Lycopene
Nutritional, Medicinal and Therapeutic Properties

Lycopene
Nutritional, Medicinal and Therapeutic Properties

Editors

Victor R. Preedy
Department of Nutrition and Dietetics
King's College
London
UK

Ronald R. Watson
Division of Health Promotion Sciences
School of Medicine
University of Arizona
Tuscon, Arizona
USA

Science Publishers

Enfield (NH)   Jersey   Plymouth

Science Publishers   *www.scipub.net*

234 May Street
Post Office Box 699
Enfield, New Hampshire 03748
United States of America

General enquiries : *info@scipub.net*
Editorial enquiries : *editor@scipub.net*
Sales enquiries   : *sales@scipub.net*

Published by Science Publishers, Enfield, NH, USA
An imprint of Edenbridge Ltd., British Channel Islands
Printed in India

ISBN: 978-1-57808-538-5

Cover illustration: Reproduced by kind permission of Dr. Zsolt Bikadi, Budapest, Hungary.

Library of Congress Cataloging-in-Publication Data

Lycopene:nutritional, medicinal and therapeutic properties/editors: Victor R. Preedy, Ronald R. Watson.--1st ed.
    p. cm.
  Includes bibliographical references and index.
  ISBN 978-1-57808-538-5 (hardcover)
  1. Lycopene. 2. Lycopene -- Health aspects. I. Preedy, Victor R. II. Watson, Ronald R. (Ronald Ross)
  QK898.L9L93 2008
  572'.592--dc22

                                                              2008016633

Dedication

Appreciation is given to the Wallace Research Foundation which has supported Dr. Ronald Watson's research. It stimulated interest in components of bioactive foods including lycopene leading to this book. The encouragement of H.B. and J. Wallace was supportive of communicating in books to the public and scientists.

Foreword

Tomatoes have become a dietary staple for humans in many parts of the world. The characteristic deep red color of the ripe tomato fruit and related products is mainly due to lycopene. Lycopene is the predominant carotenoid in tomatoes, followed by β-carotene, α-carotene, γ-carotene, and phytoene, as well as by several other minor carotenoids. Tomatoes and tomato-based foods have long been an important source of lycopene in the Western diet.

Lycopene, however, received very little attention until a decade ago. Recently, it has attracted attention for its biological and physicochemical properties, especially related to its effects as a natural antioxidant. Although it has no provitamin A activity, lycopene does exhibit a physical quenching rate constant with singlet oxygen almost twice as high as that of β-carotene. These results are due in part to lycopene's high antioxidant activity. This makes its presence in the diet of considerable interest.

There has been a growing interest in exploring the role of lycopene in the prevention of a variety of nutritional and health issues in humans, including some cancers and cardiovascular diseases. Recently, many case studies using cell cultures, animal models, and epidemiological investigations have shown a relationship between lycopene intake and a lowered risk of contracting some cancers and various chronic diseases. Increasingly, clinical evidence supports the role of lycopene as a nutrient with important health benefits, since it appears to provide protection against a broad range of epithelial cancers. The possibility that consumption of lycopene-rich foods may reduce the risk of such diseases has prompted numerous in-depth studies of the levels of lycopene in foods and of correlations between dietary lycopene and certain diseases.

Although tomatoes and lycopene have been intensively studied over the past 10 years, most of the information is scattered in many different journals and books. Professors Victor Preedy and Ronald Watson, both internationally known scholars, have edited a book titled

Lycopene: Nutritional, Medicinal and Therapeutic Properties. This book consists of three sections: characterization of lycopene from chemistry to basic physiological functions, biochemical and physiological features of lycopene's effects, and lycopene and cancer. The book offers readers a broad and comprehensive coverage through its 20 chapters.

In the five-chapter section on the characterization of lycopene from its chemistry to its basic physiological functions, the authors provide a general view of lycopene's stability during food processing and storage, lycopene metabolites, non-covalent binding of lycopene and lycophyll, and risk assessment of lycopene.

In the ten-chapter section on the biochemical and physiological features of lycopene's effects, the authors discuss lycopene and peroxynitrite modifications, lycopene and down-regulation of cyclin D1, pAKT and pBad, lycopene and chylomicrons, lycopene and chromosomal aberrations, lycopene-enriched prostasomes, lycopene's antioxidant capacity, lycopene and cardiovascular diseases, effects of lycopene and monounsaturated fat on serum lycopene and lipoprotein concentrations, cataract and oxidative stress of lycopene, and lycopene and bone tissues.

The five chapters in the section on lycopene and cancer focus on lycopene's potential preventive effects on prostate, lung, breast, and colon cancers.

The contributing authors are among the world's experts in various interdisciplinary research areas and include chemists, biochemists, molecular biologists, food technologists and engineers, and medicinal and health professionals. This multidisciplinary approach offers a comprehensive review of the most recent research findings on lycopene, cancers, and chronic diseases and provides timely information focused on tomato products and the potential benefits of lycopene for human health. This monograph will serve as a critical reference book by providing a better understanding of the role of lycopene in promoting health, and by encouraging a deeper understanding of approaches to a healthy diet and life. Possible applications of lycopene include not only traditional foods and ingredients, but also novel functional food preparations specifically formulated to promote human health, as well as applications in pharmaceutical products. Consumer demand for healthy food products provides an opportunity to develop lycopene-rich foods as new functional foods, as well as food- and pharmaceutical-grade lycopene as new nutraceutical products.

John Shi, PhD
Senior Research Scientist
Guelph Food Research Center
Agriculture and Agri-Food Canada
Canada

Preface

The concentrations of the carotenoid lycopene is particularly high in tomatoes and it has been argued by some that the health benefits of lycopene are greater than the more common antioxidants found in food items. Lycopene is found in virtually all tomato-based products though substantially higher concentrations are obtained when tomatoes have been processed. In some countries, over three quarters of the dietary lycopene is obtained from 3 food items, namely tomato sauce (ketchup), juice and pizza-base.

Lycopene reduces risk factors in a number of pathologies including cancer and cardiovascular disease. The reduced incidence of cardiovascular disease with increasing lycopene intake has been ascribed to its ability to inhibit the synthesis of cholesterol and increase LDL breakdown though its influences at the sub-cellular level are still a matter on conjecture. Other properties of lycopene include its ability to inhibit growth stimulation of cancer cells *in vitro*. This occurs via a number of molecular events such as alterations in gene expression and modulation of intracellular signalling. There is also considerable complexity in the inter-tissue distribution of lycopene after ingestion. For example there are higher concentrations of lycopene in the liver, followed by the intestine and with only low amounts in prostate or testes. All-trans- and cis-lycopene appear to be differentially regulated with the all-trans isoform being found predominantly in most mammalian tissues in contrast to the cis-isoform being found predominantly in the prostate and plasma. Gaining an insight into the metabolism, absorption and excretion of lycopene its effects at the molecular and cellular levels is one of the preludes to devising novel therapeutic strategies. Understanding this in detail requires a holistic understanding of lycopene but finding this knowledge in a single coherent volume is presently problematical. However, Lycopene: nutritional, medicinal and therapeutic properties addresses this. The book is divided into three main sections Characterisation of lycopene from chemistry to basic physiological functions; Biochemical and physiological features of lycopene's effects and Lycopene and cancer. Its coverage is extensive; from pathways in cells to whole organs.

The contributors to **Lycopene: nutritional, medicinal and therapeutic properties** are authors of international and national standing, leaders in the field and trend-setters. Emerging fields of science and important discoveries relating are also incorporated in this book. This represents "one stop shopping "of material related to lycopene in its truest sense. This book will being essential reading for nutritionists, dieticians, pharmacologists, health care professionals, research scientists, cancer workers, cardiologists, pathologists, molecular or cellular biochemists.

The Editors
Professor Victor R Preedy and Ronald R Watson

Contents

PART 1

CHARACTERIZATION OF LYCOPENE FROM CHEMISTRY TO BASIC PHYSIOLOGICAL FUNCTIONS

PART 2

BIOCHEMICAL AND PHYSIOLOGICAL FEATURES OF LYCOPENE'S EFFECTS

List of Contributors

Agarwal Renu

Delhi Institute of Pharmaceutical Sciences & Research, Pushp Vihar, Sector 3, MB Road, New Delhi 110017, India.

Agrawal Shyam Sunder

Delhi Institute of Pharmaceutical Sciences & Research, Pushp Vihar, Sector 3, MB Road, New Delhi 110017, India.

Ahuja Kiran Deep Kaur

Locked Bag 1320, School of Human Life Sciences, University of Tasmania, Launceston, 7250 TAS, Australia.

Andreassi Lucio

Department of Clinical Medicine and Immunological Sciences, Section of Dermatology, University of Siena, Italy.

Andreassi Marco

Centre of Cosmetic Science and Technology, University of Siena, Italy.

Antunes Lusânia Maria Greggi

Depto. Análises Clínicas, Toxicológicas e Bromatológicas, Faculdade de Ciências Farmacêuticas de Ribeirão Preto – USP, Av. do Café s/n, 14040-903, Ribeirão Preto - São Paulo, Brazil.

Ball Madeleine Joyce

Locked Bag 1320, School of Human Life Sciences, University of Tasmania, Launceston, 7250, TAS, Australia.

Bernard-Gallon Dominique J.

Département d'Oncogénétique, Centre Jean Perrin, 58 Rue Montalembert, BP 392, 63011 Clermont-Ferrand Cedex 01, France.

INSERM UMR 484, rue Montalembert, BP 184, 63005 Clermont-Ferrand Cedex, France.

Centre de Recherche en Nutrition Humaine (CRNH), 58 Rue Montalembert, BP 321, 63009 Clermont-Ferrand Cedex 01, France.

Bianchi Maria de Lourdes Pires

Depto. Análises Clínicas, Toxicológicas e Bromatológicas, Faculdade de Ciências Farmacêuticas de Ribeirão Preto – USP, Av. do Café s/n, 14040-903, Ribeirão Preto - São Paulo, Brazil.

Bignon Yves-Jean

Département d'Oncogénétique, Centre Jean Perrin, 58 Rue Montalembert, BP 392, 63011 Clermont-Ferrand Cedex 01, France.

INSERM UMR 484, rue Montalembert, BP 184, 63005 Clermont-Ferrand Cedex, France.

Centre de Recherche en Nutrition Humaine (CRNH), 58 Rue Montalembert, BP 321, 63009 Clermont-Ferrand Cedex 01, France.

Université d'Auvergne, 28 place Henri Dunant, BP 38, 63001 Clermont-Ferrand 1, France.

Bikadi Zsolt

Virtua Drug, Ltd., Csalogany st. 4c, H-1015 Budapest, Hungary.

Boateng Judith

P.O. Box 1628, Alabama A&M University, Department of Food and Animal Sciences, Normal, AL 35762, USA.

Botham Kathleen M.

Department of Veterinary Basic Sciences, The Royal Veterinary College, Royal College St., London NW1 0TU, UK.

Bravo Elena

Department of Haematology, Oncology and Molecular Medicine, Viale Regina Elena 299, 00161 Rome, Italy.

Chalabi Nasséra

Département d'Oncogénétique, Centre Jean Perrin, 58 Rue Montalembert, BP 392, 63011 Clermont-Ferrand Cedex 01, France.

INSERM UMR 484, rue Montalembert, BP 184, 63005 Clermont-Ferrand Cedex, France.

Centre de Recherche en Nutrition Humaine (CRNH), 58 Rue Montalembert, BP 321, 63009 Clermont-Ferrand Cedex 01, France.

Choongo Kennedy

Biomedical Sciences Department, School of Veterinary Medicine, University of Zambia, P.O. Box 32379, Lusaka, Zambia.

Chopra Mridula

Urology Research Group, Department of Pharmacy and Biomedical Sciences, University of Portsmouth, Portsmouth, UK.

Christian Mildred S.

933 Horsham Road, Horsham, PA 19044, U.S.A.

Cooper Alan

Urology Research Group, Department of Pharmacy and Biomedical Sciences, University of Portsmouth, Portsmouth, UK.

Diener Robert M.

185 Aster Court, Whitehouse Sta., NJ 08889, U.S.A.

Erdman John W.

Division of Nutritional Sciences, University of Illinois Urbana-Champaign, 448 Bevier Hall, 905 S. Goodwin Ave, Urbana, IL 61820, USA.

Ford Nikki A.

Division of Nutritional Sciences, University of Illinois Urbana-Champaign, 448 Bevier Hall, 905 S. Goodwin Ave, Urbana, IL 61820, USA.

Fujita Shoichi

Laboratory of Toxicology, Department of Environmental Veterinary Sciences, Graduate School of Veterinary Medicine, N18, W9, Kita-Ku, Sapporo 060-0818, Japan.

Goyal Anuj

Urology Research Group, Department of Pharmacy and Biomedical Sciences, University of Portsmouth, Portsmouth, UK.

Gupta S.K.

Delhi Institute of Pharmaceutical Sciences & Research, Pushp Vihar, Sector 3, MB Road, New Delhi 110017, India.

Hari Peter

Delta Elektronik, Ltd., Szentendrei st. 39 – 53, H-1033 Budapest, Hungary.

Hathcock John N.

Council for Responsible Nutrition, 1828 L St., NW, Suite 900, Washington, DC 20036-5114, USA.

Hazai Eszter

Virtua Drug, Ltd., Csalogany st. 4c, H-1015 Budapest, Hungary.

Lian Fuzhi

Nutrition and Cancer Biology Laboratory, Jean Mayer United States Department of Agriculture Human Nutrition Research Center on Aging at Tufts University, 711 Washington Street, Boston, MA 02111, USA.

Lockwood Samuel F.

Cardax Pharmaceuticals, Inc., 99-193 Aiea Heights Drive, Suite 400, Aiea, Hawaii 96701, USA.

Mackinnon E.S.

Calcium Research Laboratory, Division of Endocrinology & Metabolism & St. Michael's Hospital and Department of Medicine, University of Toronto, Ontario, Canada.

Muzandu Kaampwe

Biomedical Sciences Department, School of Veterinary Medicine, University of Zambia, P.O. Box 32379, Lusaka, Zambia.

Palozza Paola

Institute of General Pathology, Catholic University, School of Medicine, Lgo F. Vito 1, 00168 Rome, Italy.

Rao A.V.

Department of Nutritional Sciences, University of Toronto, Ontario, Canada.

Rao L.G.

Associate Professor of Medicine, University of Toronto & Director, Calcium Research Laboratory, St. Michael's Hospital, 3 Shuter St. suite 3-064, Toronto, Ontario, M5B 1W8 Canada.

Schröder F.H.

Department of Urology, ERSPC, Rochussenstraat 125, 3015 EJ Rotterdam, The Netherlands.

Sehgal Inder

Louisiana State University, School of Veterinary Medicine, Comparative Biomedical Sciences Department, Skip Bertman Drive, Baton Rouge, LA 70803.

Sestito Rosanna

Institute of General Pathology, Catholic University, School of Medicine, Lgo F. Vito 1, 00168 Rome, Italy.

Shackelford Louis

P.O. Box 1628, Alabama A&M University, Department of Food and Animal Sciences, Normal, AL 35762, USA.

Shao Andrew

Council for Responsible Nutrition, 1828 L St., NW, Suite 900, Washington, DC 20036-5114, USA.

Shi John

Guelph Food Research Center, Agriculture and Agri-Food Canada, 93 Stone Road West, Guelph, Ontario N1G 5C9 Canada.

Srivastava Sushma

Delhi Institute of Pharmaceutical Sciences & Research, Pushp Vihar, Sector 3, MB Road, New Delhi 110017, India.

Sunkara Rajitha

P.O. Box 1628, Alabama A&M University, Department of Food and Animal Sciences, Normal, AL 35762, USA.

Trion A.

Department of Urology, ErasmusMC, Dr. Molewaterplein 50, P.O. Box 2040, 3000 CA Rotterdam, The Netherlands.

van Weerden W.M.

Department of Urology, ErasmusMC, Dr. Molewaterplein 50, P.O. Box 2040, 3000 CA Rotterdam, The Netherlands.

Verghese Martha

Nutritional Biochemistry, P.O. Box 1628, Alabama A&M University, Department of Food and Animal Sciences, Normal, AL 35762, USA.

Walker Lloyd T.

P.O. Box 1628, Alabama A&M University, Department of Food and Animal Sciences, Normal, AL 35762, USA.

Wang Xiang-Dong

Nutrition and Cancer Biology Laboratory, Jean Mayer United States Department of Agriculture Human Nutrition Research Center on Aging at Tufts University, 711 Washington Street, Boston, MA 02111, USA.

Xue Sophia Jun

Guelph Food Research Center, Agriculture and Agri-Food Canada, 93 Stone Road West, Guelph, Ontario N1G 5C9 Canada.

Zsila Ferenc

Chemical Research Center of the Hungarian Academy of Science, Department of Molecular Pharmacology, PO BOX 17 H-1525 Budapest, Hungary.

PART 1

Characterization of Lycopene from Chemistry to Basic Physiological Functions

1.1

Lycopene Overview: What It Is and What It Does

[1]Robert M. Diener and [2]Mildred S. Christian

[1]185 Aster Court, Whitehouse Sta., NJ 08889, U.S.A.
[2]933 Horsham Road, Horsham, PA 19044, U.S.A.

ABSTRACT

Lycopene is the bright red carotenoid phytochemical that gives tomatoes and other red fruits their color. Because of its predominance in tomatoes, it derives its name from the Latin genus name for tomatoes. Lycopene also serves as a potent antioxidant in humans and is said to lower the risk of a variety of cancers. Information about its chemistry, safety (absence of toxicity), availability and pharmacokinetics, actions and mechanisms of action and anticarcinogenic properties, as well as results of various human clinical trials, are summarized in this overview.

INTRODUCTION

Lycopene is a fat-soluble, bright red carotenoid pigment, a phytochemical found in tomatoes and other red *fruit*. Other major sources of lycopene are watermelons, pink grapefruit, guava, papaya and rosehip.

The name is derived from the tomato's species classification, *Lycopersicon esculentum*. The importance of the tomato as a source of lycopene in Western diets is confirmed in US Department of Agriculture (USDA) data, which indicate that the average annual American per capita consumption of fresh and processed tomatoes in 2004 was 19.1 and 69.8 pounds, respectively. In the fresh state, only onions (19.3 pounds), head lettuce (21.3 pounds) and potatoes (45.6 pounds) exceeded the consumed amounts of fresh tomatoes, while processed tomato consumption far exceeded that of other processed vegetables (USDA 2005).

Unlike vitamins, lycopene content is increased in processed products, as summarized in Table 1.

Table 1 Lycopene content in raw and processed tomato products.

Tomato products	Lycopene content (µg/100 g)
Raw tomatoes	2,573
Tomato juice	9,037
Tomato sauce	15,152
Ketchup	17,007

Source: USDA data cited by Campbell et al. 2004.

Lycopene is the most common, and one of the most potent, carotenoid antioxidants in the human body. It also has been shown to be the most efficient quencher of singlet oxygen and free radicals among the common carotenoids *in vitro* (Di Mascio et al. 1989). Increased intake or blood levels of lycopene have also been associated with a lowered risk of a variety of cancers in humans, but a direct effect of lycopene has yet to be proven, for the beneficial effects have been achieved predominantly by the intake of tomatoes or tomato products rather than purified lycopene (Basu and Imrhan 2007, Giovannucci 1999).

CHEMISTRY

Lycopene is a 40 carbon atom, open-chain polyisoprenoid with 11 conjugated double bonds. Each double bond reduces the energy required for electrons to transition to higher *energy states*, thus allowing the molecule to absorb visible light of progressively longer wavelengths, which makes it appear red. Its molecular weight is 536.88 daltons. The structural formula of lycopene is illustrated in Fig. 1 (PDRhealth 2007).

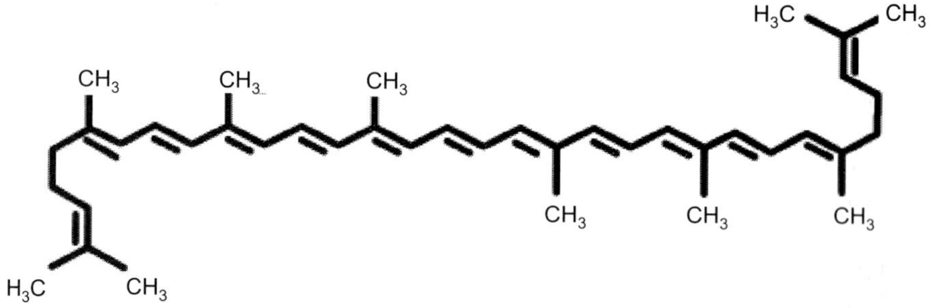

Fig. 1 Molecular formula of lycopene.

Chemically, lycopene is an acyclic isomer of beta-carotene. Beta-carotene, which contains beta-ionone rings at each end of the molecule, is formed in plants, including tomatoes, via the action of the enzyme lycopene beta-cyclase. All-*trans* lycopene is the predominant geometric isomer found in plants, although *cis* isomers of lycopene are also found in nature, including 5-*cis*, 9-*cis*, 13-*cis* and 15-*cis* isomers. However, the *trans* form of lycopene found in tomatoes is largely converted to the *cis* form by the crushing and temperature changes involved in making processed foods, such as tomato juice, ketchup, paste and sauce. As a result, the lycopene found in human plasma is a mixture of approximately 50% *cis* lycopene and 50% all-*trans* lycopene (PDRhealth 2007).

TOXICITY

A number of recent publications have reported on the safety of ingested lycopene from a variety of sources. Results from these studies are summarized in Table 2.

Table 2   Toxicity studies conducted in animals to determine the safety of lycopene.

Type of study	Results
Rat 90 d toxicity (up to 1% in diet[a])	No toxicity (NOEL was 1% in diet)
Rat 4 wk toxicity (1000 mg/kg[b])	No significant toxicity observed
Rat 14 wk toxicity (up to 500 mg/kg[b])	No significant toxicity observed
Rat 13 wk toxicity (up to 300 mg/kg[c])	NOAEL = 300 mg/kg/d
Rat 2-generation reproduction study (1000 mg/kg[b])	No teratogenic effects
Rat teratology (1000 mg/kg by gavage[b])	No teratogenic effects
Rat developmental toxicity (50-300 mg/kg[c])	No direct maternal/developmental toxicity
Rabbit developmental toxicity (50-200 mg/kg[c])	No direct maternal/developmental toxicity
Mutagenicity battery	No genotoxic effects

[a]lycopene derived from fungal biomass
[b]synthetic lycopene beadlets
[c]BASF beadlets (active lycopene product)
NOEL = no-observed-effect level, NOAEL = no-observed-adverse-effect level.

Jonker et al. (2003) fed lycopene derived from a fungal biomass (*Blakeslea trispora*) and suspended in sunflower oil (20% w/w) to groups of 20 Wistar male and female rats for 90 d. The diet contained lycopene concentrations of 0, 0.25, 0.50, and 1.0%, and lycopene intake was calculated to be 0, 145, 291 and 586 mg/kg body weight/day in male rats, and 0, 156, 312, and 616 mg/kg bw/d in female rats, respectively. Clinical and neurobehavioral observations, body weight and feed consumption measurements, ophthalmoscopic examinations, clinical pathology, organ weights, and gross pathology and histopathology failed to detect signs of toxicity in any of the groups of rats. The no-observed-effect level was 1.0% in the diet, the highest dietary concentration tested.

Lycopene from synthetic sources, which provide an alternative to extracts and can be used in dietary supplements, has been extensively tested *in vitro* and *in vivo*. Lycopene is commercially available only in formulated forms containing antioxidants, which prevent the degradation of lycopene and other excipients that provide for water dispersibility. Improperly stored, unformulated crystalline lycopene can degrade if exposed to light and air. McClain and Bausch (2003) reported on the safety of synthetic lycopene as a water-dispersible beadlet formulation containing antioxidants. In single doses, lycopene had a low order of toxicity. In repeated doses, no significant toxic effects were observed in rats treated with the lycopene beadlet formulation in the diet at doses of up to 500 mg/kg bw/d for 14 wk or 1000 mg/kg bw/d for 4 wk. The lycopene beadlets produced no teratogenic effects at dietary concentrations of 1000 ppm in a two-generation rat study or at gavaged doses of 1000 mg/kg bw/d in a rat teratology study. No genotoxic effects were observed in a comprehensive battery of mutagenicity studies. Exploratory short-term studies revealed that lycopene (in synthetic and natural form) accumulates in hepatocytes and to a lesser extent in the spleen, but the pigment accumulation was reversible in about 13 wk and produced no histopathological changes.

Results from another 13 wk toxicology study in Wistar rats were reported by Mellert et al. (2002). BASF Lycopene 10 CWD and Lyco Vit 10% formulated products, each containing approximately 10% synthetic lycopene (67% *trans*, 30.3% *cis* isomers), were evaluated for toxicological and behavioral effects at gavaged dosages of 0, 500, 1,500 and 3,000 mg/kg bw/d Lycopene 10 CWD and 3,000 mg/kg bw/d Lyco Vit 10%. No statistically significant, dose-related effects on body weight, feed consumption, hematology, urinalysis, clinical chemistry, gross and microscopic tissue examination or behavioral and sensimotor parameters were observed after 13 wk of treatment. No deaths attributed to the lycopene formulations occurred during the study; the only finding at terminal necropsy was the presence of red pigment in the feces and gastrointestinal tract, due to the red-pigmented test materials. The no-observed-adverse-effect level for this study was concluded to be 3,000 mg/kg bw/d for both Lycopene 10 CWD and Lyco Vit 10%.

The potential oral developmental toxicity of the synthetic lycopene 10 CWD and LycoVit 10% was also studied in rats and rabbits (Christian et al. 2003). Gavaged dosages of 0, 500, 1,500 or 3,000 mg/kg/d were used for 10 CWD in rats and 0, 500, 1,500 or 2,000 mg/kg/d in rabbits (maximum dose because of viscosity). LycoVit 10% was tested in rats and rabbits only at the highest dosage levels of 3,000 and 2,000 mg/kg/d, respectively. Dosages were administered on gestation days 6 through 19 (rats) or gestation days 6 through 28 (rabbits). Feed consumption and weight gain were essentially unaffected in rats and rabbits, despite intubation problems in both species and reduced gastrointestinal motility and mortality in rabbits attributable to the physical properties (high viscosity) of the gels. Neither Lycopene 10 CWD nor LycoVit 10% caused direct maternal or developmental toxicity in rats or rabbits at dosages as high as 3,000 or 2,000 mg/kg/d, respectively.

PHARMACOKINETICS

The efficiency of lycopene absorption from supplements and foods is variable, and the optimal daily intake has not been established. In foods, lycopene is part of a matrix in chloroplasts or chromoplasts within the plant, and absorption of lycopene from raw tomatoes is low, because it is mostly in the *trans* isoform and is tightly bound within the matrix. In synthetic nutritional supplements, lycopene is in the form of an oleoresin embedded in phospholipid complexes and in oils, and marketed formulations (usually capsules) range in dose from 5 to 15 mg/d. In processed tomato products, absorption is much higher for a number of reasons. First, mechanical and thermal processing releases the lycopene from crushed or ruptured plant cells. Second, heat during processing induces the *trans* isoform to change to the *cis* isomer, which is more bioavailable. Third, the addition of lipids, such as vegetable oils, increases lycopene absorption because it is lipophilic (Shi and Maguer 2000).

Lycopene's journey throughout the human body is long and complex, whether it originates from food matrices or synthetic supplements. Table 3 summarizes some of the physiological actions and chemical transformations that are given in greater detail elsewhere (PDRhealth 2007).

Studies have been conducted in rats and ferrets subjected to a 9 wk tomato oleoresin supplement to determine lycopene uptake and distribution in the body (Ferreira et al. 2000). After tissue saponification, the total lycopene levels in rats were found to be as follows: liver 14,213 nmol/kg, intestine 3,125 nmol/kg, stomach 79 nmol/kg, prostate 24 nmol/kg, and testis 3.9 nmol/kg. In ferrets, tissue content of lycopene after saponification was substantially lower than in rats, and prostate tissues had a higher content than stomach. Non-saponified extract was highest in the liver of both species. Results from this study indicate that

Table 3 Lycopene pathway from ingestion to incorporation into tissues.

Lycopene is solubilized in the lumen of the small intestine by bile salts and fat.
↓
Soluble lycopene is taken up by enterocytes in the intestinal wall.
↓
Lycopene is delivered into lymphatics as chylomicrons.
↓
Chylomicrons enter the circulation and are hydrolyzed, producing chylomicron remnants.
↓
Remnants are taken up mainly by hepatocytes.
↓
Within hepatocytes, lycopene is incorporated into lipoproteins.
↓
Low density and very low density lipoproteins enter the circulation.
↓
Lycopene incorporated mostly in low density lipoproteins enters various tissues

---

lycopene is absorbed and stored primarily in the liver of both species, but rats absorb lycopene more effectively than ferrets.

Zhao et al. (1998) studied lycopene uptake and tissue disposition in Fischer 344 rats after they had been on a diet supplemented with tomato oleoresin for 10 wk. Lycopene was present in lung (134-227 ng/g), mammary gland (174-309 ng/g) and prostate (47-97 ng/g). Serum levels ranged from 80 to 370 mg/mL in both sexes but were not reflective of dietary intake. Investigations in F344 rats by Boiliau et al. (2000) revealed that tissue and serum lycopene levels reach a plateau when rats are administered between 0.05 and 0.50 g/kg dietary lycopene provided by a water-dispersible beadlet. Rats appear to achieve tissue concentrations in the range of what is observed in humans when lycopene is consumed in high concentrations.

In humans, concentrations of lycopene in the serum are variable and range from 0.10 to 0.95 µmol/L. In the tissues, lycopene appears to concentrate particularly in those tissues with a large number of low density lipoprotein (LDL) receptors and a high rate of lipoprotein uptake, such as liver, adrenals and testes (IOM 2000).

In contrast to other carotenoids, serum values of lycopene are usually reduced not by smoking or alcohol consumption but by increasing age. The reason for this is not known. In fact, much is still unknown about the pharmacokinetics of lycopene, in particular its distribution and metabolism (Gerster 1997).

ACTION AND MECHANISM OF ACTION

The intake of tomato-based foods, especially processed tomato products, is associated with a significantly lower risk for various cancers in humans, but the exact mechanism for this possible anticarcinogenic effect has yet to be determined, although lycopene is thought to play a major role in the process.

Major beneficial actions attributed to lycopene are that it quenches singlet oxygen, traps peroxyl radicals, inhibits peroxidation, inhibits oxidative DNA damage, and stimulates gap junction communication (Sies and Stahl 1998).

The most prevalent present hypothesis is that cancer as well as several other chronic diseases, such as cardiovascular diseases, osteoporosis, and diabetes, are linked to oxidative stress, which lycopene is able to mitigate, based on *in vitro* studies that have shown it to have the highest antioxidant activity of any of the carotenoids (Di Mascio et al. 1989). However, lycopene is not the only tomato-based antioxidant with possible anticarcinogenic effects *in vitro*. Campbell et al. (2003) reported that tomato aglycone polyphenols, including quercetin, kaempherol and naringenin, inhibited cancer cell proliferation in both a human prostate cancer cell line and a mouse hepatocyte cell line. Nevertheless, lycopene also has the proven ability to act as a strong antioxidant. If oxidation proves critical to carcinogenesis, the dietary contribution to antioxidation is likely to be very complex. The benefits of tomatoes may result from the complex interaction of various carotenoids, ascorbic acid and other antioxidant polyphenolic compounds (Giovannucci 1999).

Non-antioxidant mechanisms have also been proposed. Failure of cell signaling may be a cause of cell overgrowth and eventual cancer. Lycopene may stimulate gap junction communication between cells. It is speculated that lycopene may suppress carcinogen-induced phosphorylation of regulatory proteins such as p53 and Rb antioncogenes and stop cell division at the G_0-G_1 cell cycle phase, as cited by Campbell et al. (2004) and PDRhealth (2007). Another theory is that an increase of lycopene-induced phase I liver metabolizing enzymes such as cytochrome P450-dependent enzymes, and increased hepatic quinine reductase, a phase II enzyme, may act as an underlying mechanism of protection against carcinogen-induced preneoplastic lesions in the rat liver (Breinholt et al. 2000). In addition, *in vitro* evidence suggests that lycopene may reduce proliferation of various cancer cells induced by insulin-like growth factors, which is of interest because circulating levels of IGF-I have been associated with higher risk of various cancers, including prostate cancer (Levy et al. 1995, Chan et al. 1998).

An increase in LDL oxidation is thought to be causally associated with increasing risk of atherosclerosis and coronary heart disease, and lycopene has been found to inhibit cholesterol synthesis, inhibit HMG-CoA (hydroxymethylglutaryl coenzyme A) reductase activity and up-regulate LDL receptor activity in macrophages. However, the mechanism of such an antiatherogenic activity of lycopene needs further exploration, although this effect may also be associated with lycopene's antioxidant activity. In a small preliminary study in humans, dietary supplementation of lycopene significantly increased serum lycopene levels by at least two-fold. There was no change in serum cholesterol levels (total, LDL, or high-density lipoprotein), but serum lipid peroxidation and LDL oxidation were significantly decreased. These results may have relevance for decreasing the risk of coronary heart disease (Agarwal and Rao 1998).

EPIDEMIOLOGIC STUDIES

Consumption of tomatoes or tomato-based products and plasma levels of lycopene have relatively consistently been associated with a reduced risk of a variety of cancers. This was the general conclusion at the end of an extensive review of all human studies reported in the English language of tomatoes or lycopene in relation to the risk of any cancer (Giovannucci 1999). The review covered 72 identified epidemiologic studies, of which 57 reported inverse associations between tomato intake or blood lycopene levels and the risk of cancer at a defined anatomic site; 35 of these associations were statistically significant. Evidence for a benefit was strongest for cancers of the prostate, lung and stomach. Summary data from specific sites in this comprehensive study follow. The various investigations are categorized as follows:

- Prostate cancer
- Lung and pleural cancers
- Stomach cancer
- Colorectal cancer
- Pancreatic cancer
- Esophageal cancer
- Cervical cancer and precursors
- Breast cancer
- Oral/laryngeal/pharyngeal cancer

Total Cancer

Only one prospective study with 1,271 elderly persons was reported. Individuals at the top half of tomato consumption had a lower risk of all cancers combined compared with those in the bottom half.

Prostate Cancer

Ten studies reported on the relationship between tomato/tomato products consumption or lycopene serum levels and prostate cancer risk. Of the 10 studies, four cohort studies reported statistically significant inverse associations between tomato consumption and prostate cancer risk; one of three case-control studies reported an inverse association that was not statistically significant, while the other two studies had no association to tomatoes but were flawed by the method used to report consumption. Three of the 10 reported studies were cohort investigations that compared serum lycopene levels from prediagnostic samples for prostate cancer risk. One of these studies reported a statistically significant inverse association. The second study reported a 6.2% lower median lycopene level in prostate cancer case subjects diagnosed during 13 yr, compared with age- and race-matched controls. The third study did not detect any association in Japanese-American subjects in Hawaii (ethnic differences for low lycopene levels and prostate cancer rate may have been a factor).

Lung and Pleural Cancers

Fourteen studies, mostly case-control design, reported specifically on tomato or lycopene consumption in relation to lung cancer risk. Ten of the 14 studies suggested either a statistically significant or a suggestive inverse association. Statistical significance was observed in China and Spain and in multiple US populations; non-statistically significant inverse associations were noted in the United Kingdom, Norway and Finland. Only one study reported on mesothelioma of the pleura or peritoneum, and a 40% reduction in risk was noted for those consuming tomato or tomato juice 16 or more times per month.

Stomach Cancer

Twelve case-control studies were reported from varied populations, including those of the United States, Japan, Israel, Italy, Spain, Poland and Sweden. All reported inverse associations between tomato consumption and risk of gastric cancer except Spain and Japan. These consistent results in diverse populations strongly suggest a protective effect by tomatoes/lycopene on gastric cancer.

Colorectal Cancer

Five case-control studies (from the United States, Belgium, Italy and China) were reported on tomato consumption in relation to lung cancer risk. A statistically significant inverse association was reported between tomato consumption and colon cancer risk in the US study. Two studies from Italy and one from China reported a 60% reduction in colon and rectal cancer risk with increased tomato consumption. The Belgian study found a suggestion of an inverse association only between tomato puree and colon cancer, but the study was of dubious merit because of a low consumption of tomatoes and ambiguous consumption criteria.

Pancreatic Cancer

Four small studies examined tomato or serum lycopene status in relation to risk of pancreatic cancer; all of these studies support an inverse association. A four- or five-fold increase of risk was associated with low intake of tomatoes (one of the diet studies) or a low level of serum lycopene (serum study). The other two diet studies reported an inverse trend but no documentation of relative risk, thus limiting the evaluation of the effect of tomatoes on pancreatic cancer in this review by Giovannucci (1999).

Esophageal Cancer

Only three studies were reported on the relationship of tomato intake and/or serum lycopene to esophageal cancer risk. Frequent intake of raw tomatoes produced a statistically significant risk reduction of 39% in Iranian men (but not in women), while a high intake of tomatoes in a US study produced a similar, but non-significant, reduction of 30%. In a serum bank-based study, 28 esophageal cancer patients had a 16.4% decreased mean serum level of lycopene compared to controls.

Cervical Cancer and Precursors

Five studies assessed the relationship of tomato intake and/or serum lycopene levels to cervical cancer risk. Three of the studies of pre-invasive cervical lesions reported an inverse association with tomato intake or serum lycopene level and cervical cancer risk. In one study, control subjects had higher tomato consumption than case subjects, although results were not statistically significant. No association was found in another study. A strong case for benefit cannot be made from the reviewed data.

Breast Cancer

Eight studies have been reported assessing the relationship of tomato intake and/or serum lycopene to breast cancer risk; none of four dietary-based studies supports an association between tomato intake and breast cancer risk, and lycopene levels in breast adipose tissues from case subjects were not lower than in control subjects in a small exploratory study. However, three of four studies exploring lycopene levels in serum indicated an association with reduced breast cancer risk (values were statistically significant in two studies). Further work appears to be necessary to resolve the equivocal nature of the studies to date.

Oral/Laryngeal/Pharyngeal Cancer

Only three case-control studies have been reported on the relationship of tomato intake to oral, laryngeal or pharyngeal cancer risk. Both a Chinese and an Italian study reported that high intakes of tomatoes reduced the cancer risk by approximately half. However, another Chinese study found no association between tomatoes and laryngeal cancer, which creates some doubt about overall conclusions.

Other Organs

Epidemiologic studies were also conducted on the association of tomato intake and/or serum lycopene levels to ovarian and bladder cancer risk. In none of the studies involving these organs was there any reliable relationship between tomato intake or lycopene levels and the emergence of cancer. Only one study (no association) was found involving ovarian cancer, and all four bladder cancer studies were devoid of associations between tomato intake and/or serum lycopene levels.

It is hoped that the above summary of the extensive review of the epidemiologic literature by Giovannucci (1999) will give the reader some understanding of the large amount of human clinical information that has been gathered worldwide on the possible reduced-risk or anticarcinogenic potential of tomatoes and tomato products within approximately a decade from the time that interest in lycopene started. However, despite the many studies already completed, much of the information, such as the optimal daily requirement for lycopene or tomato products in humans, is incomplete or lacking. Thus, epidemiologic research continues, and widely differing or confusing results must be evaluated as to their ultimate relevance. For instance, daily supplementation with 15, 30, 45, 60 and 90 mg/d of lycopene was well tolerated for 1 yr in a prospective trial of biochemically relapsed prostate cancer patients and produced similar plasma levels that plateaued by 3 mo, but had no discernible effect on prostate specific antigen levels (Clark et al. 2006). In contrast, Bowen et al. (2002) reported that a significant reduction of prostate specific antigen values occurred at the end of the dietary regimen in 32 patients with localized prostate adenocarcinoma who consumed tomato sauce–based pasta dishes (30 mg lycopene/d) for a 3 wk period prior to scheduled radical prostatectomy. Pretreatment serum, leukocyte and biopsy samples were compared to end-of-dietary-intervention and post-surgical tissue resections. Results from this study indicated a significant uptake of lycopene into prostate tissue and a reduction in DNA damage in both leukocytes and prostate tissue.

Finally, and most recently, Basu and Imrhan (2007), in a review of the effects of tomato product supplementation on biomarkers of oxidative stress and carcinogenesis in human clinical trials, reported that supplementation of tomato products containing lycopene has been shown to lower biomarkers of oxidative stress and carcinogenesis in healthy and type II diabetic patients and prostate cancer patients, respectively. However, the conclusion again reiterated what previous investigators have stated: (1) that consumption of processed tomato products containing lycopene is of significant health benefit, (2) that lycopene, the main tomato carotenoid, contributes to this effect, and (3) that the exact role of lycopene *per se* remains to be investigated.

References

Agarwal, S. and A.V. Rao. 1998. Tomato lycopene and low density lipoprotein oxidation: a human dietary intervention study. Lipids 33: 981-984.

Basu, A. and V. Imrhan. 2007. Tomatoes versus lycopene in oxidative stress and carcinogenesis: conclusions from clinical trials. Eur. J. Clin. Nutr. 61: 295-303.

Boiliau, T.W.M. and S.K. Clinton, and J.W. Erdman Jr. 2000. Tissue lycopene concentrations and isomer patterns are affected by androgen status and dietary lycopene concentration in male F344 rats. J. Nutr. 130: 1613-1618.

Bowen, P. and L. Chen, M. Stacewicz-Sapuntzakis, C. Duncan, R. Sharifi, L. Ghosh, H. Kim, K. Christov-Tzelkov, and R. Van Breemen. 2002. Tomato sauce supplementation and prostate cancer: lycopene accumulation and modulation of biomarkers of carcinogenesis. Exp. Biol. Med. 227: 886-893.

Breinholt, V. and S.T. Lauridsen, B. Daneshvar, and J. Jakobsen. 2000. Dose-response effects of lycopene on selected drug-metabolizing and antioxidant enzymes in the rat. Cancer Lett. 154: 201-210.

Campbell, J.K. and J.K. King, M.A. Lila, and J.W. Erdman Jr. 2003. Antiproliferation effects of tomato polyphenols in Hepa1c1c7 and LNCaP cell lines. J. Nutr. 133: 3858S-3859S.

Campbell, J.K. and K. Canene-Adams, B.L. Lindshield, T.W. Boileau, S.K. Clinton, and J. W. Erdman Jr. 2004. Tomato phytochemicals and prostate cancer risk. J. Nutr. 134(12 Suppl): 3486S-3492S.

Chan, J.M. and M.J. Stampfer, E. Giovannucci, P.H. Gann, J. Ma, P. Wilkinson. 1998. Plasma insulin-like growth factor-I and prostate cancer risk: a prospective study. Science 279: 563-566.

Christian, M.S. and S. Schulte, and J. Hellwig. 2003. Developmental (embryo-fetal toxicity/teratogenicity) toxicity studies of synthetic crystalline lycopene in rats and rabbits. Food Chem. Toxicol. 41: 773-783.

Clark, P.E. and M.C. Hall, L.S. Borden, Jr., A.A. Miller, J.J. Hu, W.R. Lee, D. Stindt, R. D'Agostino Jr., J. Lovato, M. Harmon, and F.M. Torti. 2006. Phase I-II prospective dose-escalating trial of lycopene in patients with biochemical relapse of prostate cancer after definitive local therapy. Urology 67: 1257-1261.

Di Mascio, P. and S. Kaiser, and H. Sies. 1989. Lycopene as the most efficient biological carotenoid singlet oxygen quencher. Arch. Biochem. Biophys. 274: 532-538.

Ferreira, A.L. and K-J. Yeum, C. Liu, D. Smith, N.I. Krinsky, X-D. Wang, and R.M. Russell. 2000. Tissue distribution of lycopene in ferrets and rats after lycopene supplementation. J. Nutr. 130: 1256-1260.

Gerster, H. 1997. The potential role of lycopene for human health. J. Am. Coll. Nutr. 16: 109-126.

Giovannucci, E. 1999. Tomatoes, tomato-based products, lycopene, and cancer: review of the epidemiologic literature. J. Natl. Cancer Inst. 91: 317-331.

IOM (Institute of Medicine). 2000. Dietary Reference Intakes for Vitamin C, Vitamin E, Selenium, and Carotenoids. National Academy Press, Washington, DC.

Jonker, D. and C.F. Kuper, N. Fraile, A. Estrella, and C. Rodriguez-Otero. 2003. Regul. Toxicol. Pharmacol. 37: 396-406.

Levy, J. and E. Bosin, B. Feldman, Y. Giat, A. Miinster, and M. Danilenko. 1995. Lycopene is a more potent inhibitor of human cancer cell proliferation than either α-carotene or β-carotene. Nutr. Cancer 24: 257-266.

McClain, M.R. and J. Bausch. 2003. Summary of safety studies conducted with synthetic lycopene. Regul. Toxicol. Pharmacol. 37: 274-285.

Mellert, W. and K. Deckardt, C. Gembardt, S. Schulte, B. Van Ravenzwaay, and R. Slesinski. 2002. Thirteen-week oral toxicity study of synthetic lycopene products in rats. Food Chem. Toxicol. 40: 1581-1585.

PDRhealth. 2007. Lycopene. On line. http://www.pdrhealth.com/drug_info/nmdrugprofiles/nutsupdrugs/lyc_0165.shtml (accessed 6 February 2007).

Shi, J. and M. Maguer. 2000. Lycopene in tomatoes: chemical and physical properties affected by food processing. Crit. Rev. Food Sci. Nutr. 40: 1-42.

Sies, H. and W. Stahl. 1998. Lycopene: antioxidant and biological effects and its bioavailability in the human. Proc. Soc. Exp. Biol. Med. 218: 121-124.

USDA Agricultural Statistics. 2005. Chapter on Vegetables and Melons. USDA National Agricultural Statistic Services. On line. http://www.usda.gov/nass/pubs/agstats.htm (accessed 28 February 2007).

Zhao, Z. and F. Khachik, J.P. Richie Jr., and L.A. Cohen. 1998. Lycopene uptake and tissue disposition in male and female rats. Proc. Soc. Exp. Biol. Med. 218: 109-114.

1.2

Stability of Lycopene during Food Processing and Storage

John Shi and Sophia Jun Xue
Guelph Food Research Center, Agriculture and Agri-Food Canada
93 Stone Road West, Guelph, Ontario N1G 5C9 Canada

ABSTRACT

Lycopene is the pigment principally responsible for the characteristic deep-red color of ripe tomato fruits and tomato products. Increasing clinical evidence supports the role of lycopene as a nutrient with important health benefits, since it appears to provide protection against a broad range of epithelial cancers. Thus, means of preserving lycopene during food processing and storage have drawn much attention. Tomatoes and related tomato products are the major source of lycopene compounds and are also considered an important source of carotenoids in the human diet. Lycopene belongs to the carotenoid family. Lycopene in fresh tomato fruits occurs essentially in the all-*trans* configuration. The main causes of tomato lycopene degradation during processing are isomerization and oxidation. Isomerization converts all-*trans* isomers to *cis*-isomers and results in a reduction in the biological properties of lycopene. Determination of the degree of lycopene isomerization and oxidation during processing would provide a measurement of the potential health benefits of tomato-based foods. Thermal processing generally causes some loss of lycopene in tomato-based foods. Heat induces isomerization of the all-*trans* to *cis* forms. The *cis*-isomers increase with temperature and processing time. Heat, light, oxygen and different food matrices have effects on lycopene isomerization and oxidation. Lycopene might isomerize to mono- or poly-*cis* forms with the presence of heat or oil or during dehydration. Reisomerization takes place during storage. After oxidation, the lycopene molecule splits, which causes losses of color and off-flavor. The effects of heat, oxygen, light and presence of oil on the stability of lycopene are important issues for the nutritional quality of tomato-based foods.

INTRODUCTION

Lycopene is a natural pigment that imparts red color to tomato, guava, rosehip, watermelon and pink grapefruit (Holden et al. 1999). Tomatoes (especially deep-red fresh tomato fruits) and tomato products are considered the most important source of lycopene in many human diets (Table 1); therefore, most of the research on lycopene focuses on tomato and tomato-based products.

Table 1 Lycopene content of fruits and vegetables (data from Beerh and Siddappa 1959, Gross 1987, 1991, Mangels et al. 1993).

Material	Lycopene content (mg/100 g wet basis)
Fresh tomato fruit	0.72-9.27
Watermelon	2.3-7.2
Guava (pink)	5.23-5.50
Grapefruit (pink)	0.35-3.36
Papaya	0.11-5.3
Rosehip puree	0.68-0.71
Carrot	0.65-0.78
Pumpkin	0.38-0.46
Sweet potato	0.02-0.11
Apple pulp	0.11-0.18
Apricot	0.01-0.05

Lycopene has been found in epidemiological and experimental studies to reduce the risk of macular degenerative disease, serum lipid oxidation, and cancers of the lung, bladder, cervix, and skin, especially to protect against prostate cancer, breast cancer, atherosclerosis, and associated coronary artery disease (Rao et al. 1998). Lycopene participates in a host of chemical reactions hypothesized to prevent carcinogenesis and atherogenesis by protecting critical cellular biomolecules, including lipids, proteins, and DNA. It is deposited in the liver, lungs, prostate gland, colon and skin in the human body, and its concentration in body tissues tends to be higher than that of all other carotenoids (Rao et al. 1998, Shi and Le Maguer 2000). It reduces low-density lipoprotein oxidation and helps reduce cholesterol levels in the blood. The configuration of lycopene enables it to inactivate free radicals. As an antioxidant, lycopene has a singlet-oxygen-quenching ability twice as high as that of β-carotene and 10 times that of α-tocopherol (vitamin E relative) (Weisburger 2002). According to Rao and Shen (2002), tomato juice intake showed a significant increase in serum lycopene levels for both ketchup and lycopene capsules. Lipid and protein oxidation were also observed to reduce significantly.

With increasing interest and awareness of its health benefits, lycopene has been used in functional foods and nutraceuticals. Possible applications of lycopene include not only traditional foods and ingredients, but also novel functional food preparations specifically formulated to promote human health, as well as applications in cosmetic and pharmaceutical products. Its stability during food processing and storage has drawn increasing attention. This has fueled research on the biological properties of lycopene and on the numerous factors that control these properties. This chapter provides an overview of the stability of lycopene during food processing and storage. Then it discusses some of the chemistry of lycopene and its effect on lycopene functionality.

PHYSICAL AND CHEMICAL PROPERTIES OF LYCOPENE

The basic chemical information on lycopene is fairly complete, with a research history dating back to the beginning of the 21th century (Nguyen and Schwartz 1999). The molecular formula of lycopene is $C_{40}H_{56}$. It is an acyclic open-chain polyene with 13 double bonds. There are 11 conjugated double bonds arranged in a linear array, making it longer than any other carotenoid. The acyclic structure of lycopene causes symmetrical planarity and therefore lycopene has no vitamin A activity. Lycopene is more soluble in chloroform, benzene, and other organic solvents than in water. The solubility of lycopene in vegetable oil is about 0.2 g/L at room temperature (Boskovic 1979). In aqueous systems, lycopene tends to aggregate and to precipitate as crystals. In ripe tomato fruits, lycopene exists as elongated, needle-like crystals. The molecular structure of lycopene is shown in Fig. 1.

Fig. 1   Molecular structure of lycopene.

Lycopene might be expected to undergo two changes during processing and storage: isomerization from all-*trans* to mono-*cis*-isomers or poly-*cis*-isomers and oxidation. The all-*trans* isomer of lycopene is the predominant geometrical isomer in fruits and vegetables (94-96% of total lycopene in red tomato fruits) and is the most thermodynamically stable form. In nature, lycopene exists in all-*trans* form and seven of these bonds can isomerize from the *trans* form to the mono or poly-*cis* form under the influence of heat, light, or certain chemical reactions. *Cis*-isomers of lycopene have distinct physical characteristics and chemical behaviors from their all-*trans* counterpart, including decreased color intensity, lower melting points, greater polarity, lesser tendency to crystallization, and greater solubility in oil and hydrocarbon solvents (Nguyen and Schwartz 1999).

The autoxidation of lycopene is irreversible and leads to fragmentation of the molecule, producing acetone, methylheptenone, laevulinic aldehyde and probably glyoxal, which cause apparent color loss; typically hay- or grass-like odors evolve. A reaction and degradation pathway of lycopene during production and storage of tomato-based products is shown in Fig. 2.

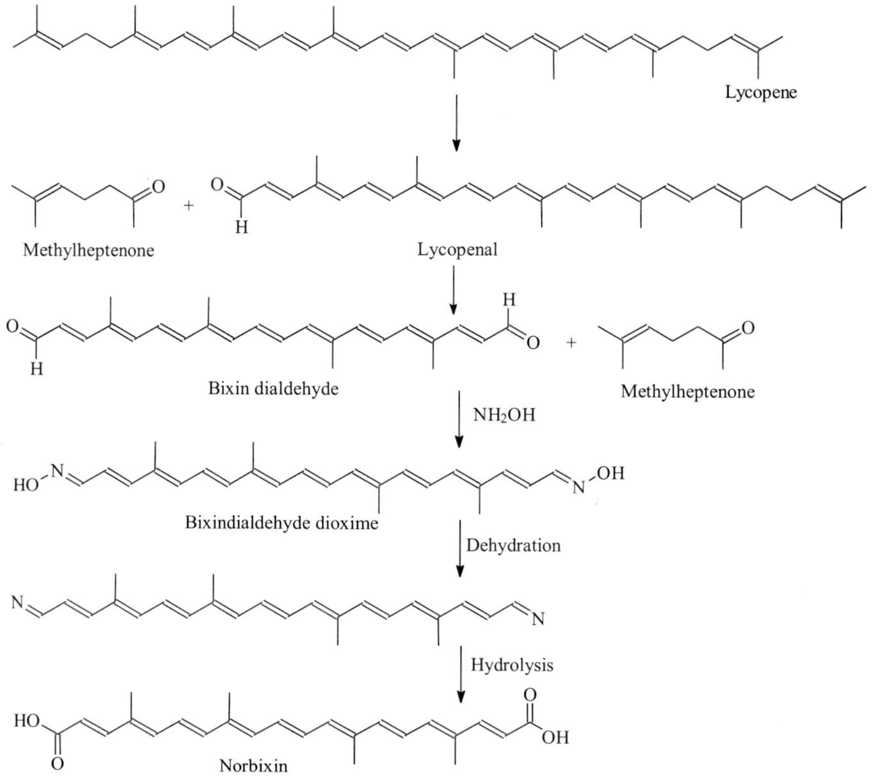

Fig. 2 Reaction pathway of lycopene during tomato powder production and storage (modified from Gould 1992).

DEGRADATION OF LYCOPENE DURING PROCESSING

Most tomato-based products are in the form of processed products such as tomato juice, paste, puree, catsup, sauce, and salsa. For juice manufacturing, sliced tomatoes undergo a hot- or cold-break method. Juice from tomatoes is usually obtained using screw or paddle extractors. In the manufacture of pulp, puree, paste, and ketchup, tomato juice is concentrated with steam coils or vacuum evaporators. For canned tomatoes, sliced or whole tomatoes are retorted. For dried tomato slices and powder, tomatoes undergo a dehydration process. Either thermal or mechanical treatments are often involved in these processes, which might affect tomato product quality. Deep-red tomato fruits that contain high concentrations of lycopene would be processed into products with a dark red color. However, the lycopene content in concentrated tomato products is generally lower than expected, because of losses during tomato processing (Tavares and Rodriguez-Amaya 1994).

The main causes of lycopene degradation during food processing are isomerization and oxidation. It is widely suspected that lycopene generally undergoes isomerization during thermal treatments. Changes in lycopene content and in the distribution of *trans-cis* isomers result in modifications of its biological properties. Determination of the degree of lycopene isomerization would provide better insights into the potential health benefits of processed food products. In processed food products, oxidation is a complex process and depends on many factors, such as processing conditions, moisture content, temperature, and the presence of pro- or autoxidants and lipids. For example, use of fine screens in juice extraction enhances the oxidation of lycopene because of the large surface exposed to air and metal. The amount of sugar, acids, and amino acids also affects lycopene degradation in processed food products by leading to the formation of brown pigments (Gould 1992). The lycopene contents of some commercial samples are listed in Table 2. The deterioration in color that occurs during the processing of various tomato products results from exposure to air at high temperatures that cause the naturally occurring all-*trans* lycopene to be isomerized and oxidized. Coupled with exposure to oxygen and light, heat treatments that disintegrate tomato tissue can result in substantial destruction of lycopene. Lycopene degradation via isomerization and oxidation affects its bioactivity and reduces its functionality with respect to health benefits. The degradation reactions of lycopene are influenced by reaction medium, temperature, physical state, and environmental conditions. The most important factors during processing are heat, light, and oxygen. The *trans*-isomer and *cis*-isomers of lycopene would degrade into oxidized fragments as shown in Fig. 3.

Table 2 A survey of lycopene content in some commercial samples.

Industrial sample	Lycopene (mg/100 g wet basis)	Sources
Tomato juice (Israel)	5.8-9.0	Lindner et al. 1984
Tomato ketchup (Finland)	9.9	Heinonen et al. 1989
Tomato puree Tomato paste Tomato ketchup (Campinas, Brazil)	19.37-8.93 18.27-6.07 10.29-41.4	Tavares and Rodriguez-Amaya 1994
Tomato soup Tomato juice Tomato sauce (USA)	8.0-13.84 9.70-11.84 6.51-19.45	Tonucci et al. 1995
Tomato pulp Pulp thin fraction (Canada)	12.09-12.83 3.98-4.08	Sharma and Le Maguer 1996

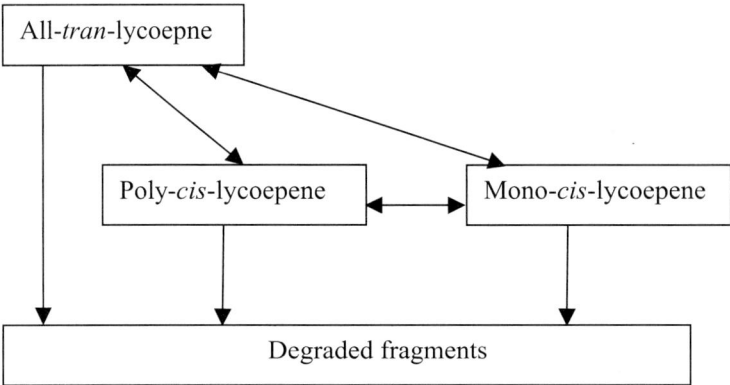

Fig. 3 Schematic of degradation pathway of all-trans-isomers and cis-isomers of lycopene.

Effects of Heating Treatments

When tomato-based products are subjected to thermal processing, the changes in lycopene content and the conversion of *trans* form to *cis*-isomers may result in a reduction in biological activity (Khachik et al. 1992, Emenhiser et al. 1995, Wilberg and Rodriguez-Amaya 1995, Stahl and Sies

1996). The results of lycopene retention in cooked tomato puree are shown in Table 3. The length of cooking time had little or no effect on the degradation of lycopene if the heating temperature was less than 100°C. These results suggest that the duration of heating, rather than cooking temperature, is the critical factor causing lycopene degradation. Temperatures higher than 120°C would result in greater lycopene loss. Significant loss can also occur when the heat treatment is too long.

Table 3   Lycopene loss rate in tomato juice during heating (Miki and Akatsu 1970).

Heating temp. (°C)	Lycopene loss (%)		
	Heating time 1 min	Heating time 3 min	Heating time 7 min
90	0.6	0.9	1.1
100	0.9	1.4	1.7
110	2.2	3.2	4.4
115	2.7	4.5	7.0
118	3.7	6.0	9.1
121	4.6	7.3	10.6
124	5.5	8.5	12.5
127	6.5	9.9	14.6
130	7.4	11.5	17.1

Lee and Chen (2002) studied the stability of lycopene by heating standard lycopene (dissolved in HPLC-grade hexane) at 50°C, 100°C and 150°C, respectively. For 50°C treatment, there was no significant change of all-*trans* lycopene found within the first 12 hr; however, the content began to decline thereafter. The levels of the mono-*cis* forms of lycopene were found to decrease with increasing heating time, which indicates that the degradation rates of mono-*cis* lycopene may be greater than the formation rate. Unlike the mono-*cis* forms of lycopene, the percentage change of two di-*cis* isomers showed increasing trends, which were probably due to conversion of mono-*cis* lycopene. There was a significant decrease of total lycopene after heating for 9 hr. The result revealed that isomerization was the main reaction during heating in the first 9 hr, after which the degradation reaction dominated. A similar result was observed for the concentration change of all-*trans* lycopene and its *cis*-isomers during heating at 100°C, the level of all-*trans* lycopene decreased by 78% after 120 min heating. The mono-*cis* forms of lycopene showed a decreasing trend. The levels of two di-*cis* isomers were found to rise in the first 60 min and then decreased, implying that di-*cis* lycopene might be converted to mono-*cis* lycopene or underwent degradation after prolonged heating. However, a large decrease was shown for the concentration of all-*trans* lycopene during heating at 150°C, and no lycopene was detected after

10 min. The levels of all the mono-*cis* forms of lycopene showed the same trend. In the first 4 min, the levels of di-*cis* isomers rose and then decreased afterward. The result indicated that the isomerization reaction was favored at the beginning and then the degradation dominated after prolonged heating at 150°C. Comparing the results, apparently with increasing temperature and heating time, degradation dominated over isomerization.

Shi et al. (2002a) dissolved extracted lycopene into canola oil and heated the sample at 25, 100 and 180°C, respectively. It was observed that increasing the temperature from 100 to 180°C or increasing thermal treatment time would increase the degradation of *trans*-isomer and *cis*-isomer of lycopene. Compared with the treatment at 25°C, the *cis*-isomers increased with thermal treatment at 100°C but dropped significantly with treatment at 180°C. The results suggested that degradation of lycopene was the main mechanism of lycopene loss at temperatures above 100°C, and lycopene in general undergoes isomerization with thermal processing.

However, the presence of certain macromolecules in fruits such as tomato might offer protection for lycopene. It was observed that lycopene loss and the rate of thermal isomerization was lower when tomato pulp was heated than when lycopene in organic solution was heated. The results from Schierle et al. (1996) showed that heating tomato-based food in oil caused greater lycopene isomerization than heating it in water, as shown in Table 4. Furthermore, Nguyen and Schwartz (1998) assessed the effects of several different heat treatments on lycopene's isomeric

Table 4 Effect of heating treatment on lycopene isomerization in aqueous and oily dispersions of tomato paste (70°C) (Schierle et al. 1996).

Heating time (min)	All-trans (%)	5-cis (%)	9-cis (%)	15-cis (%)	Other cis (%)
In water					
0	92.6	4.5	0.9	1.6	0.5
15	92.3	4.4	0.9	1.6	0.5
30	88.1	5.1	2.1	2.3	2.5
60	87.1	5.2	2.2	2.7	3.0
120	86.2	5.5	2.7	2.6	3.1
180	83.4	6.1	3.6	3.2	3.8
In olive oil					
0	87.4	4.8	4.3	3.0	0.5
30	85.2	5.8	5.5	2.9	0.5
90	83.5	6.2	5.9	3.3	1.2
120	80.3	7.0	6.9	3.2	2.6
180	76.7	8.1	8.8	3.1	3.3

distribution in a variety of tomato products, as well as in organic solvent mixtures (e.g., hexane, methanol, acetone) containing all-*trans* lycopene (Table 5). Experimental results indicated that lycopene remained relatively resistant to heat-induced geometrical conversion during typical food processing of tomatoes and related products (Table 6). The presence of fat slows the isomerization reaction and protects the *trans* and *cis* lycopene isomers against oxidation. They also found, as did Lee and Chen (2002), that lycopene in organic solvent isomerized readily as a function of time even in the absence of light and in the presence of other antioxidants.

Table 5 Lycopene isomers in various thermally processed tomato products (Nguyen and Schwartz 1998).

Sample	Total lycopene (mg/100 g, dry basis)	Cis-isomers (%)
Peeled tomato	149.89	5.37
Tomato juice (hot-break)	161.23	5.98
Tomato juice (retorted)	180.10	3.56
Tomato (whole, retorted)	183.49	3.67
Tomato paste (concentrated)	174.79	5.07
Tomato paste (retorted)	189.26	4.07
Tomato soup (retorted)	136.76	4.34
Tomato sauce (retorted)	73.33	5.13

Table 6 Total lycopene and *cis*-isomer content in dehydrated tomato samples (Shi et al. 1999).

Sample	Total lycopene (μg/g dry basis)	Loss (%)	All-trans isomers (%)	Cis-isomers (%)
Fresh tomato	75.52	0	100	0
Osmotic treatment	75.51	0	100	0
Osmotic-vacuum dried	64.09	1.51	93.49	6.51
Vacuum-dried	53.98	2.85	89.89	10.11
Air-dried	51.67	3.16	84.44	16.56

According to Sharma and Le Maguer's study (1996), heating tomato pulp at 100°C for 120 min decreased lycopene content from 185.5 to 141.5 mg/100 g total solids but was not observed to induce isomerization. Research from Agarwal et al. (2001) showed that subjecting tomato juice to cooking temperatures in the presence of corn oil resulted in the formation of the *cis*-isomeric forms. According to Ax et al. (2003), total lycopene content of oil-in-water emulsions decreased during thermal treatment with and without exposure to oxygen. Higher temperatures are directly correlated with increasing lycopene losses, and thermal treatment led to a significant decrease in the concentrations of all-*trans* and 13-*cis* isomer, while the concentration of the 9-*cis* isomer increased. Zanoni et al. (2003) and Anese et al. (2002) carried out thermal sterilization on tomato puree at temperature around 100°C and found no significant variations in the lycopene content. Takeoka et al. (2001) suggested that no consistent changes occurred in lycopene levels as fresh tomatoes were processed into hot break juice, but a statistically significant decrease in lycopene levels of 9-28% occurred as the tomatoes were processed into paste, which requires a long heating time.

In conclusion, temperatures higher than 100°C and longer heating time lead to a larger percentage of lycopene degradation; there is also controversy in the literature about whether the presence of oil during food preparation might increase the percentage of mono- or poly-*cis* lycopene in food, and whether lycopene isomerization occurs in food not cooked in oil. The molecular structural changes during heating are shown in Fig. 4, showing that high temperatures will break down molecules into small fractions (Kanasawud and Crouzet 1990).

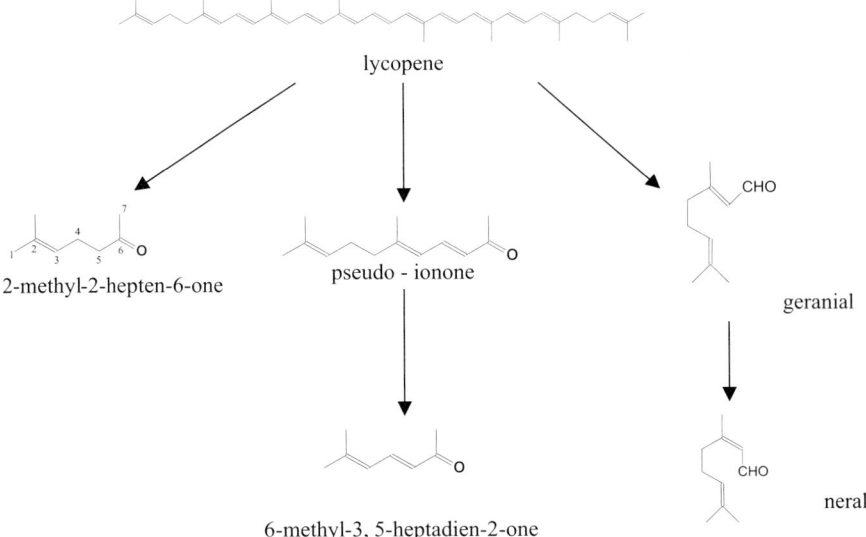

Fig. 4 Molecular structural changes and oxidated products of lycopene during heating (modified from Kanasawud and Crouzet 1990).

Effect of Light Irradiation

Lee and Chen (2002) studied lycopene by dissolving standard lycopene in hexane and illuminating it (at intensity 2000-3000 lux) at 25°C for 6 d. The content of all-*trans* lycopene was found to decrease with increasing illumination time; after 144 hr exposure to light the loss amounted to 94%. The effects of illumination on the content of total lycopene and *cis*-isomer in tomato puree are shown in Fig. 5. The increase in *cis*-isomers occurred at the onset of exposure to light. A great increase occurred in total lycopene loss as well as *cis*-isomer loss by light irradiation time and light intensity. The losses of *cis*-isomers were enhanced by exposure to light. This suggests that light irradiation causes greater loss than heating. The amount of *cis*-isomer formed during light irradiation appeared to decrease quickly as the intensity of light increased, suggesting that *cis*-isomer oxidation was the main reaction pathway. A possible explanation for this phenomenon is that the *trans*-isomer isomerizes into *cis*-isomers, then follows the oxidative pathways. The rate of *cis*-isomer oxidation was greater than rate of *cis*-isomer formation under both heat and light treatments. All the mono-*cis*-isomers of lycopene showed an inconsistent change. For instance, the level of 5-*cis* lycopene was found to increase in the beginning and then decreased after illumination time reached 2 hr. A similar trend was observed for 9-*cis*, 13-*cis*, and 15-*cis* lycopene. This result indicated that isomerization and degradation of lycopene and its *cis*-isomers, during illumination, might proceed simultaneously. The increased level of mono-*cis* lycopene is probably due to the conversion of all-*trans* lycopene, after which a decrease could occur for mono- *cis* lycopene because it might be further converted to another *cis* form of lycopene through intermediate all-*trans* lycopene or undergo degradation. Compared with concentration, the percentage changes of all-*trans* lycopene and its *cis*-isomers showed a

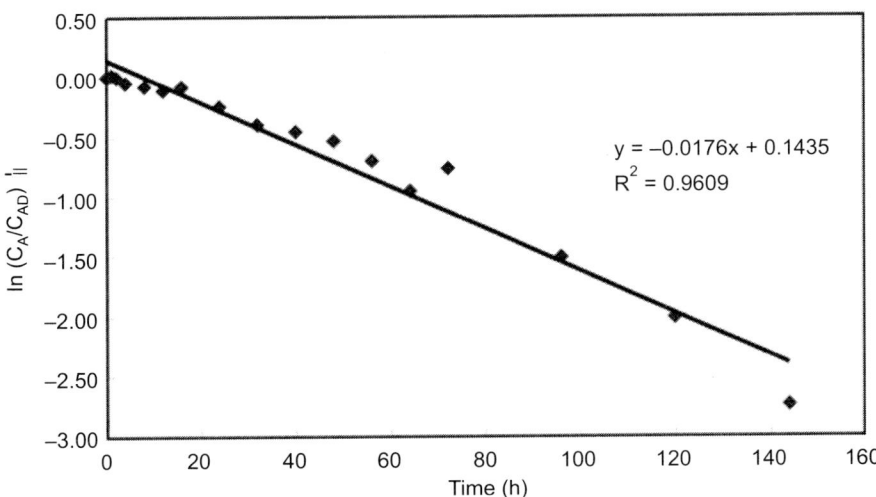

$$y = -0.0176x + 0.1435$$
$$R^2 = 0.9609$$

Fig. 5 First-order plot for the degradation of total amount of lycopene during illumination at 25°C for 144 hr (modified from Lee and Chen 2002).

different trend when the total amount of lycopene was taken into account. Prior to illumination, the total amount of lycopene was found to contain 96.4% all-*trans* lycopene, 0.99% 5-*cis* lycopene, 0.71% 9-*cis* lycopene, 0.60% 13-*cis* lycopene and 0.27% 15-*cis* lycopene. After light storage for 144 hr, a lycopene loss of 13.1% occurred for all-*trans* lycopene, while an increase of 1.47, 0.92, 5.28 and 0.44% was found for 5-*cis*, 9-*cis*, 13-*cis* and 15-*cis* lycopene, respectively. Di-*cis* lycopene also rose by 3.04%. The small percentage loss of all-*trans* lycopene during light storage revealed that all-*trans* lycopene might be isomerized to form mono-*cis* or di-*cis* lycopene. This result may account for the percentage increase of all the mono-*cis* and di-*cis* forms of lycopene during illumination.

Shi et al. (2002a) irradiated lycopene in canola oil at 2010 (outdoor), 900, 650 and 140 (indoor) $\mu mol/m^2 s$ for 1 to 6 d and found that the loss in total lycopene, *trans*-isomers and *cis*-isomers increased significantly as the intensity of the light irradiation increased. In their experiments, they found the light irradiation treatment caused more losses in total lycopene than heating treatments at 25, 100 and 180°C, while in both treatments the rate of *trans*-isomer loss was greater than the *cis*-isomer formation.

Sharma and Le Maguer (1996) studied fiber-rich fraction samples of tomato pulp under three different conditions: vacuum and dark, dark and air, and air and light, stored at –20°C, 5°C and 25°C for 60 d. The lycopene loss was highest in the presence of air and light at 25°C and lowest under vacuum and dark. According to Landers and Olson's research (1986), exposure of extracted lycopene to light should be avoided, and only gold, yellow or red lights should be used.

Effect of Oxygen

Lovric et al. (1970) found that air-packed tomato-juice powder retained the lowest lycopene levels compared with CO_2-, N_2- or vacuum-packed samples. As discussed earlier, vacuum and dark storage combination also gave the lowest lycopene loss in tomato pulp based on Sharma and Le Maguer's research (1996).

Henry et al. (1998) studied the effect of oxygen on the degradation of carotenoids in an aqueous model system. The standard lycopene was adsorbed on to a C_{18} solid phase and exposed to continuous flow of water saturated with oxygen at 30°C, and the result showed that 90% of lycopene was lost after 1 and 2 hr respectively, while other carotenoids did not achieve equivalent loss until 7 hr of exposure.

Ax et al. (2003) dissolved lycopene in the oil phase of oil-in-water emulsions, and the emulsions were filled into a glass flask with a sintered-glass frit allowing continuous flushing with either synthetic air or nitrogen gas, thus providing oxygen saturation or oxygen-free conditions. At 25°C, about 25% of lycopene was degraded within 30 hr in the oxygen-free emulsions, whereas a lycopene loss of about 80% was found for oxygen saturation. In the presence of oxygen, lycopene destabilization was about three times what it was in the absence of oxygen under inert conditions.

Nitrogen or argon headspace could be employed to keep exposure to atmospheric oxygen to a minimum (Nguyen and Schwartz 1999). As shown in Table 7, removing oxygen from the packaging improves the stability of lycopene. On the other hand, complete exclusion of oxygen using the enzyme glucose oxidase would lead to a far better stability.

Table 7 Total lycopene retention in tomato powder stored in different atmospheres, temperatures for different time lengths (Lovric et al. 1970).

Storage period (d)	Storage conditions	Total lycopene retention (%)
Fresh tomato powder		100
30	N_2, 2°C	85.5
	N_2, 20°C	90
	Air, 2°C	37
	Air, 20°C	46.3
80	N_2, 2°C	66.3
	N_2, 20°C	78.5
	Air, 2°C	11.3
	Air, 20°C	28.7
160	N_2, 2°C	54.2
	N_2, 20°C	76.5
	Air, 2°C	9.35
	Air, 20°C	25.5
210	N_2, 2°C	53.3
	N_2, 20°C	69.8
	Air, 2°C	8.55
	Air, 20°C	23
385	N_2, 2°C	53
	N_2, 20°C	65.8
	Air, 2°C	8.2
	Air, 20°C	21.8

Lycopene Isomerization

The contents of various isomers of lycopene in some commercial tomato products are shown in Table 8. It is generally accepted that the all-*trans* form of lycopene has higher stability and the *cis*-isomers have lower stability. Biological activity depends on the extent of isomerization and

Table 8 Lycopene isomers in various commercial tomato products (Schierle et al. 1996).

Sample	Total lycopene (mg/100 g wet basis)	All-trans (%)	5-cis (%)	9-cis (%)	13-cis (%)	Other cis (%)
Tomato paste (Tomatenmark, Panocchia, Italy)	52	96	4	<1	<1	<1
Tomato paste (Maracoli, Kraft, Germany)	3.7	91	5	1	2	<1
Tomato ketchup (Hot Ketchup, Del Monte, Italy)	9.5	88	7	2	3	1
Tomato ketchup (Hot Ketchup, Heinz, USA)	3.0	77	11	5	7	1
Instant meal (Eier-Ravioli, Hero, Switzerland)	0.6	76	8	5	6	5
Sauce (Hamburger Relish, Heinz, The Netherlands)	3.0	93	5	<1	3	<1
Sauce (Sauce Bolognaise, Barilla, Italy)	9.2	67	14	6	5	8
Canned tomatoes (Chris, Roger Sud, Italy)	7.1	84	5	3	5	3

oxidation as well as stability (Khachik et al. 1992, Stahl and Sies 1992, Emenhiser et al. 1995, Wilberg and Rodriguez-Amaya 1995). Although the processing of tomatoes by cooking, freezing, or canning does not usually cause substantial changes in total lycopene content, it is widely assumed that lycopene does undergo isomerization upon processing. Heat, light, acids, and other factors have been reported to cause isomerization of lycopene (Boskovic 1979, Schierle et al. 1996, Nguyen and Schwartz 1998, Shi et al. 1999). Better characterization and quantification of lycopene isomers would provide better insight into the potential nutritional quality and health benefits. The control of lycopene isomerization during production and storage could also be of benefit for improving the retention of product color and overall quality.

Lycopene isomerization increases as a function of processing time using heat. The results in Table 4 show that food processing can enhance *cis*-isomerization in tomato-based foods. Heating tomato-based foods in oil had a greater effect on lycopene isomerization than heating in water. This indicates, therefore, that not only the duration and temperature of heat treatment, but also the food matrix components such as oil or fat influence lycopene isomerization.

DEGRADATION OF LYCOPENE DURING STORAGE

During storage, besides the storage temperature, the availability of oxygen and light plays an important roles in lycopene loss. Water activity is another very important factor affecting lycoene degradation in stored tomato-based foods. While *trans*-isomers to *cis*-isomers isomerization typically occurs during processing, the storage of processed foods favors reversion from *cis*-isomers to *trans*-isomers, due to the unstable state of *cis*-isomers compared to *trans*-isomers (Shi et al. 2002b).

Sharma and Le Maguer (1996) found that, after 4 mon storage in the dark at room temperature, lycopene loss reached 97% and 79% in freeze-dried and oven-dried (25-75°C) tomato pulp, respectively. The freeze-dried samples were more voluminous and fluffy in texture compared with the thin crust of sheets of oven-dried samples, which suggested that exposure of freeze-dried fibers to air and light caused lycopene loss at a faster rate. Lovric et al. (1970) observed the effect of oxygen during the storage of lycopene powder by comparing samples stored in N_2 and air (Table 7), and results showed that oxygen is a critical factor for lycopene retention during storage.

Anguelova and Warthesen (2000) subjected tomato powder to three treatments: light exposure at room temperature, 6°C and 45°C in the dark. Differences among the storage treatments were not obvious from the data, but the amount of *cis*-isomers as a percentage of total lycopene increased to 14-18%, regardless of storage conditions (Table 9). The treatments at 6°C did not result in less 5,5′ di-*cis* lycopene than the other two treatments; this was attributed to the degradation of *cis*-isomer due to autoxidation, since *cis*-isomers were more susceptible to autoxidation than all-*trans* isomers.

Ramakrishnan and Francis (1979) indicated that the water content of the system exerted a protective influence on some carotenoid pigments above and below monolayer values. The extent of protection varied with water content, the nature of the system, and also the type of carotenoid pigments. The authors suggested that the polarity of carotenoid molecules might play a role in modifying the protective effect of moisture. Lovric et al. (1970) discussed the effect of low moisture content on the oxidative stability of dried tomato and concluded that the protective activity of water could be ascribed to a number of factors: water could inactivate

Table 9 Contents of presumptive 5,5′ di-*cis* lycopene in tomato powders (% total lycopene in the sample after a given storage period) (Abushita et al. 2000).

| Weeks | Tomato powder T1 | | | Tomato powder T2 | | |
	Light exposure	6°C	45°C	Light exposure	6°C	45°C
0	4.3	4.3	4.3	6.2	6.2	6.2
1	5.6	4.8	6.5	6.7	6.1	5.9
2	5.8	5.4	5.3	5.2	7.6	6.3
3	7	5.5	5.4	5.5	6.4	7
4	8.6	9.4	9.1	7.7	6.8	7.8
5	8.1	11	12.8	11.8	11.7	12.1
6	14.2	18.4	14.6	14.1	18.2	14.1

pro-oxidative metal ions that are present in traces; the hydrogen bonding between hydroperoxide molecules and water results in delayed chain propagation reactions in lipid autoxidation; and water can compete with oxygen for the active absorption sites.

Data from Zanoni et al. (2003) and Abushita et al. (2000) showed that very low moisture content or very low water activity seems to favor oxidative degradation in tomato products. According to Zanoni et al. (2003), adequate shelf life was obtained at relatively high moisture content (30-40%) and at high water activity (0.8-0.86). A low moisture content (≤12%) of dried tomato halves seems to promote oxidative damage. It was explained that an increase in oxidation might occur when the moisture content falls below the monolayer moisture content of the product.

Giovanelli and Paradiso (2002) used hot air to dry commercial tomato pulp to two final moisture levels—dried pulp (DP) (moisture content 9%, water activity 0.35) and intermediate moisture pulp (IMP) (moisture content 23%, water activity 0.7)—and stored them in airtight clear glass bottles in the dark without effective air exclusion for 5 mon at 4, 20, and 37°C. All-*trans* lycopene degradation was negligible in DP stored at 20°C and 37°C, while some decrease was observed in DP stored at 4°C (about 18% loss after 5 mon). All-*trans* lycopene degradation was more significant in IMP and was inversely related to storage temperature, showing about 28, 38, and 75% loss after 5 mon storage at 37, 20, and 4°C, respectively. It was concluded that moisture level in DP (approximately 9%, water activity 0.35) provides some of the protective effect, whereas at higher moisture levels (IMP moisture content 23%, water activity 0.7) the solvent effect of water prevails and reactions are favored by enhanced mobility of reaction substrates and co-substrates. Lovric et al. (1970) observed that color, total lycopene, and all-*trans* lycopene in foam-mat-dried tomato powder were better retained at 20°C than at –10, 2, and 37°C (in both cases products were stored in air and in the dark). It was explained that all-*trans* lycopene is partly converted to *cis*-isomers during drying and that re-isomerization from *cis*-isomers to the all-*trans* form is favored at 20°C (with respect to –10°C and 2°C). Since *cis*-isomers are more readily oxidized than all-*trans*-isomers, lycopene oxidation during storage occurred faster when a high *cis*-isomer ratio was present. The low storage temperatures as well as very low water activity and moisture levels have a pro-oxidative effect. Moreover, oxygen solubility increases at lower temperatures, and this could be of some importance in particulate matter such as tomato pulp pieces (Boskovic 1979).

With careful selection of storage conditions to protect the products from air (e.g., by storing in an inert atmosphere or under vacuum), it is possible to retain initial lycopene levels during storage. Low storage temperature, low oxygen content, less light irradiation, low water activity and low moisture content in storage will also limit the oxidation of lycopene. Ribeiro et al. (2003) also showed that lycopene stability strongly depends on the food system. Diluting lycopene emulsions in food system, lycopene was found very stable in orange juice; considerable lycopene loss was found in skimmed milk; and lycopene degraded very rapidly in water. Therefore, future research should pay more attention to the protective effect of different food matrices on the stability of lycopene.

In short, lycopene is prone to isomerization to a mono-*cis*-isomers or poly-*cis*-isomers configuration and degradation to oxidized products. These changes are influenced by elevated temperatures, exposure to light, oxygen, metallic ions, pH, and active surfaces. Thus, special attention must be paid to the technology used during food processing and storage so as to minimize lycopene oxidation and *trans*-isomers to *cis*-isomers isomerization. Lycopene remains relatively stable during typical food processing procedures, except at extreme conditions (e.g., very high temperature or heating for a very long time). *Trans*-isomers and *cis*-isomers isomerization may happen during processing, especially with the presence of fat, but then re-isomerization always followed during storage time. Temperature, light, oxygen, and water activity are important factors during food processing and storage of tomato-based foods and should be avoided during long storage. Autoxidation might cause final fragmentation of the lycopene molecule, which induces oxidization of the products.

Acknowledgements

The authors gratefully acknowledge the contribution of the Guelph Food Research Center, Agriculture and Agri-Food Canada (NOI 203/AAFC,

GFRC contribution No. S 277), and help from Professor Yukio Kakuda (University of Guelph).

References

Abushita, A.A. and H.G. Daood, and P.A. Biacs. 2000. Change in carotenoids and antioxidant vitamins in tomato as a function of varietal and technological factors. J. Agric. Food Chem. 48: 2075-2081.

Agarwal, A. and H. Shen, S. Agarwal, and A.V. Rao. 2001. Lycopene content of tomato products: its stability, bioavailability and in vivo anioxidant properties. J. Med. Food 4: 9-15.

Anese, M. and P. Falcone, V. Fogliano, N.C. Nicoliand, and R. Massini. 2002. Effect of equivalent thermal treatments on the color and antioxidant activity of tomato purees. Sensory and Nutritive Qualities of Food 67: 3443-3446.

Anguelova, T. and J. Warthesen. 2000. Lycopene stability in tomato powders. J. Food Sci. 65: 67-70.

Ax, K. and E. Mayer-Miebach, B. Link, H. Schuchmann, and H. Schubert. 2003. Stability of lycopene in oil-in-water emulsions. Eng. Life Sci. 4: 199-201.

Beerh, O.P. and G.S. Siddappa. 1959. A rapid spectrophotometric method for the detection and estimation of adulterants in tomato ketchup. Food Technol. 7: 414-418.

Boskovic, M.A. 1979. Fate of lycopene in dehydrated tomato products: carotenoid isomerization in food system. J. Food Sci. 44: 84-86.

Emenhiser, C. and L.C. Sander, and S.J. Schwartz. 1995. Capability of a polymeric C_{30} stationary phase to resolve cis-trans carotenoids in reversed phase liquid chromatograph. J. Chromatogr. A. 707: 205-216.

Giovanelli, G. and A. Paradiso. 2002. Stability of dried and intermediate moisture tomato pulp during storage. J. Agric. Food Chem. 50: 7277-7281.

Gould, W.V. 1992. Tomato Production, Processing and Technology. CTI Publ., Baltimore, USA.

Gross, J. 1987. Pigments in Fruits. Academic Press, London, UK.

Gross, J. 1991. Pigments in Vegetables. Van Nordstrand Reinhold, New York, USA.

Heinonen, M.I. and V. Ollilainen, E.K. Linkola, P.T. Varo, and P.E. Koivistoinen. 1989. Carotenoids in Finnish foods, vegetables, fruits, and berries. J. Agric. Food Chem. 37: 655-659.

Henry, L.K. and G.L. Catignani, and S.J. Schwartz. 1998. Oxidative degradation kinetics of lycopene, lutein, 9-cis and all-trans beta-carotene. J. Amer. Oil Chem. Soc. 75: 823-829.

Holden, J.M. and A.L. Eldridge, G.R. Beecher, I. Buzzard, B. Marilyn, C.S. Seema Davis, L.W. Douglass, H.D. Gebhardt, and S.S. Schakel. 1999. Carotenoid content of U.S. foods: an update of the database. J. Food Composition Anal. 12: 169-196.

Kanasawud, P. and J.C. Crouzet. 1990. Mechanism of formation of volatile compounds by thermal degradation of carotenoids in aqueous medium. 2. Lycopene degradation. J. Agric. Food Chem. 38: 1238-1242.

Khachik, F. and N.B. Goli, G.B. Beecher, J. Holden, W.R. Luby, M.D. Tenorio, and M.R. Barrera. 1992. Effect of food preparation on qualitative and quantitative distribution of major carotenoids constituents of tomatoes and several green vegetables. J. Agric. Food Chem. 40: 390-398.

Landers, G.M. and J.A. Olson. 1986. Absence of isomerization of retinyl palmitate, retinal, and retinal in chlorinated and unchlorinated solvents under gold light. J. AOAC 69: 50-55.

Lee, M.T. and B.H. Chen. 2002. Stability of lycopene during heating and illumination in a model system. Food Chem. 78: 425-432.

Lindner, P. and I. Shomer, and R. Vasiliver. 1984. Distribution of protein, lycopene and the elements Ca, Mg, P and N among various fractions of tomato juice. J. Food Sci. 49: 1214-1215.

Lovric, T. and Z. Sablek, and M. Boskovic. 1970. *Cis-trans* isomerisation of lycopene and colour stability of foam-mat dried tomato powder during storage. J. Sci. Food Agric. 21: 641-647.

Mangels, A.R. and J.M. Holden, G.R. Beecher, M.R. Forman, and E. Lanza. 1993. Carotenoids in fruits and vegetables: an evaluation of analytic data. J. Amer. Diet. Assoc. 93: 284-296.

Miki, N. and K. Akatsu. 1970. Effect of heating sterilization on color of tomato juice. J. Jap. Food Ind. 17: 175-181.

Nguyen, M.L. and S.J. Schwartz. 1998. Lycopene stability during food processing. Exp. Biol. Med. 218: 101-105.

Nguyen, M.L. and S.J. Schwartz. 1999. Lycopene: Chemical and biological properties. Food Technol. 53: 38-44.

Ramakrishnan, T.V. and F.J. Francis. 1979. Stability of carotenoids in model aqueous systems. J. Food Qual. 2: 177-189.

Rao, A.V. and H. Shen. 2002. Effect of low dose lycopene intake on lycopene bioavailability and oxidative stress. Nutr. Res. 22: 1125-1131.

Rao, A.V. and Z. Waseem, and S. Agarwal. 1998. Lycopene content of tomatoes and tomato products and their contribution to dietary lycopene. Food Res. Intl. 31: 737-741.

Ribeiro, H.S. and K. Ax, and H. Schubert. 2003. Stability of lycopene emulsions in food systems. J. Food Sci. 68: 2730-2734.

Schierle, J. and W. Bretzel, I. Biihler, N. Faccin, D. Hess, H. Steiner, and W. Schuep. 1996. Content and isomeric ratio of lycopene in food and human blood plasma. Food Chem. 3: 459-465.

Sharma, S.K. and M. Le Maguer. 1996. Kinetics of lycopene degradation in tomato pulp solids under different processing and storage conditions. Food Res. Intl. 29: 309-315.

Shi, J. and M. Le Maguer. 2000. Lycopene in tomatoes: Chemical and physical properties affected by food processing. Food Sci. Nutr. 40: 1-42.

Shi, J. and M. Le Maguer, Y. Kakuda, A. Liptay, and F. Niekamp. 1999. Lycopene degradation and isomerization in tomato dehydration. Food Res. Intl. 32: 15-21.

Shi, J. and Y. Wu, M. Bryan, and M. Le Maguer. 2002a. Oxidation and isomerization of lycopene under thermal treatment and light irradiation in food processing. Nutraceutical and Food 7: 179-183.

Shi, J. and M. Le Maguer, and M. Bryan. 2002b. Lycopene from Tomatoes, Functional Foods: Biochemical and Processing Aspects. CRC Press LLC, USA.

Shi, J. and M. Le Maguer, M. Bryan, and Y. Kakuda. 2003. Kinetics of lycopene degradation in tomato puree by heat and light irradiation. J. Food Proc. Eng. 25: 485-498.

Stahl, W. and H. Sies. 1992. Uptake of lycopene and its geometrical isomers is greater from heat-processed than from unprocessed tomato juice in humans. J. Nutr. 122: 2161-2166.

Stahl, W. and H. Sies. 1996. Perspectives in biochemistry and biophysics. Arch. Biochem. Biophys. 336: 1-9.

Takeoka, G.R. and L. Dao, S. Flessa, D.M. Gillespie, W.J. Jewell, B. Buebner, D. Bertwo, and S.E. Ebeler. 2001. Processing effects on lycopene content and antioxidant activity of tomatoes. J. Agric. Food Chem. 49(8): 3713-3717.

Tavares, C.A. and D.B. Rodriguez-Amaya. 1994. Carotenoid composition of Brazilian tomatoes and tomato products. Lebensm-Wiss. U-Technol. 27: 219-224.

Tonucci, L.H. and J.M. Holden, G.R. Beecher, F. Khachik, C. Davis, and G. Mulokozi. 1995. Carotenoid content of thermally processed tomato-based food products. J. Agric. Food Chem. 43: 579-586.

Weisburger, J.H. 2002. Lycopene and tomato products in health promotion. Exp. Biol. Med. 227: 924-927.

Wilberg, V.C. and B.D. Rodriguez-Amaya. 1995. HPLC quantitation of major carotenoids of fresh and processed guava, mango and papaya. Lebensm-diel-Wissenschaft und Technol. 28: 474-480.

Zanoni, B. and E. Pagliarini, G. Giovanelli, and V. Lavelli. 2003. Modeling the effect of thermal sterilization on the quality of tomato puree. J. Food Eng. 56: 203-206.

1.3

Lycopene Metabolites: Apo-lycopenals

[1]Nikki A. Ford and [2]John W. Erdman, Jr.
[1]Division of Nutritional Sciences, University of Illinois Urbana-Champaign
448 Bevier Hall, 905 S. Goodwin Ave, Urbana, IL 61820, USA
[2]Division of Nutritional Sciences, University of Illinois Urbana-Champaign
448 Bevier Hall, 905 S. Goodwin Ave, Urbana, IL 61820, USA

ABSTRACT

Epidemiological evidence suggests an inverse relationship between tomato product consumption and the risk of a number of pathologies. Lycopene is the most abundant carotenoid found in tomatoes and thus has been frequently suggested as the bioactive component responsible for the reduced risk of chronic disease. Lycopene metabolites have been recently identified and investigated *in vivo* and *in vitro*. The mechanisms of lycopene cleavage *in vivo* have not been determined although some of these metabolites are found in similar concentrations to retinoids produced *in vivo*. It is plausible that some metabolic products of lycopene are bioactive.

INTRODUCTION

Higher consumption of tomato products is associated with a reduced incidence of a number of chronic diseases (Agarwal and Rao 2000, Giovannucci 2002, Nkondjock et al. 2005). Lycopene is the most abundant carotenoid in tomatoes and therefore has received a great deal of scientific attention. Early studies suggest that an intraperitoneal injection of lycopene invoked a non-specific resistance against some strains of bacteria in mice (Lingen et al. 1959). Later, low serum lycopene levels were associated with an increased risk of pancreatic cancer (Burney et al. 1989).

A list of abbreviations is given before the references.

Lycopene has also been implicated to reduce the risk of cardiovascular disease (Martin et al. 2000). Ever since 1995, when lycopene-containing foods were reported to be associated with a reduced risk of prostate cancer, much research has been dedicated to the investigation of lycopene, tomato products, and prostate cancer (Giovannucci et al. 1995).

Two studies conducted in our laboratory indicate that intact lycopene is not as effective in disease prevention as the whole tomato. In one study, a chemically induced rat prostate cancer model was used to assess the differences in feeding lycopene or tomato powder on the progression of prostate cancer (Boileau et al. 2003). Only the tomato powder diet resulted in a significantly reduced prostate carcinogenesis and longer lifespan. The lycopene-enriched diet insignificantly reduced prostate cancer. This data suggests that lycopene contributes to a reduction in the development of prostate cancer, but other compounds within tomatoes may further contribute to cancer inhibition. In a follow-up publication from our laboratory investigating growth of transplantable prostate tumor growth in rats, tomato powder significantly reduced tumor growth while lycopene supplementation did not significantly affect growth (Canene-Adams et al. 2007). These results again suggested that although lycopene may offer some anti-tumor activity, other compounds within the whole tomato also contribute to the observed anti-tumor effect with the consumption of tomato products.

Despite the evidence insinuating a beneficial affect of lycopene against some chronic diseases, the mechanism(s) of action have yet to be elucidated. It is unclear whether lycopene itself, its metabolites, or other bioactive components of tomato products contribute to the reduced risk of some diseases. Moreover, it is plausible that metabolites of lycopene may be more biologically active than the parent compound and may be responsible for lycopene's bioactivity.

Carotenoids are cleaved into biologically active apo-carotenoids in plants, animals, and some bacteria and fungi. In plants, apo-carotenoids are important as hormones, aroma compounds, and pigments. The apo-carotenoid abscisic acid aldehyde is a plant growth regulator that acts mainly to inhibit growth, promote dormancy, and help the plant tolerate stressful conditions (Giuliano 2003). Violaxanthin is isomerized and oxidized to form xanthonin, which is spontaneously metabolized to abscisic acid aldehyde and with further oxidation results in the production of abscisic acid (Giuliano 2003). Abscisic acid is produced in plant roots in response to decreased soil water and it prevents additional water loss from the leaves during times of low water availability. The cleavage of abscisic acid aldehyde is therefore essential in plants. In addition, other plant-derived apo-carotenoids are of importance, including bixin, which is often used for food coloring, and crocin, which is the primary pigment in saffron (Giuliano 2003).

In the animal kingdom, some apo-carotenoids have vitamin A activity, can be signaling molecules, or can be part of visual pigments. The apo-carotenoid retinal is formed from the cleavage of β-carotene, α-carotene, or β-cryptoxanthin. Retinal is a necessary component of the photosensory pigment in the retina (Wald 1985). The metabolism of beta-carotene and its conversion to vitamin A has been studied extensively, while other carotenoids have received far less attention. Lycopene metabolites and metabolism have been studied *in vivo* and *in vitro* but mechanisms of lycopene cleavage *in vivo* are inconclusive (Campbell et al. 2004, Voutilainen et al. 2006, Wertz et al. 2004, Zaripheh et al. 2006).

The term "lycopenoid" has been devised by our lab to describe lycopene metabolites, including apo-lycopenals. We believe these metabolites are enzymatically produced in a similar manner to retinoids (Lindshield et al. 2007). Notably, lycopenoids have been found in similar concentrations to retinoids *in vivo* (Table 2). Since the retinoids, *all-trans* and 9-*cis* retinoic acid are ligands for nuclear receptors RAR and RXR, it is plausible that oxidation products of apo-lycopenals may also be biologically active as agonists or antagonists for the genes controlled by these or other nuclear receptors.

PROPERTIES OF LYCOPENALS

Structure of Apo-lycopenals

Apo-lycopenals are poly-isoprenoid compounds of 40 carbons or less in length derived from the parent compound, lycopene. They contain at least one aldehyde end group. The molecular weight of apo-8′-lycopenal, apo-10′-lycopenal, and apo-12′-lycopenal are 416, 376, and 350, respectively (Fig. 1).

Physical Properties of Apo-lycopenals

Lycopene and apo-8′-lycopenal maximally absorb light at 472 and 473 nm in hexane, respectively. The maximum absorbance of apo-10′-lycopenal is 460 nm, while apo-12-lycopenal is absorbed at 445 nm. Chain-shortened aldehydes of lycopene lose their classical, "three-fingered" spectra, as shown in Fig. 2.

Color is an obvious observable difference between lycopene and some of its metabolites. The perceived color of these compounds is affected by

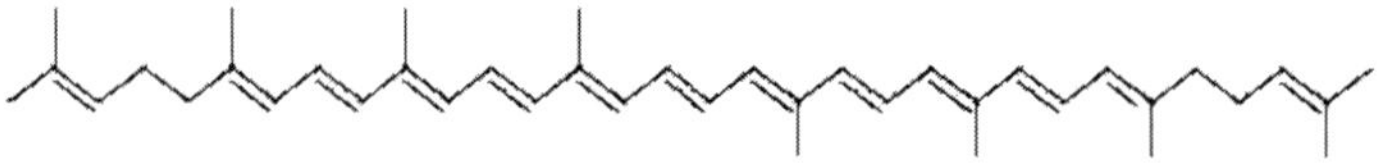

Lycopene
MW: 537

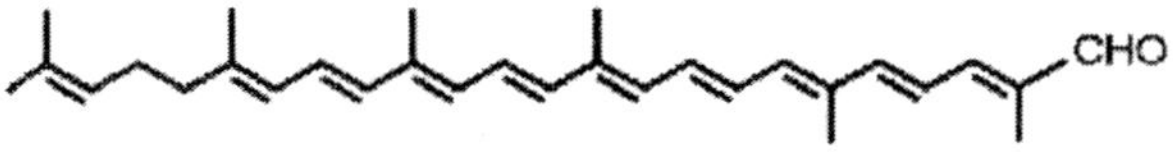

Apo-8'-lycopenal
MW: 416

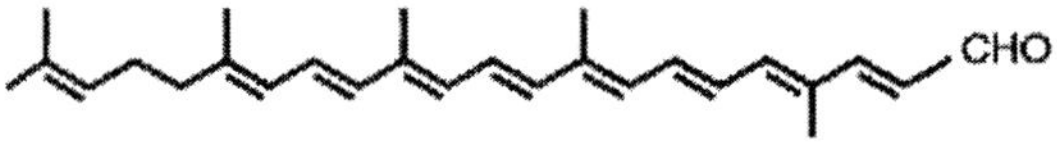

Apo-10'-lycopenal
MW: 376

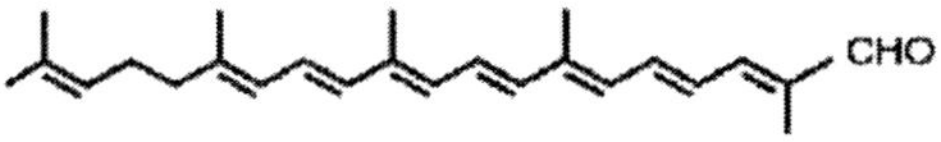

Apo-12'-lycopenal
MW: 350

Fig. 1 Lycopene and three apo-lycopenals. Apo-lycopenals are chain-shortened aldehydes with lower molecular weights than lycopene.

the maximum wavelength absorbance where lycopene and apo-8'-lycopenal are a deeper red than the other two apo-lycopenals.

Absorption and Transport

Currently there is no data available about the absorption and transport of apo-lycopenals. Lycopene is transported by chylomicrons from the gut to the liver in the mesenteric lymph and portal systems and then transported from the liver to peripheral tissues in lipoproteins. In the fasted state, serum lycopene is mostly found in low density lipoprotein (LDL) or high density lipoprotein (HDL) fractions. Lycopene is thought to be absorbed into the enterocyte by passive or facilitative diffusion. Intestinal binding proteins facilitate absorption of other carotenoids and could also bolster lycopene or lycopenoid absorption. The study of ATP binding cassette G5 genetic variants suggests that this binding protein facilitates lutein absorption and could be responsible for the transport of other carotenoids into the enterocyte (Clark et al. 2006). Scavenger Receptor class B type I protein (SR-BI) and cluster determinant 36 (CD36) are responsible for the transport of lipids and cholesterol from lipoproteins into tissues. In

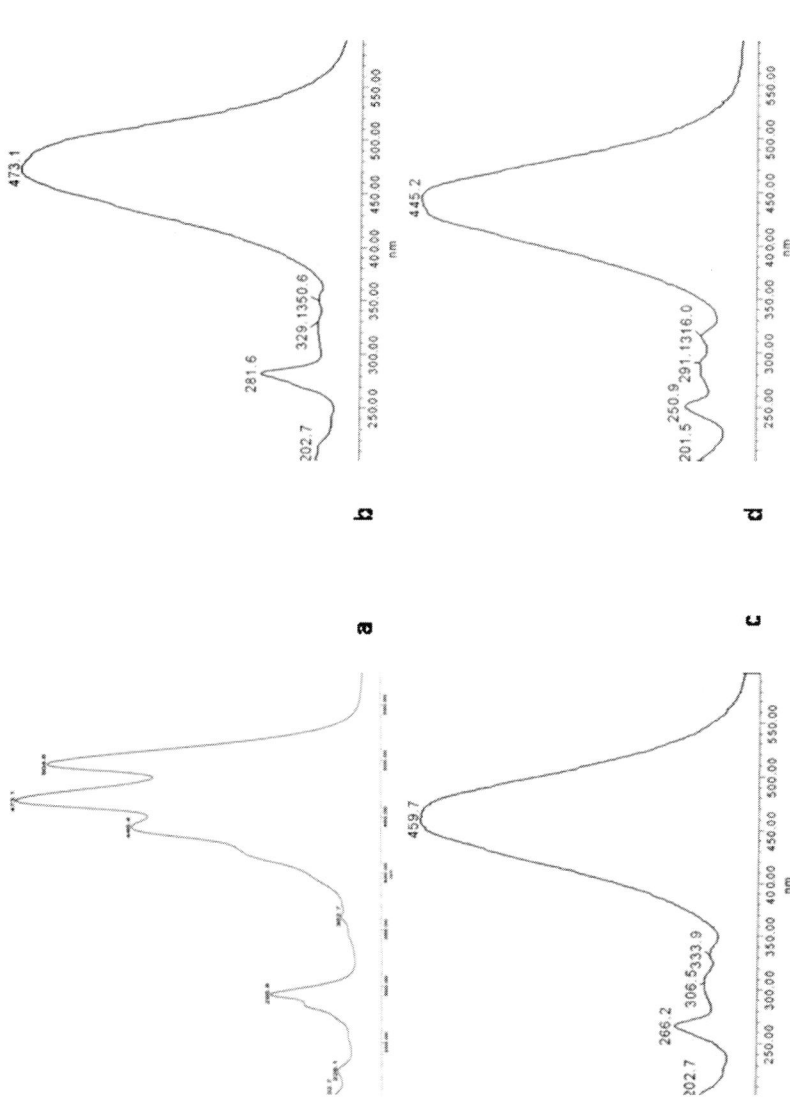

Fig. 2 UV spectrum of apo-lycopenals. Lycopene (a) optimally absorbs at 472 nm with a three-fingered spectrum. Apo-8'-lycopenal (b) optimally absorbs at 473 nm with one peak while apo-10'-lycopenal (c) and apo-12'-lycopenal (d) maximally absorb at 460 and 445 nm, respectively. All compounds were dissolved in hexane (Ford et al., unpublished data).

addition, they facilitate the absorption of β-carotene and lutein (Reboul et al. 2005, van Bennekum et al. 2005). However, SR-BI has a greater influence on carotenoid transport than CD36. Carotenoids may also be taken up into tissues upon contact with chylomicrons or LDL particles. Therefore, it has been suggested that the differential tissue uptake of lycopene and other carotenoids is dependent on the amount of LDL receptors and SR-BI (Borel et al. 2005).

Retinoids are solubilized, transported, and protected by retinoid-binding proteins (Noy 2000). Retinoid-binding proteins bind different retinoids with varying affinities to regulate disposition, metabolism, and bioactivity. However, no apo-lycopenal binding proteins have currently been identified.

Isomers of lycopene differentially accumulate in tissues and differ in bioavailability. Moreover, while approximately 90% of dietary lycopene is in the all-*trans* conformation, *cis*-isomers are predominately found in human tissues. Cooking tomatoes usually increases the bioavailability of lycopene through its release from the matrix into the lipid portion of the meal. Additionally, through tomato processing, some lycopene is modestly converted from its all-*trans*-isomer to its *cis*-isomer (Nguyen and Schwartz 1998, Shi and Le Maguer 2000). Data from our and other labs indicates *cis*-lycopene isomers are more readily absorbed by enterocytes and preferentially accumulate in tissues than all-*trans* lycopene (Boileau et al. 2002, Stahl and Sies 1992). The mechanisms accountable for favored uptake of *cis*-isomers may be explained by two theories; the reduced length of *cis*-isomers more readily allows them to be incorporated into micelles, and/or all-*trans* isomers more readily form crystals, severely reducing their uptake by micelles (Boileau et al. 2002). Short-chain apo-lycopenals may be more bioavailable than some isomers of lycopene. Further work is necessary to determine the bioavailability of lycopene metabolites.

Like other carotenoids, apo-lycopenals are hydrophobic molecules with little to no solubility in water although they are significantly more polar than the parent compound, lycopene. It can be predicted that they would be restricted to lipid portions of cells, mainly membranes and lipid droplets, and may be bound to protein for transport through the body. In addition, the aldehyde portion of these molecules may readily interact with neighboring compounds. It may also be expected that apo-lycopenals might be easily reduced to apo-lycopenols and perhaps oxidized to apo-lycopenoic acids. It is of vital interest to identify the biological function of these compounds *in vivo*.

PRODUCTION OF APO-LYCOPENALS

Apo-carotenoids may be produced *in vivo* by cleavage enzymes or oxidation or may even be of dietary origin. The study of carotenoid metabolites is often difficult because of their instability and the high vulnerability to oxidation of the parent compound. The production of apo-lycopenals has not yet been clearly defined.

In vitro Oxidation

Carotenoids are readily oxidized because of the highly reactive double bonds with reactive oxygen species (ROS). Lycopene, because of its high number of conjugated double bonds, is most susceptible to oxidation. Oxygenated lycopene has been found in human blood including *cis* and all-*trans* 5,6-dihydroxy-5,6-dihydrolycopene (Khachik et al. 1992). The production of lycopene oxidation products has been studied extensively *in vitro* and some of the oxidative products formed are apo-lycopenals (Ferreira et al. 2003, Kim et al. 2001, Zhang et al. 2003). It has been hypothesized that lycopene can be indiscriminately cleaved by non-enzymatic reactions in oxidatively stressed systems. Therefore, it appears that apo-lycopenals may form enzymatically or non-enzymatically.

Many *in vitro* studies using mild oxidative catalysts have produced a spectrum of oxidative and cleavage products including apo-lycopenals by reaction with singlet oxygen or radical oxidation attack (Handelman et al. 1991, Stratton et al. 1993). It has been suggested that these cleavage products are produced and then further metabolized for excretion by the body (Nagao 2004). Unfortunately, this method of apo-lycopenal formation is random and unpredictable.

Enzymatic Cleavage

Lipoxygenase

Lipoxygenases, found in a variety of plants and animals, catalyze the dioxygenation of polyunsaturated fatty acids. They are often used *in vitro* to induce oxidation of substrates. One study investigated the metabolism of lycopene in post-mitochondrial fractions of rat intestinal mucosa with added soybean lipoxygenase at 37°C for up to 90 min (Ferreira et al. 2003). Incubation of lycopene with lipoxygenase significantly increased the creation of lycopene cleavage and oxidation products. Two of the proposed cleavage products were 3,4-dehydro-5,6-dihydro-15-apo-lycopenal and 2-apo-5,8-lycopenal-furanoxide.

Excentric cleavage products of β-carotene have also been identified by incubating human gastric mucosal homogenates with lipoxygenase, including β-apo-8′, -10′, -12′, and 14′-carotenals (Yeum et al. 1995). Furthermore, a specific inhibitor of lipoxygenase, NDGA, inhibited β-apo-carotenoid production.

In conclusion, lipoxygenase appears to produce cleavage products of lycopene and β-carotene. It is evident that lipoxygenase-type enzymes are capable of metabolizing lycopene and other carotenoids.

Cytochrome P450

Cytochrome P450s (cyp 450) are membrane-associated proteins found in the endoplasmic reticulum or the inner mitochondrial membrane. They are found throughout the body and involved in numerous metabolic reactions. Cyp P450s are often associated with the metabolism of toxins, drugs, or waste products, although they are also important in the synthesis and degradation of hormones. Oxygenase reactions are the most common reactions catalyzed by P450 enzymes. CYP26 is specific to the metabolism of retinoids (White et al. 1997). Although much research has been dedicated to this diverse superfamily, many cyp 450s have unknown functions.

P450 enzymes may regulate the metabolism of carotenoids *in vivo*. The production of biologically active metabolites must be regulated to prevent toxicity yet maintain a balance of essential compounds. The cytochrome P450RAI, also called CYP26a, is responsible for the inactivation of all-trans retinoic acid (White et al. 1997). This enzyme is predominantly expressed in the cerebellum and pons of the human brain, which investigators suggest must serve to protect these tissues from excess exposure to retinoids (White et al. 2000). Furthermore, the specificity of P450RAI for different retinoids was investigated in monkey kidney fibroblast cells. This cyp 450 was most specific for all-trans retinoic acid followed by the apo-carotenoid, retinal, and least specific for retinol. At 10 μM, retinal competitively inhibited P450RAI-mediated all-trans retinoic acid metabolism by 25%. Other retinoids may also be metabolized by this cyp 450 family. Therefore, it is plausible that apo-lycopenals are metabolized by cyp 450s.

In our lab, we investigated the effects of feeding a lycopene-enriched diet on expression of rat hepatic P450 enzymes (Zaripheh et al. 2006). Rats were fed an AIN-93G based diet supplemented with 0.25 g lycopene/kg diet for 30 d. The expression of phase I and phase II detoxification enzymes was not enhanced by the lycopene-enriched diet. The results of this study suggest that a moderately high concentration of lycopene in the diet was not sufficient to induce P450 enzymes.

Converse to the previous study, three hepatic cyp 450-dependent enzymes were induced when female rats were orally administered 0.001 to 0.1 g lycopene per kg body weight per day for 2 wk (Breinholt et al. 2000). Induction of these enzymes by a relatively low dose of lycopene suggests that human lycopene consumption may modulate drug-metabolizing enzymes.

The conflicting results from studies investigating carotenoids and induction of P450 enzyme activity may be due to differing carotenoid substrates, different models, or most importantly varying dose concentrations of carotenoid. Induction of these detoxification enzymes by high doses of carotenoid is the most plausible explanation as P450 enzymes are often involved in the metabolism of toxins, drugs, or high levels of specific food components.

Carotenoid cleavage enzymes

Three carotenoid cleavage oxygenases have been identified in mammals, CMOI, CMOII, and RPE65. Many other members of this enzyme family are found in bacteria and plants with a wide substrate specificity (Kloer et al. 2005). These enzymes belong to a superfamily of iron-containing oxygenases that encompass a wide array of physiological roles. It is suggested that RPE65 is a binding protein for all-trans-retinyl esters but is not responsible for apo-carotenoid production (von Lintig et al. 2005).

Carotenoid Oxygenase I and Carotenoid Oxygenase II. Carotenoid Oxygenase I (CMOI) is also known in the literature as β,β-carotene 15,15′-Dioxygenase, BCOI, and βCMOOX (Wyss et al. 2000), while Carotenoid Oxygenase II (CMOII) is also known as β,β-carotene-9′,10′-oxygenase and BCOII (Kiefer et al. 2001, Wyss 2004). Further characterization of these enzymes and their functions will help to clarify appropriate nomenclature. Figure 3 demonstrates the action of CMOI on β-carotene and the proposed mechanism of CMOII on lycopene.

Carotenoid cleavage enzymes were first proposed in the 1930s as a mechanism for vitamin A formation from β-carotene cleavage at the central 15,15′-double bond (Moore 1930). The central cleavage of β-carotene to form retinal was established in rat liver and intestine in the laboratories of Goodman and Olson in 1965 (Goodman and Huang 1965, Olson and Hayaishi 1965). Carotenoid oxygenase I symmetrically cleaves β-carotene to form two molecules of retinal. Additionally, through the use of lycopene-producing *E. coli*, it was demonstrated that lycopene was not oxidatively cleaved by CMOI (Kiefer et al. 2001). This cytoplasmic enzyme was first cloned in 2000 by Wyss et al. (2000). It is primarily found in the intestinal mucosa of mammals but has also been found in numerous other tissues (Lakshman et al. 1989, Nagao et al. 1996, Parker 1996).

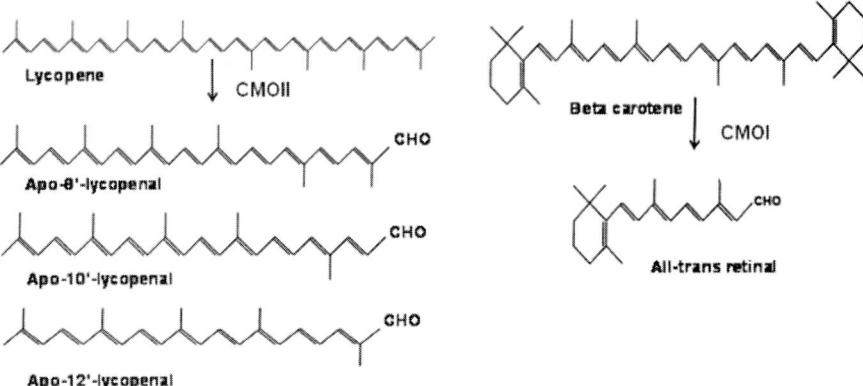

Fig. 3 Proposed carotenoid metabolism pathway. Carotenoid Oxygenase I (CMOI) is the primary enzyme for the central cleavage of β-carotene to two molecules of all-*trans* retinal. Carotenoid Oxygenase II (CMOII) is proposed to be the primary enzyme for the asymmetric cleavage of lycopene in the formation of apo-lycopenals (Lindshield et al. 2007).

A second carotenoid cleavage enzyme, CMOII, was later identified and is proposed to be responsible for the excentric cleavage of carotenoids by Wang and colleagues. This group then identified two β-carotene products of this excentric cleavage as β-apo-13-carotenone and β-apo-14′-carotenal (Wang et al. 1991). However, β-carotene was not cleaved by CMOII in *E. coli*. Carotenoid oxygenase II is expressed in many mammalian tissues including those insensitive to vitamin A deficiency and also where no CMOI has been found; this suggests that CMOII has a function outside of vitamin A synthesis (Lindqvist et al. 2005). The two enzymes are similarly distributed and expressed in the human small intestine, testes, liver, and adrenal glands (Table 1). Both are independently expressed in many other tissues. Carotenoid oxygenase II appears to be less specific, accepting a wider range of substrates, including non-cyclic lycopene, although it has been reported to prefer cis-isomers over trans-isomers (Hu et al. 2006, Kiefer et al. 2001).

Recently, our laboratory measured the expression of CMOI and CMOII in F344 rats fed a lycopene-enriched diet. Expression of CMOI was decreased in the kidney and adrenal with lycopene consumption, while that of CMOII was only significantly reduced in the kidney (Zaripheh et al. 2003, 2006). In another study, the expression of CMOII mRNA increased 4-fold in the lungs of ferrets fed lycopene when compared to control ferrets (Hu et al. 2006).

Table 1 Carotenoid oxygenase tissue expression

Tissue	CMOI	CMOII
Adrenals	+ +	+ +
Small intestine	+ + +	+ + +
Liver	+ + +	+ + +
Skeletal muscle	+	+ + +
Heart	–	+ + +
Eye	+ + +	+ + +
Prostate	+	+ +
Pancreas	+ + +	+ +

An expression profile of tissues of moderate to high expression of CMOII compared to the expression of CMOI in the same tissues. (–) zero, (+) low, (+ +) moderate, (+ + +) high expression levels (Kiefer et al. 2001).

Other *in vivo* factors may influence carotenoid cleavage by CMOI and CMOII. As detailed below, α-tocopherol and ascorbic acid may affect the enzymatic cleavage of β-carotene. Ascorbic acid and α-tocopherol are potent antioxidants *in vivo* and *in vitro*. There appears to be an interaction among these vitamins and carotenoids that facilitates valuable defense against oxidative stress. Lycopene is a potent antioxidant *in vitro*, but little data is available about its possible interactions with other antioxidants in the prevention of disease.

It was found that in the presence of α-tocopherol, β-carotene was only symmetrically cleaved to retinal in the rat intestinal mucosa; in the absence of α-tocopherol, β-carotene was additionally converted to apo-carotenoids (Yeum et al. 2000). The same group then investigated the effects of α-tocopherol and ascorbic acid on cleavage of β-carotene in cigarette smoke-exposed ferret lungs *in vitro* (Liu et al. 2004). In contrast to the previous study, α-tocopherol reduced the production of apo-carotenals. Ascorbic acid had little effect on the production of apo-carotenals, while the combination of α-tocopherol and ascorbic acid significantly reduced the cleavage of β-carotene to apo-carotenals. Whether the contradiction in results is due to the extreme oxidative differences between tissue models or other factors, further research is necessary to elucidate the effects of other anti-oxidants on the metabolism of carotenoids.

It has been suggested that apo-carotenals are further shortened through a β-oxidation-like mechanism (Wolf 2001). These β-apo-carotenals could be shortened to produce retinal and then oxidized to retinoic acid or reduced to retinol. This identifies a secondary mechanism for the cleavage of β-carotene to retinol. However, this mechanism is estimated to contribute only about 5% of retinoid production *in vivo* (Wolf 2001).

Apocarotenoid-14',13'-Dioxygenase. Another carotenoid cleavage enzyme, β-apocarotenoid-14',13'-dioxygenase (ADO) was identified in rat and rabbit intestinal mucosa extracts (Dmitrovskii et al. 1997). This enzyme is involved in the oxidation of apo-8'-carotenol to apo-14'-carotenal. It was distinguished from the other oxygenase enzymes by differences in optimal pH, heat inactivity, and required activators. Researchers suggest that metabolism of apo-carotenols to retinal and other apo-carotenals is mediated by ADO through a different metabolic pathway than the conversion of β-carotene to retinal by CMOI (Gessler et al. 1998).

Carotenoid Cleavage in Plants. The pathways for production of apo-lycopenals in plants have two things in common; they originate from a C40 carotenoid and the first cleavage yields an aldehyde. A family of carotenoid cleavage dioxygenases (CCD) has been identified in plants in recent years (Simkin et al. 2004). VP14 is a member of this family and is also the enzyme responsible for synthesis of abscisic acid. These enzymes are dioxygenases that cleave a variety of double bond substrates into apo-carotenoids. Little is known about the secondary products formed by these cleavage enzymes beyond their contribution to flavor and aroma of the fruit.

A carotenoid cleavage dioxygenase, AtCCD1, was identified that cleaves the 9',10' double bond of many carotenoids to form C14 dialdehyde and C13 cyclohexone *in vitro* (Schwartz et al. 2001). Two homologous enzymes in tomatoes, LeCCD1A and LeCCD1B, also were determined *in vitro* to cleave multiple carotenoid substrates at the 9',10' double bond (Simkin et al. 2004). Conversely, gene silencing of LeCCD1 in transgenically altered plants did not significantly affect the production of phytoene, lycopene, β-carotene, or lutein in the fruit.

CCD7/MAX3 and CCD8/MAX4 are two enzymes with broad carotenoid substrate specificity that putatively synthesize a novel apo-carotenoid hormone that controls lateral shoot growth. CCD7/MAX4 cleaves at the 9',10' position to form two aldehydes. Specifically, it cleaves β-carotene to generate β-ionone and β-apo-10'-carotenal (Schwartz et al. 2004). CCD8/MAX4 was then shown to further cleave β-apo-10'-carotenal into two dialdehydes. Investigation of CCD8/MAX4 induction in an *E. coli* strain engineered to accumulate lycopene indicated an increased cleavage due to reduced lycopene accumulation, but researchers were unable to identify any metabolites to confirm cleavage products (Auldridge et al. 2006).

Another lycopene oxygenase initiates the first step in the production of bixin from the tropical plant *Bixa orellana* at the 5',6' bond. Bixin is widely used in foods and cosmetics and is also known as annatto. Some have reported that bixin and norbixin induce CYP1A1 and antigenotoxic activity, respectively (De-Oliveira et al. 2003, Junior et al. 2005).

The identification of plant carotenoid cleavage enzymes raises the question of whether we consume apo-lycopenals in foods. One group identified apo-8'-lycopenal and apo-6'-lycopenal in ripe tomatoes (Ben-Aziz et al. 1973). They indicated that more of these compounds were present in over-ripe tomatoes. It was suggested that they were formed in response to stress, injury, or during senescence of the tomatoes. It has also been proposed that the processing of tomato products exposes lycopene to high temperatures and oxygen, thus allowing for an increase in lycopene oxidative products (Shi and Le Maguer 2000).

Comparison of Plant and Animal Carotenoid Cleavage Enzymes. One study compared carotenoid cleavage enzymes in plants and animals (Fleischmann 2002). Commercially available star fruit and quince fruit and blowflies, locusts, Wistar rats and birdwing butterflies were used in this investigation. Plant and animal carotenoid cleavage enzymes were different in all but two of the characteristics studied. Plant enzymes had a much smaller molecular weight and isoelectric point than animal carotenoid cleavage enzymes. Additionally, plant enzymes cleaved beta-carotene optimally at higher temperatures and pH than animal enzymes. All enzymes studies were first order and had differing Vmax although the Km was similar between plants and animals. The time constant was also similar between plant and animal carotenoid cleavage enzymes. This analysis concludes there are clearly two separate groups; animal and plant carotenases.

In conclusion, the production of apo-lycopenals may involve free radical oxidation, lipoxygenase activity, cyp 450s, or a host of carotenoid cleavage enzymes currently known or yet to be identified. The structure of apo-lycopenals may be similar to other apo-carotenoids or very distinct as observed by the two different functions and specificities of CMOI and CMOII. Additionally, production of apo-lycopenals may require co-factors or inhibitors such as that seen in many *in vitro* studies using lipoxygenase. Through the future use of knock-out models and other *in vivo* work, the formation of apo-lycopenals should be further elucidated.

APO-LYCOPENALS IDENTIFIED *IN VITRO*

Many studies have reported the identification and characterization of *in vitro* lycopene metabolites and oxidative products. Some of these compounds have also shown anti-cancer activity *in vitro* through inhibition of proliferation, enhancement of gene expression of connexin 43 (which enhances communication between cells), or induction of apoptosis (King et al. 1997, Zhang et al. 2003, Ford et al. 2007). Thus, some of these compounds may be physiologically bioactive and further identification and characterization of lycopene metabolites *in vitro* is warranted.

One study investigated the *in vitro* autoxidation of lycopene under different conditions. Lycopene was solubilized in tween 40 or dissolved in toluene and incorporated into a liposomal phospholipid membrane and extracted and analyzed after 72 hr. It was concluded that lycopene was sequentially cleaved into a series of apo-lycopenals and short-chain carbonyl compounds *in vitro* under oxidative conditions (Kim et al. 2001). Among other oxidation products formed, apo-14'-lycopenal, apo-12'-lycopenal, apo-10'-lycopenal and apo-6'-lycopenal were identified. The author also suggested that apo-6'-lycopenal may have been oxidized further to shorter chain apo-lycopenals. This data suggests that, like retinoic acid, tissues exposed to oxidative stress may induce lycopene metabolite formation and these metabolites may be essential for the biological effects observed with fruit and vegetable consumption (Kim et al. 2001).

Numerous other studies have investigated lycopene metabolite formation *in vitro* (Caris-Veyrat et al. 2003, Khachik et al. 1998). For example, Caris-Veyrat and colleagues studied metabolite formation by two different oxidation systems; potassium permanganate and oxygen catalyzed by metalloporphyrin. Eleven apo-lycopenals were found and characterized and a novel compound, apo-11-lycopenal, was identified. The authors proposed that apo-lycopenals were oxidatively cleaved from epoxides of lycopene.

There are some problems associated with the use of *in vitro* systems for the investigation of carotenoids. The partial pressure of oxygen during *in vitro* studies is often very high in comparison to most tissues in the body, while the concentration of carotenoids is usually supraphysiological in these studies. Conversely, these models may be important representatives of carotenoid cleavage in smokers, cancer patients, or other high oxidative stress conditions.

Many lycopene metabolites or oxidative products have been identified *in vitro*, but it is unknown whether many of these products are biologically produced. Further studies are necessary to identify whether these compounds are produced *in vivo* and to elucidate their biological function.

METABOLITES IDENTIFIED *IN VIVO*

A few metabolites of lycopene have been identified *in vivo*. Interestingly, those lycopene metabolites identified are of comparable biological concentrations as β-carotene metabolites (Table 2).

Table 2 *In vivo* carotenoid metabolite concentrations.

Metabolite	Animal	Tissue	In vivo conc.	Source
Apo-8'-lycopenal	F344 rats	liver	~250 ng/g	Gajic et al. 2006
Apo-12'-lycopenal	F344 rats	liver	≥ 250 ng/g	Gajic et al. 2006
Apo-10'-lycopenol	ferrets	lung	2-4 ng/g	Hu et al. 2006
Retinal	Wistar rats	small intestine	8 µg/g	Thomas 2005
All-*trans*-retinoic acid	humans	plasma	3.7 ng/ml	Napoli 1985
All-trans-retinoic acid	Sprague-Dawley rats	plasma	1.2 ng/ml	Napoli 1985
β-apo-8'-carotenal	humans	plasma	245 ng/ml	Ho 2007

Apo-8'-lycopenal and Putatively Apo-12'-lycopenal

Apo-lycopenals were identified in the liver of rats fed lycopene-enriched diets (Gajic et al. 2006). Male F344 rats consumed a diet with a total lycopene content equivalent to humans consuming 2.7 tomatoes daily for 30 d (Zaripheh and Erdman 2005). A single oral dose of $(6,7,6'7')$- ^{14}C labeled lycopene was administered 24 hr before sacrifice. Apo-8'-lycopenal and putatively apo-12'-lycopenal were identified using HPLC-PDA, radioactivity detection, and either ESI-MS or ESI/LC-MS (Fig. 4, see arrow). Many other polar metabolites or oxidative products were observed but were not identified. No apo-10'-lycopenal, apo-10'lycopenol, apo-10'-lycopenoic acid, apo-8'-lycopenol, or apo-12'-lycopenol was detected in rat liver. The lycopenoic acids may have been present in the more polar fractions that could not be separated for analysis. It was estimated that apo-8'-lycopenal comprised approximately 2% of total lycopene metabolites in the liver at approximately 250 ng per gram of liver tissue (Gajic et al. 2006). Further analysis of the polar lycopene metabolites may have revealed additional short-chain apo-lycopenals produced in rat liver. Figure 4 shows an HPLC and radioactivity profile of a rat fed lycopene for 30 d and then gavaged 24 hr before sacrifice with ^{14}C-lycopene.

Apo-10'-lycopenal

Another lycopene metabolite, apo-10'-lycopenol, was identified in the lungs of ferrets consuming a lycopene supplemented diet for 9 wk (Hu et al. 2006). Ferrets consumed an equivalent of 60 mg lycopene per day for a 70 kg human. Apo-10'-lycopenal was not detected in ferret lung tissue. It was determined that after supplementation with lycopene, ferret lungs contain 5-11 pmol apo-10'-lycopenol per gram of wet weight tissue. Figure 5 shows the HPLC profile of apo-10'-lycopenol identified in lung extract of ferrets supplemented with lycopene for 9 wk.

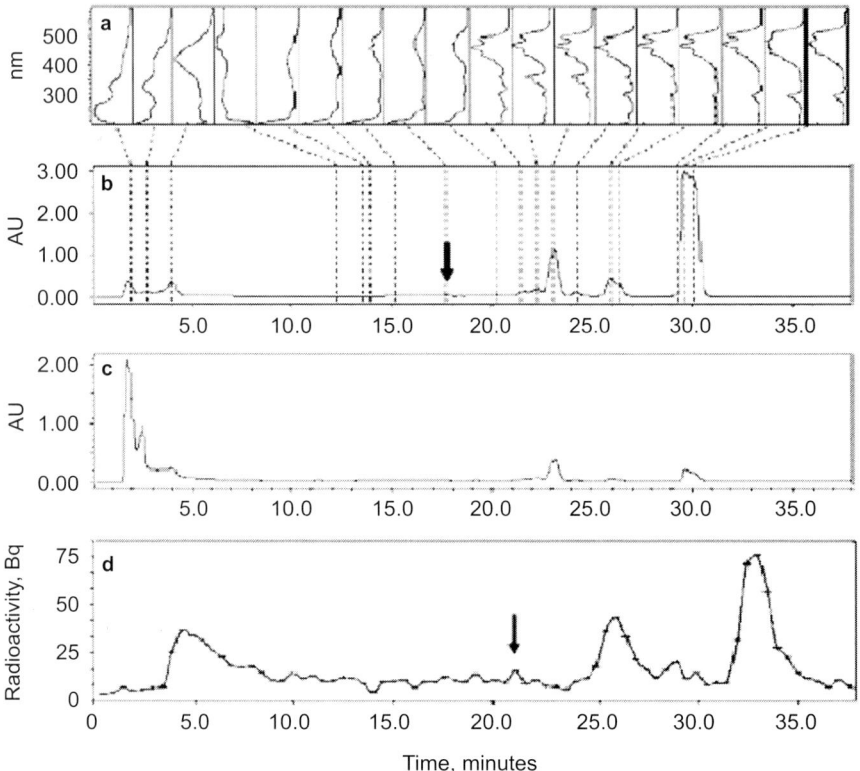

Fig. 4 HPLC and radioactivity profile of liver carotenoids from a rat fed lycopene for 30 d and gavaged with ^{14}C-lycopene 24 hr before sacrifice. Selected UV spectra (a) reveal many isomers and metabolites of lycopene. HPLC retention times, observed at 472 nm (b) and 340 nm (c), show lycopene, its *cis*-isomers, and more polar metabolites. All-*trans* and 5-*cis* lycopene elute at about 30 min. Additional *cis*-isomers of lycopene are seen between 21 and 30 min. Radioactive profile of the liver extract (d) was monitored using a β-ram radioactivity detector in tandem with an HPLC-PDA. There was a delay between the HPLC-PDA detection and the radioactive detector (Gajic et al. 2006). The arrows indicate the peak of apo-8′-lycopenal detected by UV-vis and β-ram radioactivity.

Although apo-10′-lycopenal is theoretically formed by CMOII, it has not been identified *in vivo*. It is suggested that apo-10′-lycopenal is an *in vivo* intermediate that may be rapidly reduced to its alcohol form or is present in undetectable levels. Researchers have confirmed the metabolism of apo-10′-lycopenal to its alcohol form *in vitro* (Hu et al. 2006).

Other Important *in vivo* Metabolites

The first lycopene metabolite identified in human plasma was 5,6-dihydroxy-5′,6′-dihydrolycopene by Khachik and coworkers at the US

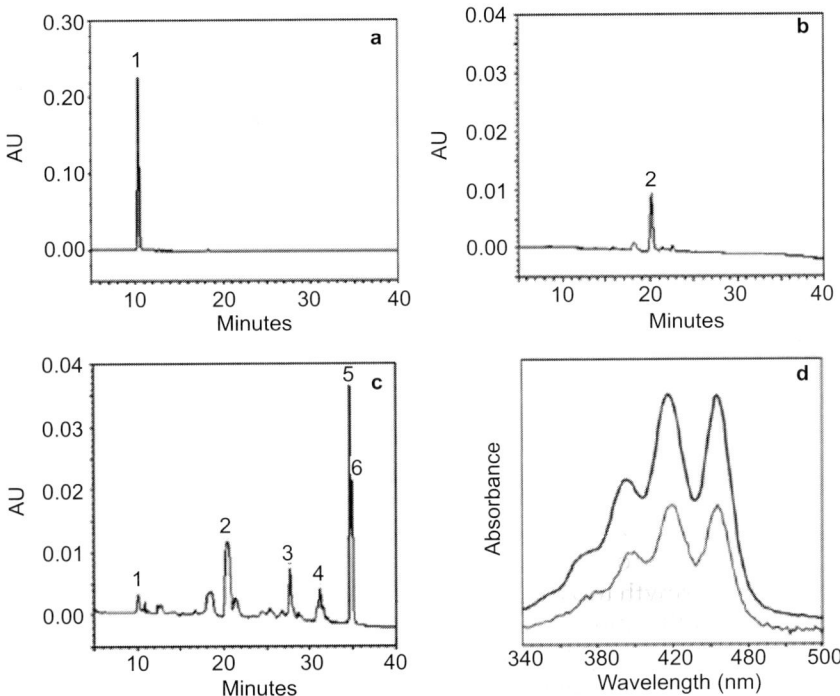

Fig. 5 Detection of apo-10′-lycopenol *in vivo*. Apo-10′-lycopenol retention time (a) is observed using HPLC. Representative lung extracts from ferrets supplemented with placebo is shown (b). HPLC analysis (c) of lung extracts from ferrets supplemented with lycopene for 9 wk shows a peak eluting at the same time as the respective apo-10′-lycopenol standard. UV spectral analysis (d) matches that of standard, apo-10′-lycopenol (Hu et al. 2006).

Department of Agriculture (Khachik et al. 1992). This group then reported finding 2,6-cyclolycopene-1,5-diol in human serum and breast milk (Khachik et al. 1997).

Recently, β-apo-8′-carotenal was identified in human plasma (calculated at ~245 ng/ml) (Ho et al. 2007). A healthy man was given an oral dose of ^{14}C labeled-all-trans-β-carotene in a banana milkshake. Serum, feces, and urine were collected over time and measured by accelerated mass spectrometry. Retinyl palmitate and retinol appeared in the plasma within 24 hr, while β-apo-8′-carotenal appeared only after 72 hr. This data suggest that the ^{14}C labeled-all-trans-β-carotene was absorbed into the body and then excentrally cleaved.

POTENTIAL FUNCTIONS OF APO-CAROTENOIDS

Cancer

Epidemiology studies establish that tomato products are associated with reduced incidence of prostate cancer (Giovannucci 2002). Other studies have shown positive effects of lycopene treatment with other forms of cancer including, colon, leukemia, and breast cancer (Nkondjock et al. 2005). Lycopene oxidative products are widely studied in cancer research as they have been found to inhibit cancer growth, induce apoptosis, and enhance cell-cell communication.

Autoxidation products of lycopene have been shown to inhibit the growth of human leukemia cells *in vitro* (Nara et al. 2001). Cells were treated with either an acyclic carotenoid or the products of its oxidation. The oxidation products were prepared by incubation of either lycopene, phytoene, phytofluene, or ζ-carotene in toluene at 37°C for 24 hr. Forty-two percent of intact lycopene remained in the oxidative mixture after treatment. The growth of leukemia cells was not affected by 6 μM of lycopene, while the oxidative mixture of lycopene at 6 μM most effectively inhibited cell growth to a dramatic 3.4% of the control. Researchers further report the identification of a novel cleavage product, (E,E,E)-4-methyl-8-oxo-2,4-6-nonatrienal (MON), formed by autoxidation of lycopene. Cell viability was significantly reduced by 5 μM of MON. Moreover, at 15 μM this metabolite induced apoptosis and DNA fragmentation in human promyelocytic leukemia cells (HL-60), while lycopene had no effect. Other oxidized lycopene products were present in this mixture and therefore could also be responsible for these anti-cancer effects (Zhang et al. 2003). This *in vitro* study suggests that oxidation products of lycopene, which may include apo-lycopenals, significantly reduce proliferation while inducing apoptosis of leukemia cells.

Loss of gap junction communication is often a characteristic of carcinogenesis. Carotenoids and their metabolites have been shown to stimulate the important gap junction communication between cells that may prove to be essential in the prevention of cancer. Aust and colleagues exposed lycopene to hydrogen peroxide and osmium tetroxide oxidation and further separated oxidation products from intact lycopene (Aust et al. 2003). These oxidation products were then used in the investigation of gap junction communication in rat liver epithelial cells. Gap junction communication was stimulated with exposure to the equivalent of 1 μM lycopene oxidative products. Researchers identified the biologically active compound as a dialdehyde, although they suggest that other compounds present may be responsible for the observed effect (Aust et al. 2003).

Lycopene oxidative products, and therefore potentially apo-lycopenals, induce apoptosis and/or inhibit the growth of prostate cancer cells. Three different prostate cancer cell lines (PC-3, DU145, and LNCaP) were treated *in vitro* with 20 µM of the lycopene oxidative product acyclo-retinoic acid (Kotake-Nara et al. 2002). This supraphysiological treatment significantly reduced cell viability in two of the prostate cancer cell lines, PC-3 and DU145, more effectively than geranylgeranoic acid, all-*trans*-retinoic acid or 9-*cis*-retinoic acid. Acyclo-retinoic acid did not affect LNCaP cell proliferation. The cell viability was reduced through induction of apoptosis as noted by increased DNA nuclear fragmentation.

The effects of apo-lycopenals on prostate cancer have also recently been investigated in our laboratory (Ford 2007). Apo-10′-lycopenal and apo-12′-lycopenal inhibited proliferation of DU145 cells but increased proliferation of androgen-dependent LNCaP cells in a dose-dependent manner. Reduced cell viability was observed at 1 µM but viability was significantly inhibited only at concentrations of 15 µM or greater. Apo-8′-lycopenal did not affect the proliferation of either prostate cancer cell line. Further investigation determined that apo-lycopenals and additional metabolites accumulated in both prostate cancer cell lines treated with apo-lycopenals.

Through a host of evidence, it appears that lycopene metabolites may be important in the reduction or prevention of cancer. Identification of bioactive lycopene metabolites and further characterization of their *in vivo* function(s) is essential for future cancer research.

In contrast, another apo-carotenal may cause harmful affects on the body by possibly stimulating factors that lead to cancer. Human alveolar epithelial cells (A549) were treated *in vitro* with a high level (20 µM) of Apo-8′-carotenal for 1 hr and then exposed to 20 µM benzopyrene, a cigarette-associated carcinogen, for 24 hr (Yeh and Wu 2006). This treatment resulted in DNA strand breaking and lipid peroxidation in the cells. Additionally, apo-8′-carotenal induced the same detrimental effects without initiation by benzopyrene. This induction was mediated through increased expression of CYP1A2.

The effects of other metabolites of lycopene on cancer have also been investigated. The other product of CMOII β-carotene cleavage, β-ionone, was given in a very high dose (1 g/kg bw) to male Wistar rats by intraperitoneal injection for 4 d (Robottom-Ferreira, Aquino et al. 2003). Beta-ionone was shown to induce CYP2B1 in rat liver and increase the expression of CYP2A3 in rat lungs. CYP2A enzymes are involved in the metabolism of many drugs and hormones but activate different carcinogens. In contrast, β-ionone at a concentration of 100 µM has been shown by another lab to inhibit proliferation, cell cycle progression, and the expression of cyclin-dependent kinase 2 activity *in vitro* in breast cancer cells (Duncan et al. 2004). Cyclin-dependent kinase 2 is essential for cell cycle progression, while methods of inhibiting cancer progression can involve suppression of the cell cycle. This very high concentration of β-ionone effectively slowed the G2/M phase and at higher concentrations also significantly inhibited the G0/G1 phase, thus resulting in significant inhibition of proliferation. More research is clearly needed on the carotenoid metabolic product, β-ionone.

Nuclear Receptor Activation

There is a paucity of research on the function of apo-lycopenals, but through the *in vivo* and *in vitro* investigation of other apo-carotenoids, researchers may predict some of the potential roles of apo-lycopenals. Central cleavage of β-carotene and subsequent oxidation results in the production of bioactive ligands, all-trans and 9-cis-retinoic acids. These ligands can activate the nuclear receptors RAR and RXR. The RXR receptor heterodimerizes a variety of nuclear receptors including peroxisome proliferator-activated receptors (PPARs). Thus, RXR and PPAR-V are essential for lipogenesis. Therefore, researchers were interested in whether asymmetric β-carotene cleavage products, apo-carotenals, also affected these receptors or transcription factors (Ziouzenkova et al. 2007). Less than 1 µM of apo-14-carotenal inhibited lipogenesis *in vitro*. Other longer apo-carotenals had no effect in this model. Apo-14-carotenal at concentrations greater than 10 µM mildly induced RARα activation. This effect may reduce lipogenesis, but activation of RARα was not found at lower concentration of apo-14-carotenal. Therefore, investigators looked at the ligand binding domain of RXR, PPARα and PPARγ. Apo-14-carotenal at a 10 µM concentration effectively inhibited activation of the ligand binding domain of RXR by 70% and also reduced the activation of PPARα and PPARγ binding domains. Again, longer apo-carotenals showed no effects. Apo-14-carotenal was also found to inhibit activation of a PPAR response element construct (Ziouzenkova et al. 2007). Together, these results suggest that formation of apo-14-carotenal *in vivo* may be important as it may play a role in lipid metabolism. Investigation of lipogenesis with other apo-carotenals may elucidate further essential metabolites.

Basal Cellular Functions

The metabolites produced from CMOII cleavage of β-carotene play a role in a variety of physiological processes. Oxidation products of β-carotene, including apo-10'-carotenal, potently inhibited Na^+/K^+-ATPase at low concentrations *in vitro* (Siems et al. 2000). The oxidized carotenoid mixture mostly comprising aldehydes inhibited Na+-K+-ATPase activity by 50% in equivalent concentrations to 10 μM β-carotene. Function of Na^+/K^+-ATPase could not be restored and therefore it was suggested that the oxidation products irreversibly bind the enzyme *in vitro*. The function of Na^+/K^+-ATPase was also inhibited by apo-10'-carotenal and retinal at the same concentration. Therefore, apo-10'-carotenal and other oxidation products of β-carotene may be detrimental to cellular life.

Work by the same authors also suggests that 5-20 μM of *in vitro* carotenoid oxidative products induced apoptosis of human neutrophils (Salerno et al. 2005). This could promote a pro-carcinogenic effect as neutrophils are essential for acute inflammation and host defense. This could be due to premature activation of neutrophil apoptosis, which could increase the possible risk of infection by bacteria or interfere with the removal of arrested neoplastic cells. It was further elucidated that β-carotene and its oxidation products were more cytotoxic than lycopene.

Some lycopene oxidation and cleavage products may prove to be highly active *in vivo*. Lycopene metabolites may potentially inhibit some carcinogenic factors while inducing other. Overall, one must be cautious in reviewing the data because most of the *in vitro* effects are seen only at supraphysiological concentrations and therefore would not be expected to be encountered in human tissues. However, preliminary work with lower levels of lycopene metabolites proves interesting and potentially beneficial for human health.

FUTURE DIRECTIONS WITH APO-LYCOPENALS

Although a large number of carotenoid cleavage products have been identified and described, we are only beginning to understand the biological importance of these compounds. Apo-lycopenals may prove to be physiologically important, only useful in certain disease states, potentially detrimental to the body, or simply just inactive waste products. It is therefore essential to identify apo-lycopenals in human tissues and further investigate apo-lycopenals *in vivo*. As many theories have been proposed, it is also necessary to elucidate the metabolic pathways for the production of apo-lycopenals and further clarify dose concentration concerns. It is also of great importance to then determine the roles of apo-lycopenals in human health.

ABBREVIATIONS

RAR: retinoic acid receptor; RXR: retinoid x receptor; LDL: low density lipoproteins; HDL: high density lipoproteins; SR-B1: scavenger receptor class B type 1; CD36: cluster determinant 36; ROS: reactive oxygen species;

Cyp P450: cytochrome P450; CMOI: carotenoid oxygenase I; CMOII: carotenoid oxygenase II; ADO: β-apocarotenoid-14′,13′-dioxygenase; CCD: carotenoid cleavage dioxygenase; HPLC-PDA: high performance liquid chromatography with photo-diode array; ESI-MS: electrospray ionization mass spectrometry; MON: (E,E,E,)-4-methyl-8-oxo-2,4-6 nonatrienal; PPAR: peroxisome proliferators-activated receptor

References

Agarwal, S. and A. Rao. 2000. Tomato lycopene and its role in human health and chronic diseases. Can. Med. Assoc. J. 163: 739-744.

Auldridge, M.E. and A. Block, J.T. Vogel, C. Dabney-Smith, I. Mila, M. Bouzayen, M. Magallanes-Lundback, D. DellaPenna, D.R. McCarty, and H.J. Klee. 2006. Characterization of three members of the Arabidopsis carotenoid cleavage dioxygenase family demonstrates the divergent roles of this multifunctional enzyme family. Plant J. 45: 982-993.

Aust, O. and N. Ale-Agha, L. Zhang, H. Wollersen, H. Sies, and W. Stahl. 2003. Lycopene oxidation product enhances gap junctional communication. Food Chem. Toxicol. 41: 1399-1407.

Ben-Aziz, A. and G. Britton, and T.W.Goodwin. 1973. Carotene epoxides of *Lycopersicon esculentum*. Phytochemistry 12: 2759-2764.

Boileau, T.W. and A C. Boileau, and J.W. Erdman Jr. 2002. Bioavailability of all-trans- and cis-isomers of lycopene. Exp. Biol. Med. (Maywood) 227: 914-919.

Boileau, T.W. and Z. Liao, S. Kim, S. Lemeshow, J.W. Erdman Jr., and S.K. Clinton. 2003. Prostate carcinogenesis in N-methyl-N-nitrosourea (NMU)-testosterone-treated rats fed tomato powder, lycopene, or energy-restricted diets. J. Natl. Cancer Inst. 95: 1578-1586.

Borel, P. and J. Drai, H. Faure, V. Fayol, C. Galabert, M. Laromiguiere, and G. Le Moel. 2005. Recent knowledge about intestinal absorption and cleavage of carotenoids. Ann. Biol. Clin. (Paris) 63: 165-177.

Breinholt, V. and S.T. Lauridsen, B. Daneshvar, and J. Jakobsen. 2000. Dose-response effects of lycopene on selected drug-metabolizing and antioxidant enzymes in the rat. Cancer Lett. 154: 201-210.

Burney, P.G. and G.W. Comstock, and J.S. Morris. 1989. Serologic precursors of cancer: serum micronutrients and the subsequent risk of pancreatic cancer. Amer. J. Clin. Nutr. 49: 895-900.

Campbell, J.K. and K. Canene-Adams, B.L. Lindshield, T.W. Boileau, S.K. Clinton, and J.W. Erdman Jr. 2004. Tomato phytochemicals and prostate cancer risk. J. Nutr. 134: 3486S-3492S.

Canene-Adams, K. and B.L. Lindshield, S. Wang, E.H. Jeffery, S.K. Clinton, and J.W. Erdman Jr. 2007. Combinations of tomato and broccoli enhance antitumor activity in dunning r3327-h prostate adenocarcinomas. Cancer Res. 67: 836-843.

Caris-Veyrat, C. and A. Schmid, M. Carail, and V. Bohm. 2003. Cleavage products of lycopene produced by in vitro oxidations: characterization and mechanisms of formation. J. Agric. Food Chem. 51: 7318-7325.

Clark, R.M. and K.L. Herron, D. Waters, and M.L. Fernandez. 2006. Hypo- and hyperresponse to egg cholesterol predicts plasma lutein and beta-carotene concentrations in men and women. J. Nutr. 136: 601-607.

De-Oliveira, A.C. and I.B. Silva, D.A. Manhaes-Rocha, and F.J. Paumgartten. 2003. Induction of liver monooxygenases by annatto and bixin in female rats. Braz. J. Med. Biol. Res. 36: 113-118.

Dmitrovskii, A.A. and N.N. Gessler, S.B. Gomboeva, V. Ershov Yu, and V. Bykhovsky. 1997. Enzymatic oxidation of beta-apo-8'-carotenol to beta-apo-14'-carotenal by an enzyme different from beta-carotene-15,15'-dioxygenase. Biochemistry (Mosc.) 62: 787-792.

Duncan, R.E. and D. Lau, A. El-Sohemy, and M.C. Archer. 2004. Geraniol and beta-ionone inhibit proliferation, cell cycle progression, and cyclin-dependent kinase 2 activity in MCF-7 breast cancer cells independent of effects on HMG-CoA reductase activity. Biochem. Pharmacol. 68: 1739-1747.

Ferreira, A.L. and K.J. Yeum, R.M. Russell, N.I. Krinsky, and G. Tang. 2003. Enzymatic and oxidative metabolites of lycopene. J. Nutr. Biochem. 14: 531-540.

Fleischmann, P. and A. Lutz-Roder, P. Winterhalter, and N. Watanabe. 2002. Carotenoid cleavage enzymes in animals and plants. Carotenoid-Derived Aroma Compounds 802: 76-88.

Ford, N.A. and J.W. Erdman Jr. 2007. Investigation of apo-lycopenals in DU145 and LNCaP Cells. FASEB J. 21.

Gajic, M. and S. Zaripheh, F. Sun, and J.W. Erdman Jr. 2006. Apo-8'-lycopenal and apo-12'-lycopenal are metabolic products of lycopene in rat liver. J. Nutr. 136: 1552-1557.

Gessler, N.N. and A.A. Dmitrovskii, S.B. Gomboeva, and V. Bykhovskii. 1998. Conversion of beta-carotene and some beta-apocarotinoids under the effect of substances from the intestinal mucosa. Prikl. Biokhim. Mikrobiol. 34: 645-649.

Giovannucci, E. 2002. A review of epidemiologic studies of tomatoes, lycopene, and prostate cancer. Exp. Biol. Med. (Maywood) 227: 852-859.

Giovannucci, E. and A. Ascherio, E.B. Rimm, M.J. Stampfer, G.A. Colditz, and W.C. Willett. 1995. Intake of carotenoids and retinol in relation to risk of prostate cancer. J. Natl. Cancer Inst. 87: 1767-1776.

Giuliano, G. and A.-B. Salim, and J. von Lintig. 2003. Carotenoid oxygenases: cleave it or leave it. Trends Plant Sci. 8.

Goodman, D.S. and H.S. Huang. 1965. Biosynthesis of vitamin A with rat intestinal enzymes. Science 149: 879-880.

Handelman, G.J. and F.J. van Kuijk, A. Chatterjee, and N.I. Krinsky. 1991. Characterization of products formed during the autoxidation of beta-carotene. Free Radic. Biol. Med. 10: 427-437.

Hessel, S. and A. Eichinger, et al. 2007. CMO1 Deficiency Abolishes Vitamin A Production from beta-Carotene and Alters Lipid Metabolism in Mice. J Biol Chem 282: 33553-33561.

Ho, C.C. and F.F. de Moura, S.H. Kim, and A.J. Clifford. 2007. Excentral cleavage of {beta}-carotene in vivo in a healthy man. Amer. J. Clin. Nutr. 85: 770-777.

Hu, K.Q. and C. Liu, H. Ernst, N.I. Krinsky, R.M. Russell, and X.D. Wang. 2006. The biochemical characterization of ferret carotene-9',10'-monooxygenase catalyzing cleavage of carotenoids in vitro and in vivo. J. Biol. Chem. 281: 19327-19338.

Junior, A.C. and L.M. Asad, E.B. Oliveira, K. Kovary, N.R. Asad, and I. Felzenszwalb. 2005. Antigenotoxic and antimutagenic potential of an annatto pigment (norbixin) against oxidative stress. Genet. Mol. Res. 4: 94-99.

Khachik, F. and G.R. Beecher, M.B. Goli, W.R. Lusby, and J.C. Smith Jr. 1992. Separation and identification of carotenoids and their oxidation products in the extracts of human plasma. Anal. Chem. 64: 2111-2122.

Khachik, F. and C.J. Spangler, J.C. Smith Jr., L.M. Canfield, A. Steck, and H. Pfander. 1997. Identification, quantification, and relative concentrations of carotenoids and their metabolites in human milk and serum. Anal. Chem. 69: 1873-1881.

Khachik, F. and A. Steck, U. Niggli, and H. Pfander. 1998. Partial synthesis and structural elucidation of the oxidative metabolites of lycopene identified in tomato paste, tomato juice, and human serum. J. Agric. Food Chem. 46: 4874-4884.

Kiefer, C. and S. Hessel, J.M. Lampert, K. Vogt, M.O. Lederer, D.E. Breithaupt, and J. von Lintig. 2001. Identification and characterization of a mammalian enzyme catalyzing the asymmetric oxidative cleavage of provitamin A. J. Biol. Chem. 276: 14110-14116.

Kim, S.J. and E. Nara, H. Kobayashi, J. Terao, and A. Nagao. 2001. Formation of cleavage products by autoxidation of lycopene. Lipids 36: 191-199.

King, T. and F. Khachik, H. Bortkiewicz, L. Fukushima, S. Morioka, and J. Bertram. 1997. Metabolites of dietary carotenoids as potential cancer preventive agents. Pure Appl. Chem. 69: 2135-2140.

Kloer, D.P. and S. Ruch, S. Al-Babili, P. Beyer, and G.E. Schulz. 2005. The structure of a retinal-forming carotenoid oxygenase. Science 308: 267-269.

Kotake-Nara, E. and S.J. Kim, M. Kobori, K. Miyashita, and A. Nagao. 2002. Acyclo-retinoic acid induces apoptosis in human prostate cancer cells. Anticancer Res. 22: 689-695.

Lakshman, M.R. and I. Mychkovsky, and M. Attlesey. 1989. Enzymatic conversion of all-trans-beta-carotene to retinal by a cytosolic enzyme from rabbit and rat intestinal mucosa. Proc. Natl. Acad. Sci. USA 86: 9124-9128.

Lindqvist, A. and Y.G. He, and S. Andersson. 2005. Cell type-specific expression of beta-carotene 9',10'-monooxygenase in human tissues. J. Histochem. Cytochem. 53: 1403-1412.

Lindshield, B.L. and K. Canene-Adams, and J.W. Erdman Jr. 2007. Lycopenoids: are lycopene metabolites bioactive? Arch. Biochem. Biophys. 458: 136-140.

Lingen, C. and L. Ernster, and O. Lindberg. 1959. The promoting effect of lycopene on the non-specific resistance of animals. Exp. Cell. Res. 16: 384-393.

Liu, C. and R. Russell, and X. Wang. 2004. alpha-Tocopherol and ascorbic acid decrease the production of beta-apo-carotenals and increase the formation of retinoids from beta-carotene in the lung tissues of cigarette smoke-exposed ferrets in vitro. J. Nutr. 134: 426-430.

Martin, K.R. and D. Wu, and M. Meydani. 2000. The effect of carotenoids on the expression of cell surface adhesion molecules and binding of monocytes to human aortic endothelial cells. Atherosclerosis 150: 265-274.

Moore, T. 1930. Vitamin A and carotene: The absence of the liver oil vitamin A from carotene. VI. The conversion of carotene to vitamin A in vivo. Biochem. J. 24: 692-702.

Nagao, A. 2004. Oxidative conversion of carotenoids to retinoids and other products. J. Nutr. 134: 237S-240S.

Nagao, A. and A. During, C. Hoshino, J. Terao, and J.A. Olson. 1996. Stoichiometric conversion of all trans-beta-carotene to retinal by pig intestinal extract. Arch. Biochem. Biophys. 328: 57-63.

Napoli, J.L. and B.C. Pramanik, et al. 1985. Quantification of retinoic acid by gas-liquid chromatography-mass spectrometry: total versus all-trans-retinoic acid in human plasma. J Lipid Res 26: 387-392.

Nara, E. and H. Hayashi, M. Kotake, K. Miyashita, and A. Nagao. 2001. Acyclic carotenoids and their oxidation mixtures inhibit the growth of HL-60 human promyelocytic leukemia cells. Nutr. Cancer 39: 273-283.

Nguyen, M.L. and S.J. Schwartz. 1998. Lycopene stability during food processing. Proc. Soc. Exp. Biol. Med. 218: 101-105.

Nkondjock, A. and P. Ghadirian, K.C. Johnson, and D. Krewski. 2005. Dietary intake of lycopene is associated with reduced pancreatic cancer risk. J. Nutr. 135: 592-597.

Noy, N. 2000. Retinoid-binding proteins: mediators of retinoid action. Biochem. J. 348 Pt 3: 481-495.

Olson, J.A. and Hayaishi, O. 1965. The enzymatic cleavage of beta-carotene into vitamin A by soluble enzymes of rat liver and intestine. Proc. Natl. Acad. Sci. USA 54: 1364-1370.

Parker, R.S. 1996. Absorption, metabolism, and transport of carotenoids. Faseb J. 10: 542-551.

Reboul, E. and L. Abou, C. Mikail, O. Ghiringhelli, M. Andre, H. Portugal, D. Jourdheuil-Rahmani, M.J. Amiot, D. Lairon, and P. Borel. 2005. Lutein transport by Caco-2 TC-7 cells occurs partly by a facilitated process involving the scavenger receptor class B type I (SR-BI). Biochem. J. 387: 455-461.

Salerno, C. and C. Crifo, E. Capuozzo, O. Sommerburg, C. Langhans, and W. Siems. 2005. Effect of carotenoid oxidation products on neutrophil viability and function. Biofactors 24: 185-192.

Schwartz, S.H. and X. Qin, and J.A. Zeevaart. 2001. Characterization of a novel carotenoid cleavage dioxygenase from plants. J. Biol. Chem. 276: 25208-25211.

Schwartz, S. and X. Qin, and M. Loewen. 2004. The biochemical characterization of two carotenoid cleavage enzymes from Arabidopsis indicates that a carotenoid-derived compound inhibits lateral branching. J. Biol. Chem. 279: 46940-46945.

Shi, J. and M. Le Maguer. 2000. Lycopene in tomatoes: chemical and physical properties affected by food processing. Crit. Rev. Biotechnol. 20: 293-334.

Siems, W.G. and O. Sommerburg, J.S. Hurst, and F.J. van Kuijk. 2000. Carotenoid oxidative degradation products inhibit Na+-K+-ATPase. Free Radic. Res. 33: 427-435.

Simkin, A.J. and S.H. Schwartz, M. Auldridge, M.G. Taylor, and H.J. Klee. 2004. The tomato carotenoid cleavage dioxygenase 1 genes contribute to the formation of the flavor volatiles beta-ionone, pseudoionone, and geranylacetone. Plant J. 40: 882-892.

Stahl, W. and H. Sies. 1992. Uptake of lycopene and its geometrical isomers is greater from heat-processed than from unprocessed tomato juice in humans. J. Nutr. 122: 2161-2166.

Stratton, S.P. and W.H. Schaefer, and D.C. Liebler. 1993. Isolation and identification of singlet oxygen oxidation products of beta-carotene. Chem. Res. Toxicol. 6: 542-547.

Thomas, S. and R. Prabhu, et al. 2005. Retinoid metabolism in the rat small intestine. Br J Nutr 93: 59-63.

van Bennekum, A. and M. Werder, S.T. Thuahnai, C.H. Han, P. Duong, D.L. Williams, P. Wettstein, G. Schulthess, M.C. Phillips, and H. Hauser. 2005. Class B scavenger receptor-mediated intestinal absorption of dietary beta-carotene and cholesterol. Biochemistry 44: 4517-4525.

Vogel, J.T. and B.C. Tan, et al. 2008. The carotenoid cleavage dioxygenase 1 enzyme has broad substrate specificity, cleaving multiple carotenoids at two different bond positions. J. Biol. Chem. 283: published online before print.

von Lintig, J. and S. Hessel, A. Isken, C. Kiefer, J.M. Lampert, O. Voolstra, and K. Vogt. 2005. Towards a better understanding of carotenoid metabolism in animals. Biochim. Biophys. Acta 1740: 122-131.

Voutilainen, S. and T. Nurmi, J. Mursu, and T.H. Rissanen. 2006. Carotenoids and cardiovascular health. Amer. J Clin. Nutr. 83: 1265-1271.

Wald, G. 1985. Vitamin-A in the eye tissues. Nutr. Rev. 43: 244-246.

Wang, X.D. and G.W. Tang, J.G. Fox, N.I. Krinsky, and R.M. Russell. 1991. Enzymatic conversion of beta-carotene into beta-apo-carotenals and retinoids by human, monkey, ferret, and rat tissues. Arch. Biochem. Biophys. 285: 8-16.

Wertz, K. and U. Siler, and R. Goralczyk. 2004. Lycopene: modes of action to promote prostate health. Arch. Biochem. Biophys. 430: 127-134.

White, J.A. and B. Beckett-Jones, Y.D. Guo, F.J. Dilworth, J. Bonasoro, G. Jones, and M. Petkovich. 1997. cDNA cloning of human retinoic acid-metabolizing enzyme (hP450RAI) identifies a novel family of cytochromes P450. J. Biol. Chem. 272: 18538-18541.

White, J.A. and H. Ramshaw, M. Taimi, W. Stangle, A. Zhang, S. Everingham, S. Creighton, S.P. Tam, G. Jones, and M. Petkovich. 2000. Identification of the human cytochrome P450, P450RAI-2, which is predominantly expressed in the adult cerebellum and is responsible for all-trans-retinoic acid metabolism. Proc. Natl. Acad. Sci. USA 97: 6403-6408.

Wolf, G. 2001. The enzymatic cleavage of beta-carotene: end of a controversy. Nutr. Rev. 59: 116-118.

Wyss, A. 2004. Carotene oxygenases: a new family of double bond cleavage enzymes. J. Nutr. 134: 246S-250S.

Wyss, A. and G. Wirtz, W. Woggon, R. Brugger, M. Wyss, A. Friedlein, H. Bachmann, and W. Hunziker. 2000. Cloning and expression of beta,beta-carotene 15,15'-dioxygenase. Biochem. Biophys. Res. Commun. 271: 334-336.

Yeh, S.L. and S.H. Wu. 2006. Effects of quercetin on beta-apo-8'-carotenal-induced DNA damage and cytochrome P1A2 expression in A549 cells. Chem. Biol. Interact. 163: 199-206.

Yeum, K.J. and Y.C. Lee-Kim, S. Yoon, K.Y. Lee, I.S. Park, K.S. Lee, B.S. Kim, G. Tang, R.M. Russell, and N.I. Krinsky. 1995. Similar metabolites formed from beta-carotene by human gastric mucosal homogenates, lipoxygenase, or linoleic acid hydroperoxide. Arch. Biochem. Biophys. 321: 167-174.

Yeum, K.J. and A.L. dos Anjos Ferreira, D. Smith, N.I. Krinsky, and R.M. Russell. 2000. The effect of alpha-tocopherol on the oxidative cleavage of beta-carotene. Free Radic. Biol. Med. 29: 105-114.

Zaripheh, S. and J.W. Erdman Jr. 2005. The biodistribution of a single oral dose of [14C]-lycopene in rats prefed either a control or lycopene-enriched diet. J. Nutr. 135: 2212-2218.

Zaripheh, S. and T.W. Boileau, M.A. Lila, and J.W. Erdman Jr. 2003. [14C]-lycopene and [14C]-labeled polar products are differentially distributed in tissues of F344 rats prefed lycopene. J. Nutr. 133: 4189-4195.

Zaripheh, S. and T.Y. Nara, M.T. Nakamura, and J.W. Erdman Jr. 2006. Dietary lycopene downregulates carotenoid 15,15'-monooxygenase and PPAR-gamma in selected rat tissues. J. Nutr. 136: 932-938.

Zhang, H. and E. Kotake-Nara, H. Ono, and A. Nagao. 2003. A novel cleavage product formed by autoxidation of lycopene induces apoptosis in HL-60 cells. Free Radic. Biol. Med. 35: 1653-1663.

Ziouzenkova, O. and G. Orasanu, G. Sukhova, E. Lau, J.P. Berger, G. Tang, N.I. Krinsky, G.G. Dolnikowski, and J. Plutzky. 2007. Asymmetric cleavage of beta-carotene yields a transcriptional repressor of retinoid X receptor and peroxisome proliferator-activated receptor responses. Mol. Endocrinol. 21: 77-88.

1.4

Non-covalent Binding of Lycopene and Lycophyll

[1]Zsolt Bikadi, [2]Peter Hari, [1]Eszter Hazai, [3]Samuel F. Lockwood and [4]Ferenc Zsila

[1]Virtua Drug, Ltd., Csalogany st. 4c, H-1015 Budapest, Hungary
[2]Delta Elektronik, Szentendrei st. 39 – 53, H-1033 Budapest, Hungary
[3]Cardax Pharmaceuticals, Inc., 99-193 Aiea Heights Drive, Suite 400, Aiea Hawaii 96701, USA
[4]Chemical Research Center, Hungarian Academy of Sciences
Institute of Biomolecular Chemistry
PO BOX 17 H-1525 Budapest, Hungary

ABSTRACT

The consumption of a diet rich in carotenoids – in particular lycopene – has been epidemiologically correlated with a lower risk for different types of cancer and cardiovascular diseases. The antioxidant effect of lycopene has been well established, and it is thought to be a major mechanism of lycopene action. Recently, evidence has accumulated for other mechanisms of lycopene action involving non-covalent (**covalent**: a chemical link between two atoms in which the electrons are shared between them) interactions with proteins and cell membranes, such as the role of lycopene in regulation of cell proliferation, growth factor signaling, modulation of hormone and immune response, and enhancing gap junctional intercellular communication (GJIC). This chapter focuses on the growing body of evidence that carotenoids specifically interact with proteins, and that this interaction affects their mode of action. There remains a large gap of knowledge concerning the effect of carotenoids on proteins at the molecular level. When the availability of experimental data is limited, molecular modeling studies can be used for exploring the potential

---

A list of abbreviations is given before the references.

mechanistic functions of carotenoids. In this review we discuss the applicability of docking studies in the exploration of the mode of action of lycopene and lycophyll on lipoxygenase enzymes.

INTRODUCTION

Carotenoids comprise a class of natural pigments including more than 750 structurally distinct compounds that are found in fruits and vegetables, as well as fish, crustaceans, and birds (Britton et al. 2004, Delgado-Vargas et al. 2000). Tomato fruit contains a related set of acyclic C40 carotenoids, the carotene lycopene and the xanthophylls lycophyll and lycoxanthin (Fig. 1). Lycopene is by far the most abundant among them and is responsible for the fruit's deep red color. Together with α- and β-carotene, lycopene is a prominent member of the dietary carotene group and contains only carbon and hydrogen atoms. Lycophyll (dihydroxylycopene) is an intermediate between lycopene and the methoxylated carotenoids. The carbon chain of alternating single and double bonds in the lycopene series consists of 11 conjugated bonds and possesses the greatest singlet oxygen quenching ability among tested carotenoids (Di Mascio et al. 1989). The all-*trans* form of lycopene predominates in nature and is also the most abundant geometric isomer in human food sources (up to 90% in tomato products) (Clinton 1998).

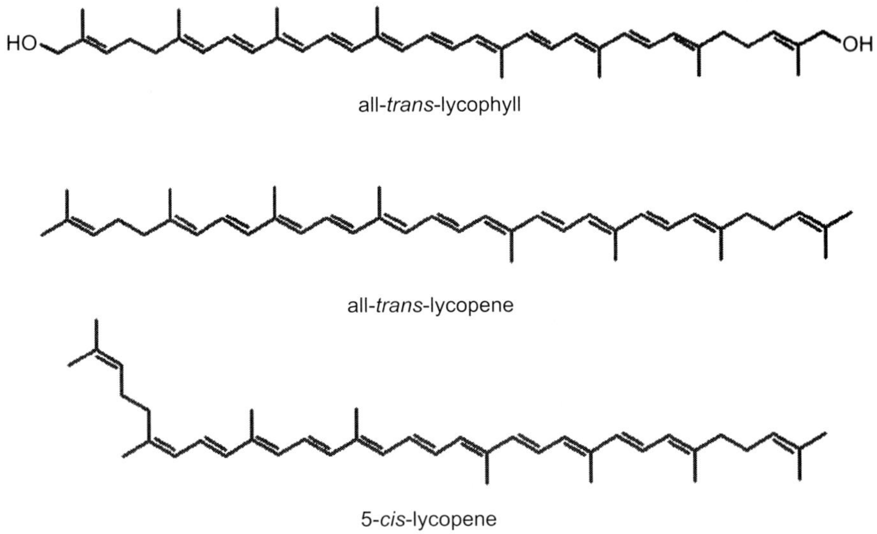

all-*trans*-lycophyll

all-*trans*-lycopene

5-*cis*-lycopene

Fig. 1 Structures of all-*trans* lycophyll and lycopene (most common geometric isomers in nature); 5-*cis*-lycopene is found in elevated concentration in human prostate.

However, *cis*-isomeric forms of lycopene are the predominant carotenes in human blood, with the all-*trans* fraction being below 50%, strongly suggesting metabolic conversion, preferential dietary absorption of *cis* forms, or radical interactions converting the all-*trans* form to other geometric isomers in oxidative tissues.

Lycopene exerts its multiple biological effects through different mechanisms. The dietary intake of food rich in oxo- and hydrocarbon carotenoids—in particular lycopene—has been epidemiologically correlated with a lower risk of several chronic diseases (Ben-Dor et al. 2001, Clinton 1998, Tapiero et al. 2004). Lycopene acts as an antioxidant *in vivo* and this mechanism has been considered the major important effect in biological systems (Heber and Lu 2002, Lindshield et al. 2007). In addition, it reduces oxidative DNA damage and inhibits prostatic IGF-I signaling, IL-6 expression, and androgen signaling. Moreover, lycopene improves communication among cells. Although extensive research has been carried out on carotenoids, other mechanisms of lycopene action (involving non-covalent interactions with proteins and cell membranes) have only recently become the focus of investigation. The present chapter reviews our current knowledge in the field of non-covalent interactions of lycopene and lycophyll and its relevance to human health.

NON-COVALENT BINDING OF LYCOPENE TO PROTEINS

The beneficial effects of lycopene and related carotenoid tomato fruit compounds on human health can be divided into two parts. Since lycopene is an excellent scavenger of singlet oxygen and other reactive oxygen and nitrogen species, it is very effective in protecting membrane lipids (McNulty et al. 2007), lipoproteins, proteins (including important enzymes), and DNA against oxidative damage (Kun et al. 2006). Spatial proximity can be sufficient for exerting the observed antioxidant effect; however, it does not require the direct binding of lycopene and hydroxylycopenes to target objects that need to be protected. Recently, a growing body of evidence has indicated that several properties of lycopene do not correlate with its effects on reactive oxygen and nitrogen species, including the regulation of cell proliferation, growth factor signaling, modulation of hormone and immune response, and enhancing GJIC (Elliott 2005, Kun et al. 2006). These non-oxidative mechanisms preferentially involve non-covalent binding interactions between lycopene and proteins as well as biomembranes (Sharoni et al. 2004).

Early contact between lycopene and macromolecules of the human body occurs in the small intestine during absorption. As a highly lipophilic plant pigment, lycopene and its congeners require solubilization for effective transfer across the intestinal barrier after the release from the food matrix. The uptake of lycopene is greatly facilitated by the presence of dietary lipids such as triacylglycerols, cholesterol and phospholipids that form micelles of variable size and composition (Baskaran et al. 2003). The micelles possess a triacylglycerol core surrounded by a monomolecular layer of partially digested proteins, polysaccharides and lipids, especially phospholipid and partially ionized fatty acids (Furr and Clark 1997). In contrast to the more polar dietary carotenoids (e.g., astaxanthin, lutein, and zeaxanthin), lycopene is incorporated into the lipophilic core of these particles where it forms a large number of hydrophobic interactions (e.g., van der Waals, CH-π) with the constituents of the micelle (Borel et al. 1996). After digestion of the triacylglycerol core by pancreatic lipase, lycopene is transferred into mixed bile salt micelles. It has been demonstrated that both the transfer of cis isomers of lycopene to micelles and the subsequent uptake of cis isomers by human intestinal Caco-2 intestinal epithelial cells were more efficient than that of all-trans lycopene (Failla et al. 2008). A possible explanation for these findings is the more favorable binding of the cis configuration of lycopene to membrane transporters responsible for the uptake of carotenoids across the brush border membrane. These data suggest that the greater bioaccessibility of cis compared with all-trans isomers of lycopene contributes to the enrichment of the cis isomers in tissues. According to the conventional view, the mixed bile salt micelles move through the aqueous phase of the intestinal lumen to the membranes of enterocytes and subsequently diffuse passively into the cytoplasm. This scenario has been augmented by the demonstration that lutein and β-carotene directly interact with class B scavenger receptors on the apical surface of the epithelial cells, which mediate their intracellular absorption (Reboul et al. 2005, During and Harrison 2005, van Bennekum et al. 2005).

In enterocytes, carotenoid molecules are assembled into lymphatic lipoproteins called chylomicrons. It is postulated that due to the poor aqueous solubility of polyenic compounds, a specific carotenoid binding protein(s) exists in the cytosol in order to facilitate the formation of chylomicrons (Diwadkar-Navsariwala et al. 2003, Furr and Clark 1997). Indeed, a specific carotenoid-binding protein has recently been characterized in ferret liver, which was highly specific for carotenoids with at least one unsubstituted β-ionone ring (Rao et al. 1997).

In blood plasma, the evidence suggests that carotenoids are transported almost exclusively by lipoproteins. Lycopene is predominantly associated with the low-density lipoprotein (LDL) fraction (Bhosale and Bernstein 2007, Segrest et al. 2001). In the case of more hydrophilic, semi-synthetic carotenoid derivatives that can be given parenterally (disodium disuccinate astaxanthin; tetrahydrochoride dilysinate astaxanthin) (Zsila et al. 2003, 2004) and natural carboxylic acid apocarotenoids (crocetin, bixin) (Zsila et al. 2001, 2002, 2005a, b), serum albumin and the acute-phase component serum α_1-acid glycoprotein may also be involved in plasma transport through formation of non-covalent carotenoid-protein complexes. The exact structure of LDL is unknown, and localization of lycopene molecules inside the particles is still debated. According to the classical model, carotenes are nonspecifically dissolved in the hydrophobic core of LDL, which is composed primarily of esterified cholesterol molecules (Borel et al. 1996). However, novel results obtained by Raman spectroscopy indicated the presence of lycopene and β-carotene in the LDL shell formed by phospholipids, unesterified cholesterol, and the apolipoprotein B-100 (apoB-100) (Lin et al. 2000). In relation to β-carotene, lycopene was found in closer spatial proximity to apoB-100, suggesting a specific lycopene-protein interaction that could provide an explanation for the distinct tissue uptake of the two carotenes (Lin et al. 2000).

LDL not only function as carotenoid transporters but also supply cells with cholesterol, an essential component of all animal cell membranes. Interestingly, lycopene possesses a moderate hypocholesterolemic action, explained by its *in vitro* inhibitory effect on the cellular HMG-CoA reductase enzyme of macrophages, the rate-limiting enzyme in cholesterol synthesis (Fuhrman et al. 1997). An additional impact of lycopene on cardiovascular health and thrombo-embolic disorders is inhibition of platelet aggregation. This effect could be mediated by the interference of lycopene with the phospholipase C pathway, or more directly on phosphodiesterases involved with nitric oxide signaling (Hsiao et al. 2005, Lazarus and Garg 2004).

The organ- and tissue-specific distribution of lycopene is likely to be governed, at least in part, by non-covalent interactions with specialized proteins. Recently, absorption and subcellular localization studies of lycopene in human prostate cancer cells suggested that rapid cellular intake of lycopene is mediated by a receptor or binding protein (Liu et al. 2006). A similar conclusion was drawn from [14C]-lycopene biodistribution studies in rats, which showed the preferential uptake of lycopene into some tissues (Zaripheh and Erdman 2005).

Metabolism of Lycopene

Besides intact incorporation into lipoprotein particles, carotenoids may also undergo enzymatic conversion inside enterocytes. The enzymes most likely play a major role in the metabolism of lycopene are carotenoid monooxygenase I and II (CMO I, CMO II), the two most important carotenoid-cleaving enzymes in mammals. The enzymatic cleavage of β-carotene into two retinal molecules is catalyzed by β-carotene 15,15′-monooxygenase (CMO I). Lycopene, lutein and astaxanthin are competitive inhibitors of this enzyme (Ershov Yu et al. 1994). It was concluded from experiments conducted with several substrate analogs that the 15,15′-monooxygenase possesses a hydrophobic, barrel-like substrate binding pocket (Wirtz et al. 2001). This observation received support from the reported crystal structure of the *Synechocystis* carotenoid

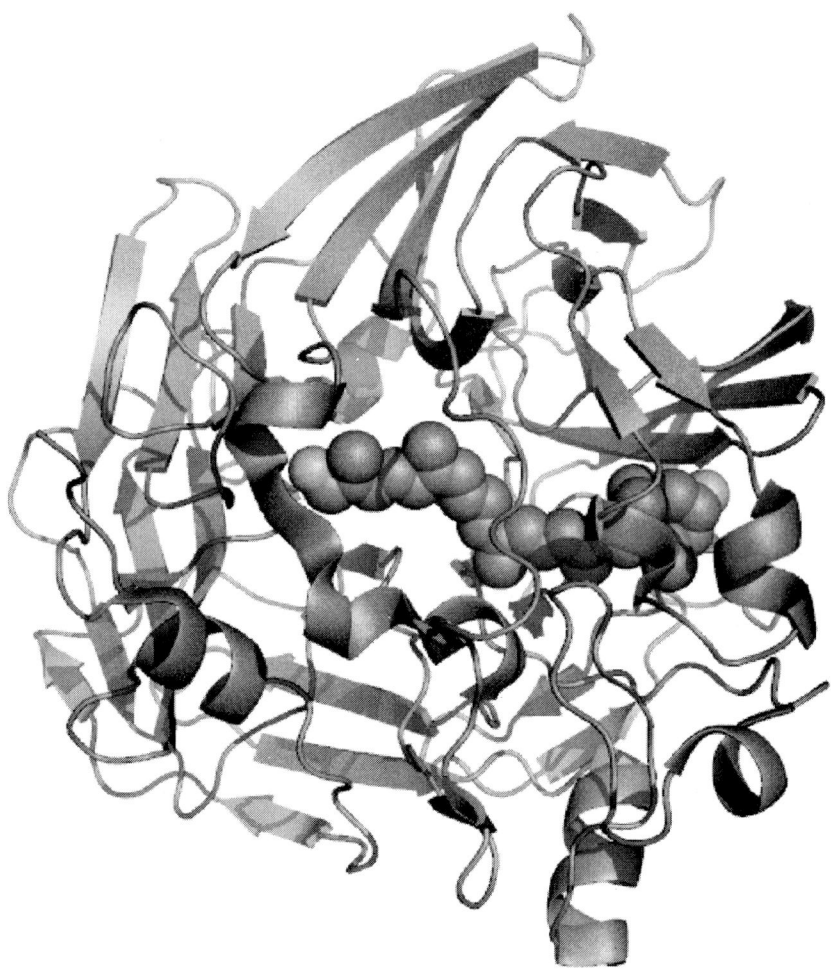

Fig. 2 Apocarotenoid cleavage oxygenase complexed with lycopene. When the carotenoid became trapped in the enzyme, three consecutive double bonds of the carotenoid changed from an all-*trans* to a *cis-trans-cis* conformation (Kloer et al. 2005).

oxygenase (Fig. 2), showing a long tunnel inside the protein lined with numerous nonpolar residues (mainly with aromatic sidechains), suitable to stabilize the polyene chain via π-π interactions (Kloer et al. 2005). Recently, an additional carotenoid oxygenase named β-carotene 9′,10′-monooxygenase was discovered in mice. This enzyme cleaves lycopene at the 9′,10′ position, resulting in apolycopenals (CMO II) (Kiefer et al. 2001).

Although little is known about the *in vivo* action of CMO I and II on lycopene, in theory, lycopene products are suggested to be acycloretinal or apo-lycopenals for CMO I and II, respectively. In addition, it seems possible that cytochrome P450 enzymes could be involved in the cleavage of lycopene. Recent advances in characterization of carotenoid cleavage enzymes may also provide clues to the possibility of the enzymatic formation of active lycopene-series metabolites. Novel 9′,10′-dioxygenases have been added to the growing family of double bond cleavage enzymes characterized from both plants and animals (Wyss 2004).

Health Effects of the Protein Binding of Lycopene

Many epidemiological studies have indicated that lycopene significantly reduces the risk for several malignant disorders including breast tumors, bladder, pancreas, lung, and prostate cancer, as well as cervical intraepithelial neoplasia (Guns and Cowell 2005, Wane and Lengacher 2006, Rao and Rao 2007). The understanding of the molecular events behind this protection is still in an embryonic state, but accumulating evidence suggests that the direct binding interaction of lycopene with protein members of the cell transcription system is involved (Barber et al. 2006, Sharoni et al. 2004).

The proliferation of the HL-60 promyelocytic leukemia cell line was synergistically inhibited upon co-exposure of lycopene with 1,25-dihydroxyvitamin D_3 or retinoic acid (Amir et al. 1999). The latter compounds are ligands of the nuclear receptor super-family, suggesting that lycopene or one of its derivatives may also interact with members of this family of receptors.

It is now well established that intact GJIC is an important limiting factor in the growth inhibition of precancerous and malignant cells. The chemopreventive properties of carotenoids and retinoids are partly assigned to their GJIC stimulating effects. Lycopene, pro-vitamin A carotenoids, and retinoids are able to up-regulate the expression of connexin proteins (Cx43 and others) that are assembled into channels connecting the cytoplasm of adjacent cells (Bertram and Vine 2005). The macromolecular target of lycopene, through which it enhances the synthesis of Cx43, remains unknown. A likely candidate for this function is the peroxisome proliferator activated receptor γ (PPAR-γ). This receptor possesses a large T-shaped, mainly hydrophobic pocket suitable for accommodation of lipophilic ligands, including long-chain saturated/ polyunsaturated fatty acids and arachidonic acid metabolites (Willson et al. 2000). Additionally, the oxidative and/or enzymatic degradation of lycopene into smaller, biologically active derivatives (e.g., acyclo-retinoic acid) could produce molecules capable of binding to the sterically more restricted binding site of retinoid nuclear receptors (Ben-Dor et al. 2001, Bertram and Vine 2005). Recently, apo-8'-lycopenal and apo-12'-lycopenal have been identified as metabolic products of lycopene in rat liver (Gajic et al. 2006). In addition, the lycopene metabolite apo-10'-lycopenoic acid detected in mammalian tissues has been shown to inhibit lung cancer cell growth in vitro and to suppress lung tumorigenesis in animal mouse model (Lian et al. 2007). This derivative transactivated the retinoic acid receptor beta promoter and induced the expression of the receptor.

It should be noted that inhibition of multidrug transporters might also be involved in the anticancer activity of lycopene and other carotenoids. Over-expression of P-glycoprotein, the prominent member of the ABC transporter family, is a major obstacle to successful cancer chemotherapy (Teodori et al. 2006). Residing in the cytoplasmic membrane of malignant cells, this protein actively pumps out a broad array of chemotherapeutic agents from the cell, resulting in multidrug resistance. It has been found that lycopene and its oxidized derivative lycophyll are able to reverse multidrug resistance *in vitro* by inhibiting the efflux pump function of P-glycoprotein (Molnar et al. 2006).

Association and Aggregation Properties

The biological effects of carotenoids can be mediated not only by association with proteins, but also by interaction with cell membranes (McNulty et al. 2007). It has been suggested that carotenoids could influence mitochondrial function via the modulation of mitochondrial membrane fluidity (Elliott 2005). In mitochondria, ATP is synthesized by the F_1F_0ATPase complex in which the F_0 portion is embedded in the inner mitochondrial membrane. The F_0 portion moves during ATP synthesis. Since carotenoids — and especially lycopene — strongly reduce the fluidity of biomembranes (Gruszecki and Strzalka 2005, Suwalsky et al. 2002), they could alter the activity of the F_1F_0ATPase by hindering the movement of the F_0 component.

Non-covalent interactions can be exploited in drug formulation procedures to reduce or eliminate unfavorable physicochemical properties of pharmaceutical agents. Due to their chemical composition (carbons and

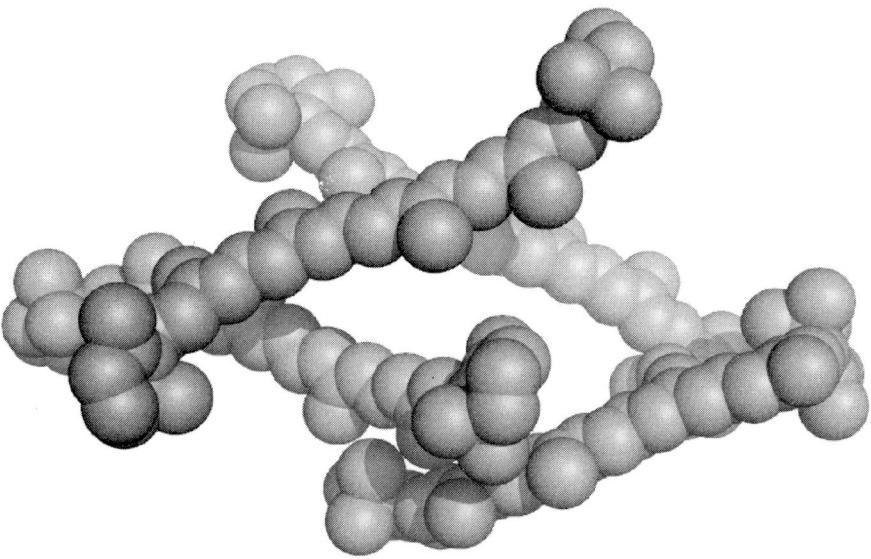

Fig. 3 Tetramer structure formed by β-carotene molecules. Many carotenoids form supramolecular assemblies ("aggregates") in aqueous as well as other solutions.

hydrogens only), lycopene and the majority of carotenoids are highly hydrophobic and are therefore practically insoluble in aqueous solution. Carotenoids form aggregates in aqueous solution (Fig. 3). Lycopene molecules promptly form H-type (or "card pack") supramolecular assemblies in aqueous media in which the molecules are held together by non-covalent forces (such as π-π stacking and van der Waals interactions) (Wang et al. 2005). By shielding the hydrophobic polyene chain from water molecules, solubilization is a frequently used method to enhance the water solubility of lycopene and other carotenoids for investigations of their effects in *in vitro* or *in vivo* test systems and in therapeutic applications.

Possessing a hydrophobic cavity, cyclodextrins (CDs) are well-suited host molecules for the formation of inclusion complexes with a variety of lipophilic species by admitting them, at least partially, into the cavity in aqueous media (Davis and Brewster 2004). Water-solubilization and stabilization of lycopene were successfully achieved by using α-, β- and γ-cyclodextrins and their derivatives (Mele et al. 2002, Ukai et al. 1996, Vertzoni et al. 2006). Among the non-covalent lycopene-cyclodextrin associations, both true 1:1 inclusion complexes and large supramolecular aggregates were observed and characterized (Mele et al. 2002). In the case of γ-CDs, a molar ratio of 1:200 of lycopene to the host was required for 100% complexation of the carotene molecules (Ukai et al. 1996). Our recent study on the complexation behavior of carotenoids with CDs using solubility experiments and molecular modeling methods demonstrated the importance of the self-association properties of CD derivatives on their carotenoid complexation behavior (Bikadi et al. 2006).

NON-COVALENT PROTEIN BINDING OF LYCOPENE: A MOLECULAR MODELING STUDY

When the availability of experimental data is limited, molecular modeling studies are valuable initial tools for the determination of the potential binding affinities of carotenoids to proteins. Molecular docking is the most common technique for the *in silico* investigation of protein-ligand interactions. The "blind docking" method is generally used for the prediction of possible ligand binding sites on the whole protein (Hetenyi and van der Spoel 2002). When the binding site of a protein was unknown, or other sites than the catalytic site were shown to modulate enzymatic activity (i.e., allosteric mechanisms), this method successfully revealed possible sites of interactions. In our recent article (Hazai et al. 2006), docking studies were used to predict the binding affinities of lycopene and lycophyll to lipoxygenase (LOX) enzymes and to explain the results at the molecular level.

In vitro experiments showed that lycopene was a substrate of soybean LOX (Biacs and Daood 2000). The presence of this enzyme significantly increased the production of lycopene oxidative metabolites (Biacs and Daood 2000). In order to gain a deeper insight into the molecular mechanisms by which lycopene and lycophyll might exert its anti-cancer effects, docking calculations were carried out with human 5-LOX. The intact tomato carotenoids were found to bind with high affinity to a putative allosteric site on the 5-LOX enzyme (the "cleavage site"), while their potential oxidative metabolites were capable of high-affinity direct competitive binding at the 5-LOX active (catalytic) site (Fig. 4).

The long, linear groove at the interface of the N- and C-terminal domains of 5-LOX renders the rigid carotenoids ideal ligands for this site. The shape of both all-*trans* lycophyll and all-*trans* lycopene fits almost exactly to the cleft formed by the surface residues at this site. The interaction is formed with carotenoids by numerous hydrophobic residues (Val109, Val110 and Leu603), as well as with hydrophilic and charged residues near the polyene chain. The hydroxyl (OH) groups of lycophyll were fixed in the cleft and formed hydrogen bonds with Q129, K161, and D162, respectively. The molecular lengths of the investigated carotenoids (about 30 Å) did not allow their binding to the catalytic site in both *trans* and *cis* geometric conformations. Carotenoid binding at this cleavage site

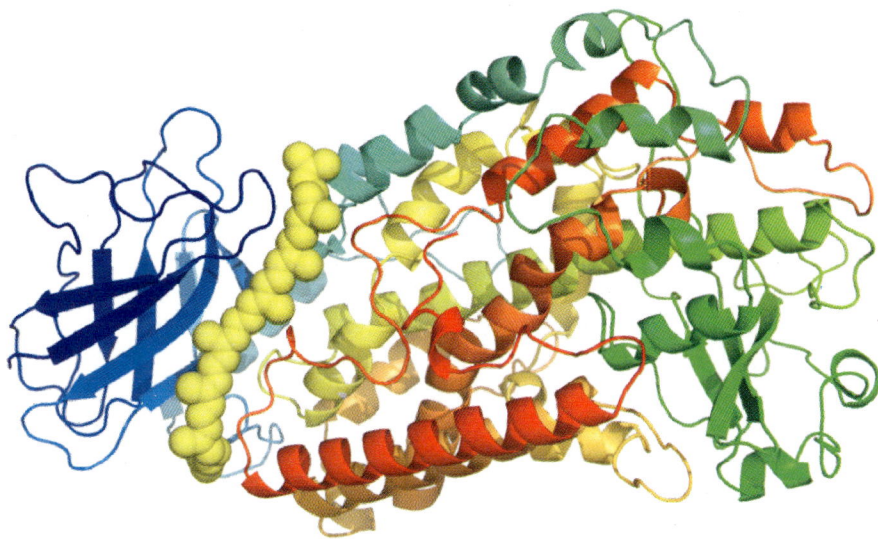

Fig. 4 Lycopene docked to the "cleavage" site of 5-LOX (5-lipoxygenase) model (see text as well as Hazai et al. 2006).

could explain the effect of polyenic compounds on the 5-LOX enzymatic function via an allosteric mechanism, or by radical scavenging in proximity to the active center.

Two bioactive metabolites of lycopene were calculated to bind to the catalytic site with high affinity, suggesting potential direct competitive inhibition of 5-LOX activity that should be shared by both lycopene and lycophyll (and lycoxanthin as well) after *in vivo* supplementation, particularly in the case of the dial metabolite.

The results suggested at least three possible mechanisms by which tomato carotenoids could modulate 5-LOX enzyme activity:

1. Binding of all-*trans* lycopene or lycophyll into the cleft could modulate enzyme activity by an allosteric mechanism and also could affect the interdomain movement of the N-terminal and C-terminal domains, which has been shown to have functional importance for regulation of the catalytic activity (Hammel et al. 2004, Kühn et al. 2005). The presence of an allosteric binding site in both soybean and human 15-LOX has been demonstrated experimentally (Mogul et al. 2000, Ruddat et al. 2003).

2. Binding of natural tomato carotenoid molecule(s) in the cleavage site could place the carotenoids in reasonable proximity for trapping free radical intermediates generated during the enzyme-

substrate reaction, or altering the ion state of the active center. It was shown that antioxidant activity in the vicinity of the activity center can result in the inhibition of LOX-catalyzed reactions (Serpen and Gokmen 2006).

3. Recent advances in characterization of carotenoid cleavage enzymes may also provide clues to the possibility of enzymatic formation of active lycopene-series metabolites. Novel 9',10'-dioxygenases have been added to the growing family of double bond cleavage enzymes characterized from plants and animals (Wyss 2004). If an excentric dioxygenase is capable of utilizing lycophyll as well as lycopene as a substrate *in vivo*, then supplementation with either compound could produce direct enzyme inhibitory activity of the 5-LOX catalytic site through "metabolic activation". Such enzymes have already been identified in the tomato genome; they produce a symmetric C14 dial only slightly shorter than 2,7,11-trimethyl-tetradecahexaene-1,14-dial utilized in the current study (Simkin et al. 2004). Oxidative metabolites of astaxanthin have been identified by high performance liquid chromatography in humans, which suggest the presence of such an excentric cleavage enzyme (Kistler et al. 2002).

Future Directions for Lycopene Series Work

Lycopene is easily isolated from tomato fruit for nutraceutical production; as the predominant C40 carotenoid therein, it is also easily absorbed from dietary tomato sauce, paste, juice, and ketchup, particularly after appropriate processing. The direct chemical synthesis was described in 1972 (Kjosen and Liaaen-Jensen 1972), allowing for cost-effective industrial production today. Recently, the direct chemical synthesis and isolation of all-*trans* lycophyll was described (Jackson et al. 2005; Braun et al. 2006). As lycoxanthin and lycophyll have been shown to have distinct chemical and biological properties from lycopene, such synthesis should allow for appropriate small-scale and eventual large-scale *in vitro* and *in vivo* testing of these important carotenoids.

ACKNOWLEDGMENTS

F. Zsila is grateful for the support of research grants of OTKA K69213 and NKFP1-00012/2005 (Jedlik Anyos project).

ABBREVIATIONS

Å: angstrom; apoB-100: apolipoprotein B-100; ATP: adenosine triphosphate; ^{14}C: carbon 14 isotope; C40: carotenoids containing 40 carbon atoms; CDs: cyclodextrins; CMO: carotenoid monooxygenase; Cx43: connexin 43; GJIC: gap junctional intercellular communication; HMG-CoA: 3-hydroxy-3-methyl-glutaryl-coenzyme A; IGF-1: insulin-like growth factor-1; IL-6: interleukin 6; LDL: low-density lipoprotein; LOX: lipoxygenase (as in 5-LOX)

References

Amir, H. and M. Karas, J. Giat, M. Danilenko, R. Levy, T. Yermiahu, J. Levy, and Y. Sharoni. 1999. Lycopene and 1,25-dihydroxyvitamin D3 cooperate in the inhibition of cell cycle progression and induction of differentiation in HL-60 leukemic cells. Nutr. Cancer 33: 105-112.

Barber, N.J. and X. Zhang, G. Zhu, R. Pramanik, J.A. Barber, F.L. Martin, J.D. Morris, and G.H. Muir. 2006. Lycopene inhibits DNA synthesis in primary prostate epithelial cells in vitro and its administration is associated with a reduced prostate-specific antigen velocity in a phase II clinical study. Prost Cancer Prost Diseases 9: 407-413.

Baskaran, V. and T. Sugawara, and A. Nagao. 2003. Phospholipids affect the intestinal absorption of carotenoids in mice. Lipids 38: 705-711.

Ben-Dor, A. and A. Nahum, M. Danilenko, Y. Giat, W. Stahl, H.D. Martin, T. Emmerich, N. Noy, J. Levy, and Y. Sharoni. 2001. Effects of acyclo-retinoic acid and lycopene on activation of the retinoic acid receptor and proliferation of mammary cancer cells. Arch. Biochem. Biophys. 391: 295-302.

Bertram, J.S. and A.L. Vine. 2005. Cancer prevention by retinoids and carotenoids: independent action on a common target. Biochim. Biophys. Acta 1740: 170-178.

Bhosale, P. and P.S. Bernstein. 2007. Vertebrate and invertebrate carotenoid-binding proteins. Arch. Biochem. Biophys. 458: 121-127.

Biacs, P.A. and H.G. Daood. 2000. Lipoxygenase-catalysed degradation of carotenoids from tomato in the presence of antioxidant vitamins. Biochem. Soc. Trans. 28: 839-845.

Bikadi, Z. and R. Kurdi, S. Balogh, J. Szeman, and E. Hazai. 2006. Aggregation of cyclodextrins as an important factor to determine their complexation behavior. Chem. Biodiversity 3: 1266-1278.

Borel, P. and P. Grolier, M. Armand, A. Partier, H. Lafont, D. Lairon, and V. AzaisBraesco. 1996. Carotenoids in biological emulsions: Solubility, surface-to-core distribution, and release from lipid droplets. J. Lipid Res. 37: 250-261.

Braun, C.L. and H.L. Jackson, S.F. Lockwood, and G. Nadolski. 2006. Purification of synthetic all-E lycophyll (psi,psi-carotene-16,16'-diol). J. Chromatogr. B. 834: 208-212.

Britton, G. and S. Liaaen-Jensen, H. Pfander, A.Z. Mercadante, and E.S. Egeland. [eds.] 2004. Carotenoids Handbook. Birkhäuser, Basel.

Clinton, S.K. 1998. Lycopene: chemistry, biology, and implications for human health and disease. Nutr. Rev. 56: 35-51.

Davis, M.E. and M.E. Brewster. 2004. Cyclodextrin-based pharmaceutics: past, present and future. Nat. Rev. Drug Discovery 3: 1023-1035.

Delgado-Vargas, F. and A.R. Jimenez, and O. Paredes-Lopez. 2000. Natural pigments: carotenoids, anthocyanins, and betalains—characteristics, biosynthesis, processing, and stability. Crit. Rev. Food Sci. Nutr. 40: 173-289.

Di Mascio, P. and S. Kaiser, and H. Sies. 1989. Lycopene as the most efficient biological carotenoid singlet oxygen quencher. Arch. Biochem. Biophys. 274: 532-538.

Diwadkar-Navsariwala, V. and J.A. Novotny, D.M. Gustin, J.A. Sosman, K.A. Rodvold, J.A. Crowell, M. Stacewicz-Sapuntzakis, and P.E. Bowen. 2003. A physiological pharmacokinetic model describing the disposition of lycopene in healthy men. J. Lipid Res. 44: 1927-1939.

During, A. and E.H. Harrison. 2005. An in vitro model to study the intestinal absorption of carotenoids. Food Res. Int. 38: 1001-1008.

Elliott, R. 2005. Mechanisms of genomic and non-genomic actions of carotenoids. Biochim. Biophys. Acta - Mol. Basis Dis. 1740: 147-154.

Ershov, Yu. V. and V. Bykhovsky, and A.A. Dmitrovskii. 1994. Stabilization and competitive inhibition of beta-carotene 15,15'-dioxygenase by carotenoids. Biochem. Mol. Biol. Int. 34: 755-763.

Failla, M.L. and C. Chitchumroonchokchai, and B.K. Ishida. 2008. In vitro micellarization and intestinal cell uptake of *cis* isomers of lycopene exceed those of all-*trans* lycopene. J Nutr. 138: 482-486.

Fuhrman, B. and A. Elis, and M. Aviram. 1997. Hypocholesterolemic effect of lycopene and beta-carotene is related to suppression of cholesterol synthesis and augmentation of LDL receptor activity in macrophages. Biochem. Biophys. Res. Commun. 233: 658-662.

Furr, H.C. and R.M. Clark. 1997. Intestinal absorption and tissue distribution of carotenoids. J. Nutr. Biochem. 8: 364-377.

Gajic, M. and S. Zaripheh, and F. Sun, and J.W. Erdman. 2006. Apo-8'-lycopenal and apo-12'-lycopenal are metabolic products of lycopene in rat liver. J. Nutr. 136: 1552-1557.

Gruszecki, W.I. and K. Strzalka. 2005. Carotenoids as modulators of lipid membrane physical properties. Biochim. Biophys. Acta - Mol. Basis Dis. 1740: 108-115.

Guns, E.S. and S.P. Cowell. 2005. Drug Insight: lycopene in the prevention and treatment of prostate cancer. Nat. Clin. Practice Urol. 2: 38-43.

Hammel, M. and M. Walther, R. Prassl, and H. Kuhn. 2004. Structural flexibility of the N-terminal beta-barrel domain of 15-lipoxygenase-1 probed by small angle X-ray scattering. Functional consequences for activity regulation and membrane binding. J. Mol. Biol. 343: 917-929.

Hazai, E. and Z. Bikadi, F. Zsila, and S.F. Lockwood. 2006. Molecular modeling of the non-covalent binding of the dietary tomato carotenoids lycopene and lycophyll, and selected oxidative metabolites with 5-lipoxygenase. Bioorg. Med. Chem. 14: 6859-6867.

Heber, D. and Q.Y. Lu. 2002. Overview of mechanisms of action of lycopene. Exp. Biol. Med. (Maywood) 227: 920-923.

Hetenyi, C. and D. van der Spoel. 2002. Efficient docking of peptides to proteins without prior knowledge of the binding site. Protein Sci. 11: 1729-1737.

Hsiao, G. and Y. Wang, N.H. Tzu, T.H. Fong, M.Y. Shen, K.H. Lin, D.S. Chou, and J.R. Sheu. 2005. Inhibitory effects of lycopene on in vitro platelet activation and in vivo prevention of thrombus formation. J. Lab. Clin. Med. 146: 216-226.

Jackson, H.L. and G.T. Nadolski, C. Braun, and S.F. Lockwood. 2005. Efficient total synthesis of lycophyll (psi,psi-carotene-16,16 '-diol). Org. Process Res. Dev. 9: 830-836.

Kiefer, C. and S. Hessel, J.M. Lampert, K. Vogt, M.O. Lederer, D.E. Breithaupt, and J. von Lintig. 2001. Identification and characterization of a mammalian enzyme catalyzing the asymmetric oxidative cleavage of provitamin A. J. Biol. Chem. 276: 14110-14116.

Kistler, A. and H. Liechti, L. Pichard, E. Wolz, G. Oesterhelt, A. Hayes, and P. Maurel. 2002. Metabolism and CYP-inducer properties of astaxanthin in man and primary human hepatocytes. Arch. Toxicol. 75: 665-675.

Kjosen, H. and S. Liaaen-Jensen. 1972. Carotenoids of higher plants. 6. Total synthesis of lycoxanthin and lycophyll. Acta Chem. Scand. [A] 26: 4121-4129.

Kloer, D.P. and S. Ruch, S. Al-Babili, P. Beyer, and G.E. Schulz. 2005. The structure of a retinal-forming carotenoid oxygenase. Science 308: 267-269.

Kühn, H. and J. Saam, S. Eibach, H.G. Holzhutter, I. Ivanov, and M. Walther. 2005. Structural biology of mammalian lipoxygenases: enzymatic consequences of targeted alterations of the protein structure. Biochem. Biophys. Res. Commun. 338: 93-101.

Kun, Y. and U.S. Lule, and D. Xiao-Lin. 2006. Lycopene: Its properties and relationship to human health. Food Rev. Int. 22: 309-333.

Lazarus, S.A. and M.L. Garg. 2004. Tomato extract inhibits human platelet aggregation in vitro without increasing basal cAMP levels. Int. J. Food Sci. Nutr. 55: 249-256.

Lian, F. and D.E. Smith, and H. Ernst, and R.M. Russell and X.D. Wang. 2007. Apo-10'-lycopenoic acid inhibits lung cancer cell growth in vitro, and suppresses lung tumorigenesis in the A/J mouse model in vivo. Carcinogenesis 28: 1567-1574.

Lin, S. and L. Quaroni, W.S. White, T. Cotton, and G. Chumanov. 2000. Localization of carotenoids in plasma low-density lipoproteins studied by surface-enhanced resonance Raman spectroscopy. Biopolymers 57: 249-256.

Lindshield, B.L. and K. Canene-Adams, and J.W. Erdman, Jr. 2007. Lycopenoids: are lycopene metabolites bioactive? Arch. Biochem. Biophys. 458: 136-140.

Liu, A. and N. Pajkovic, Y. Pang, D.W. Zhu, B. Calamini, A.L. Mesecar, and R.B. van Breemen. 2006. Absorption and subcellular localization of lycopene in human prostate cancer cells. Mol. Cancer Ther. 5: 2879-2885.

McNulty, H.P. and J. Byun, S.F. Lockwood, R.F. Jacob, and R.P. Mason. 2007. Differential effects of carotenoids on lipid peroxidation due to membrane interactions: X-ray diffraction analysis. Biochim. Biophys. Acta 1768: 167-174.

Mele, A. and R. Mendichi, A. Selva, P. Molnar, and G. Toth. 2002. Non-covalent associations of cyclomaltooligosaccharides (cyclodextrins) with carotenoids in water. A study on the alpha- and beta-cyclodextrin/psi,psi-carotene (lycopene) systems by light scattering, ionspray ionization and tandem mass spectrometry. Carbohydr. Res. 337: 1129-1136.

Mogul, R. and E. Johansen, and T.R. Holman. 2000. Oleyl sulfate reveals allosteric inhibition of soybean lipoxygenase-1 and human 15-lipoxygenase. Biochemistry 39: 4801-4807.

Molnar, J. and N. Gyemant, M. Tanaka, J. Hohmann, E. Bergmann-Leitner, P. Molnar, J. Deli, R. Didiziapetris, and M.J.U. Ferreira. 2006. Inhibition of multidrug resistance of cancer cells by natural diterpenes, triterpenes and carotenoids. Curr. Pharm. Des. 12: 287-311.

Rao, A.V. and L.G. Rao. 2007. Carotenoids and human health. Pharmacol. Res. 55: 207-216.

Rao, M.N. and P. Ghosh, and M.R. Lakshman. 1997. Purification and partial characterization of a cellular carotenoid-binding protein from ferret liver. J. Biol. Chem. 272: 24455-24460.

Reboul, E. and L. Abou, and C. Mikail, and O. Ghiringhelli, and M. André, and H. Portugal, and D. Jourdheuil-Rahmani, and M.J. Amiot, and D. Lairon, and P. Borel. 2005. Lutein transport by Caco-2 TC-7 cells occurs partly by a facilitated process involving the scavenger receptor class B type I (SR-BI). Biochem. J. 387: 455-461.

Ruddat, V.C. and S. Whitman, T.R. Holman, and C.F. Bernasconi. 2003. Stopped-flow kinetic investigations of the activation of soybean lipoxygenase-1 and the influence of inhibitors on the allosteric site. Biochemistry 42: 4172-4178.

Segrest, J.P. and M.K. Jones, H. De Loof, and N. Dashti. 2001. Structure of apolipoprotein B-100 in low density lipoproteins. J. Lipid Res. 42: 1346-1367.

Serpen, A. and V. Gokmen. 2006. A proposed mechanism for the inhibition of soybean lipoxygenase by β-carotene. J. Sci. Food Agric. 86: 401-406.

Sharoni, Y. and M. Danilenko, N. Dubi, A. Ben-Dor, and J. Levy. 2004. Carotenoids and transcription. Arch. Biochem. Biophys. 430: 89-96.

Simkin, A.J. and S.H. Schwartz, M. Auldridge, M.G. Taylor, and H.J. Klee. 2004. The tomato carotenoid cleavage dioxygenase 1 genes contribute to the formation of the flavor volatiles beta-ionone, pseudoionone, and geranylacetone. Plant J. 40: 882-892.

Suwalsky, M. and P. Hidalgo, K. Strzalka, and A. Kostecka-Gugala. 2002. Comparative X-ray studies on the interaction of carotenoids with a model phosphatidylcholine membrane. Zeitschrift fur Naturforschung C - a J. Biosci. 57: 129-134.

Tapiero, H. and D.M. Townsend, and K.D. Tew. 2004. The role of carotenoids in the prevention of human pathologies. Biomed. Pharmacother. 58: 100-110.

Teodori, E. and S. Dei, C. Martelli, S. Scapecchi, and F. Gualtieri. 2006. The functions and structure of ABC transporters: implications for the design of new inhibitors of Pgp and MRP1 to control multidrug resistance (MDR). Curr. Drug Targets 7: 893-909.

Ukai, N. and H. Nakamura, Y. Lu, H. Etoh, K. Ina, S. Ohshima, F. Ojima, H. Sakamoto, and Y. Ishiguro. 1996. The water-solubilization and stabilization of lycopene by cyclodextrin and addition of protein. J. Jap. Soc. Food Sci. Technol. - Nippon Shokuhin Kagaku Kogaku Kaishi 43: 247-250.

van Bennekum, A. and M. Werder, S.T. Thuahnai, C.H. Han, P. Duong, D.L. Williams, P. Wettstein, G. Schulthess, M.C. Phillips, and H. Hauser. 2005. Class B scavenger receptor-mediated intestinal absorption of dietary ss-carotene and cholesterol. Biochemistry (Moscow) 44: 4517-4525.

Vertzoni, M. and T. Kartezini, C. Reppas, H. Archontaki, and G. Valsami. 2006. Solubilization and quantification of lycopene in aqueous media in the form of cyclodextrin binary systems. Int. J. Pharm. 309: 115-122.

Wane, D. and C.A. Lengacher. 2006. Integrative review of lycopene and breast cancer. Oncol. Nurs. Forum 33: 127-134.

Wang, L.X. and Z.L. Du, R.X. Li, and D.C. Wu. 2005. Supramolecular aggregates of lycopene. Dyes Pig. 65: 15-19.

Willson, T.M. and P.J. Brown, D.D. Sternbach, and B.R. Henke. 2000. The PPARs: from orphan receptors to drug discovery. J. Med. Chem. 43: 527-550.

Wirtz, G.M. and C. Bornemann, A. Giger, R.K. Muller, H. Schneider, G. Schlotterbeck, G. Schiefer, and W.D. Woggon. 2001. The substrate specificity of beta,beta-carotene 15,15'-monooxygenase. Helv. Chim. Acta 84: 2301-2315.

Wyss, A. 2004. Carotene oxygenases: A new family of double bond cleavage enzymes. J. Nutr. 134: 246S-250S.

Zaripheh, S. and J.W. Erdman, Jr. 2005. The biodistribution of a single oral dose of [^{14}C]-lycopene in rats prefed either a control or lycopene-enriched diet. J. Nutr. 135: 2212-2218.

Zsila, F. and Z. Bikadi, and M. Simonyi. 2002. Further insight into the molecular basis of carotenoid-albumin interactions: circular dichroism and electronic absorption study on different crocetin-albumin complexes. Tetrahedron-Asymmetry 13: 273-283.

Zsila, F. and Z. Bikadi, and M. Simonyi. 2001. Induced chirality upon crocetin binding to human serum albumin: origin and nature. Tetrahedron-Asymmetry 12: 3125-3137.

Zsila, F. and M. Simonyi, and S.F. Lockwood. 2003. Interaction of the disodium disuccinate derivative of meso-astaxanthin with human serum albumin: From chiral complexation to self-assembly. Bioorg. Med. Chem. Lett. 13: 4093-4100.

Zsila, F. and P. Molnar, and J. Deli. 2005a. Analysis of binding interaction between the natural apocarotenoid bixin and human serum albumin by circular dichroism and fluorescence spectroscopy. Chem. Biodiversity 2: 758-772.

Zsila, F. and P. Molnar, J. Deli, and S.F. Lockwood. 2005b. Circular dichroism and absorption spectroscopic data reveal binding of the natural cis-carotenoid bixin to human alpha(1)-acid glycoprotein. Bioorg. Chem. 33: 298-309.

Zsila, F. and I. Fitos, Z. Bikadi, M. Simonyi, H.L. Jackson, and S.F. Lockwood. 2004. In vitro plasma protein binding and aqueous aggregation behavior of astaxanthin dilysinate tetrahydrochloride. Bioorg. Med. Chem. Lett. 14: 5357-5366.

Risk Assessment of Lycopene

Andrew Shao and John N. Hathcock
Council for Responsible Nutrition, 1828 L St., NW, Suite 900
Washington, DC 20036-5114

ABSTRACT

Lycopene is one of the most prevalent carotenoids in the human diet and has become an increasingly popular ingredient in dietary supplements and functional food products. A large body of human and animal research suggests that supplemental lycopene may provide benefits in the areas of prostate, skin and cardiovascular health. The increased awareness and use of this ingredient in dietary supplements and functional foods warrants a comprehensive review of its safety. Systematic evaluation of the research designs and data provide a basis for risk assessment and the usual tolerable upper intake level (UL) derived from it if the newer methods described as the Observed Safe Level (OSL) or Highest Observed Intake (HOI) are used (terms are defined in Table 1). The OSL risk assessment method indicates that the evidence of safety is strong at intakes up to 75 mg/d for lycopene, and this level is identified as the OSL. Although much higher levels have been tested without adverse effects and may be safe, the data for intakes above this level are not sufficient for a confident conclusion of long-term safety.

INTRODUCTION

Lycopene is among the most prevalent carotenoids in the human diet and serum (Table 2) (Beecher and Khachik 1992, Khachik et al. 1992). Its popularity in dietary supplements and functional foods and beverages has

A list of abbreviations is given before the references.

Table 1 Definitions of terms.

Term	Definition
Dietary Reference Intakes (DRI)	Series of reports on dietary reference values for the intake of essential nutrients by Americans and Canadians. Includes details on the application of statistically valid methods and reviews the roles that macronutrients play in traditional deficiency diseases as well as chronic diseases (Institute of Medicine, National Academies of Science).
Highest Observed Intake (HOI)	An estimate of a safe nutrient intake level only when no adverse health effects have been identified. It is the highest level of intake observed or administered as reported within a study or studies of acceptable quality. A Model for Establishing Upper Levels of Intake for Nutrients and Related Substances (FAO/WHO 2006).
Lowest observed adverse effect level (LOAEL)	The lowest tested dose of a substance that has been reported to cause harmful health effects on people or animals (FNB 1998b).
Observed Safe Level (OSL)	The highest intake level of a nutrient with sufficient and appropriate evidence to affirm safety, even if there are no established adverse effects at any level (Hathcock 2004).
No observed adverse effect level (NOAEL)	The highest tested dose of a substance that has been reported to have no harmful health effects on people or animals (FNB 1998b).
Tolerable Upper Intake Level (UL)	The highest level of daily nutrient intake that is likely to pose no risk of adverse health effects to almost all individuals in the general population (FNB 1998b).
Uncertainty Factor (UF)	A correction factor used in the calculation of the UL to account for uncertainty involved in the extrapolation of data from specific study populations to the general population (FNB 1998b).
Upper Level for Supplements (ULS)	Represents the amounts of nutrients known to be safe for supplemental intake by healthy adults who consume typical diets.

soared in recent years, partly because of increase in research and submissions of petitions to the US Food and Drug Administration for the approval of Qualified Health Claim language (Food and Drug Administration 2006).

Lycopene lacks oxygen atoms and is therefore a member of the hydrocarbon family of carotenoids (Khachik et al. 2002). It is present in the human diet primarily in dark red fruits and vegetables such as tomatoes (Holden et al. 1999). Increased consumption and higher serum levels of lycopene have been linked to a reduced risk of cardiovascular disease (Arab and Steck 2000, Willcox et al. 2003) and prostate cancer (Chan et al. 2005, Giovannucci 2005). Findings from a growing collection of placebo-

Table 2 Prevalence of various carotenoids in human serum*.

Carotenoid	% Distribution in serum
Lutein	20
Lycopene	20
α-carotene	10
ζ-carotene	10
Phytofluene	8
β-cryptoxanthin	8
β-carotene	6
α-cryptoxanthin	4
Phytoene	4
Anhydrolutein	3
Zeaxanthin	3
γ-carotene	2
Neurosporene	2

*Derived from Khachik et al. (1992).

controlled intervention trials suggest that consumption of lycopene (either as a dietary supplement or in the form of processed tomatoes) can reduce DNA damage (Astley et al. 2004, Zhao et al. 2006) and may have beneficial effects on prostate cancer (Kucuk et al. 2001, 2002, Ansari and Gupta 2003, van Breemen 2005) and lung cancer (Liu et al. 2003, 2006, Wang 2005).

The increase in both public awareness and use of this ingredient in dietary supplements and functional food products warrants a comprehensive safety evaluation and identification of a tolerable upper intake level (UL) or equivalent. The UL is the highest level of daily nutrient intake that is likely to pose no risk of adverse health effects to almost all individuals in the general population (FNB 1998b). The UL is not a threshold value above which toxicity occurs, but rather serves as an important guideline for industry regulators, policy makers, product formulators and consumers to protect against excessive intakes. The ULs for most nutrients and related substances are based on widely applicable risk assessment models used by the US Food and Nutrition Board (FNB) in its establishment of the Dietary Reference Intakes (Refer to Fig. 1A) (FNB 1997, 1998a, 1998b, 2000, 2001). Risk assessment is a scientific approach to characterizing the nature and likelihood of harm resulting from human exposure to any substance (FNB 1998b). Traditionally used in toxicology to assess the safety of exposure to non-carcinogenic chemicals (National Research Council 1983), risk assessment has become one of the most well-accepted formal and quantitative scientific methods for determining the safety of orally ingested nutrients. Because of the systematic, comprehensive and authoritative character of the FNB risk assessment

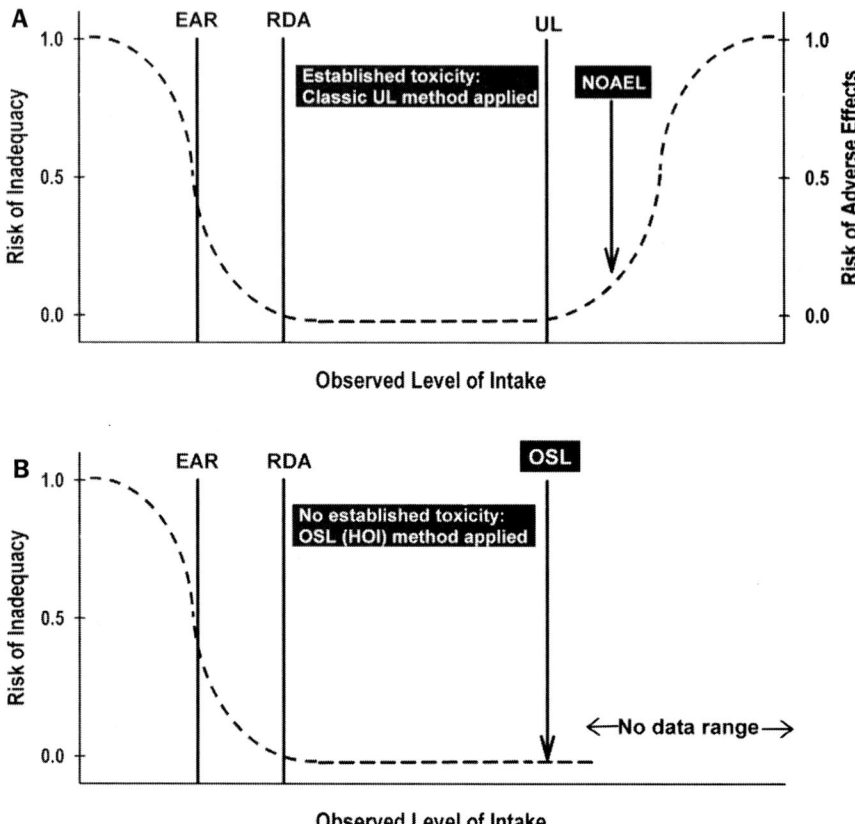

Fig. 1 Theoretical description of health effects of a nutrient as a function of level of intake. The upper graph (A) is a classic figure used by the US Food and Nutrition Board to illustrate the concept of the Dietary Reference Intakes (DRI) for a given nutrient. The Estimated Average Requirement (EAR), Recommended Dietary Allowance (RDA) and Tolerable Upper Intake Level (UL) are three of the population-based values. The UL is determined based on an established adverse (or critical) effect and is calculated from the NOAEL. The lower graph (B) is a derivative of the upper illustrating the OSL concept. In the absence of established adverse effects, a NOAEL and therefore a UL cannot be identified. In such cases the OSL (or HOI) approach may be applied to determine an upper intake level with sufficient confidence.

method for nutrients, this approach, with some slight modifications, is now widely supported and adopted by others such as the European Commission Scientific Committee on Food (Scientific Committee on Food 2001), the United Kingdom Expert Group on Vitamins and Minerals (Expert Group on Vitamins and Minerals 2003) and more recently the Food and Agriculture Organization/World Health Organization in its project report *A Model for Establishing Upper Levels of Intake for Nutrients and Related Substances* (FAO/WHO 2006).

A drawback of the UL method developed by the FNB is that for nutrients for which there are no established adverse effects or toxicity, a UL is not set, regardless of the depth, quality or scope of available data relevant to safety of those nutrients (FNB 1998b). The absence of evidence of toxicity in the face of extensive data has led to the decision by the FNB not to establish a UL for several nutrients, including thiamin, riboflavin, biotin, and vitamin B12 (FNB 1998a). Unfortunately, the absence of a UL has been erroneously misinterpreted to indicate that there is insufficient evidence for the safety of these substances. Vitamin B12 is perhaps the best example of this error. Despite a large body of data indicating safety at doses as high as several milligrams a day (Hathcock 2004), the absence of a UL from the FNB has been used by some international policy makers to support an illogically low UL (9 μg/d proposed in Germany, Domke et al. 2006; 3 μg/d in France, Ministre de l'economie 2006). To reduce this type of misinterpretation of the absence of a UL, use of an alternative approach termed the Observed Safe Level (OSL) method is advocated (Refer to Fig. 1B). The 2006 FAO/WHO report adopted the same concept under the name Highest Observed Intake (HOI) (FAO/WHO, 2006), although the report did not include examples of the application of the method.

METHODS

The safety evaluation method applied to orally administered lycopene is from the Council for Responsible Nutrition. The Council's *Vitamin and Mineral Safety, 2nd Edition* (Hathcock 2004), contains the basic features of the FNB method and also the OSL modification, defined as the highest level of a nutrient studied with convincing evidence of safety for those nutrients where no toxicity can be identified at any level. This approach was recently adopted as a Highest Observed Intake (HOI) in the FAO/WHO report (FAO/WHO 2006) and was recently applied to the lycopene database (Shao and Hathcock 2006a).

Overall, this risk analysis was derived from the available published human clinical trials obtained from the Medline database (www.pubmed.gov) between 1966 and March 2007. The major steps typically used in nutrient risk assessment are outlined in Fig. 2 (FNB 1998b). Briefly, the first step in the UL process is to identify a hazard or adverse effect related to excessive intake, followed by assessment of the dose-response relationship for the identified hazard. A no observed adverse effect level (NOAEL) or lowest observed adverse effect level (LOAEL) is then selected to which an uncertainty factor (UF) is then applied to account for uncertainty that may pertain to extrapolation of the data to the generally healthy population. The UL is then derived by

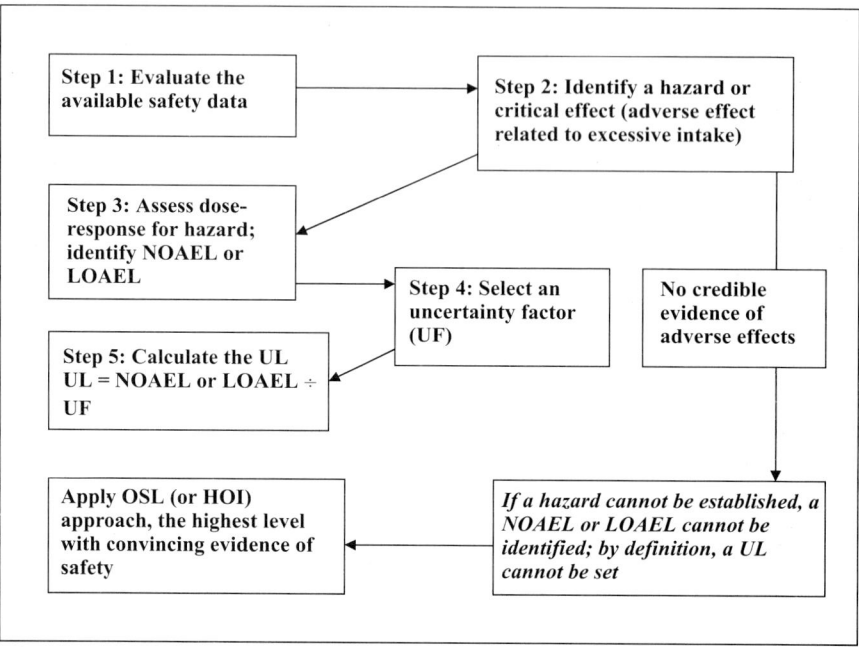

Fig. 2 Basic steps involved in nutrient risk assessment. The classic UL method involves several basic steps: (1) Evaluation of the available safety data. (2) Identification of a hazard or adverse effect related to excessive intake. (3) Assessment of the dose-response relationship for the identified hazard and selection of a no observed adverse effect level (NOAEL) or lowest observed adverse effect level (LOAEL). (4) Selection and application of an uncertainty factor (UF). (5) Calculation of the UL by dividing the NOAEL or LOAEL by the UF.

If no data establish adverse effects in humans, the classic approach cannot be used and a UL cannot be derived. In these circumstances, the highest intake level with sufficient evidence of safety is identified as the OSL (or HOI).

dividing the NOAEL or LOAEL by the UF. If no data establish adverse effects in humans, the above procedure cannot be used. In these circumstances, the highest intake level with sufficient evidence of safety is identified as the OSL (Hathcock 2004) and the HOI by FAO/WHO (FAO/WHO 2006). Uncertainty is considered in selection of the OSL value, although using an approach that differs from that used by the FNB (see section on Uncertainty Evaluation below).

We applied the first procedure to the published lycopene human clinical trial data and found no basis for a NOAEL or LOAEL and thus could not derive a UL. Consequently, we applied the OSL procedure, with the results described in the sections below.

No consistent efforts were made in any of the clinical trials to control dietary lycopene, and therefore the assumption is that the study subjects were consuming usual dietary levels of this substance. Thus, the OSL value identified from the trials does not require correction for dietary intakes, and the OSL can be identified as a safe Upper Level for Supplements (ULS).

SCIENTIFIC EVIDENCE RELATED TO SAFETY

Human Studies

There have been more than 50 peer-reviewed, published human intervention trials involving various forms of lycopene. Of these, the 17 most relevant studies regarding safety are presented in this chapter (Table 3). Criteria for study inclusion were study duration of at least one week and lycopene dose used greater than 8 mg/d, and studies had to be randomized, placebo-controlled intervention trials. Studies that were uncontrolled and unblinded, those investigating acute bioavailability, pharmacokinetics or post-prandial responses from single bolus doses were excluded from this analysis and are used solely as supportive information. Also excluded were studies that did not quantify the dosage of lycopene being administered (such as certain feeding studies). Lycopene exists in nature in several isomeric forms (Khachik et al. 2002), with the majority of both natural and synthetic sources of supplemental lycopene occurring in the *trans* configuration (Hoppe et al. 2003). For the purposes of this risk assessment, human trial data are limited to the use of total or *trans*-lycopene, as specified. Only a few of the studies undertaken to assess the beneficial effects of lycopene monitored possible adverse effects, and then primarily through self-reporting. There are no human studies that have focused specifically on the safety aspects of lycopene supplementation. Two comprehensive reviews have been published examining the safety of natural (Matulka et al. 2004) and synthetic (McClain and Bausch 2003) sources of supplemental lycopene. Both reviews concluded, from the available human and animal toxicity data, that there is no indication of significant adverse effect of oral lycopene.

The highest lycopene dosage used in a randomized, controlled human clinical trial was 150 mg/d for 7 d (Rao and Agarwal 1998). The longest duration trial was 20 wk (140 d) at a lycopene dose of 13.3 mg/d in healthy adults (Olmedilla et al. 2002). Due to differences in study duration and variations in bioavailability from different lycopene sources, serum lycopene levels did not follow a consistent dose-dependent response in the 17 reviewed trials. Although supplemental forms of lycopene derived from synthetic and natural sources appear to have comparable bioavailability (Hoppe et al. 2003), this is not the case with other dietary lycopene sources, such as raw or processed tomatoes or tomato juice

Table 3 Published safety observations* for randomized, controlled trials examining lycopene supplementation.

Study population	Dosage and study design	Key observations	OSL considerations	Reference
Healthy adults $n = 20$	75, 150 mg/d lycopene (as tomato oleoresin); randomized, controlled, crossover; 7 d	Significant 2-, and 2.4-fold increase, respectively, in serum lycopene (to 661.9 and 793.8 nmol/L) vs. placebo; no adverse effects reported	150 mg, but relatively short duration and small sample size argue against use of this study for identification of an OSL	Rao and Agarwal (1998)
Healthy adults $n = 15$	70 mg/d lycopene (as beadlets), 75 mg/d (as oleoresin and tomato juice, respectively); randomized, controlled, crossover; 4 wk	Significant 41% increase in serum lycopene (to approx. 825 nmol/L) for all three treatments; significant 2-fold increase in buccal mucosa cell lycopene level for beadlet and oleoresin forms, respectively; no adverse effects reported	75 mg in healthy adults, substantial duration of exposure to multiple forms at a high dosage; based on the findings of the study above, this study is chosen to serve as the basis for the human OSL	Paetau et al. (1998, 1999)
Healthy elderly $n = 29$	47 mg/d lycopene (as tomato juice); randomized, controlled; 8 wk	Significant 4-fold increase in serum lycopene (to 540 nmol/L); no adverse effects reported	47 mg, supports the OSL selected	Watzl et al. (2000)
Healthy women $n = 10$	45 mg/d lycopene as tomato juice; randomized, controlled; 29 d	Significant 2.8-fold increase in serum lycopene (to 1842 nmol/L); no adverse effects reported	45 mg, supports the OSL selected	Maryama et al. (2001)
Prostate cancer patients $n = 15$	30 mg/d lycopene (as tomato oleoresin); randomized, controlled; 3 wk	No significant changes for lycopene in serum or tumor tissues; no adverse effects observed	30 mg, supports the OSL selected	Kucuk et al. (2001, 2002)
Exercise-induced asthma patients $n = 20$	30 mg/d lycopene (as tomato oleoresin); randomized, controlled, crossover; 1 wk	Significant 2-fold increase in serum lycopene (to 149 nmol/L); no adverse effects reported	30 mg, supports the OSL selected	Neuman et al. (2000)
Healthy adults $n = 22$	25 mg/d lycopene (as tomato paste or oleoresin, respectively); randomized, controlled; 8 wk	Significant 2.7 and 2.5-fold increases in serum lycopene in the two treatment groups, respectively; significant 3.7-fold increase in buccal mucosa cell lycopene level; no adverse effects reported	25 mg, supports the OSL selected	Richelle et al. (2002)

(Table Contd.)

Table Contd.

Subjects	Dose and design	Outcome	OSL	Reference
Healthy adults n = 9	20 mg/d lycopene (as tomato paste); randomized, controlled, crossover; 11 d	Significant 2.4-fold increase in serum lycopene (to 3430 nmol/L); no adverse effects reported	20 mg, supports the OSL selected	Horvitz et al. (2004)
Oral submucous fibrosis patients n = 21	16 mg/d lycopene; randomized, controlled; 2 mon (60 d)	Significant improvement in mouth-opening values; no adverse effects reported	16 mg, supports the OSL selected	Kumar et al. (2007)
Healthy smoking and non-smoking adults n = 27	15 mg/d lycopene (as tomato paste); randomized, controlled, crossover; 2 wk	Significant 3.3-fold increase in serum lycopene (to 800 nmol/L); no adverse effects observed	15 mg, supports the OSL selected	Briviba et al. (2004)
Healthy men n = 52	15 mg/d lycopene (as tomato oleoresin); randomized, controlled; 12 wk	Significant 1.9-fold increase in serum lycopene (to 1170 nmol/L); no adverse effects reported	15 mg, supports the OSL selected	Hininger et al. (2001)
Healthy adults n = 24	15 mg/d lycopene (as tomato oleoresin or synthetic lycopene, respectively); randomized, controlled; 28 d	Significant 2.5 and 2.8-fold increase in serum lycopene, respectively (to 950 and 910 nmol/L, respectively); no adverse effects observed	15 mg, supports the OSL selected	Hoppe et al. (2003)
Healthy men n = 28	15 mg/d lycopene (as tomato oleoresin); randomized, controlled, crossover; 4 wk	Significant 4-fold increase in serum lycopene (to 750 nmol/L); no adverse effects reported	15 mg, supports the OSL selected	Astley et al. (2004)
Healthy adults n = 23	15 mg/d lycopene (as tomato oleoresin); randomized, controlled; 26 d	Significant 44% increase in serum lycopene (to 1010 nmol/L); no adverse effects reported	15 mg, supports the OSL selected	Hughes et al. (2000)
Healthy elderly n = 17	13.3 mg/d lycopene (as tomato oleoresin); randomized, controlled; 12 wk	Significant 2.7-fold increase in serum lycopene (to 270 nmol/L); no adverse effects reported	13.3 mg, supports the OSL selected	Corridan et al. (2001)

(Table Contd.)

Table Contd.

Healthy adults n = 93	13.3 mg/d lycopene (as tomato oleoresin); randomized, controlled; 20 wk	Significant 2-fold increase in serum lycopene (to 1200 nmol/L); no adverse effects reported	13.3 mg, supports the OSL selected	Olmedilla et al. (2002)
Healthy postmenopausal women n = 8	12 mg/d lycopene (as synthetic lycopene); randomized, controlled; 8 wk	Significant 2.6-fold increase in serum lycopene (to 1300 nmol/L); significant decrease in endogenous DNA damage vs. placebo; no adverse effects reported	12 mg, supports the OSL selected	Zhao et al. (2006)

*No adverse effects observed = investigators report observing no adverse effects and/or study subjects reported no adverse effects
No adverse effects reported = safety or adverse effects not addressed in study publication

(Gartner et al. 1997, Cohn et al. 2004, Tang et al. 2005). Other potential confounders include the inconsistencies in the amount of time allowed for depletion and/or wash-out phases, which varied considerably between studies and the baseline lycopene status of the subjects being tested. However, irrespective of these inter-study inconsistencies, there were no adverse effects observed or reported at any intake in any of the reviewed studies.

Mean level of lycopene consumption from foods by Americans has been estimated to be just over 8 mg/d (Matulka et al. 2004), suggesting that the doses used in the reviewed trials (up to many times higher than what is typically consumed in the diet) are adequate to assess safety of supplementation. The absence of any pattern of adverse effects related to lycopene consumption in any of the published human trials provides support for a high level of confidence in the safety of this substance.

The only documented side effect of lycopene supplementation is carotenodermia, a condition characterized by a yellowish discoloration of the skin resulting from elevated dermal concentrations. The condition is most often associated with high β-carotene intake from foods or supplements (> 30 mg/d) (Bendich, 1988), has been reported in only a few instances involving lycopene (Reigh 1960, La Placa et al. 2000, Olmedilla et al. 2002), and is recognized as a harmless, reversible nuisance effect (FNB, 2000). In contrast to β-carotene, emerging research suggests that lycopene may have a protective effect against smoking-induced lung cancer (Liu et al. 2003, 2006, Wang 2005).

Animal and *in vitro* Studies

Animal and *in vitro* studies addressing the safety as well as the metabolism and metabolic effects of lycopene have been reviewed in detail (McClain and Bausch 2003, Matulka et al. 2004). For natural lycopene (derived from tomato oleoresin), no LD_{50} in rats was established but it was determined to be more than 5000 mg/kg body weight, and the NOAEL was established to be more than 4500 mg/kg body weight (based on a 13 wk study) (Matulka et al. 2004). For synthetic lycopene, no adverse effects were detected up to the highest dose of 1000 mg/kg body weight for 4 wk or 500 mg/kg body weight for 13 wk (McClain and Bausch 2003). The results of teratogenicity and mutagenicity studies conducted on synthetic and natural lycopene also showed no irreversible adverse effects at comparable dosages. Such a large margin of safety from animal studies provides a high level of confidence that supplemental lycopene, regardless of the form, can be consumed safely at relatively high doses by humans.

HUMAN NOAEL OR OSL (HOI)

Human NOAEL

None of the clinical trials found any adverse effects related to lycopene administration, and therefore there is, by definition, no basis for identifying a NOAEL or LOAEL. Without either of these two values a UL usually is not identified (FNB 1998b).

Human OSL

Published relevant human clinical trials involved lycopene doses of up to 150 mg/d (Rao and Agarwal 1998). All human studies reviewed were double-blind, randomized, controlled trials. A series of non-randomized, open-label clinical trials has also been published. The dosages involved in these studies ranged from 12.5 to 37 mg/d, the results of which are consistent with respect to safety, showing no observed or reported adverse effects (Chen et al. 2001, Watzl et al. 2003, Cohn et al. 2004, van Breemen 2005, Frohlich et al. 2006).

Although the absence of adverse effects with 150 mg/d in the study by Rao and Agarwal (1998) suggests this dose may be appropriate for identifying the OSL, the relatively short duration (1 wk) and small sample size (n = 20) argue against use of this study for identification of an OSL for healthy adults.

The next highest dose tested in a published, double-blind, placebo-controlled trial is 75 mg/d for 4 wk (28 d) in 15 healthy adults (Paetau et al. 1998, 1999). Subjects receiving lycopene experienced a significant 41% increase in serum lycopene and significant 2-fold increase in buccal mucosa cell lycopene level, with no adverse effects reported. The longer duration and the absence of adverse effects in this study, as well as that of Rao and Agarwal (1998), which also implemented a 75 mg/d dose, provides sufficient support for the designation of this study as the basis for the human OSL of 75 mg/d.

The remainder of the reviewed trials in Table 3 include lycopene doses ranging from 47 down to 12 mg/d for durations ranging from 1 to 20 wk (Hughes et al. 2000, Neuman et al. 2000, Watzl et al. 2000, Corridan et al. 2001, Hininger et al. 2001, Kucuk et al. 2001, 2002, Maruyama et al. 2001, Olmedilla et al. 2002, Richelle et al. 2002, Hoppe et al. 2003, Astley et al. 2004, Briviba et al. 2004, Horvitz et al. 2004, Zhao et al. 2006, Kumar et al. 2007). The complete absence of adverse effects in all the published human trials using lycopene doses above, at, and below the 75 mg/d level provides sufficient support for this value to confidently be designated as the OSL.

The quantities of lycopene involved in these trials are supplemental amounts well above the estimated average amount consumed in foods consumed in the United States (8 mg/d; Matulka et al. 2004). Therefore, this risk assessment represents a direct approach to a ULS and no correction is needed for the lycopene present in the US food supply.

Uncertainty Evaluation

The traditional UL method used by the FNB and others involves selection of the highest possible NOAEL value and assignment and use of a UF to correct for uncertainty related to the extrapolation of the data to the general population (FNB 1998b). This process itself carries much uncertainty, as there are no well-established or accepted quantitative approaches to uncertainty. The result is a seemingly arbitrary assignment of UF values in the calculation of UL values (Table 4). Whether the UF values chosen for past risk assessments are an accurate indication of uncertainty is a matter of debate.

Table 4 Examples of uncertainty factors (UF)* applied in the UL method.

Nutrient	UF	Data type	Rationale
Vitamin E	36	Animal LOAEL	Species difference
Folic acid	5	Human LOAEL	No NOAEL
Vitamin B6	2	Human NOAEL	Small sample sizes, uncontrolled clinical trials
Iron	1.5	Human LOAEL	Large dataset
Nicotinic acid	1.5	Human LOAEL	Transient, non-toxic effect (flushing)
Vitamin D	1.2	Human NOAEL	Short duration, small sample size
Fluoride	1.0	Human NOAEL	LOAEL identified as a cosmetic effect (mottling of teeth)
Manganese	1.0	Human NOAEL	Clinical trials and epidemiology consistent
Vitamins B1, B2, B12	—	Human data	No toxicological basis for UL
Coenzyme Q10	1.0	Human OSL	No adverse effects established, higher dose clinical trials increase confidence in selected OSL

*UF values for vitamin E, folic acid, vitamin B1, B2, B6, B12, nicotinic acid, iron, vitamin D, fluoride, manganese derived from their respective DRI reports (1997, 1998a, 2000, 2001a). UF value for coenzyme Q10 derived from Hathcock and Shao (2006b).

A new approach to risk assessment assesses uncertainty in a different manner. All relevant clinical trial data are analyzed in decreasing order of intake (as illustrated above). Possible NOAEL or OSL values are identified and the strength of the data is evaluated. A NOAEL or OSL value is selected that provides high enough confidence to justify UF of unity. For lycopene, the highest dose used in a randomized controlled trial is

150 mg/d (Rao and Agarwal 1998) and could be considered as the OSL. However, the study has several limitations, including small sample size and short duration. Use of this study to serve as the basis of the OSL would require some level of correction for uncertainty (i.e., UF > 1.0). Exactly what UF value should be assigned is unclear. Therefore, the approach of choosing the next lowest intake level studied (75 mg/d; Paetau et al. 1998, 1999), while more conservative, also promotes a higher level of confidence and less uncertainty (i.e., UF = 1.0). This approach to uncertainty has been the basis of many other recently published risk assessments on both essential and nonessential nutrients (Hathcock and Shao 2006a, b, 2007, Shao and Hathcock 2006a, b, Hathcock et al. 2007).

The highest lycopene dose from published animal toxicity studies is reported by Matulka et al. (2004) using lycopene derived from tomato oleoresin. An acute dose of 5000 mg/kg in rats failed to produce clinical signs of toxicity or death and a sub-chronic dose of up to 4500 mg/kg body weight per day for 13 wk also caused no adverse effects. This NOAEL in rats equates to approximately 270 g/d in a healthy 60 kg adult. There are important species-specific differences that exist between rats and humans with respect to the absorption and metabolism of carotenoids that preclude direct extrapolation of the research from one species to the other (Castenmiller and West 1998, West and Castenmiller 1998, Borel 2003). Application of a conservative UF of 1000 would result in an OSL of 270 mg.

While acute and sub-chronic animal toxicity studies may provide some further insight into the safety profile of lycopene, these studies cannot be confidently extrapolated to a quantitative risk assessment for an orally administered substance in humans. Rather, such studies help to provide further confidence in the already chosen OSL value.

NOAEL and LOAEL: >150 mg/d lycopene

OSL:

- 75 mg/d based on randomized, controlled human trials
- 270 mg lycopene based on extrapolation from animal data

CONCLUSIONS

In summary, lycopene is one of the most prevalent carotenoids in the human diet and serum. The continuously emerging research on the health benefits of this compound, combined with its increased use in dietary supplement and functional food products, warranted the present risk assessment evaluation. The absence of clear adverse effects from the available published human clinical and animal data on lycopene provides a high level of confidence in its safe use in dietary supplements and functional food products. The absence of a well-defined critical effect precludes the selection of a NOAEL and therefore requires use of the observed safe level (OSL) or highest observed intake (HOI) approach established by FAO/WHO to conduct this risk assessment. Evidence from well-designed randomized, controlled human clinical trials indicates that the Upper Level for Supplements (ULS) for lycopene is 75 mg/d (270 mg based on extrapolation from animal data).

ACKNOWLEDGEMENTS

We would like to acknowledge Ingrid Lebert for her assistance in the copying and acquisition of the relevant research articles.

ABBREVIATIONS

DRI: Dietary Reference Intakes; FAO/WHO: Food and Agriculture Organization/World Health Organization; HOI: Highest Observed Intake; LOAEL: Lowest observed adverse effect level; OSL: Observed Safe Level; NOAEL: No observed adverse effect level; UF: Uncertainty factor; UL: Upper intake level; ULS: Upper Level for Supplements; FNB: US Food and Nutrition Board

References

Ansari, M.S. and N.P. Gupta. 2003. A comparison of lycopene and orchidectomy vs. orchidectomy alone in the management of advanced prostate cancer. BJU Int. 92: 375-378; discussion 378.

Arab, L. and S. Steck. 2000. Lycopene and cardiovascular disease. Amer. J. Clin. Nutr. 71: 1691S-1695S; discussion 1696S-1697S.

Astley, S.B. and D.A. Hughes, A.J. Wright, R.M. Elliott, and S. Southon. 2004. DNA damage and susceptibility to oxidative damage in lymphocytes: effects of carotenoids in vitro and in vivo. Br. J. Nutr. 91: 53-61.

Beecher, G.R. and F. Khachik. 1992. Qualitative relationship of dietary and plasma carotenoids in human beings. Ann. NY Acad. Sci. 669: 320-321.

Bendich, A. 1988. The safety of beta-carotene. Nutr. Cancer 11: 207-214.

Borel, P. 2003. Factors affecting intestinal absorption of highly lipophilic food microconstituents (fat-soluble vitamins, carotenoids and phytosterols). Clin. Chem. Lab. Med. 41: 979-994.

Briviba, K. and S.E. Kulling, J. Moseneder, B. Watzl, G. Rechkemmer, and A. Bub. 2004. Effects of supplementing a low-carotenoid diet with a tomato extract for 2 weeks on endogenous levels of DNA single strand breaks and immune functions in healthy non-smokers and smokers. Carcinogenesis 25: 2373-2378.

Castenmiller, J.J. and C.E. West. 1998. Bioavailability and bioconversion of carotenoids. Annu. Rev. Nutr. 18: 19-38.

Chan, J.M. and P.H. Gann, and E.L. Giovannucci. 2005. Role of diet in prostate cancer development and progression. J. Clin. Oncol. 23: 8152-8160.

Chen, L. and M. Stacewicz-Sapuntzakis, C. Duncan, R. Sharifi, L. Ghosh, R. van Breemen, D. Ashton, and P.E. Bowen. 2001. Oxidative DNA damage in prostate cancer patients consuming tomato sauce-based entrees as a whole-food intervention. J. Natl. Cancer Inst. 93: 1872-1879.

Cohn, W. and P. Thurmann, U. Tenter, C. Aebischer, J. Schierle, and W. Schalch. 2004. Comparative multiple dose plasma kinetics of lycopene administered in tomato juice, tomato soup or lycopene tablets. Eur. J. Nutr. 43: 304-312.

Corridan, B.M. and M. O'Donoghue, D.A. Hughes, and P.A. Morrissey. 2001. Low-dose supplementation with lycopene or beta-carotene does not enhance cell-mediated immunity in healthy free-living elderly humans. Eur. J. Clin. Nutr. 55: 627-635.

Domke, A. and R. Grossklaus, B. Niemann, H. Przyrembel, K. Richter, E. Schmide, A. Weissenborn, B. Wörner, and R. Ziegenhagen. 2006. Use of Vitamins in Foods. Federal Institute for Risk Assessment (Germany). Berlin, Germany.

Expert Group on Vitamins and Minerals. 2003. Safe Upper Levels for Vitamins and Minerals. Food Standards Agency, UK.

FAO/WHO. 2006. A Model for Establishing Upper Levels of Intake for Nutrients and Related Substances. FAO/WHO Technical Workshop on Risk Assessment. Food and Agriculture Organization, World Health Organization, Geneva, Switzerland.

Food and Nutrition Board. Institute of Medicine. 1997. Dietary reference intakes for calcium, phosphorus, magnesium, vitamin D, and fluoride. National Academy Press, Washington, D.C.

Food and Nutrition Board. Institute of Medicine. 1998a. Dietary Reference Intakes for Thiamin, Riboflavin, Niacin, Vitamin B6, Folate, Vitamin B12, Pantothenic Acid, Biotin and Choline. National Academy Press, Washington, D.C.

Food and Nutrition Board. Institute of Medicine. 1998b. Dietary Reference Intakes: A Risk Assessment Model for Establishing Upper Intake Levels for Nutrients. National Academy Press, Washington, D.C.

Food and Nutrition Board. Institute of Medicine. 2000. Dietary Reference Intakes for Vitamin C, Vitamin E, Selenium and Carotenoids. National Academy Press, Washington, D.C.

Food and Nutrition Board. Institute of Medicine. 2001. Dietary Reference Intakes for Vitamin A, Vitamin K, Arsenic, Boron, Chromium, Copper, Iodine, Iron, Manganese, Molybdenum, Nickel, Silicon, Vanadium and Zinc. National Academy Press, Washington, D.C.

Frohlich, K. and K. Kaufmann, R. Bitsch, and V. Bohm. 2006. Effects of ingestion of tomatoes, tomato juice and tomato puree on contents of lycopene isomers, tocopherols and ascorbic acid in human plasma as well as on lycopene isomer pattern. Br. J. Nutr. 95: 734-741.

Gartner, C. and W. Stahl, and H. Sies. 1997. Lycopene is more bioavailable from tomato paste than from fresh tomatoes. Amer. J. Clin. Nutr. 66: 116-122.

Giovannucci, E. 2005. Tomato products, lycopene, and prostate cancer: a review of the epidemiological literature. J. Nutr. 135: 2030S-2031S.

Hathcock, J. 2004. Vitamin and Mineral Safety. Council for Responsible Nutrition, Washington, D.C.

Hathcock, J.N. and A. Shao. 2006a. Risk assessment for carnitine. Regul. Toxicol. Pharmacol. 46: 23-28.

Hathcock, J.N. and A. Shao. 2006b. Risk assessment for coenzyme Q10 (Ubiquinone). Regul. Toxicol. Pharmacol. 45: 282-288.

Hathcock, J.N. and A. Shao. 2007. Risk assessment for glucosamine and chondroitin sulfate. Regul. Toxicol. Pharmacol. 47: 78-83.

Hathcock, J.N. and A. Shao, R. Vieth, and R. Heaney. 2007. Risk assessment for vitamin D. Amer. J. Clin. Nutr. 85: 6-18.

Hininger, I.A. and A. Meyer-Wenger, U. Moser, A. Wright, S. Southon, D. Thurnham, M. Chopra, H. Van Den Berg, B. Olmedilla, A.E. Favier, and A.M. Roussel. 2001. No significant effects of lutein, lycopene or beta-carotene supplementation on biological markers of oxidative stress and LDL oxidizability in healthy adult subjects. J. Amer. Coll. Nutr. 20: 232-238.

Holden, J.M. and A.L. Eldridge, G.R. Beecher, I.M. Buzzard, S. Bhagwat, C.S. Davis, L.W. Douglass, S. Gebhardt, D. Haytowitz, and S. Schakel. 1999. Carotenoid content of U.S. foods: An update of the database. J. Food Comp. Anal. 12: 169-196.

Hoppe, P.P. and K. Kramer, H. van den Berg, G. Steenge, and T. van Vliet. 2003. Synthetic and tomato-based lycopene have identical bioavailability in humans. Eur. J. Nutr. 42: 272-278.

Horvitz, M.A. and P.W. Simon, and S.A. Tanumihardjo. 2004. Lycopene and beta-carotene are bioavailable from lycopene 'red' carrots in humans. Eur. J. Clin. Nutr. 58: 803-811.

Hughes, D.A. and A.J. Wright, P.M. Finglas, A.C. Polley, A.L. Bailey, S.B. Astley, and S. Southon. 2000. Effects of lycopene and lutein supplementation on the expression of functionally associated surface molecules on blood monocytes from healthy male nonsmokers. J. Infect. Dis. 182 (Suppl 1): S11-15.

Khachik, F. and G.R. Beecher, M.B. Goli, W.R. Lusby, and C.E. Daitch. 1992. Separation and quantification of carotenoids in human plasma. Meth. Enzymol. 213: 205-219.

Khachik, F. and L. Carvalho, P.S. Bernstein, G.J. Muir, D.Y. Zhao, and N.B. Katz. 2002. Chemistry, distribution, and metabolism of tomato carotenoids and their impact on human health. Exp. Biol. Med. (Maywood). 227: 845-851.

Kucuk, O. and F.H. Sarkar, Z. Djuric, W. Sakr, M.N. Pollak, F. Khachik, M. Banerjee, J.S. Bertram, and D.P. Wood Jr. 2002. Effects of lycopene supplementation in patients with localized prostate cancer. Exp. Biol. Med. (Maywood) 227: 881-885.

Kucuk, O. and F.H. Sarkar, W. Sakr, Z. Djuric, M.N. Pollak, F. Khachik, Y.W. Li, M. Banerjee, D. Grignon, J.S. Bertram, J.D. Crissman, E.J. Pontes, and D.P. Wood Jr. 2001. Phase II randomized clinical trial of lycopene supplementation before radical prostatectomy. Cancer Epidemiol. Biomarkers Prev. 10: 861-868.

Kumar, A. and A. Bagewadi, V. Keluskar, and M. Singh. 2007. Efficacy of lycopene in the management of oral submucous fibrosis. Oral Surg. Oral Med. Oral Pathol. Oral Radiol. Endod. 103: 207-213.

La Placa, M. and M. Pazzaglia, and A. Tosti. 2000. Lycopenaemia. J. Eur. Acad. Dermatol. Venereol. 14: 311-312.

Liu, C. and F. Lian, D.E. Smith, R.M. Russell, and X.D. Wang. 2003. Lycopene supplementation inhibits lung squamous metaplasia and induces apoptosis via up-regulating insulin-like growth factor-binding protein 3 in cigarette smoke-exposed ferrets. Cancer Res. 63: 3138-3144.

Liu, C. and R.M. Russell, and X.D. Wang. 2006. Lycopene supplementation prevents smoke-induced changes in p53, p53 phosphorylation, cell proliferation, and apoptosis in the gastric mucosa of ferrets. J. Nutr. 136: 106-111.

Maruyama, C. and K. Imamura, S. Oshima, M. Suzukawa, S. Egami, M. Tonomoto, N. Baba, M. Harada, M. Ayaori, T. Inakuma, and T. Ishikawa. 2001. Effects of tomato juice consumption on plasma and lipoprotein carotenoid concentrations and the susceptibility of low density lipoprotein to oxidative modification. J. Nutr. Sci. Vitaminol. (Tokyo) 47: 213-221.

Matulka, R.A. and A.M. Hood, and J.C. Griffiths. 2004. Safety evaluation of a natural tomato oleoresin extract derived from food-processing tomatoes. Regul. Toxicol. Pharmacol. 39: 390-402.

McClain, R.M. and J. Bausch. 2003. Summary of safety studies conducted with synthetic lycopene. Regul. Toxicol. Pharmacol. 37: 274-285.

Le Ministre de l'économie, des finances et de l'industrie. Décrets, arêtés, circulaires—textes généraux. Arrêté de 9 mai 2006 relatif aux nutriments pouvant être employés dan la fabrication de compléments alimentaires. Journal Officiel de la Républic Française. Texte 7 sur 62, 2006.

Neuman, I. and H. Nahum, and A. Ben-Amotz. 2000. Reduction of exercise-induced asthma oxidative stress by lycopene, a natural antioxidant. Allergy 55: 1184-1189.

Olmedilla, B. and F. Granado, S. Southon, A.J. Wright, I. Blanco, E. Gil-Martinez, H. van den Berg, D. Thurnham, B. Corridan, M. Chopra, and I. Hininger. 2002. A European multicentre, placebo-controlled supplementation study with alpha-tocopherol, carotene-rich palm oil, lutein or lycopene: analysis of serum responses. Clin. Sci. (Lond.) 102: 447-456.

Paetau, I. and F. Khachik, E.D. Brown, G.R. Beecher, T.R. Kramer, J. Chittams, and B.A. Clevidence. 1998. Chronic ingestion of lycopene-rich tomato juice or lycopene supplements significantly increases plasma concentrations of lycopene and related tomato carotenoids in humans. Amer. J. Clin. Nutr. 68: 1187-1195.

Paetau, I. and D. Rao, E.R. Wiley, E.D. Brown, and B.A. Clevidence. 1999. Carotenoids in human buccal mucosa cells after 4 wk of supplementation with tomato juice or lycopene supplements. Amer. J. Clin. Nutr. 70: 490-494.

Qualified Health Claims. Center for Food Safety and Applied Nutrition. 2006. Food and Drug Administration., Rockville, MD. http://www.cfsan.fda.gov/~dms/lab-qhc.html. Accessed on February 26, 2007.

Rao, A.V. and A. Agarwal. 1998. Bioavailability and in vivo antioxidant properties of lycopene from tomato products and their possible role in the prevention of cancer. Nutr. Cancer 31: 199-203.

Reigh, P. 1960. Lycopenemia: A variant of carotenemia. New England J. Med. 80: 353-361.

Richelle, M. and K. Bortlik, S. Liardet, C. Hager, P. Lambelet, M. Baur, L.A. Applegate, and E.A. Offord. 2002. A food-based formulation provides lycopene with the same bioavailability to humans as that from tomato paste. J. Nutr. 132: 404-408.

National Research Council Risk Assessment in the Federal Government: Managing the Process, 1983. National Academy Press, Washington, DC.

Scientific Committee on Food. 2001. Guidelines of the Scientific Committee on Food for the Development of Tolerable Upper Intake Levels for Vitamins and Minerals. European Commission, Brussels, Belgium.

Shao, A. and J.N. Hathcock. 2006a. Risk assessment for creatine monohydrate. Regul. Toxicol. Pharmacol. 45: 242-251.

Shao, A. and J.N. Hathcock. 2006b. Risk assessment for the carotenoids lutein and lycopene. Regul. Toxicol. Pharmacol. 45: 289-298.

Tang, G. and A.L. Ferreira, M.A. Grusak, J. Qin, G.G. Dolnikowski, R.M. Russell, and N.I. Krinsky. 2005. Bioavailability of synthetic and biosynthetic deuterated lycopene in humans. J. Nutr. Biochem. 16: 229-235.

van Breemen, R.B. 2005. How do intermediate endpoint markers respond to lycopene in men with prostate cancer or benign prostate hyperplasia? J. Nutr. 135: 2062S-2064S.

Wang, X.D. 2005. Can smoke-exposed ferrets be utilized to unravel the mechanisms of action of lycopene? J. Nutr. 135: 2053S-2056S.

Watzl, B. and A. Bub, M. Blockhaus, B.M. Herbert, P.M. Luhrmann, M. Neuhauser-Berthold, and G. Rechkemmer. 2000. Prolonged tomato juice consumption has no effect on cell-mediated immunity of well-nourished elderly men and women. J. Nutr. 130: 1719-1723.

Watzl, B. and A. Bub, K. Briviba, and G. Rechkemmer. 2003. Supplementation of a low-carotenoid diet with tomato or carrot juice modulates immune functions in healthy men. Ann. Nutr. Metab. 47: 255-261.

West, C.E. and J.J. Castenmiller. 1998. Quantification of the "SLAMENGHI" factors for carotenoid bioavailability and bioconversion. Int. J. Vitam. Nutr. Res. 68: 371-377.

Willcox, J.K. and G.L. Catignani, and S. Lazarus. 2003. Tomatoes and cardiovascular health. Crit. Rev. Food Sci. Nutr. 43: 1-18.

Zhao, X. and G. Aldini, E.J. Johnson, H. Rasmussen, K. Kraemer, H. Woolf, N. Musaeus, N.I. Krinsky, R.M. Russell, and K.J. Yeum. 2006. Modification of lymphocyte DNA damage by carotenoid supplementation in postmenopausal women. Amer. J. Clin. Nutr. 83: 163-169.

PART 2

Biochemical and Physiological Features of Lycopene's Effects

2.1

Lycopene and Peroxynitrite Modifications

[1]Kaampwe Muzandu, [1]Kennedy Choongo and
[2]Shoichi Fujita

[1]Biomedical Sciences Department, School of Veterinary Medicine
University of Zambia, P.O. Box 32379, Lusaka, Zambia
[2]Laboratory of Toxicology, Department of Environmental Veterinary Sciences
Graduate School of Veterinary Medicine, N18, W9, Kita-Ku
Sapporo 060-0818, Japan

ABSTRACT

Reactive oxygen species (e.g., superoxide and hydroxyl radicals) and reactive nitrogen species (e.g., peroxynitrite and nitric oxide) generated during oxidative/nitrosative stress have been implicated in the initiation and/or progression of various ailments including chronic inflammation and cancer, and cardiovascular and neurological diseases. Peroxynitrite, formed from the near diffusion-limited reaction of superoxide and nitric oxide, has gained much attention in recent years for its now recognized role in the pathophysiology of various conditions. Upon formation at, for example, sites of infection or inflammation, it initiates oxidative or nitrosative reactions with various cellular biomolecules such as proteins, lipids and nucleic acids potentially leading to toxicity. Early and on-going epidemiological studies suggest that lycopene and tomato products could reduce the risk of development of ailments associated with oxidative stress. A multitude of effects or mechanisms of lycopene in the prevention of disease is unfolding. Some of these mechanisms are associated with its antioxidant capacity, such as scavenging of peroxynitrite itself, as well as other reactive nitrogen species (e.g., nitrogen dioxide). Other protective mechanisms of lycopene include antiproliferative and prodifferentiation activities and modulation of various mediators of inflammation and the immune function.

A list of abbreviations is given before the references.

INTRODUCTION

Carotenoids are attracting immense interest as possible deterrents to chronic diseases, including cardiovascular disease, neurological conditions and cancers of the prostate, breast and gastrointestinal tract (Giovannucci 2002, Mayne 1996, Tapiero et al. 2004). These compounds are ubiquitous in nature as they are synthesized in plants, microorganisms and algae but must be obtained from the diet in animals and humans. They are responsible for the pigmentation of various fruits and vegetables. For example, the predominant carotenoids found in human tissues, lycopene and β-carotene, impart a red color to tomatoes and an orange color to carrots, respectively. Besides their well-recognized function as antioxidants, carotenoids have a wide range of other biological effects such as modulation of the immune response and induction of gap-junction communications (Khachick et al. 1995, Stahl and Sies 1996). Consumption of fruits and vegetables has been associated with a reduced risk of development of a number of chronic diseases. Most of these diseases probably arise as a result of oxidative and/or nitrosative stress. Normal cellular metabolism is a source of reactive oxygen species (ROS) and reactive nitrogen species (RNS). Oxidative or nitrosative stress refers to the excessive formation of these species. Thus, "oxidative stress" is an imbalance between oxidants and antioxidants in favor of the oxidants, potentially leading to damage (Sies et al. 2005). Reactive oxygen species, i(ncluding superoxide, hydrogen peroxide, and hypochlorite ions) and RNS (including nitric oxide or NO and peroxynitrite) mediate host defense systems against intruding microorganisms. Reactive oxygen species are also produced by ionizing or ultraviolet radiation and by redox cycling during cellular metabolism of certain chemicals. Superoxide plays a central role as it is the source of various ROS and RNS. For instance, it may combine with NO to generate the potent oxidant and nitrating species peroxynitrite (Burney et al. 1999). Peroxynitrite ($ONOO^-$/$ONOOH$) is both a one-electron and two-electron oxidant. It can oxidize ascorbate, thiol groups, methionine, tryptophan, lipids and DNA. Peroxynitrite also nitrates free and protein-associated tyrosines and other phenolic compounds. In addition to its role in pathologic processes, peroxynitrite also serves a physiological function as a cell-signaling molecule.

Of the carotenoids, lycopene is the most efficient quencher of singlet oxygen ($^1O_2$). It is also a potent quencher of peroxyl radicals. There is a paucity of data on the interactions of carotenoids with RNS. However, β-carotene has been shown to be a strong scavenger of RNS in solution, including nitrogen dioxide (Everett et al. 1996, Mortensen et al. 1997), and peroxynitrite (Kikugawa et al. 1997). Lycopene and β-carotene have also been shown to be consumed in the process of scavenging peroxynitrite in low density lipoproteins (Panasenko et al. 2000, Pannala et al. 1998). This chapter provides an overview of peroxynitrite-induced modifications to biomolecules and the modulatory role of lycopene in some of these modifications.

FORMATION AND DECOMPOSITION OF PEROXYNITRITE IN BIOLOGICAL SYSTEMS

Peroxynitrite can be produced by a variety of cells including endothelial cells, Kupffer cells, macrophages, and neutrophils. During phagocytosis, these cells undergo a vigorous respiratory burst in which they consume oxygen and convert it enzymatically to $O_2^{\bullet-}$, which is rapidly transformed to H_2O_2, hypochlorous acid, hydroxyl radical, and other potent oxidants (Fig. 1). The conversion of O_2 to $O_2^{\bullet-}$ is catalyzed by enzymes such as the membrane-associated NADPH oxidase and the cytosolic xanthine oxidase. Superoxide is also produced by the mitochondrial respiratory chain. Superoxide dismutase catalyzes the dismutation of $O_2^{\bullet-}$ to H_2O_2 and triplet (ground state) oxygen. Superoxide can also dismutate spontaneously to H_2O_2 and $^1O_2$ (Afanas'ev 1999). The very toxic hypohalous acids (HOCl and HOBr) are produced via reactions catalyzed by myeloperoxidase (MPO) or eosinophil peroxidase (EPO) in the presence of hydrogen peroxide and chloride or bromide ions to cause tissue injury during which thiols, amines, and aromatic compounds are halogenated (van der Vliet and Cross 2000). Nitrogen dioxide can also be formed from reactions catalyzed by MPO or EPO using H_2O_2 and nitrite as substrates (Ohshima et al. 2003). Nitric oxide is a free radical produced from L-arginine via the enzyme nitric oxide synthase (NOS). Three isoforms of NOS are recognized, namely, neuronal (nNOS), endothelial (eNOS), and inducible (iNOS).

Peroxynitrite, a potent oxidizing and nitrating species (Beckman 1996, Koppenol et al. 1992), is the product of the near diffusion-limited reaction of NO and superoxide anion ($O_2^{\bullet-}$). The rate constant for this reaction is commonly reported as 6.7×10^9 M^{-1} s^{-1} (Burney et al. 1999). Other reports estimate it to be even faster, up to 2.0×10^{10} M^{-1} s^{-1} (Kissner et al. 1997). The formation of peroxynitrite will occur near sites of superoxide generation owing to the short half-life of superoxide (milliseconds) and limited diffusion, which occurs via anion channels (Fridovich 1995). Nitric oxide, on the other hand, has a longer half-life (seconds) and readily diffuses across biological membranes (Liu et al. 1998). Peroxynitrous acid diffuses passively across membranes, but the anion traverses membranes via anion channels (Denicola et al. 1998). Peroxynitrite is relatively long-lived with a half-life of approximately 1 s under physiological conditions (Beckman

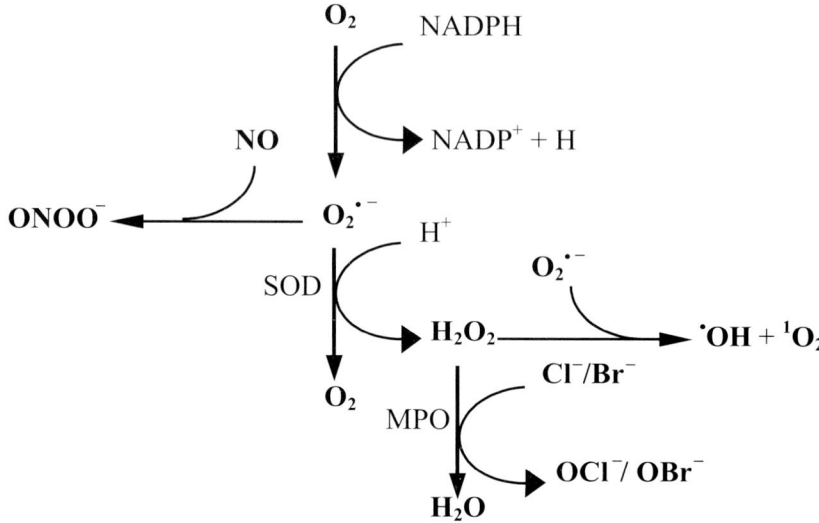

Fig. 1 Formation of cytotoxic oxidants by inflammatory cells. Nitric oxide generated from various types of cells, e.g., macrophages, reacts with superoxide formed from enzyme systems, e.g., NADPH oxidase or xanthine oxidase, to yield peroxynitrite. Superoxide may lead to the generation of various reactive oxygen species such as hydrogen peroxide, hydroxyl radicals, singlet oxygen and hypohalous acids.

Cl^-/Br^-, chloride/bromide; MPO, myeloperoxidase; $NADP^+$, oxidized nicotinamide adenine dinucleotide phosphate; NADPH, reduced nicotinamide adenine dinucleotide phosphate; NO, nitric oxide; $O_2^{\cdot-}$, superoxide; $^1O_2$, singlet oxygen; OCl^-/OBr^-, hypochlorous/hypobromous ions; $^{\cdot}OH$, hydroxyl radical; $ONOO^-$, peroxynitrite anion; SOD, superoxide dismutase.

et al. 1990, Koppenol et al. 1992, Hughes and Nicklin 1968). At physiological pH and temperature, peroxynitrite will, following protonation, exist as 20% peroxynitrous acid (ONOOH, pK_a 6.8) (Beckman 1996), which in turn homolyzes to form nitrogen dioxide ($^{\cdot}NO_2$) and the hydroxyl radical ($^{\cdot}OH$). These radicals are initially formed in a solvent cage, in which 70% of them recombine to form nitrate and the remaining 30% escape from the cage as free $^{\cdot}NO_2$ and $^{\cdot}OH$. This route of peroxynitrite decay is considered less important *in vivo*. It is most likely limited to acidic environments such as those obtaining in ischemic tissues, in the phagolysosome, at the cell surface and around the negatively charged DNA (Lamm and Pack 1990, Olmos et al. 2007). An important biological reaction of peroxynitrite is with CO_2 to yield nitrosoperoxycarbonate ($ONOO\ CO_2^-$) (Dedon and Tannenbaum, 2004), which also eventually decomposes to yield NO_3^-. The carbonate ($CO_3^{\cdot-}$) and $^{\cdot}NO_2$ radicals are intermediates in this decomposition process. The reaction of peroxynitrite with CO_2 (4.6×10^4 M^{-1} s^{-1}) (Denicola et al. 1996) is more plausible *in vivo* considering the high concentrations of bicarbonate and CO_2 in equilibrium in the body fluid compartments. The intracellular and intravascular concentrations of CO_2 are approximately 1 and 25 mM, respectively. The nitrosoperoxycarbonate adduct is unstable, homolyzing to 30-35% into $CO_3^{\bullet-}$ and $^\bullet NO_2$ (Kirsch and Lehnig, 2005). Both these radicals have potent reactivity towards various substances, including thiols, amines, purines and ascorbate (Fig. 2). The reaction of peroxynitrite with CO_2 limits its half-life from approximately 1 s to 10-20 ms.

MODIFICATIONS OF BIOMOLECULES BY PEROXYNITRITE

Reactions of peroxynitrite with biomolecules are complex and numerous and include oxidations, such as DNA damage leading to sugar and base modifications, mutations, and also single- and double-strand breaks

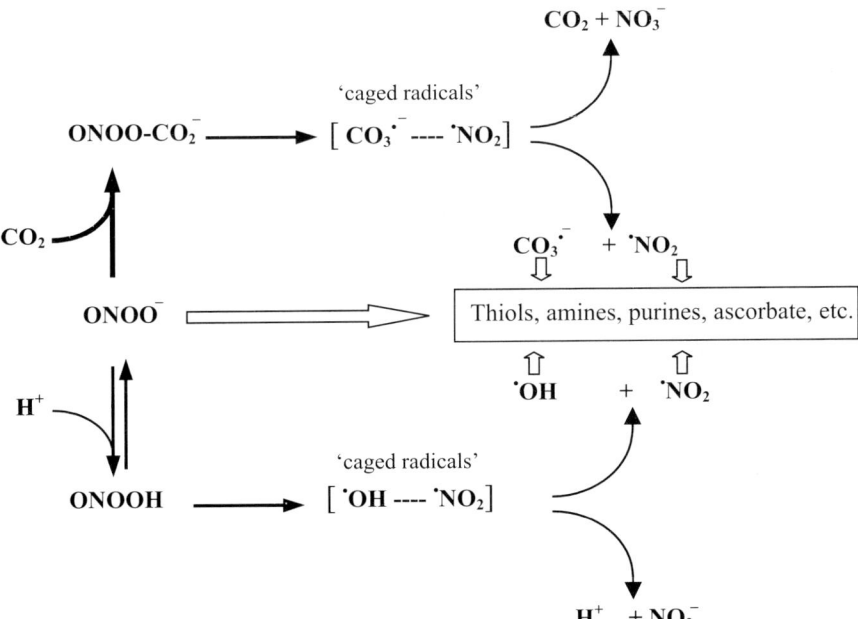

Fig. 2 Direct and radical reactions of peroxynitrite. Peroxynitrite reacts directly with various biomolecules. However, it also reacts indirectly via secondary radicals formed either following the reaction of peroxynitrite with carbon dioxide to yield carbonate and nitrogen dioxide radicals or the protonation of peroxynitrite into peroxynitrous acid which in turn decays to hydroxyl and nitrogen dioxide radicals.

CO_2, carbon dioxide; $CO_3^{\bullet-}$, carbonate radical; $^\bullet OH$, hydroxyl radical; $ONOO^-$, peroxynitrite anion; $ONOO\text{-}CO_2^-$ nitrosoperoxycarbonate; $ONOOH$, peroxynitrous acid; $^\bullet NO_2$, nitrogen dioxide; NO_3^-, nitrate.

(Burney et al. 1999, Dedon and Tannenbaum 2004). Peroxynitrite may also nitrate DNA, though the extent and importance of this reaction is unclear. Other than DNA, peroxynitrite also produces oxidative and nitrosative damage to lipids and proteins. Protein tyrosine nitration, the addition of a nitro group ($-NO_2$) on to one of the two equivalent *ortho* carbons of the aromatic ring of tyrosine residues (Gow et al. 2004), has been associated with a number of pathological states (Pietraforte et al. 2003).

Interaction of Peroxynitrite with Metal Centers

In addition to the fast and direct one- or two-electron oxidation reactions with CO_2 and thiols (e.g., glutathione and cysteine), peroxynitrite reacts rapidly and directly with transition metal centers. The rate constants for these reactions range from 10^3 in the case of thiols to $10^6 \ M^{-1} \ s^{-1}$ for metal centers. Peroxynitrite also mediates indirect oxidative reactions or secondary radical reactions ($CO_3{}^{\bullet -}$, $^\bullet NO_2$ and $^\bullet OH$) (Alvarez and Radi 2003). The reaction of peroxynitrite with metal centers proceeds in much the same way that it reacts with other Lewis acids (A), electron-accepting compounds such as H^+ and CO_2, that favor the cleavage of the $O-O$ bond. An adduct is formed that homolyzes to produce $^\bullet NO_2$ and the corresponding oxyradical ($^\bullet O-A^-$) (Alvarez and Radi 2003, Alvarez et al. 2004). The oxyradical so formed may rearrange to produce the corresponding radical of the oxo-compound via oxidation of the Lewis acid (Alvarez and Radi 2003, Eq. 1).

$$ONOO^- + A \rightarrow ONOO-A \rightarrow {}^\bullet NO_2 + {}^\bullet O-A^- \rightarrow {}^\bullet NO_2 + O{=}{=}A^{\bullet -} \qquad (1)$$

Thus, interaction of peroxynitrite with metals (M) leads to the formation of peroxynitrite-metal adduct that homolyzes to yield $^\bullet NO_2$ and the corresponding oxy-metal radical ($^\bullet O- M^n$), which in turn converts to an oxo-metal complex (Alvarez and Radi 2003, Eq. 2).

$$ONOO^- + M^n \rightarrow ONOO-M^n \rightarrow {}^\bullet NO_2 + {}^\bullet O-M^n \ {}^\bullet NO_2 + O{=}{=} M^{(n+1)+} \qquad (2)$$

The reaction of peroxynitrite with metal centers can lead to reversible redox modification (e.g., iron in cytochrome *c*) or disruption of the metal center and release of the metal (e.g., iron-sulfur cluster in aconitase and zinc-sulfur cluster in alcohol dehydrogenase). Metals that are liberated from metal centers may bind to phosphate, low molecular weight chelators or macromolecules (Radi 2004). In addition, these liberated metals may become involved in the generation of more ROS. Reduced metal centers can also be oxidized by peroxynitrite yielding an oxyradical or oxo-compound and nitrite, instead of nitrogen dioxide (e.g., one-electron oxidation of Fe^{2+} to Fe^{3+} in cytochrome *c*) (Alvarez and Radi 2003).

Interaction of Peroxynitrite with Proteins

Protein oxidation is commonly measured by the detection of carbonyls. Carbonyl groups are introduced into proteins by direct oxidation of proline, arginine, lysine, and threonine side chains or by Micheal addition reactions with products of lipid peroxidation or glyco-oxidation (Levine and Stadtman 2001, Choi et al. 2005). The major products of metal-catalyzed oxidation of proteins are α-aminoadipic semialdehyde from lysine and glutamic semialdehyde from arginine and proline (Stadtman 2006). Protein carbonylation often leads to loss of function or structural changes that may be deleterious (Levine and Stadtman 2001). Carbonyl groups are elevated during the process of aging, in neurodegenerative diseases such as Alzheimer's disease and Parkinson's disease, and in cancer, cataractogenesis, diabetes and sepsis (Levine 2002, Dalle-Donne et al. 2003). Protein carbonylation is an irreversible process in which the carbonylated proteins are marked for degradation by proteases and the Lon proteosome (Nystrom 2005). Carbonylated proteins are normally targeted for degradation by the Lon proteosome. However, heavily oxidized proteins may escape the degradation pathway and form high molecular weight aggregates that can inhibit proteosome activity and become cytotoxic (Nystrom 2005). Peroxynitrite has been demonstrated to lead to the formation of protein carbonyls in cells (Jung et al. 2007).

Peroxynitrite can oxidize as well as nitrate proteins. In fact, protein nitration has often been used as a biomarker of peroxynitrite formation *in vivo*. However, other mechanisms can also lead to the nitration of proteins as discussed below.

Peroxynitrite can react directly with prosthetic groups of proteins (e.g., hemeproteins) or directly with the peptide chain resulting in conformational and functional changes (Denicola and Radi 2005). Cysteine, methionine and tryptophan react directly with peroxynitrite, whereas other protein-forming amino acids, such as tyrosine, phenylalanine and histidine, react via secondary radical mechanisms (e.g., hydroxyl, carbonate and nitrogen dioxide) (Alvarez and Radi 2003). The reaction of peroxynitrite with the thiol group–containing cysteine is a major reaction. The products of the reaction of peroxynitrite with thiols are corresponding disulfides and nitrite (Alvarez and Radi 2003) (Fig. 3). An intermediate in this reaction, sulfenic acid (RSOH), reacts with another thiol to form a corresponding disulfide (RSSR). Peroxynitrite oxidizes thiols of low molecular weight (e.g., albumin and glutathione) as well as protein-bound thiols (e.g., cysteines occurring in enzymes such as glyceraldehyde-3-phosphate dehydrogenase and creatine kinase) leading to protein inactivation (Konorev et al. 1998, Souza and Radi 1998, and Buchczyk et al. 2003).

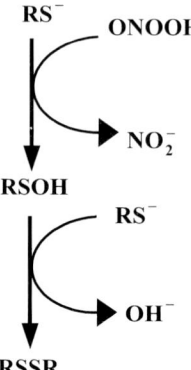

Fig. 3 Direct oxidation of thiols by peroxynitrite. Peroxynitrite can react directly with thiols to yield sulfenic acid and nitrite. The thiol further reacts with sulfenic acid to form the corresponding disulfide and hydroxide anion.

NO_2^-, nitrite; RS^-, thiol; OH^-, hydroxide anion; ONOOH, peroxynitrous acid; RSOH, sulfenic acid; RSSR, disulfide.

Peroxynitrite may also interact with thiols via secondary radical mechanisms (e.g., carbonate radical and nitrogen dioxide) (Fig. 4). The thiyl radical ($RS^•$) is produced upon the reaction of $CO_3^{•-}$ and $^•NO_2$ with thiols (RSH). Bicarbonate and nitrite are produced during this process from $CO_3^{•-}$ and $^•NO_2$, respectively. The thiyl radical can further be oxidized to nitrosothiols (RSNO) by NO. Nitrosothiols are also formed by the reaction of thiols with nitrous anhydride (N_2O_3, dinitrogen trioxide). Nitrous anhydride is formed by the reaction of excess $^•NO_2$ with NO. A consequence of thiyl radical formation is the generation and propagation of free radical chain reactions in the presence of oxygen that lead to oxidative stress.

An important reaction of peroxynitrite is nitration of proteins. This occurs by radical mechanisms as peroxynitrite does not react directly with tyrosine (Fig. 5). The hydroxyl radical plays a minor role as its reaction with cellular macromolecules, though widespread and non-selective, proceeds at a rate that is non-competitive with that of carbonate and nitrogen dioxide. Nitrogen dioxide is considered an inefficient nitrating species owing to the fact that it must first react with tyrosine to form the tyrosyl radical, which then yields 3-nitrotyrosine upon further reaction with nitrogen dioxide (Radi 2004). Thus, the more plausible and important nitrating species *in vivo* are the carbonate radical and the oxo-metal complexes, which enhance protein nitration (e.g., in heme proteins, and manganese and copper,zinc superoxide dismutase) (Alvarez and Radi 2003, Radi 2004). Alternative mechanisms for the *in vivo* nitration of

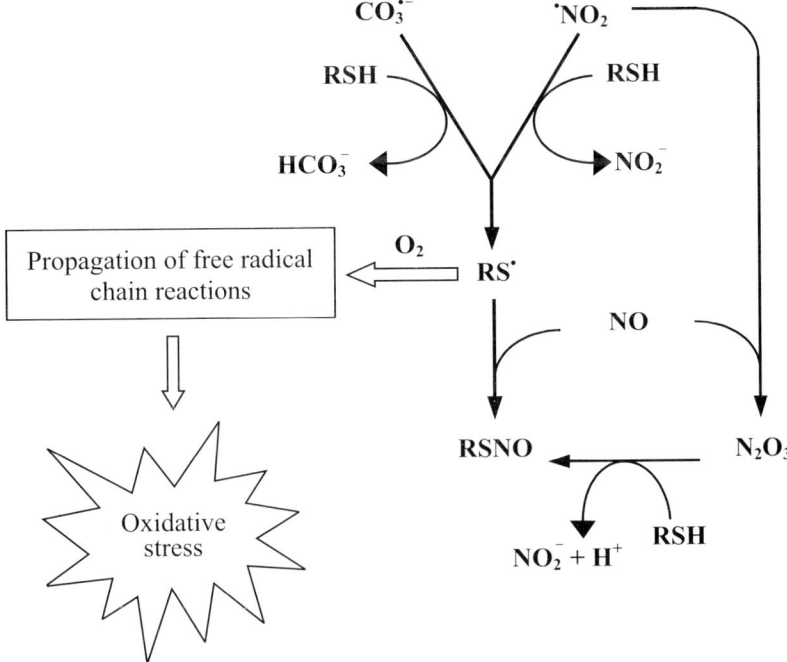

Fig. 4 Peroxynitrite-mediated radical mechanisms of thiol oxidation. Peroxynitrite can also react with thiols via secondary radical mechanisms (e.g., carbonate radical and $^\bullet NO_2$). The carbonate radical can react with thiols to yield the thiyl radical. Similarly, $^\bullet NO_2$ reacts with thiols to yield the thiyl radical. Thiyl radicals in the presence of oxygen lead to the propagation of free radical chain reactions that may contribute to oxidative stress. Nitrosothiols can be formed by the reaction of NO with thiyl radicals or by the reaction of thiols with nitrous anhydride, generated from the interaction of excess $^\bullet NO_2$ with NO.

$CO_3^{\bullet -}$, carbonate radical; HCO_3^-, bicarbonate; NO, nitric oxide; $^\bullet NO_2$, nitrogen dioxide; NO_2^-, nitrite; N_2O_3, nitrous anhydride (dinitrogen trioxide); $RS^\bullet$, thiyl radical; RSH, thiol; RSNO, nitrosothiol.

tyrosine exist. One such mechanism is the MPO-catalyzed oxidation of NO_2^-, a catabolic end product of NO (Hurst 2002). Another mechanism of protein tyrosine nitration, though of doubtful significance *in vivo*, is the formation of nitryl chloride (NO_2Cl) by the reaction of MPO-derived HOCl with NO_2^- (Fig. 5) (Whiteman et al. 2003). Neutrophils are known to produce large amounts of HOCl and NO_2^- at sites of inflammation during the respiratory burst. The formation of 3-nitrotyrosine can also occur with other peroxidases such as EPO (released by activated eosinophils in inflamed tissues), and horseradish peroxidase and heme proteins, such as hemoglobin and catalase, in the presence of NO_2^-. The MPO-catalyzed oxidation of NO_2^- in the presence of H_2O_2 and chloride ions (Cl^-) yields HOCl and this occurs predominantly in neutrophils (van der Vliet et al.

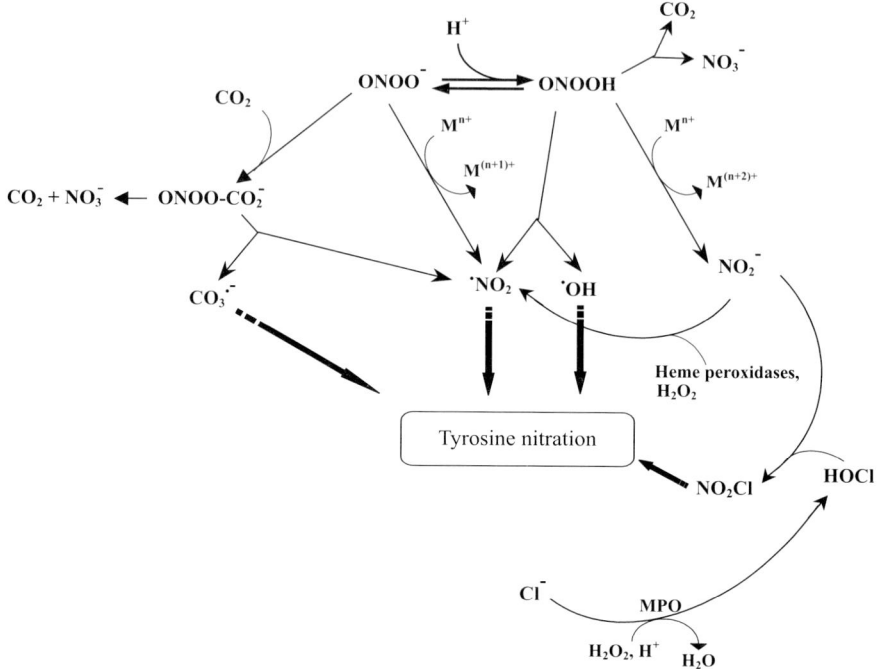

Fig. 5 Pathways of protein nitration. Peroxynitrite nitrates tyrosine residues in proteins via radical mechanisms only. The scheme above outlines the formation of various radicals involved in protein nitration, mainly carbonate, nitrogen dioxide, hydroxyl, and nytryl chloride. One-electron oxidations of metal centers by peroxynitrite lead to the formation of $^{\bullet}NO_2$, whereas two-electron oxidations lead to the formation of nitrite.

CO_2, carbon dioxide; $CO_3^{\bullet-}$, carbonate radical; H_2O_2, hydrogen peroxide; HOCl, hypochlorous acid; M^{n+}, reduced metal center; $M^{(n+1)+}$ and $M^{(n+2)+}$, one electron- and two electron-oxidized metal centers, respectively; $^{\bullet}OH$, hydroxyl radical; $ONOO^-$, peroxynitrite anion; $ONOO\text{-}CO_2^-$ nitrosoperoxycarbonate; ONOOH, peroxynitrous acid; NO, nitric oxide; $^{\bullet}NO_2$, nitrogen dioxide; NO_2^-, nitrite; MPO, myeloperoxidase, NO_2Cl, nitryl chloride.

1997, Gaut et al. 2002). The EPO-catalyzed oxidation of NO_2^- in the presence of H_2O_2 and bromide ions (Br^-) yielding HOBr occurs predominantly in eosinophils during parasitic infections (Wu et al. 1999a, b). Carbon dioxide modulates the reactivity and cytotoxicity of peroxynitrite. Nitration of proteins is enhanced in the presence of CO_2. High concentrations of CO_2 have been reported to protect bacteria (*Escherichia coli*) and parasites (*Trypanosoma cruzi*) from the cytotoxic effects of peroxynitrite (Alvarez et al. 2004, Denicola et al. 1993, Zhu et al. 1992). In contrast, Linares et al. (2001) attributed the toxicity of peroxynitrite to *Leishmania amazonensis* parasites to nitration of the parasite's proteins following interaction of peroxynitrite with CO_2 to generate secondary radical species. As mentioned earlier, protein nitration is also enhanced in the presence of certain low molecular weight metal compounds, such as copper, iron and manganese compounds, as well as proteins containing these metals (e.g., hemeproteins, manganese and copper,zinc superoxide dismutase (Alvarez and Radi 2003). Protein nitration may be a reversible and regulated process with a role in cellular signaling processes. This is in light of the discovery of "denitrase" enzyme activity (Gow et al. 1996, Kamisaki et al. 1998, Aulak et al. 2004, Koeck et al. 2004).

Interaction of Peroxynitrite with Lipids

There are numerous putative mechanisms for the generation of oxidized lipids *in vivo* such as metal-dependent Fenton oxidation, enzyme-catalyzed oxidation by lipoxygenase or MPO, reaction with HOCl, cell-dependent oxidation via a diversity of $O_2^{\bullet-}$ and H_2O_2-generating oxidases, and oxidation by nitric oxide-derived reactive species (e.g., $^{\bullet}NO_2$, nitryl chloride and ONOO') (O'Donnell and Freeman, 2001). Peroxynitrite can, *in vitro*, oxidize via hydrogen abstraction a diversity of lipids that include membranes, liposomes, and lipoproteins yielding lipid hydroperoxides, conjugated dienes, and aldehydes (Denicola and Radi 2005, Radi et al. 1991, O'Donnell and Freeman 2001). The formation of these radicals leads to propagation of free radical attacks on neighboring polyunsaturated fatty acids. A consequence of this is damage to membrane lipids resulting in permeability and fluidity changes culminating in deleterious biological effects. In addition to oxidation, peroxynitrite also mediates nitration of lipids. Nitration occurs either by hydrogen abstraction by $^{\bullet}NO_2$ or addition mechanisms involving the nitronium cation (NO_2^+) (O'Donnell and Freeman 2001).

Interaction of Peroxynitrite with Nucleic Acids

The reactivity of peroxynitrite with DNA and the consequent biological implications have been reviewed in detail elsewhere (Szabo and Ohshima 1997, Burney et al. 1999, Dedon and Tannenbaum 2004). The following is a summary of peroxynitrite-induced DNA modifications. Peroxynitrite reacts with DNA to yield mainly DNA strand breaks. Damage to both the base and the sugar in DNA can occur. Of the nitrogen bases, peroxynitrite reacts preferentially with guanine to yield 8-oxoguanine and 8-nitroguanine. It also reacts with nucleosides to form various modifications. The reaction with deoxyguanosine (dG) forms a number of oxidized products, including 8-nitrodG and 8-oxodG that may further react with peroxynitrite to yield other products.

Treatment of plasmid DNA with peroxynitrite causes strand breaks (Salgo et al. 1995, Epe et al. 1996, Yermilov et al. 1996, Kennedy et al. 1997, Yoshie and Ohshima 1997, Muzandu et al. 2006) (Fig. 6). Peroxynitrite also induces strand breaks in mammalian cells (Delaney et al. 1996, Szabo et al. 1996, 1998, Zingarelli et al. 1996, Spencer et al. 1996, Muzandu et al. 2006).

Several lines of evidence suggest that DNA oxidation is mutagenic and is a major contributor to human cancer (Beckman and Ames 1997). As earlier mentioned, damage to nucleotide base in DNA leads to various modifications. These modifications may result in blocking of replication

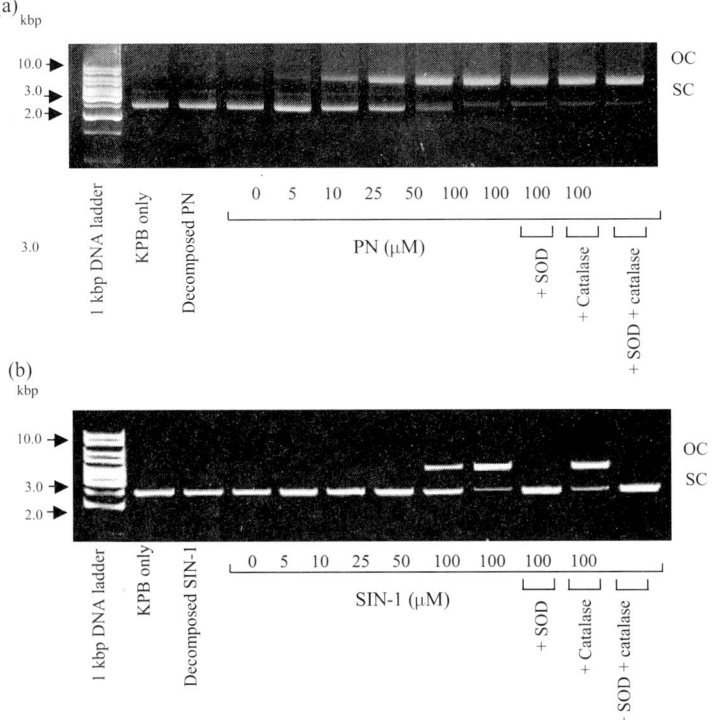

Fig. 6 Induction of strand breaks in pUC18 DNA by peroxynitrite. Both authentic peroxynitrite and SIN-1, the peroxynitrite generator, induced dose-dependent increase in DNA strand breaks as observed by the conversion of supercoiled plasmid DNA to the open circular form. SIN-1 simultaneously releases superoxide and nitric oxide to form peroxynitrite. In the case of SIN-1 treatment, superoxide dismutase (as well as the combination of superoxide dismutase and catalase) prevented damage to DNA, indicating a scavenging of superoxide. (a) DNA was incubated with 0-100 μM PN alone for 30 min at 37°C. The concentration at 0 μM contained 10 mM NaOH, the vehicle used for PN. (b) DNA was incubated with 0-100 μM SIN-1 alone for 30 min at 37°C. The concentration at 0 μM contained 10 mM NaOH, the vehicle used for SIN-1.

PN, authentic peroxynitrite; SIN-1, 3-morpholinosydnonimine; SOD, superoxide dismutase; KPB, potassium phosphate buffer; OC, open circular DNA; SC, supercoiled DNA.

and frequent mispairing, leading to base substitutions. Damage to the sugar moiety in DNA leads to modification and the release of the free base, yielding abasic sites. Abasic sites are mutagenic because of the preferential incorporation of adenine opposite abasic sites by DNA polymerases during replication (Jackson and Loeb 2001). Peroxynitrite has been implicated in causing G:C to A:T mutations (Souici et al. 2000) and G:C to T:A mutations (Juedes and Wogan 1996, Kim et al. 2005).

Besides oxidation of DNA, RNA can also be oxidized by various oxidants potentially leading to a disruption of cellular homeostasis (Tanaka et al. 2007). The oxidation of RNA has been associated with neurodegenerative diseases (Nunomura et al. 1999, Zhang et al. 1999, Nunomura et al. 2002, Martinet et al. 2004, Tanaka et al. 2007). Peroxynitrite derived from SIN-1 (3-morpholinosydnonimine), a metabolite of the nitrovasodilator molsidomine, causes oxidative damage to total RNA isolated from cultured human THP-1 monocytic cells (Martinet et al. 2004, 2005). Oxidative damage to mRNA may lead to abnormal protein translation, and damage to tRNA and rRNA could result in dysfunction of protein synthesis (Martinet et al. 2005).

PEROXYNITRITE AND DISEASE

Reactive oxygen and nitrogen species have been implicated in causing diseases associated with aging, chronic inflammation and cancer. Normal cellular metabolism generates ROS, including hydrogen peroxide, hydroxyl radical, superoxide anion and singlet oxygen. Mitochondria are an important cellular source of ROS. Reactive oxygen and nitrogen species oxidize cellular macromolecules, such as nucleic acids, proteins, and lipids, leading to deleterious effects in the host. Organisms have mechanisms by which they can combat these reactive species through antioxidants (e.g., glutathione, uric acid, tocopherol, ascorbate), enzymes (e.g., superoxide dismutase, catalase, glutathione peroxidase) and repair systems. However, in certain situations such as in chronic infection and chronic inflammation or exposure to exogenous chemicals, the amount of ROS produced may overwhelm the capacity of the endogenous antioxidants and enzymes. During inflammation, cells such as neutrophils, eosinophils and macrophages produce an arsenal of toxic substances (e.g., NO, $O_2^{\bullet-}$, H_2O_2, $^\bullet OH$, HOCl) to combat various pathogens including bacteria, parasites, viruses and foreign bodies. In chronic infection and inflammation there is overproduction of these toxic substances leading to mutagenesis and tissue damage.

The discovery of peroxynitrite formation from NO and superoxide has generated immense interest in its possible involvement in physiological and pathophysiological processes. Indeed, peroxynitrite has been implicated in over 80 diseases (Reiter et al. 2000). The involvement of peroxynitrite has been inferred from the detection of nitrotyrosine in tissues. However, peroxynitrite is not the only RNS capable of nitration of tyrosine. The most prominent diseases in which peroxynitrite has been implicated include Parkinson's disease, amyotrophic lateral sclerosis, Alzheimer's disease, infectious bowel disease and various cardiovascular pathologies. For a detailed review, see Pacher et al. (2007). Peroxynitrite is also suspected to have a role in cancer as it can produce oxidative and nitrative modifications to DNA that potentially lead to mutagenesis and ultimately carcinogenesis. The presence of 8-nitroguanine in certain inflamed tissues has provided evidence that peroxynitrite is involved *in vivo* in cancers associated with chronic infection and inflammation (e.g., cholangiocarcinoma, gastric carcinoma, and colon carcinoma). Lycopene has received particular attention as a possible chemopreventive agent for prostate cancer. Though not a disease, aging is believed to be accelerated by ROS and RNS, including peroxynitrite, that damage proteins, lipids and nucleic acids. For instance, in progeria and Werner Syndrome, the rare human diseases characterized by premature aging, there is an abnormally high number of oxidized proteins (Stadtman 1992). Despite the various endogenous antioxidant defenses, it has been estimated that a rat will accumulate twice as many oxidative lesions by the time it is old (Ames et al. 1995).

EFFECT OF LYCOPENE ON PEROXYNITRITE MODIFICATIONS

Modulatory Effect of Lycopene on Peroxynitrite-induced DNA Modifications

Damage to macromolecules such as DNA may lead to tumorigenesis in the event that the damage is unrepaired, misrepaired or tolerated. Lycopene protects against DNA damage from various genotoxic agents.

Hydrogen peroxide causes DNA damage to different cells, possibly through the generation of the hydroxyl radical. Park et al. (2005) demonstrated the dose-dependent inhibitory effect of lycopene on H_2O_2-induced DNA damage in Hep3B human hepatoma cells using the comet assay. This assay detects DNA strand breaks, DNA-DNA or DNA-protein cross-links and abasic sites (Collins et al. 1997, Tice et al. 2000). A more recent study shows that lycopene administered to Chinese Hamster Ovary cells (CHO K-1) either one hour prior to or simultaneously with 0.6 mM H_2O_2 had a protective effect on DNA (Scolastici et al. 2007). Park et al. (2005) had similar findings. In contrast to the results of some other researchers (Woods et al. 1999), β-carotene had no protective effects in HepG2 cells exposed to H_2O_2 treatment. In a xanthine/xanthine oxidase system, low lycopene concentrations afforded some level of protection but this was rapidly lost at higher concentrations (4-10 mM) (Lowe et al. 1999). These observations could be due to the pro-oxidant nature of carotenoids under certain conditions, e.g., high carotenoid concentration and high oxygen tension.

In a study using plasmid DNA and the catechol estrogen, 4-hydroxy-estradiol (4-OHE_2), plus copper sulfate system (Muzandu et al. 2005), physiological concentrations of lycopene, as well as β-carotene, strongly inhibited 4-OHE_2/$CuSO_4$-induced strand breakage. Lycopene and β-carotene also ameliorated DNA damage in Chinese hamster fibroblasts incubated with 4-OHE_2 (Muzandu et al. 2005). The putative mechanism of tumor initiation and/or progression by endogenous or exogenous agents is the generation of toxic radicals during the redox cycling of metabolites, e.g., catechol estrogens, which are also directly genotoxic (Cavalieri et al. 1997), and cadmium (Yang et al. 1996).

Little research has been carried out on the interactions of carotenoids with peroxynitrite-induced modifications to nucleic acids. Hiramoto et al. (1999) found that β-carotene prevented DNA strand breaks in human promyelocytic leukemia HL-60 cells that were exposed to either authentic peroxynitrite or SIN-1-derived peroxynitrite. In a more recent study (Muzandu et al. 2006), peroxynitrite-induced DNA damage was shown to be ameliorated by both lycopene and β-carotene (Table 1, Figs. 7 and 8). In this study, plasmid DNA (pUC 18) and Chinese Hamster fibroblasts (V79 cells) were employed. Lycopene and β-carotene prevented strand breaks in DNA caused by SIN-1-derived peroxynitrite (Table 1). Treatment of cells with authentic peroxynitrite or SIN-1 caused a dose- and time-dependent increase in the level of DNA damage as detected with the comet assay. DNA damage was preventable by lycopene when administered 24 hr prior to SIN-1 (Fig. 8).

Other toxic agents or substances that lycopene has been found to protect against DNA damage include iron (or ferric nitrilotriacetate) (Matos et al. 2000), methyl methanesulfonate (MMS) and 4-nitroquinoline 1-oxide (4-NQO) (Scolastici et al. 2007), catechol estrogens (Muzandu et al. 2005) and gamma-irradiation (Srinivasan et al. 2007).

Other than protecting DNA from damage via an antioxidant mechanism, it has been suggested that lycopene may also modulate DNA repair mechanisms (Astley and Elliot 2005).

Table 1 Effect of lycopene and β-carotene on strand breaks induced by the peroxynitrite donor, SIN-1, in pUC18 plasmid DNA.[†]

Concentration (μM)[§]	Lycopene		β-Carotene	
	SSB/10^4 bp DNA	% control[#]	SSB/10^4 bp DNA	% control[#]
0.00 (control)	1.37 ± 0.12 [a]	100.0	1.42 ± 0.13[a]	100.0
0.05	1.36 ± 0.03 [a]	99.6	1.44 ± 0.19[a]	101.0
0.10	1.04 ± 0.07 [b]	76.0	1.05 ± 0.14[ab]	74.0
0.25	0.72 ± 0.07[c]	52.8	0.77 ± 0.16[bc]	54.2
0.50	0.55 ± 0.04[cd]	40.3	0.41 ± 0.07 [cd]	28.8
1.00	0.47 ± 0.02[cd]*	34.2	0.19 ± 0.02 [d]	13.3
2.50	0.38 ± 0.02[d]*	27.9	0.16 ± 0.01 [d]	11.4
5.00	0.32 ± 0.03[d]*	23.3	0.14 ± 0.02 [d]	9.5
10.00	0.31 ± 0.02[d]*	22.6	0.13 ± 0.02 [d]	9.5

[†]DNA was incubated with carotenoids or vehicle (THF) and/or SIN-1 for 30 min at 37°C. DNA damage was expressed as single strand breaks (SSB) per 10^4 base pairs (bp) DNA. Data are means ± SEM of three experiments. [§]Concentration of carotenoid + 50 μM SIN-1. [#]Percentage of SSB in pUC18 plasmid DNA calculated relative to control. Differences were evaluated by two-way ANOVA and post hoc tests. A common letter indicated no difference among data in each carotenoid-treated group (post hoc Tukey's-test, $P \leq 0.05$). *Different from same concentration of μ-carotene treatment (post hoc Student's t-test with Bonferroni correction, $P \leq 0.05$).

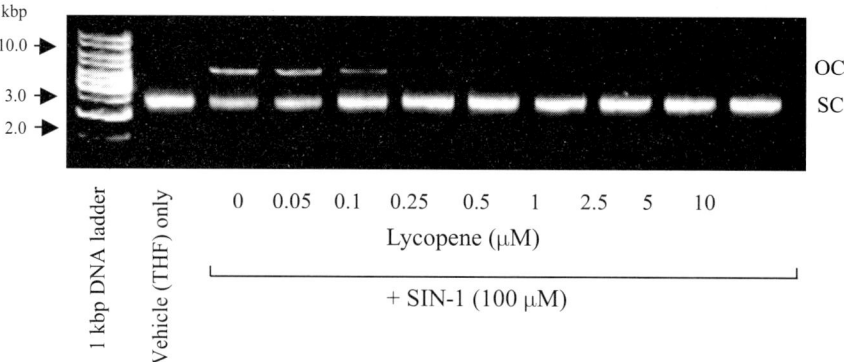

Fig. 7 Lycopene protects DNA against damage by peroxynitrite. The carotenoid lycopene inhibits damage to SIN-1-treated pUC18 plasmid DNA in a dose-dependent manner. DNA was incubated for 30 min at 37°C with either vehicle only (0.1 % THF) or 0.25-10 μM lycopene plus 100 μM SIN-1.

SIN-1, 3-morpholinosydnonimine; THF, tetrahydrofuran; OC, open circular DNA; SC, supercoiled DNA.

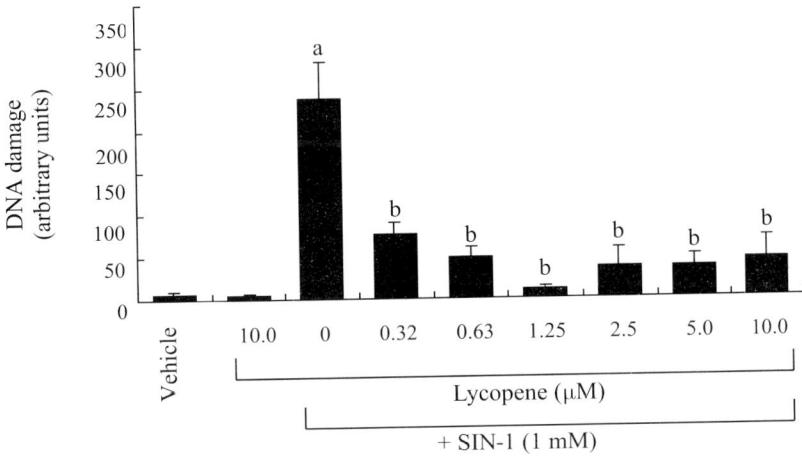

Fig. 8 Effect of lycopene on cellular DNA damage caused by SIN-1. Lycopene protects V79 cells against SIN-1-mediated DNA damage. V79 cells were incubated with media containing 0 to 10 μM lycopene (n=3) for 24 hr and then exposed to 1 mM SIN-1 for 1 hr. DNA damage by 1 mM of SIN-1 was assessed with the comet assay. Data are expressed as means ± SEM. A common letter indicates no difference among data (Tukey's test, $P \leq 0.05$).

Modulatory Effect of Lycopene on Peroxynitrite-induced DNA Mutations

No studies have been reported on the effect of lycopene on DNA mutations specifically caused by peroxynitrite. However, there are numerous studies concerning the effect of lycopene or tomato products on DNA mutations and chromosome aberrations caused by other agents, some believed to generate ROS. Cisplastin is one of the main antineoplastic drugs used in testicular, ovarian, bladder, head and neck tumors and malignant lymphomas with high efficacy. However, its clinical use is limited because of various side effects, some of which may be caused by its ability to generate ROS (Masuda et al. 1994, Weijl et al. 1997). Low doses of lycopene administered to rats reduce cisplastin-induced chromosome damage in bone marrow (Sendao et al. 2006). Lycopene is also very effective in reducing spontaneous mutagenesis in mismatch repair-deficient colon carcinoma cells (Mure and Rossman 2001).

Modulatory Effect of Lycopene on Peroxynitrite-induced Protein Modifications

Peroxynitrite has been implicated in the pathology of various diseases because of the detection of 3-nitrotyrosine in tissues, which has traditionally been the biomarker of this nitrating agent. It is now appreciated that several mechanisms that do not involve peroxynitrite exist that could lead to the nitration of proteins. Nevertheless, peroxynitrite is thought to mediate the toxic effects seen in diseases associated with aging, neurodegeneration, cardiovascular diseases, chronic infection and inflammation, and cancer. At the same time, lycopene (or the consumption of tomato products) has been linked to a reduced risk to the development of these same ailments.

In a recent study, the ability of lycopene and β-carotene to modulate protein nitration in V79 cells was investigated (Muzandu et al. 2006). SIN-1 increased nitration of proteins in cells above basal levels in a dose-dependent manner as determined by western blotting (Fig. 9a). It was revealed that at a concentration of 5 mM, lycopene and, less so, β-carotene, were strong inhibitors of protein nitration (Fig. 9). As has been pointed out earlier, lycopene and β-carotene are potent quenchers of singlet oxygen and peroxyl radicals. It has been determined that they have a role in scavenging RNS. Beta-carotene strongly scavenges RNS in solution, including nitrogen dioxide (Everett et al. 1996, Mortensen et al. 1997) and peroxynitrite (Kikugawa et al. 1997). Lycopene and b-carotene are also consumed in the process of scavenging peroxynitrite in low density lipoproteins (Pannala et al. 1998, Panasenko et al. 2000, Terao et al. 2001).

Modulatory Effect of Lycopene on Peroxynitrite-induced Lipid Modifications

Lycopene is a strong inhibitor of lipid peroxidation, both *in vitro* (Matos et al. 2000, Shi et al. 2004, Srinivasan et al. 2007) and *in vivo* (Atessahin et al. 2005, 2006, Matos et al. 2006).

A study by Velmurugan et al. (2004) evaluated the effect of tomato paste on clastogenesis induced by the carcinogen MNNG (N-methyl-N-Nitro-N-Nitrosoguanidine) in rats. These researchers found that tomato paste (0.5, 1.0 and 2.0 g/kg bw) administered for 5 consecutive days significantly reduced clastogenicity as well as lipid peroxidation. It is worth noting that even though some research studies show lycopene or tomato products can afford some protection against lipid peroxidation induced by various agents, other studies have failed to demonstrate a similar effect (Breinholt et al. 2000, Frederiksen et al. 2007).

Peroxynitrite is well known to oxidize lipids. There are indications that lycopene is a scavenger of peroxynitrite and nitrogen oxide species (e.g., $^{•}NO_2$) (Bohm et al. 1995, 2001, Pannala et al. 1998, Panasenko et al. 2000, Terao et al. 2001). However, it has yet to be experimentally established whether or not lycopene can inhibit peroxynitrite-induced lipid oxidation.

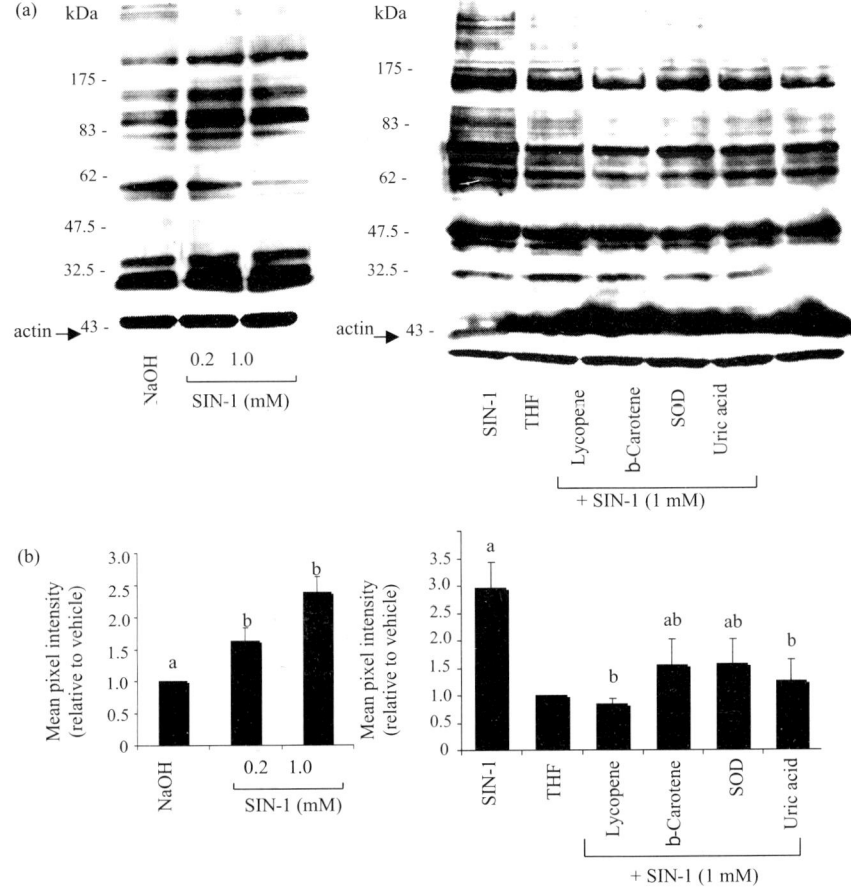

Fig. 9 Carotenoids reduce nitrotyrosine formation. (a) Western blotting reveals an increase in nitration of proteins with SIN-1 treatment of cells (left panel). Nitration by SIN-1 (1 mM) is inhibited by antioxidants uric acid (1 mM) as well as by lycopene (5 μM) (right panel). (b) Densitometric analysis of each lane from ~47.5 kDa to ~100 kDa showing differences in nitrotyrosine formation is presented. Total pixel intensity was taken in the said region of each lane. Data are expressed as mean pixel intensity normalized to actin and relative to the control (vehicle; NaOH or THF). Error bars represent standard error of the mean from four experiments. A common letter indicated no difference among data (Tukey's test, $P \leq 0.05$).

Modulatory Effect of Lycopene on General Peroxynitrite-induced Cellular Functions

There is strong evidence for the formation of peroxynitrite during chronic infection and inflammation, which is thought to lead to carcinogenesis. Lycopene may inhibit inflammatory processes or modulate immunity by its antioxidant mechanisms (e.g., direct interaction with peroxynitrite or its derived radicals) or by influencing various mediators of inflammation. Lycopene inhibits lipopolysaccharide (LPS)-induced inducible nitric oxide synthase (iNOS) protein and mRNA expression in mouse macrophage cells (Rafi et al. 2007). This would have the effect of reducing NO available to react with superoxide for the formation of additional peroxynitrite. The transcription factor nuclear factor kappaB (NF-κB) that regulates the expression of iNOS has also been shown to be down-regulated by lycopene (Kim et al. 2004).

ACKNOWLEDGEMENTS

The authors wish to express their appreciation to Dr. Boniface Namangala for his helpful comments and discussions in preparation of the manuscript.

ABBREVIATIONS

$CO_3^{\bullet-}$: carbonate anion; eNOS: endothelial nitric oxide synthase; EPO: eosinophil peroxidase; H_2O_2: hydrogen peroxide; HOBr: hypobromous acid; HOCl: hypochlorous acid; iNOS: inducible nitric oxide synthase; $^{\bullet}$OH: hydroxyl radical; $ONOO\ CO_2^-$: nitrosoperoxycarbonate; MPO: myeloperoxidase; nNOS: neuronal nitric oxide synthase; NO_2^-: nitrite; NO_3^-: nitrate; NO: nitric oxide; $^{\bullet}NO_2$: nitrogen dioxide; NO^+: nitrosonium/nitrosyl cation; NO^-: nitroxyl anion; NO_2Cl: nitryl chloride; ROS: reactive oxygen species; RNS: reactive nitrogen species; $RS^{\bullet}$: thiyl radical; RSH: thiol; RSNO: nitrosothiol; RSOH: sulfenic acid; RSSR: disulfide; $ONOO^-$: peroxynitrite anion; ONOOH: peroxynitrous acid; ROOH: peroxyl radical; $^1O_2$: singlet oxygen; $O_2^{\bullet-}$: superoxide anion; SIN-1: 3-morpholinosydnonimine

References

Afanas'ev, I.B. 1999. Superoxide Ion: Chemistry and Biological Implications. CRC Press, Inc., Florida, USA.

Alvarez, B. and V. Demicheli, R. Duran, M. Trujillo, C. Cervenansky, B.A. Freeman, and R. Radi. 2004. Inactivation of human Cu,Zn Superoxide dismutase by peroxynitrite and formation of histidinyl radical. Free Radic. Biol. Med. 37: 813-822.

Alvarez, B. and R. Radi. 2003. Peroxynitrite reactivity with amino acids and proteins. Amino Acids 25: 295-311.

Ames, B.N. and W.C. Gold, and W.C. Willet. 1995. The causes and prevention of cancer. Proc. Natl. Acad. Sci. USA 92: 5258-5265.

Astley, S.B. and R.M. Elliott. 2005. How strong is the evidence that lycopene supplementation can modify biomarkers of oxidative damage and DNA repair in human lymphocytes? J. Nutr. 135: 2071S-2073S.

Atessahin, A., and I. Karahan, G. Turk, S. Gur, S. Yilmoz, and A. Osman Ceribasi. 2006. Protective role of lycopene on cisplatin-induced changes in sperm characteristics, testicular damage and oxidative stress in rats. Reprod. Toxicol. 21: 42-47.

Atessahin, A. and S. Yilmaz, I. Karahan, A.O. Ceribasi, and A. Karaoglu. 2005. Effects of lycopene against cisplatin-induced nephrotoxicity and oxidative stress in rats. Toxicology 212: 116-123.

Aulak, K.S. and T. Koeck, J.W. Crabb, and D.J. Stuehr. 2004. Dynamics of protein nitration in cells and mitochondria. Am. J. Physiol. Heart Circ. Physiol. 286: H30-H38.

Beckman, J.S. 1996. Oxidative damage and tyrosine nitration from peroxynitrite. Chem. Res. Toxicol. 9: 836-844.

Beckman, J.S. and T.W. Beckman, J. Chen, P.A. Marshall, and B.A. Freeman. 1990. Apparent hydroxyl radical production by peroxynitrite: Implications for endothelial injury from nitric oxide and superoxide. Proc. Natl. Acad. Sci. USA 87: 1620-1624.

Beckman, K.B. and B.N. Ames. 1997. Oxidative decay of DNA. J. Biol. Chem. 272: 19633-19636.

Bohm, F. and R. Edge, M. Burke, and T.G. Truscott. 2001. Dietary uptake of lycopene protects human cells from singlet oxygen and nitrogen dioxide — ROS components from cigarette smoke. J. Photochem. Photobiol. B. 64: 176-178.

Bohm, F. and J.H. Tinkler, and T.G. Truscott. 1995. Carotenoids protect against cell membrane damage by the nitrogen dioxide radical. Nat. Med. 1: 98-99.

Breinholt, V. and S.T. Lauridsen, B. Daneshvar, and J. Jakobsen. 2000. Dose-response effects of lycopene on selected drug-metabolizing and atioxidant enzymes in the rat. Cancer Lett. 154: 201-210.

Buchczyk, D.P. and T. Grune, H. Sies, and L.O. Klotz. 2003. Modifications of glyceraldehyde-3-phosphate dehydrogenase induced by increasing concentrations of peroxynitrite: early recognition by 20S proteasome. Biol. Chem. 384: 237-241.

Burney, S. and J.L. Caulfield, J.C. Niles, J.S. Wishnok, and S.R. Tannenbaum. 1999. The chemistry of DNA damage from nitric oxide and peroxynitrite. Mutat. Res. 424: 37-49.

Cavalieri, E.L. and D.E. Stack, P.D. Devanesan, R. Todorovic, I. Dwivedy, S. Higginbotham, S.L. Johansson, K.D. Patil, M.L. Gross, J.K. Gooden, R. Ramanathan, R.L. Cerny, and E.G. Rogan. 1997. Molecular origin of cancer: catechol estrogen-3,4-quinones as endogenous tumor initiators. Proc. Natl. Acad. Sci. USA 94: 10937-10942.

Choi, J. and H.D. Rees, S.T. Weintraub, A.I. Levey, L.S. Chin, and L. Li. 2005. Oxidative modifications and aggregation of Cu,Zn-superoxide dismutase associated with Alzheimer and Parkinson diseases. J. Biol. Chem. 280: 11648-11655.

Collins, A.R. and V.L. Dobson, M. Dusinska, G. Kennedy, and R. Stetina. 1997. The comet assay: what can it really tell us? Mutat. Res. 375: 183-193.

Dalle-Donne, I. and D. Giustarini, R. Colombo, R. Rossi, and A. Milzani. 2003. Protein carbonylation in human diseases. Trends Mol. Med. 9: 169-176.

Dedon, P.C. and S.R. Tannenbaum. 2004. Reactive nitrogen species in the chemical biology of inflammation. Arch. Biochem. Biophys. 423: 12-22.

Delaney, C.A. and B. Tyrberg, L. Bouwens, H. Vaghef, B. Hellman, and D.L. Eizirik. 1996. Sensitivity of human pancreatic islets to peroxynitrite-induced cell dysfunction and death. FEBS Lett. 394: 300-306.

Denicola, A. and B.A. Freeman, M. Trujillo, and R. Radi. 1996. Peroxynitrite reaction with carbon dioxide/bicarbonate: kinetics and influence on peroxynitrite-mediated oxidations. Arch. Biochem. Biophys. 333: 49-58.

Denicola, A. and R. Radi. 2005. Peroxynitrite and drug-dependent toxicity. Toxicology 208: 273-288.

Denicola, A. and H. Rubbo, D. Rodriguez, and R. Radi. 1993. Peroxynitrite-mediated cytotoxicity to *Trypanosoma cruzi*. Arch. Biochem. Biophys. 304: 279-286.

Denicola, A. and J.M. Souza, and R. Radi. 1998. Diffusion of peroxynitrite across erythrocyte membranes. Proc. Natl. Acad. Sci. USA 95: 3566-3571.

Epe, B. and D. Ballmaier, I. Roussyn, K. Briviba, and H. Sies. 1996. DNA damage by peroxynitrite characterized with DNA repair enzymes. Nucleic Acids Res. 24: 4105-4110.

Everett, S.A. and M.F. Dennis, K.B. Patel, S. Maddix, S.C. Kundu, and R.L. Willson. 1996. Scavenging of nitrogen dioxide, thiyl, and sulfonyl free radicals by the nutritional antioxidant beta-carotene. J. Biol. Chem. 271: 3988-3994.

Frederiksen, H. and S.E. Rasmussen, M. Schroder, A. Bysted, J. Jakobsen, H. Frandsen, G. Ravn-Haren, and A. Mortensen. 2007. Dietary supplementation with an extract of lycopene-rich tomatoes does not reduce atherosclerosis in Watanabe Heritable Hyperlipidemic rabbits. Br. J. Nutr. 97: 6-10.

Fridovich, I. 1995. Superoxide radical and superoxide dismutases. Annu. Rev. Biochem. 64: 97-112.

Gaut, J.P. and J. Byun, H.D. Tran, W.M. Lauber, J.A. Carroll, R.S. Hotchkiss, A. Bellouaj, and J.W. Heinecke. 2002. Myeloperoxidase produces nitrating oxidants *in vivo*. J. Clin. Invest. 109: 1311-1319.

Giovannucci, E. 2002. A review of epidemiologic studies of tomatoes, lycopene, and prostate cancer. Exp. Biol. Med. 227: 852-859.

Gow, A. and D. Duran, S. Malcolm, and H. Ischiropoulos. 1996. Effects of peroxynitrite-induced protein modifications on tyrosine phosphorylation and degradation. FEBS Lett. 385: 63-66.

Gow, A.J. and C.R. Farkouh, D.A. Munson, M.A. Posencheg, and H. Ischiropoulos. 2004. Biological significance of nitric oxide-mediated protein modifications. Amer. J. Physiol. Lung Cell Mol. Physiol. 287: 262-268.

Hiramoto, K. and S. Tomiyama, and K. Kikugawa. 1999. Effective inhibition by beta-carotene of cellular DNA breaking induced by peroxynitrous acid. Free Radic. Res. 30: 21-27.

Hughes, M.N. and H.G. Nicklin. 1968. The chemistry of pernitrites. Part I: Kinetics of decomposition of pernitrous acid. J. Chem. Soc., A 450-452.

Hurst, J.K. 2002. Whence nitrotyrosine? J. Clin. Invest. 109: 1287-1289.

Jackson, A.L. and L.A. Loeb. 2001. The contribution of endogenous sources of DNA damage to the multiple mutations in cancer. Mutat. Res. 477: 7-21.

Juedes, M.J. and G.N. Wogan. 1996. Peroxynitrite-induced mutation spectra of pSP189 following replication in bacteria and in human cells. Mutat. Res. 349: 1-61.

Jung, T. and M. Engels, L.O. Klotz, K.D. Kroncke, and T. Grune. 2007. Nitrotyrosine and protein carbonyls are equally distributed in HT22 cells after nitrosative stress. Free Radic. Biol. Med. 42: 773-786.

Kamisaki, Y. and K. Wada, K. Bian, B. Balabanli, K. Davis, E. Martin, F. Behbod, Y.C. Lee, and F. Murad. 1998. An activity in rat tissues that modifies nitrotyrosine-containing proteins. Proc. Natl. Acad. Sci. USA 95: 11584-11589.

Kennedy, L.J. and K. Moore Jr., J.L. Caulfield, S.R. Tannenbaum, and P.C. Dedon. 1997. Quantitation of 8-oxoguanine and strand breaks produced by four oxidizing agents. Chem. Res. Toxicol. 10: 386-392.

Khachick, F. and G.R. Beecher, and J.C. Smith Jr. 1995. Lutein, lycopene, and their oxidative metabolites in chemoprevention of cancer. J. Cell. Biochem. 22 (suppl): 236-246.

Kikugawa, K. and K. Hiramoto, S. Tomiyama, and Y. Asano. 1997. Beta-Carotene effectively scavenges toxic nitrogen oxides: nitrogen dioxide and peroxynitrous acid. FEBS Lett. 404: 175-178.

Kim, G.Y. and J.H. Kim, S.C. Ahn, H.J. Lee, D.O. Moon, C.M. Lee, and Y.M. Park. 2004. Lycopene suppresses the lipopolysaccharide-induced phenotypic and functional maturation of murine dendritic cells through inhibition of mitogen-activated protein kinases and nuclear factor-κB. Immunology 113: 203-211.

Kim, M.Y. and M. Dong, P.C. Dedon, and G.N. Wogan. 2005. Effects of peroxynitrite dose and dose rate on DNA damage and mutation in the *supF* shuttle vector. Chem. Res. Toxicol. 18: 76-86.

Kirsch, M. and M. Lehnig. 2005. Generation of peroxynitrite from reaction of N-acetyl-N-nitrosotryptophan with hydrogen peroxide over a wide range of pH values. Org. Biomol. Chem. 3: 2085-2090.

Kissner, R. and T. Nauser, P. Bugnon, P.G. Lye, and W.H. Koppenol. 1997. Formation and properties of peroxynitrite as studied by laser flash photolysis, high-pressure stop-flow technique, and pulse radiolysis. Chem. Res. Toxicol. 10: 1285-1292.

Koeck, T. and X. Fu, S.L. Hazen, J.W. Crabb, D.J. Stuehr, and K.S. Aulak. 2004. Rapid and selective oxygen-regulated protein tyrosine denitration and nitration in mitochondria. J. Biol. Chem. 279: 27257-27262.

Konorev, E.A. and N. Hogg, and B. Kalyanaraman. 1998. Rapid and irreversible0inhibition of creatine kinase by peroxynitrite. FEBS. Lett. 427: 171-174.

Koppenol, W.H. and J.J. Moreno, W.A. Pryor, H. Ischiropoulos, and J.S. Beckman. 1992. Peroxynitrite, a cloaked oxidant formed by nitric oxide and superoxide. Chem. Res. Toxicol. 5: 834-842.

Lamm, G. and G.R. Pack. 1990. Acidic domains around nucleic acids. Proc. Natl. Acad. Sci. USA 87: 9033-9036.

Levine, R.L. 2002. Carbonyl modified proteins in cellular regulation, aging, and disease. Free Radic. Biol. Med. 32: 790-796.

Levine, R.L. and E.R. Stadtman. 2001. Oxidative modification of proteins during aging. Exp. Gerontol. 36: 1495-1502.

Linares, E. and S. Giorgio, R.A. Mortara, C.X.C. Santos, A.T. Yamada, and O. Augusto. 2001. Role of peroxynitrite in macrophage microbicidal mechanisms *in vivo* revealed by protein nitration and hydroxylation. Free Radic. Biol. Med. 30: 1234-1242.

Liu, X. and M.J. Miller, M.S. Joshi, H. Sadowska-Krowicka, D.A. Clark, and J.R. Lancaster. 1998. Diffusion-limited reaction of free nitric oxide with erythrocytes. J. Biol. Chem. 273: 18709-18713.

Lowe, G.M. and L.A. Booth, A.J. Young, and R.F. Bilton. 1999. Lycopene and beta-carotene protect against oxidative damage in HT29 cells at low concentrations but rapidly lose this capacity at higher doses. Free Radic. Res. 30: 141-151.

Martinet, W. and G.R. de Meyer, A.G. Herman, and M.M. Kockx. 2004. Reactive oxygen species induce RNA damage in human atherosclerosis. Eur. J. Clin. Invest. 34: 323-327.

Martinet, W. and G.R. de Meyer, A.G. Herman, and M.M. Kockx. 2005. RNA damage in human atherosclerosis: pathophysiological significance and implications for gene expression studies. RNA Biol. 2: 4-7.

Masuda, H. and T. Takaka, and U. Takahama. 1994. Cisplatin generates superoxide anion by interaction with DNA in a cell-free system. Biochem. Biophys. Res. Commun. 203: 1175-1180.

Matos, H.R. and P. Di Mascio, and M.H. Medeiros. 2000. Protective effect of lycopene on lipid peroxidation and oxidative DNA damage in cell culture. Arch. Biochem. Biophys. 383: 56-59.

Matos, H.R. and S.A. Marques, O.F. Gomes, A.A. Silva, J.C. Heimann, P. Di Mascio, and M.H. Medeiros. 2006. Lycopene and beta-carotene protect in vivo iron-induced oxidative stress damage in rat prostate. Braz. J. Med. Biol. Res. 39: 203-210.

Mayne, S.T. 1996. Beta-carotene, carotenoids, and disease prevention in humans. FASEB J. 10: 690-701.

Mortensen, A. and L.H. Skibsted, J. Sampson, C. Rice-Evans, and S.A. Everett. 1997. Comparative mechanisms and rates of free radical scavenging by carotenoid antioxidants. FEBS Lett. 418: 91-97.

Mure, K. and T.G. Rossman. 2001. Reduction of spontaneous mutagenesis in mismatch repair-deficient and proficient cells by dietary antioxidants. Mutat. Res. 480-481: 85-95.

Muzandu, K. and K. El Bohi, Z. Shaban, M. Ishizuka, A. Kazusaka, and S. Fujita. 2005. Lycopene and beta-carotene ameliorate catechol estrogen-mediated DNA damage. Jap. J. Vet. Res. 52: 173-184.

Muzandu, K. and M. Ishizuka, K.Q. Sakamoto, Z. Shaban, K. El Bohi, A. Kazusaka, and S. Fujita. 2006. Effect of lycopene and beta-carotene on peroxynitrite-mediated cellular modifications. Toxicol. Appl. Pharmacol. 215: 330-340.

Nunomura, A.S. and S. Chiba, K. Kosaka, A. Takeda, R.J. Castellani, M.A. Smith, and G. Perry. 2002. Neuronal RNA oxidation is a prominent feature of dementia with Lewy bodies. NeuroReport 13: 2035-2039.

Nunomura, A. and G. Perry, M.A. Pappolla, R. Wade, K. Hirai, S. Chiba, and M.A. Smith. 1999. RNA oxidation is a prominent feature of vulnerable neurons in Alzheimer's disease. J. Neurosci. 19: 1959-1964.

Nystrom, T. 2005. Role of oxidative carbonylation in protein quality control and senescence. EMBO J. 24: 1311-1317.

O'Donnell, V.B. and B.A. Freeman. 2001. Interactions between nitric oxide and lipid oxidation pathways: implications for vascular disease. Circ. Res. 88: 12-21.

Ohshima, H. and M. Tatemichi, and T. Sawa. 2003. Chemical basis of inflammation-induced carcinogenesis. Arch. Biochem. Biophys. 417: 3-11.

Olmos, A. and R.M. Giner, and S. Manez. 2007. Drugs modulating the biological effects of peroxynitrite and related nitrogen species. Med. Res. Rev. 27: 1-64.

Pacher, P. and J.S. Beckman, and L. Liaudet. 2007. Nitric oxide and peroxynitrite in health and disease. Physiol. Rev. 87: 315-424.

Panasenko, O.M. and V.S. Sharov, K. Briviba, and H. Sies. 2000. Interaction of peroxynitrite with carotenoids in human low density lipoproteins. Arch. Biochem. Biophys. 373: 302-305.

Pannala, A.S. and C. Rice-Evans, J. Sampson, and S. Singh. 1998. Interaction of peroxynitrite with carotenoids and tocopherols within low density lipoprotein. FEBS Lett. 423: 297-301.

Park, Y.O. and E.S. Hwang, and T.W. Moon. 2005. The effect of lycopene on cell growth and oxidative DNA damage of Hep3B human hepatoma cells. BioFactors 23: 129-139.

Pietraforte, D. and A.M. Salzano, G. Marino, and M. Minetti. 2003. Peroxynitrite-dependent modifications of tyrosine residues in hemoglobin. Formation of tyrosyl radical(s) and 3-nitrotyrosine. Amino Acids 25: 341-350.

Radi, R. 2004. Nitric oxides, oxidants, and protein tyrosine nitration. Proc. Natl. Acad. Sci. USA 101: 4003-4008.

Radi, R. and J.S. Beckman, K.M. Bush, and B.A. Freeman. 1991. Peroxynitrite-induced membrane lipid peroxidation: the cytotoxic potential of superoxide and nitric oxide. Arch. Biochem. Biophys. 288: 481-487.

Rafi, M.M. and P.N. Yadav, and M. Reyes. 2007. Lycopene inhibits LPS-induced proinflammatory mediator inducible nitric oxide synthase in mouse macrophage cells. J. Food Sci. 72: S69-S74

Reiter, C.D. and R.J. Teng, and J.S. Beckman. 2000. Superoxide reacts with nitric oxide to nitrate tyrosine at physiological pH via peroxynitrite. J. Biol. Chem. 275: 32460-32466.

Salgo, M.G. and K. Stone, G.L. Squadrito, J.R. Battista, and W.A. Pryor. 1995. Peroxynitrite causes DNA nicks in plasmid pBR322. Biochem. Biophys. Res. Commun. 210: 1025-1030.

Scolastici, C. and R.O. de Lima, L.F. Barbisan, A.L. Ferreira, D.A. Ribeiro, and D.M. Salvadori. 2007. Lycopene activity against chemically induced DNA damage in Chinese hamster ovary cells. Toxicol. In Vitro 21: 840-845.

Sendao, M.C. and E.B. Behling, R.A. dos Santos, L.M. Antunes, and M. de Lourdes Pires Bianchi. 2006. Comparative effects of acute and subacute lycopene administration on chromosomal aberrations induced by cisplatin in male rats. Food Chem. Toxicol. 44: 1334-1339.

Shi, J. and Y. Kakuda, and D. Yeung. 2004. Antioxidative properties of lycopene and other carotenoids from tomatoes: Synergistic effects. BioFactors 21: 203-210.

Sies, H. and W. Stahl, and A. Sevanian. 2005. Nutritional, dietary and postprandial oxidative stress. J. Nutr. 135: 969-972.

Souici, A.C. and J. Mirkovitch, P. Hausel, K. Keefer, and E. Felley-Bosco. 2000. Transition mutation in codon 248 of the p53 tumor suppressor gene induced by reactive oxygen species and a nitric oxide-releasing compound. Carcinogenesis 21: 281-287.

Souza, J.M. and R. Radi. 1998. Glyceraldehyde-3-phosphate dehydrogenase inactivation by peroxynitrite. Arch. Biochem. Biophys. 360: 187-194.

Spencer, J.P.E. and J. Wong, A. Jenner, O.I. Aruoma, C.E. Cross, and B. Halliwell. 1996. Base modification and strand breakage in isolated calf thymus DNA and in DNA from human skin epidermal keratinocytes exposed to peroxynitrite or 3-morpholinosydnonimine. Chem. Res. Toxicol. 9: 1152-1158.

Srinivasan, M. and A.R. Sudheer, K.R. Pillai, P.R. Kumar, P.R. Sudhakaran, and V.P. Menon. 2007. Lycopene as a natural protector against gamma-radiation induced DNA damage, lipid peroxidation and antioxidant status in primary culture of isolated rat hepatocytes in vitro. Biochim. Biophys. Acta. 1770: 659-665.

Stadtman, E.R. 1992. Protein oxidation and aging. Science 257: 1220-1224.

Stadtman, E.R. 2006. Protein oxidation and aging. Free Radic. Res. 40: 1250-1258.

Stahl, W. and H. Sies. 1996. Lycopene: a biologically important carotenoid for humans? Arch. Biochem. Biophys. 336: 1-9.

Szabo, C. and H. Ohshima. 1997. DNA damage induced by peroxynitrite: subsequent biological effects. Nitric Oxide 1: 373-385.

Szabo, C. and B. Zingarelli, M. O'Connor, and A.L. Salzman. 1996. DNA strand breakage, activation of poly (ADP-ribose) synthetase, and0cellular energy depletion are involved in the cytotoxicity of macrophages and smooth muscle cells exposed to peroxynitrite. Proc. Natl. Acad. Sci. USA 93: 1753-1758.

Tanaka, M. and P.B. Chock, and E.R. Stadtman. 2007. Oxidized messenger RNA induces translation errors. Proc Natl. Acad. Sci. USA 104: 66-71.

Tapiero, H. and D.M. Townsend, and K.D. Tew. 2004. The role of carotenoids in the prevention of human pathologies. Biomed. Pharmacother. 58: 100-110.

Terao, J. and S. Yamaguchi, M. Shirai, M. Miyoshi, J.H. Moon, S. Oshima, T. Inakuma, T. Tsushida, and Y. Kato. 2001. Protection by quercetin and quercetin 3-O-beta-D-glucuronide of peroxynitrite-induced antioxidant consumption in human plasma low-density lipoprotein. Free Radic. Res. 35: 925-931.

Tice, R.R. and E. Agurell, D. Anderson, B. Burlinson, A. Hartmann, H. Kobayashi, Y. Miyamae, E. Rojas, J.C. Ryu, and Y.F. Sasaki. 2000. Single cell gel/comet assay: guidelines for in vitro and in vivo genetic toxicology testing. Environ. Mol. Mutagen. 35: 206-221.

van der Vliet, A. and J.P. Eiserich, B. Halliwell, and C.E. Cross. 1997. Formation of reactive nitrogen species during peroxidase-catalyzed oxidation of nitrite. A potential additional mechanism of nitric oxide-dependent toxicity. J. Biol. Chem. 272: 7617-7625.

van der Vliet, A. and C.E. Cross. 2000. Oxidants, nitrosants and the lung. Amer. J. Med. 109: 398-421.

Velmurugan, B. and V. Bhuvaneswari, S. Abraham, and S. Nagini. 2004. Protective effect of tomato against N-methyl-N-nitro-N-nitrosoguanidine-induced in vivo clastogenicity and oxidative stress. Basic Nutr. Invest. 20: 812-816.

Weijl, N.I. and F.J. Cleton, and S. Osanto. 1997. Free radicals and antioxidants in chemotherapy-induced toxicity. Cancer Treat. Rev. 23: 209-240.

Whiteman, M. and H.L. Siau, and B. Halliwell. 2003. Lack of tyrosine nitration by hypochlorous acid presence of physiological concentrations of nitrite. Implications for the role of nitryl chloride in tyrosine nitration *in vivo*. J. Biol. Chem. 278: 8380-8384.

Woods, J.A. and R.F. Bilton, and A.J. Young. 1999. Beta-carotene enhances hydrogen peroxide-induced DNA damage in human hepatocellular HepG2 cells. FEBS Lett. 449: 255-258.

Wu, W. and Y. Chen, and S.L. Hazen. 1999a. Eosinophil peroxidase nitrates protein tyrosyl residues. Implications for oxidative damage by nitrating intermediates in eosinophilic inflammatory disorders. J. Biol. Chem. 274: 25933-25944.

Wu, W. and Y. Chen, A. d'Avignon, and S.L. Hazen. 1999b. 3-Bromotyrosine and 3,5-dibromotyrosine are major products of protein oxidation by eosinophil peroxidase: potential markers for eosinophil-dependent tissue injury *in vivo*. Biochemistry 38: 3538-3548.

Yang, J.L. and J.I. Chao, and J.G. Lin. 1996. Reactive oxygen species may participate in the mutagenicity and mutational spectrum of cadmium in Chinese hamster ovary-K1 cells. Chem. Res. Toxicol. 9: 1360-1367.

Yermilov, V. and Y. Yoshie, J. Rubio, and H. Ohshima. 1996. Effects of carbon dioxide/bicarbonate on induction of DNA single-strand breaks and formation of 8-nitroguanine, 8-oxoguanine and base-propenal mediated by peroxynitrite. FEBS Lett. 399: 67-70.

Yoshie, Y. and H. Ohshima. 1997. Nitric oxide synergistically enhances DNA strand breakage induced by polyhydroxyaromatic compounds, but inhibits that induced by the Fenton reaction. Arch. Biochem. Biophys. 342: 13-21.

Zhang, J. and G. Perry, M.A. Smith, D. Robertson, S.J. Olson, D.G. Graham, and T.J. Montine. 1999. Parkinson's disease is associated with oxidative damage to cytoplasmic DNA and RNA in substantia nigra neurons. Amer. J. Pathol. 154: 1423-1429.

Zhu, L. and C. Gunn, and J.S. Beckman. 1992. Bactericidal activity of peroxynitrite. Arch. Biochem. Biophys. 298: 452-457.

Zingarelli, B. and M. O'Connor, H. Wong, A.L. Salzman, and C. Szabo. 1996. Peroxynitrite-mediated DNA strand breakage activates polyadenosine diphosphate ribosyl synthetase and causes cellular energy depletion in macrophages stimulated with bacterial lipopolysaccharide. J. Immunol. 156: 350-358.

2.2

Lycopene and Down-regulation of Cyclin D1, pAKT and pBad

Rosanna Sestito and Paola Palozza

Institute of General Pathology, Catholic University, School of Medicine
Lgo F. Vito 1, 00168 Rome

ABSTRACT

Although there is epidemiological evidence to support the beneficial effects of lycopene, tomato and tomato products in the prevention of chronic diseases, including cancer and cardiovascular diseases, the mechanism(s) responsible for such beneficial effects has not yet been completely clarified. The up-regulation of cyclin D1 and the deregulation of PI3k/Akt pathway have been largely associated with the pathogenesis and development of cancer and cardiovascular diseases. This review focused on the main evidence showing interference of lycopene with cyclin D1 and PI3K/Akt pathway. The available data indicate that lycopene may arrest cell cycle progression at the G0/G1 phase by acting as a potent down-regulator of cyclin D1 in both normal and tumor cultured cells. Moreover, the carotenoid may modulate cyclin D1 levels in cells exposed *in vitro* and *in vivo* to cigarette smoke. On the other hand, lycopene interference with PI3K/Akt pathway has been reported to suppress PI3K-dependent proliferative and survival signaling and to activate apoptosis pathways in tumor cells or in normal cells exposed to cigarette smoke. In particular, lycopene has been reported to promote apoptosis by decreasing the phosphorylation of Bad. In addition, recent data suggest a role for lycopene in modulating tissue factor activity and expression through PI3K/Akt pathway in endothelial cells. Although further work is needed, the available data seem to indicate in the modulation of cyclin D1 and PI3K/Akt pathway a key mechanism in the anticancer and cardio-protective effects of lycopene.

A list of abbreviations is given before the references.

INTRODUCTION

Lycopene is a natural pigment synthesized by plants and microorganisms but not by animals. It is a carotenoid, anacyclic isomer of β-carotene and has no vitamin A activity (Rao and Agarwal 2000, Stahl and Sies 1996). It is a highly unsaturated, straight chain hydrocarbon containing 11 conjugated and two non-conjugated double bonds.

Increasing evidence suggests that lycopene, the main carotenoid present in tomato and responsible for its red color, may have protective effects in chronic diseases, including cardiovascular diseases (CVD) and cancer.

Dietary intake of tomatoes and tomato products has been found to be associated with a lower risk of a variety of cancers in a number of epidemiological studies (Giovannucci 1999a). The most impressive results came from the US Health Professionals Follow-up Study evaluating the intake of various carotenoids and retinols, from a food frequency questionnaire, in relation to risk of prostate cancer. The estimated intake of lycopene from various tomato products, and not any other carotenoid, was inversely related to the risk of prostate cancer (Giovannucci et al. 1995). Similarly, serum and tissue levels of lycopene were inversely associated with prostate cancer risk in recent case-control and cohort studies (Rao et al. 1999, Gann et al. 1999). Moreover, high intake of tomatoes was consistently associated with a reduced risk of cancers of the digestive tract (especially stomach, colon and rectal), cervical intraepithelial cancer (Van Eewyck et al. 1991), and breast and bladder cancers (Helzlsouer et al. 1989). Giovannucci (1999a) reviewed 72 epidemiological studies on tomatoes, tomato-based products, lycopene and cancer. A significant number of such studies demonstrated an inverse relationship between intakes of tomatoes or plasma lycopene levels and cancer. The strongest associations were observed for cancers of the prostate, lung and stomach. However, for cancers of pancreas, colon and rectum, esophagus, oral mucosa, breast and cervix the associations were only suggestive. These results were consistent across numerous diverse populations and with the use of several different study designs. None of the studies analyzed indicated increased risk of cancer (Giovannucci 1999a).

The association between dietary intake or blood or tissue concentrations of lycopene and CVD and atherosclerosis has been studied over the last ten years. One of the first reports to suggest a possible relationship between lycopene and CVD risk was the nested case-control study published in 1994 by Street and coworkers. During a 14 yr follow-up, they found an increased risk of events in smokers with lower serum levels of β-carotene, lycopene, lutein and zeaxanthin, but not in non-smokers (Street et al.

1994). A later multicenter case-control study (EURAMIC or European Study of Antioxidants, Myocardial Infarction and Cancer of the Breast) evaluated the relationship between adipose tissue antioxidant status (α- and β-carotene and lycopene) and acute myocardial infarction (Kohlmeir et al. 1997). Subjects were recruited from 10 European countries to maximize the variability in exposure within the study. After adjusting for a range of dietary variables, only lycopene, and not β-carotene, levels were found to be protective. The protective potential of lycopene was maximal among individuals with highest polyunsaturated fat stores. Similarly, lower blood lycopene levels were found to be associated with increased risk and mortality from coronary heart disease in a concomitant cross-sectional study comparing Lithuanian and Swedish populations showing diverging mortality rates from coronary heart disease (Kristenson et al. 1997). In the Finnish prospective Kuopio Ischemic Hearth Disease Risk Factor (KIHD) study involving 1038 middle-aged men, serum levels of lycopene were inversely and independently associated with the risk-acute coronary event or stroke (Rissanen et al. 2001). The Antioxidant Supplementation in Atherosclerosis Prevention (ASAP) study examined cardiovascular risk factors and intake of nutrients in Finnish men and women. An inverse association was also seen among women although the difference did not remain statistically significant after the adjustments (Rissanen et al. 2000). In a small clinical trial involving six male subjects dietary supplementation of lycopene (60 mg/d for 3 mon) resulted in 14% reduction of plasma low density lipoprotein levels and thus acted as a moderate hypocholesterolemic agent (Fuhramn et al. 1997). In addition, a significant association between plasma lycopene and a reduced risk of CVD in a nested case-control study of middle-aged and older women has been reported (Sesso et al. 2004).

Although several mechanisms have been proposed to explain the beneficial effects of lycopene on CVD and cancer, including its antioxidant properties (Rao and Agarwal 2000, Stahl and Sies 1996, Nguyen and Schwartz 1999), its ability to enhance intercellular gap junction communication (Zhang et al. 1991, 1992), hormonal and immune system (Levy et al. 1995, Nagasawa et al. 1995, Kobayashi et al. 1996), and metabolic pathways (Astrog et al. 1997, Fuhramn et al. 1997), the exact mechanism(s) responsible for such benefits is still under investigation. This review focuses on the growing body of evidence that lycopene may be protective in human health by modulating cellular signaling of proliferation and apoptosis through changes in cyclin D1 and PI3K/Akt pathways.

MODULATION OF CELL CYCLE AND APOPTOSIS BY LYCOPENE

Several studies demonstrate that lycopene is able to inhibit cell cycle progression of different cell lines, as shown in Table 1.

Table 1   Evidence for an arrest of cell cycle progression by lycopene in different cell lines.

Lycopene conc.	Cell line	Target	Reference
0.6-3 µM	HL-60	–	Amir et al. 1999
0.75-3 µM	MCF-7	IGF-1	Karas et al. 2000
1-20 µM	MCF-7	AP-1	Prakash et al. 2001
	ER-Hs578T	–	
	MDA-MB-231		
2-3 µM	MCF-7	Cyclin D	Nahum et al. 2001
	T-47D		
	ECC-1		
1-10 µM	LCNaP	–	Kim et al. 2002a
up to 5 µM	PrEC	Cyclin D1	Obermuller-Jevic et al. 2003
0.1-5 µM	LCNaP	–	Hwang and Bowen 2004
0.5-2 µM	RAT-1	Cyclin D	Palozza et al. 2005
	Mv1LU		
10 nM-1 µM	LCNaP	Cyclin D1	Ivanov et al. 2007
	PC-3	Cyclin E	
		CdK-4	
		Rb	
		IFG-1	
		AKT	

On the other hand, increasing evidence underlines the role of lycopene in apoptosis in different cell lines, including LNCaP prostate cancer (Hantz et al. 2005, Kanagaraj et al. 2007, Ivanov et al. 2007), lymphoblast (Muller et al. 2002) and mammary cancer cells (Chalabi et al. 2006, 2007).

Similarly, tomato and tomato-derived products, very rich in lycopene, have been reported to induce an arrest of proliferation and apoptosis in *in vitro* (Hwang and Bowen 2005) and *in vivo* (Canene-Adams et al. 2007) models as well as in an *in vivo* model of prostate cancer.

Although these studies demonstrated a clear function of lycopene in inhibiting cell cycle progression and inducing apoptosis, at the moment, not much data are available on the mechanism(s) implicated. The most important targets/pathways implicated in the regulation of cell growth by lycopene, as reported from the recent data of the literature, seem to be cyclin D1 and PI3K/Akt pathway.

Cyclin D1

Functions

Several proteins are known to regulate the timing of the events in the cell cycle. Major control switches of the cell cycle are the cyclins and the cyclin-dependent kinases. Among them, cyclin D1, a component subunit of cyclin-dependent kinase (Cdk)4 and Cdk6, is a rate limiting factor in progression of cells through the first gap (G1) phase of the cell cycle. It is the regulatory subunit of the holoenzymes that phosphorylate and, together with sequential phosphorylation by cyclin E/CDK2, inactivate the cell-cycle inhibiting function of the retinoblastoma protein. Retinoblastoma protein serves as a gatekeeper of the G_1 phase, and passage through the restriction point leads to DNA synthesis.

In addition to their CDK-binding function, a body of evidence now indicates that D-type cyclins have CDK-independent properties (Wang et al. 2004, Dey and Li 2000, Pestell et al. 1999). As previously proposed (Pestell et al. 1999), the role of these properties in cellular growth, metabolism, and cellular differentiation is substantial. Cyclin D1 forms physical associations with more than 30 transcription factors or transcriptional co-regulators (Wang et al. 2004, Zhang et al. 1999, Horstmann et al. 2000).

Regulation

Cyclin D1 is induced by growth factors, including epithelial growth factor and insulin-like growth factors (IGF) I (Albanese et al. 1999) and II (Holnthoner et al. 2002), amino acids (Nelsen et al. 2003), lysophosphatidic acid (Hu et al. 2003), hormones (Fu et al. 2003), retinoic acid (Suzui et al. 2002), and peroxisome proliferator-activated receptor (PPAR)γ ligand (Wang et al. 2003), secreted factors from adipocytes (Iyengar et al. 2003), and gastrointestinal hormones such as gastrin. Each of them regulates cyclin D1 expression in specific cell types (Song et al. 2003, Pradeep et al. 2004). Many oncogenic signals induce cyclin D1 expression and do so through distinct DNA sequences in the cyclin D1 promoter including Ras (Albanese et al. 1995), Src (Lee et al. 1999), ErbB2 (Lee et al. 2000), β-catenin (Shtutman et al. 1999, Lin et al. 2000), oncogenic signal transducer and activator of transcriptions (Bromberg et al. 1999, Kim et al. 2002b, Matsumura et al. 1999), and simian virus 40 small t antigen, a particularly crucial oncogene in transformation of human cells (Watanabe et al. 1996, Baldin et al. 1993, Sherr 2003). Deregulation of the cell cycle check points and over-expression of growth-promoting cell cycle factors are associated with tumorigenesis. Cyclin D1 has been shown to be over-expressed in many cancers including cancers of breast, esophagus, head, neck and prostate (Aggarwal and Shishodia 2006).

The levels of cyclin D1 are regulated by phosphorylation-dependent proteolytic degradation through PI3K pathway. Over-expression of constitutively active Akt (MyrAkt) was shown to extend the half-life of cyclin D1 protein, whereas treatment with the PI3K inhibitor Wortmannin accelerated cyclin D1 degradation (Diehl et al. 1998). Akt also promotes transcriptional activation of cyclin D1 levels (Ahmed et al. 1997). Moreover, translation of cyclin D1 mRNA has been shown to be dependent on the activation of PI3K/AKT/eIF4e pathway.

Modulation of Cyclin D1 by Lycopene

Although these studies demonstrated a clear inhibition of cell cycle progression by lycopene, at the moment, not much data are available on the mechanism(s) implicated. The available evidence seems to indicate that lycopene may act as a potent modulator of cyclin D1 in both normal and tumor cultured cells. Moreover, a modulation of cyclin D1 by lycopene in cells exposed *in vitro* and *in vivo* to cigarette smoke has been recently demonstrated.

Lycopene was able to inhibit the growth of mammary (MCF-7 and T-47D) and endometrial (ECC-1) cancer cells through down-regulation of cyclin D1 and cyclin D3. It has been suggested that the reduction in cyclin D1 levels by lycopene has two consequences. The first is a direct effect causing reduction in cyclin D-cdk4 complexes resulting in a decrease of both cdk4 and cdk2 kinase activity and in a reduction of the hypophosphorylation of retinoblastoma protein. The second is a retention of p27 in cyclin E-cdk2 complexes, an indirect effect that leads to the inhibition of cdk2 activity (Nahum et al. 2001).

In a recent study, LNCaP and PC3 prostate cancer cells treated with lycopene-based agents have been reported to undergo mitotic arrest. Immunoblot screening indicated that lycopene's antiproliferative effects are likely achieved through a block in G1/S transition mediated by decreased levels of cyclins D1 and E and cyclin-dependent kinase4 and suppressed retinoblastoma phosphorylation (Ivanov et al. 2007).

We recently reported that tomato added to cultured colon (HT-29 and HCT-116) cancer cells by an *in vitro* digestion procedure was able to induce an arrest of cell cycle progression at the G0/G1 phase. Such an effect was accompanied by a dose-dependent decrease in the expression of cyclin D1.

Although tomato digestates contain a complex mix of compounds besides lycopene, including a large variety of micronutrients and microconstituents, such as polyphenols and other non pro-vitamin A carotenoids, this observation seems to support the notion that lycopene may be a molecule extremely important in the regulation of intracellular levels of cyclin D (Palozza, unpublished data).

Similarly, lycopene was also able to inhibit cell cycle progression at the G0/G1 phase and to reduce cell proliferation by a mechanism involving cyclin D1 in normal cells. It has been reported that, after stimulation of synchronized human normal prostate epithelial cells (PrEC) with growth factors, cyclin D1 protein expression increased in lycopene-untreated cells. Such an increase was lower or even absent following treatment with lycopene at the concentration of 0.5 mmol/L and 5.0 mmol/L, respectively. Interestingly, such an effect was specific for cyclin D1, since cyclin E levels remained constant and were unaffected by lycopene treatment (Obermüller-Jevic et al. 2003).

Moreover, we recently reported that lycopene was able to enhance the arrest of cell cycle progression induced by cigarette smoke condensate (TAR) in RAT-1 immortalized fibroblasts. TAR-exposed cells treated with lycopene showed a delay in cell cycle at the G0/G1 phase and a concomitant reduction in S phase. Such effects were accompanied by a dose-dependent decrease in cyclin D1 levels. On the other hand, fibroblasts treated with lycopene alone showed the same effects, although to a lower extent. The down-regulation of cyclin D1 observed in this study was dose-dependent and occurred at lycopene concentration achievable *in vivo* after carotenoid supplementation (Palozza et al. 2005).

According with these *in vitro* studies, treatment with lycopene *in vivo* has also been reported to induce modulatory effects on cyclin D1 expression. It has been reported that smoke exposure substantially decreased the levels of p21[Waf1/Cip1] and increased those of cyclin D1 and proliferating cellular nuclear antigen (PCNA) in gastric mucosa from ferrets. Supplementation of ferrets with either low or high doses of lycopene prevented the changes in p21[Waf1/Cip1], cyclin D1 and PCNA caused by smoke exposure in a dose-dependent fashion (Liu et al. 2006).

Although further studies are needed to clarify the mechanism(s) of lycopene interference with cell signaling leading to down-regulation of cyclin D1 and ultimately to cell cycle arrest in both normal and tumor cells, these reports suggest that the reduction in cyclin D1 by lycopene treatment may be a key event in the ability of the carotenoid to arrest cell cycle progression.

PI3K/Akt Pathway

Activation

The serine/threonine protein kinase Akt is the cellular homolog of the viral oncogene v-Akt (Bellacosa et al. 1991). Tyrosine kinase and G-protein coupled receptors activate Class 1A and Class 1B PI3-kinases (Wymann et al. 2003), which phosphorylate PIP2 at the 3-OH group, generating the second messenger PIP3. PIP3 levels are tightly regulated by the action of phosphatases such as phosphatase and tensin homolog (PTEN), which removes phosphate from the 3-OH position (Simpson and Parsons 2001). PIP3 does not activate Akt directly but instead appears to recruit Akt to the plasma membrane and to alter its conformation to allow subsequent phosphorylation at Thr308 by the phosphoinositide-dependent kinase-1 (PDK1) or at Ser473 by PDK2 (Alessi et al. 1997, Stokoe et al. 1997, Stephens et al. 1998). It has been reported that Akt is also activated by cellular stress and heat shock through association with Hsp27 (Konishi et al. 1997). Many receptors, including those for cytokines [interleukin-3 (IL-3), IL-2], neurotrophic factors (nerve growth factor), brain-derived neurotrophic factor, and growth factors (IGF-1, platelet-derived growth factor), transmit survival signals through the Akt pathway (Talapatra and Thompson 2001).

Deregulation

Increasing evidence shows that Akt deregulation is associated with the development of several chronic diseases, including CVD and cancer (Vivanco and Sawyers 2002, Shiojima and Walsh 2006).

It has been demonstrated that Akt-mediated growth-promoting signals in cardiac muscle cells are able to enhance angiogenesis in the heart in a paracrine manner (Shiojima and Walsh 2006). Akt activation in skeletal muscle cells also results in myofiber growth associated with enhanced vascular endothelial growth factor secretion and induces blood vessel recruitment (Takahashi et al. 2006).

The over-expression of Akt isoforms has been described in breast, colon, ovarian, pancreatic, prostate and bile duct cancers (Vivanco and Sawyers 2002, Kandel and Hay 1999, Roy et al. 2002, Altomare et al. 2003, Tanno et al. 2004). Samuels et al. (2004) described also gene mutations of PIK3CA, which encodes one of the P110 α catalytic subunits of PI3K, in human colon, gastric, breast, and lung cancer and glioblastoma. Moreover, several studies have shown gene amplification of PIK3C in human ovarian cancer (Shayasteh et al. 1999), cervical cancer (Ma et al. 2000), head and neck cancer (Woenckhaus et al. 2002), gastric cancer (Byun et al. 2004), and glioblastoma (Knobbe and Reifenberger 2003). In addition, mutation of

PTEN, which dephosphorylates PI(3,4,5)P3 and down-regulates PI3K activation, has been reported in primary cancers in thyroid, breast, colon, prostate, uterus, central nervous system and soft tissue (Nassif et al. 2004, Frisk et al. 2002, Garcia et al. 1999). Finally, Akt activity promotes resistance to chemo- and radiotherapy (Tanno et al. 2004, Clark et al. 2002, Brognard et al. 2001). Therefore, the mechanism of Akt activation remains a target for researchers in the hunt for suitable options for cancer therapy.

Role in Cell Survival and in Apoptosis

The control of Akt on cell survival can occur through two mechanisms. The first involves a regulation of cell cycle-related proteins and the second involves the modulation of apoptosis-related proteins, as summarized in Fig. 1. It has been reported that Akt can inhibit cell cycle arrest by phosphorylating p21 (Zhou et al. 2001). Furthermore, Akt may enhance cell cycle progression by suppression of AFX forkhead transcription factor activity resulting in diminished expression of AFX target genes such as p27kipi (Medema et al. 2000). Also, Akt may potentiate cell cycle progression through increased translation of cyclin D (Muise-Helmericks et al. 1998). Interestingly, Akt indirectly regulates the tumor suppressor p53 protein, which act as a sensor of cellular stress, and transduces stress signals into apoptotic signals (Evan and Vousden 2001).

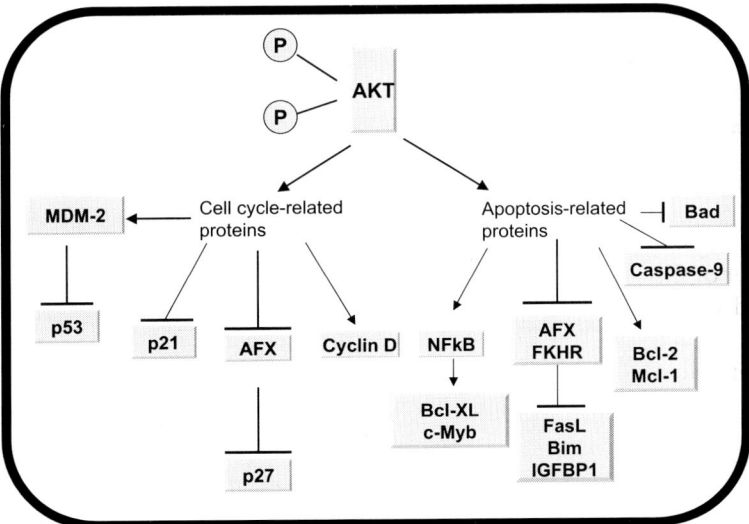

Fig. 1 Effects of AKT phosphorylation on cell cycle- and apoptosis-related proteins. ↓ = activation; ⊥ = inhibition.

Akt regulates the apoptotic machinery by regulating transcription factors and pro-apoptotic proteins. Phosphorylation of the transcription factors (AFX, FKHR and FKHRL1) by Akt inhibits transcription of pro-apoptotic genes such as FasL, IGFBP-1 and Bim (Datta et al. 1999, Nicholson and Anderson 2002). Akt can regulate the activation of nuclear factor-kappaB (NFkappaB) (Ozes et al. 1999, Kane et al. 1999), which, in turn, regulates the expression of pro-survival genes, including Bcl-XL, caspase inhibitors and c-Myb (Barkett and Gilmore 1999, Lauder et al. 2001). In addition, Akt phosphorylates and activates the cyclic AMP-response element-binding protein (Du and Montminy 1998), which increases the transcription of anti-apoptotic genes, such as those for Bcl-2 (Pugazhenthi et al. 2000), Mcl-1 (Wang et al. 1999, Osaki et al. 2004) and Akt itself (Reusch and Klemm 2002). Recent results suggest that Akt also promotes cell survival through the inactivation of caspase 9, downstream of cytochrome C release (Franke et al. 1998). However, one of the most known components of the apoptotic machinery found to be phospho-rylated by AKT is Bad .

Bad

The BCL-2 family member Bad is similar to Bcl-2 exclusively within its BH3 domain (Yang et al. 1995). It directly interacts via its BH3 domain with pro-survival Bcl-2 family members such as Bcl-XL and when over-expressed blocks Bcl-XL-dependent cell survival (Yang et al. 1995, Ottilie et al. 1997, Zha et al. 1997). Disruption of the Bad BH3 domain by a point mutation or deletion abrogates the ability of Bad, both to interact with Bcl-XL and to induce cell death (Zha et al. 1997). Korsmeyer and colleagues demonstrated that the activity of Bad was modulated by the addition of growth factors to GF-deprived cells (Zha et al. 1996). In the absence of survival factors, Bad is phosphorylated and associates with Bcl-XL, as shown in Fig. 2. The exposure of cell lines to GFs induces the phosphorylation of Bad at two sites, Ser-112 and Ser-136. Phosphorylation of either of these sites causes Bad to dissociate from Bcl-XL and to associate instead with cytoplasmic 14-3-3 proteins, adapter proteins that interact with a variety of signaling molecules in a phosphorylation-dependent manner (Muslin et al. 1996, Yaffe et al. 1997). Association of Bad with 14-3-3 proteins may protect Bad from dephosphorylation or sequester Bad away from its targets at the mitochondria. Thus, phosphorylation of Bad inactivates its ability to cause cell death and promotes cell survival. Conversely, dephosphorylation of Bad results in targeting of Bad to mitochondrial membranes where Bad has been proposed to interact with and inactivate survival-promoting Bcl-2 family members and thereby induce cell death.

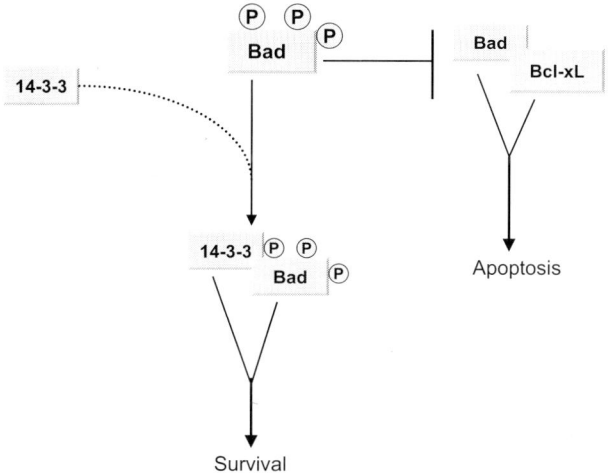

Fig. 2 Scheme representing the effects of Bad phosphorylation on cell survival and apoptosis. ↓ = activation; ⊥ = inhibition.

Modulation of PI3K/Akt Pathway and Bad by Lycopene

Cancer studies

Among the factors that activate PI3K/Akt pathway, one of the most important is the family of insulin-like growth factors. The IGFs play a pivotal role in regulating mitogenic and apoptotic pathways (Yu and Rohan 2000). Several lines of evidence implicate IGF-1 and its receptor, IGF-1R, in lung cancer and other malignancies (Yu and Rohan 2000, Giovannucci 1999b). Up-regulation of IGF-1 accompanies tumor progression in the TRAMP mouse prostate cancer model (Kaplan et al. 1999). Also, IGF-1 over-expression in the prostate epithelium causes prostatic intraepithelial neoplasia in transgenic mice (Di Giovanni et al. 2000). It is known that IGFBP-3, one of the six members of the IGBP family and a major circulating protein in human plasma (Giovannucci 1999b, Rechler 1997), regulates bioactivity of IGF-1 by sequestering it away from its receptor in the extracellular milieu, thereby inhibiting the mitogenic and anti-apoptotic action of IGF-1. IGFBP-3 also inhibits cell growth and induces apoptosis independent of IGF-1 and its receptors (Liu et al. 2000, Oh et al. 1993). It also acts as a potent inhibitor of both PI3K/AKT and MAPK signalling pathways (Lee et al. 2002). Low blood IGFBP-3 concentrations and high IGF-1 concentrations and IGF-1/IGFBP-3 ratios have been implicated in a variety of cancers, including lung cancer (Karas et al. 2000, Hochscheid et al. 2000). Serum levels of IGFBP-3 have been also negatively correlated with the number of cigarettes smoked per day or pack-year history of smoking (Kaklamani et al. 1999).

Available data suggest that lycopene and tomato products are able to modulate IGF-1 pathway *in vitro* as well as in *in vivo* models. Clinical data show that higher intake of cooked tomatoes or lycopene is significantly associated with lower circulating levels of IGF-1 (Mucci et al. 2001) and higher levels of IGFBP-3 (Holmes et al. 2002). Moreover, it has been demonstrated that, in the MatLyLu Dunning prostate cancer model, IGF-1 expression was decreased locally in prostate tumors by lycopene supplementation (Siler et al. 2004). In addition, lycopene treatment has been reported to strongly reduce the IGF-1 stimulation of activator protein 1 binding in MCF-7 mammary cancer cells and such an effect was associated with a delayed G1-S cell cycle progression (Karas et al. 2000). Lycopene has also been reported to significantly increase the levels of IGFBP-3 in PC-3 prostate cancer cells. Such an inhibition was accompanied by a decrease in cell proliferation and by an increase in apoptosis induction (Kanagaraj et al. 2007). Interestingly, in a recent study, the arrest of cell cycle and the decrease in cyclins D1 and E induced by lycopene in androgen-responsive LNCaP and androgen-independent PC3 prostate cancer cells correlated with decreased IGF-1R expression and activation, increased IGFBP-2 expression and decreased Akt activation, further confirming that lycopene can suppress PI3K-dependent proliferative and survival signaling (Du and Montminy 1998). A scheme of the modulatory effects of lycopene on IGF-1 and AKT pathways is reported in Fig. 3.

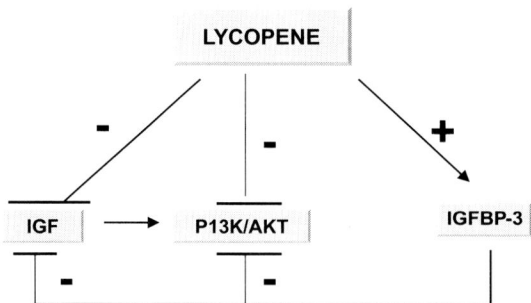

Fig. 3 Modulatory effects of lycopene on IGF, IGFBP-3 and PI3K/AKT pathway. ↓ = activation; ⊥ = inhibition.

Two recent studies seem to demonstrate that lycopene is able to counteract the dangerous effects of smoke by acting through a mechanism involving IGF-1 and/or AKT pathways. In the lung of ferrets, it has been shown that cigarette smoke–induced lesions (e.g., squamous metaplasia,

PCNA over-expression, and diminished apoptosis) were associated with reduced plasma IGFBP-3 concentrations and increased IGF-1/IGFBP-3 ratios. Such changes significantly affected the status of cell proliferation and apoptosis in the lung of ferrets. Smoke exposure significantly decreased cleaved caspase-3 protein and increased PCNA. Furthermore, smoke exposure suppressed Bad-mediated apoptosis by inducing the phoshorylation of Bad at both Ser136 and Ser112. These smoke-induced changes were prevented by lycopene supplementation in a dose-dependent manner. The carotenoid was able to increase IGFBP-3 levels and to decrease IGF-1/IGFBP-3 ratio. Moreover, it decreased Bad phosphorylation at both Ser136 and Ser112 and increased cleaved caspase-3, preventing cigarette smoke–induced squamous metaplasia and the increase in PCNA (Liu et al. 2003).

A recent *in vitro* study also suggests that the modulation of AKT pathway may have a key role in the pro-apoptotic effects of lycopene under smoke conditions (Palozza et al. 2005). In fact, while RAT-1 fibroblasts exposed to cigarette smoke condensate (tar) exhibited high levels of phosphorylated AKT, cells exposed to a combination of tar and lycopene strongly decreased them. Moreover, the exposition of RAT-1 fibroblasts to tar alone suppressed Bad-mediated apoptosis by inducing the phosphorylation of Bad at Ser136. Conversely, lycopene was able to completely prevent the phosphorylation of Bad induced by tar, confirming *in vitro* the results obtained *in vivo* by Liu et al. (Liu et al. 2003). In our laboratory, similar results have been recently found in the human prostate DU-145 cancer cells exposed to lycopene in association with tar. The carotenoid was able to prevent tar-induced AKT (Fig. 4) and Bad (Fig. 5) phosphorylation (unpublished data). Moreover, in the same study, the expression of the heat shock protein (hsp)90 was increased following tar exposure (Palozza et al. 2005). Such an increase was counteracted by lycopene. This finding is particularly interesting in view of a previous report showing that hsp90 maintains Akt activity by binding to Akt and by preventing PP2A-dependent dephosphorylation of Akt (Sato et al. 2000). Moreover, hsp90 has been reported to prevent proteasome-dependent degradation of PDK1, which is known to activate Akt (Fujita et al. 2002). On the other hand, the finding that lycopene is able to counteract the effect of tar on hsp90 is not surprising in view of the fact that heat shock proteins increase as a consequence of oxidative stress, including smoke (Pinot et al. 1997, Hunt 1986) and that lycopene acts as a potent antioxidant (Conn et al. 1991, Di Mascio et al. 1989). The modulation of hsp90 by lycopene under smoke conditions could be a further suggestive intracellular mechanism to explain the modulatory activity of lycopene on Bad.

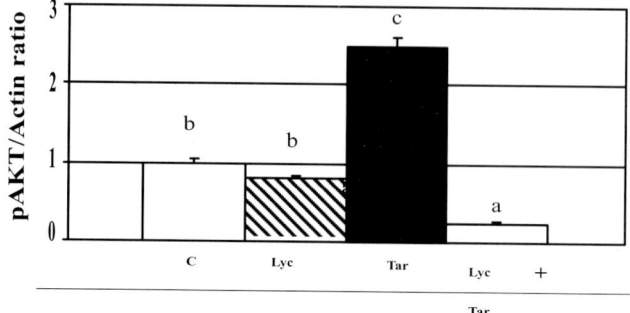

Fig. 4 Effects of a 12 hr treatment with lycopene and cigarette smoke condensate (tar), alone and in combination, on the expression of pAKT in DU-145 cancer cells. Tar was added at the concentration of 25 µg/ml and the carotenoid at the concentration of 1 µM. The values were the means ± SEM, n = 5. Values not sharing the same letter were significantly different (P < 0.05, Fisher test).

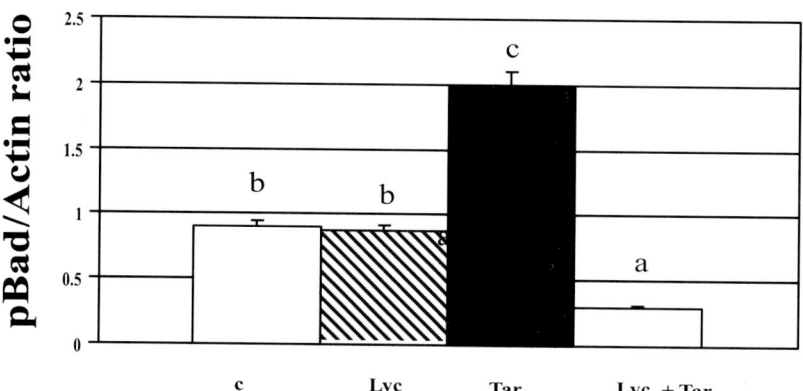

Fig. 5 Effects of a 12 hr treatment with lycopene and cigarette smoke condensate (tar), alone and in combination, on the expression of pBad in DU-145 cancer cells. Tar was added at the concentration of 25 µg/ml and the carotenoid at the concentration of 1 µM. The values were the means ± SEM, n = 5. Values not sharing the same letter were significantly different (P < 0.05, Fisher test).

CVD studies

Very little is known about the specific biological mechanisms through which lycopene can protect against atherosclerosis and CVD. Although many of the biological effects and health benefits of lycopene are hypothesized to occur via protection against oxidative damage, the direct modulation of AKT pathway might be one mechanism through which the carotenoid exerts its protective effects (Lee et al. 2006). It has been recently reported that the activation of tissue factor plays a key role in vascular thrombosis (Levi et al. 1999) and that vessel injury results in an increase of tissue factor expression and activation. Moreover, previous authors demonstrated that the activation of PI3K pathway suppressed tissue factor expression in endothelial and monocytic cells (Blum et al. 2001, Guha et al. 2002). In a recent study, water-dispersible forms of various carotenoids (β-carotene, lutein and lycopene) from natural sources in microemulsion were used to study effects of carotenoids on tissue factor activity in human endothelial cells. All carotenoids studied, including lycopene, suppressed tissue factor activity and gene expression in human endothelial cells. Interestingly, such a study also demonstrated that addition of an Akt-specific inhibitor reversed the effect of carotenoids on tissue factor activity, indicating that carotenoids enhanced phosphorylation of Akt and suppressed tissue factor activity in endothelial cells by this mechanism.

CONCLUSIONS

Although there is epidemiological evidence to support the beneficial effects of lycopene, tomato and tomato products in the prevention of chronic diseases, the mechanism(s) responsible for such beneficial effects has not yet been completely clarified. The up-regulation of cyclin D1 and the deregulation of PI3k/Akt pathway have been largely associated with the pathogenesis and development of cancer and CVD. As discussed throughout this review, increasing evidence shows interference of lycopene with cyclin D1 and PI3K/Akt pathway. Lycopene interference with cell signaling leading to down-regulation of cyclin D1 and ultimately to cell cycle arrest may be a key event by which the carotenoid arrests tumor cell growth. Moreover, lycopene interference with PI3K/Akt pathway may suppress PI3K-dependent proliferative and survival signaling in tumor cells or in cells exposed to oxidative stress, such as cigarette smoke, or may modify the expression of tissue factors regulating endothelial functions. Despite these promising reports, further work is needed to better understand the interactions between lycopene and cyclin D1/PI3K/Akt pathway, in particular, knowledge of dose-effects, tissue-specific effects and lycopene metabolites effects. This information is critically needed to improve the knowledge of strategies in the prevention of chronic diseases.

ABBREVIATIONS

AFX: forkhead transcription factor; (C/EBP)β: CCAAT/enhancer binding protein; CDK: cyclin-dependent kinase; CVD: cardiovascular diseases; FKHR: forkhead homologue in rhabdomyosarcoma; FKHRL1: FKHR-like protein 1; hsp: heat shock protein; IGF: insulin growth factor;

IGFBP1: insulin-like growth factor binding protein 1; PCNA: proliferating cellular nuclear antigen; PDK: phosphoinositide-dependent kinase; PH: pleckstrin homology; PI(3,4,5)P3: phosphatidylinositol 3,4,5-trisphosphate; PI3K: phosphatidil inositol-3-kinase; PP2A: type 2 protein serine/threonine phosphatase; PPARγ: peroxisome proliferator-activated receptor γ; PTEN: phosphatase and tensin homolog; TAR: tobacco smoke condensate; TGFβ: tumor growth factor β.

References

Aggarwal, B.B. and S. Shishodia. 2006. Molecular targets of dietary agents for prevention and therapy of cancer. Bioc. Pharm. 71: 1397-1421.

Ahmed, N.N. and H.L. Grimes, A. Bellacosa, T.O. Chan, and P.N. Tsichlis. 1997. Transduction of interleukin-2 antiapoptotic and proliferative signals via Akt protein kinase. PNAS USA 94: 3627-3632.

Albanese, C. and J. Johnson, G. Watanabe, N. Eklund, D. Vu, A. Arnold, and R.G. Pestell. 1995. Transforming p21^{ras} mutants and c-Ets-2 activate the cyclin D1 promoter through distinguishable regions. J. Biol. Chem. 270: 23589-23597.

Albanese, C. and M. D'Amico, A.T. Reutens, M. Fu, G. Watanabe, R.J. Lee, R.N. Kitsis, B. Henglein, M. Avantaggiati, K. Somasundaram, B. Thimmapaya, and R.G. Pestell. 1999. Activation of the cyclin D1 gene by the E1A-associated protein p300 through AP-1 inhibits cellular apoptosis. J. Biol. Chem. 274: 34186-34195.

Alessi, D.R. and M. Deak, A. Casamayor, F.B. Caudwell, N. Morrice, D.G. Norman, P. Gaffney, C.B. Reese, C.N. MacDougall, D. Harbison, A. Ashworth, and M. Bownes. 1997. 3-Phosphoinositide-dependent protein kinase-1 (PDK1): structural and functional homology with the Drosophila DSTPK61 kinase. Curr. Biol. 7: 776-789.

Altomare, D.A. and S. Tanno, A. De Rienzo, A.J. Klein-Szanto, S. Tanno, K.L. Skele, J.P. Hoffman, and J.R. Testa. 2003. Frequent activation of AKT2 kinase in human pancreatic carcinomas. J. Cell. Biochem. 88: 470-476.

Amir, H. and M. Karas, G. Giat, M. Danilenko, R. Levy, T. Yermiahu, J. Levy, and Y. Sharoni. 1999. Lycopene and 1,25-dihydroxyvitamin D3 cooperate in the inhibition of cell cycle progression and induction of differentiation in HL-60 leukemic cells. Nutr. Cancer 33: 105-112.

Astrog, P. and S. Gradelet, R. Berges, and M. Suschetet. 1997. Dietary lycopene decreases initiation of liver preneoplastic foci by diethylnitrosamine in rat. Nutr. Cancer 29: 60-68.

Baldin, V. and J. Lukas, M.J. Marcote, M. Pagano, and G. Draetta. 1993. Cyclin D1 is a nuclear protein required for cell cycle progression in G1. Genes Dev. 7: 812-821.

Barkett, M. and T.D. Gilmore. 1999. Control of apoptosis by Rel/NF-κB transcription factors. Oncogene 18: 6910-6924.

Bellacosa, A. and J.R. Testa, S.P. Staal, and P.N. Tsichlis. 1991. A retroviral oncogene, akt, encoding a serine-threonine kinase containing an SH2-like region. Science 254: 274-277.

Blum, S. and K. Issbrüker, A. Willuweit, S.Hehlgans, M. Lucerna, D. Mechtcheriakova, K. Walsh, D. von der Ahe, E. Hofer, and M. Clauss. 2001. An inhibitory role of the PI3K signalling pathway in vascular endothelial growth factor-induced tissue factor expression. J. Biol. Chem. 276: 33428-33434.

Brognard, J. and A.S. Clark, Y. Ni, and P.A. Dennis. 2001. Akt/protein kinase b is constitutively active in non-small cell lung cancer cells and promotes cellular survival and resistance to chemotherapy and radiation. Cancer Res. 61: 3986-3997.

Bromberg, J.F. and M.H. Wrzeszczynska, G. Devgan, Y. Zhao, R.G. Pestell, C. Albanese, and J.E. Darnell Jr. 1999. Stat3 as an oncogene. Cell 98: 295-303.

Byun, D.S. and K. Cho, B.K. Ryu, M.G. Lee, J.I. Park, K.S. Chae, H.J. Kim, and S.G. Chi. 2003. Frequent monoallelic deletion of PTEN and its reciprocal association with PIK3CA amplification in gastric carcinoma. Int. J. Cancer 104: 318-327.

Canene-Adams, K. and B.L. Lindshield, S. Wang, E.H. Jeffery, S.K. Clinton, and J.W. Erdman Jr. 2007. Combinations of tomato and broccoli enhance antitumor activity in dunning r3327-h prostate adenocarcinomas. Cancer Res. 67: 836-843.

Chalabi, N. and L. Delort, L. Le Corre, S. Satih, Y.J. Bignon, and D. Bernard-Gallon. 2006. Gene signature of breast cancer cell lines treated with lycopene. Pharmacogenomics 7: 663-672.

Chalabi, N. and S. Satih, L. Delort, Y.J. Bignon, and D.J. Bernard-Gallon. 2007. Expression profiling by whole-genome microarray hybridization reveals differential gene expression in breast cancer cell lines after lycopene exposure. Biochim. Biophys. Acta 1769: 124-130.

Clark, A.S. and K. West, S. Streicher, and P.A. Dennis. 2002. Constitutive and inducible Akt activity promotes resistance to chemotherapy, trastuzumab, or tamoxifen in breast cancer cells. Mol. Cancer Ther. 1: 707-717.

Conn, P.F. and W. Schalch, and T.G. Truscott. 1991. The singlet oxygen and carotenoid interaction. Photochem. Photobiol. 11B: 41-47.

Datta, S.R. and A. Brunet, M.E. Greenberg. 1999. Cellular survival: A play in three Akts. Genes Dev. 13: 2905-2927.

Dey, A. and W. Li. 2000. Cell cycle-independent induction of D1 and D2 cyclin expression, but not cyclin-Cdk complex formation or Rb phosphorylation, by IFNγ in macrophages. Biochim. Biophys. Acta 1497: 135-147.

Di Giovanni, J. and K. Kiguchi, A. Frijhoff, E. Wilker, D.K. Bol, L. Beltran, S. Moats, A. Ramirez, J. Jorcano, and C. Conti. 2000. Deregulated expression of insulin-like growth factor 1 in prostate epithelium leads to neoplasia in transgenic mice. Proc. Natl. Acad. Sci. USA 97: 3455-3460.

Di Mascio, P. and S. Kaiser, and H. Sies. 1989. Lycopene as the most efficient biological carotenoid singlet oxygen quencher. Arch. Biochem. Biophys. 274: 532-538.

Diehl, J.A. and M. Cheng, M.F. Roussel, and C.J. Sherr. 1998. Glycogen synthase kinase-3beta regulates cyclin D1 proteolysis and subcellular localization. Genes Dev. 12: 3499-3511.

Du, K. and M. Montminy. 1998. CREB is a regulatory target for the protein kinase Akt/PKB. J. Biol. Chem. 273: 32377-32379.

Evan, G.I. and K.H. Vousden. 2001. Proliferation, cell cycle and apoptosis in cancer. Nature 411: 342-348.

Franke, T.F. and E. Stanbridge, S. Frisch, and J.C. Reed. 1998. Regulation of cell death protease caspase-9 by phosphorylation. Science 282: 1318-1321.

Frisk, T. and T. Foukakis, T. Dwight, J. Lundberg, A. Hoog, G. Wallin, C. Eng, J. Zedenius, and C. Larsson. 2002. Silencing of the PTEN tumor-suppressor gene in anaplastic thyroid cancer. Genes Chromosomes Cancer 35: 74-80.

Fu, M. and M. Rao, C. Wang, T. Sakamaki, J. Wang, D. Di Vizio, X. Zhang, C. Albanese, S. Balk, C. Chang, S. Fan, E. Rosen, J.J. Palvimo, O.A. Janne, S. Muratoglu, M.L. Avantaggiati, and R.G. Pestell. 2003. Acetylation of androgen receptor enhances coactivator binding and promotes prostate cancer cell growth. Mol. Cell. Biol. 23: 8563-8575.

Fuhramn, B. and A. Elis, and M. Aviram. 1997. Hypocholesterolemic effect of lycopene and β-carotene is related to suppression of cholesterol synthesis and augmentation of LDL receptor activity in macrophage. Biochem. Biophys. Res. Commun. 233: 658-662.

Fujita, N. and S. Sato, A. Ishida, and T. Tsuruo. 2002. Akt-dependent phosphorylation of p27Kip1 promotes binding to 14-3-3 and cytoplasmic localization. J. Biol. Chem. 277: 10346-10353.

Gann, P. and J. Ma, E. Giovannucci, W. Willett, F.M. Sacks, C.H. Hennekens, and M.J. Stampfer. 1999. Lower prostate cancer risk in men with elevated plasma lycopene levels: results of a prospective analysis. Cancer Res. 59: 1225-1230.

Garcia, J.M. and J.M. Silva, G. Dominguez, R. Gonzalez, A. Navarro, L. Carretero, M. Provencio, P. Espana, and F. Bonilla. 1999. Allelic loss of the PTEN region (10q23) in breast carcinomas of poor pathophenotype. Breast Cancer Res Treat. 57: 237-243.

Giovannucci, E. 1999a. Tomatoes, tomato-based products, lycopene, and cancer: review of the epidemiologic literature. J. Natl. Cancer Inst. 91: 317-331.

Giovannucci, E. 1999b. Insulin-like growth factor-I and binding protein-3 and risk of cancer. Horm. Res. 51 (Suppl 3): 34-41.

Giovannucci, E. and A. Ascherio, E.B. Rimm, M.J. nd retinol in relation to risk of prostate cancer. J. Natl. Cancer Inst. 87: 1767-1776.

Guha, M. and N. Mackman. 2002. The PI3K-Akt pathway limits lipopolysaccharide activation of signaling pathway and expression of inflammatory mediators in human monocytic cells. J. Biol. Chem. 277: 32124-32132.

Hantz, H.L. and L.F. Young, and K.R. Martin. 2005. Physiologically attainable concentrations of lycopene induce mitochondrial apoptosis in LNCaP human prostate cancer cells. Exp. Biol. Med. (Maywood) 230: 171-179.

Helzlsouer, K.J. and G.W. Comstock, and J.S. Morris. 1989. Selenium, lycopene, α-tocopherol, β-carotene, retinol, and subsequent bladder cancer, Cancer Res. 49: 6144-6148.

Hochscheid, R. and G. Jaques, and B. Wegmann. 2000. Transfection of human insulin-like growth factor-binding protein 3 gene inhibits cell growth and tumorigenicity: a cell culture model for lung cancer. J. Endocrinol. 166: 553-563.

Holmes, M.D. and M.N. Pollak, W.C. Willett, and S.E. Hankinson. 2002. Dietary correlates of plasma insulin-like growth factor I and insulin-like growth factor-binding protein 3 concentrations. Cancer Epidemiol. Biomark. Prev. 11: 852-861.

Holnthoner, W. and M. Pillinger, M. Groger, K. Wolff, A.W. Ashton, C. Albanese, P. Neumeister, R.G. Pestell, and P. Petzelbauer. 2002. Fibroblast growth factor-2 induces Lef/Tcf-dependent transcription in human endothelial cells. J. Biol. Chem. 277: 45847-45853.

Horstmann, S. and S. Ferrari, and K.H. Klempnauer. 2000. Regulation of B-Myb activity by cyclin D. Oncogene 19: 298-306.

Hu, Y.L. and C. Albanese, R.G. Pestell, and R.B. Jaffe. 2003. Dual mechanisms for lysophosphatidic acid stimulation of human ovarian carcinoma cells. J. Natl. Cancer Inst. 95: 733-740.

Hunt, L.A. 1986. Sidestream cigarette smoke-exposure of mouse cells induces stress/heat shock-like proteins. Toxicology 39: 259-273.

Hwang, E.S. and P.E. Bowen. 2004. Cell cycle arrest and induction of apoptosis by lycopene in LNCaP human prostate cancer cells. J. Med. Food. 7: 284-289.

Hwang, E.S. and P.E. Bowen. 2005. Effects of tomato paste extracts on cell proliferation, cell-cycle arrest and apoptosis in LNCaP human cancer cells. Biofactors 23: 75-84.

Ivanov, N.I. and S.P. Cowell, P. Brown, P.S. Rennie, E.S. Guns, and M.E. Cox. 2007. Lycopene differentially induces quiescence and apoptosis in androgen-responsive and -independent prostate cancer cell lines. Clin. Nutr. 26: 2522-2563.

Iyengar, P. and T.P. Combs, S.J. Shah, V. Gouon-Evans, J.W. Pollard, C. Albanese, L. Flanagan, M.P. Tenniswood, C. Guha, M.P. Lisanti, R.G. Pestell, and P.E. Scherer. 2003. Adipocyte-secreted factors synergistically promote mammary tumorigenesis through induction of anti-apoptotic transcriptional programs and proto-oncogene stabilization. Oncogene 22: 6408-6423.

Kaklamani, V.G. and A. Linos, E. Kaklamani, I. Markaki, and C. Mantzoros. 1999. Age, sex, and smoking are predictors of circulating insulin-like growth factor1 and insulin-like growth factor-binding protein 3. J. Clin. Oncol. 17: 813-817.

Kanagaraj, P. and M.R. Vijayababu, B. Ravisankar, J. Anbalagan, M.M. Aruldhas, and J. Arunakaran. 2007. Effect of lycopene on insulin-like growth factor-I, IGF binding protein-3 and IGF type-I receptor in prostate cancer cells. J. Cancer Res. Clin. Oncol. 133: 351-359.

Kandel, E.S. and N. Hay. 1999. The regulation and activities of the multifunctional serine/threonine kinase Akt/PKB. Exp. Cell. Res. 253: 210-229.

Kane, L.P. and V.S. Shapiro, D. Stokoe, and A. Weiss. 1999. Induction of NF-κB by the Akt/PKB kinase. Curr. Biol. 9: 601-604.

Kaplan, P.J. and S. Mohan, P. Cohen, B.A. Foster, and N.M. Greenberg. 1999. The insulin-like growth factor axis and prostate cancer: lessons from the transgenic adenocarcinoma of mouse prostate (TRAMP) model. Cancer Res. 59: 2203-2209.

Karas, M. and H. Amir, D. Fishman, M. Danilenko, S. Segal, A. Nahum, A. Koifmann, Y. Giat, J. Levy, and Y. Sharoni. 2000. Lycopene interferes with cell cycle progression and Insulin-like Growth Factor I signaling in mammary cancer cells. Nutr. Cancer 36: 101-111.

Kim, L. and A.V. Rao, and L.G. Rao. 2002a. Effect of lycopene on prostate LNCaP cancer cells in culture. J. Med. Food 5: 181-187.

Kim, M.O. and Q. Si, J.N. Zhou, R.G. Pestell, C.F. Brosnan, J. Locker, and S.C. Lee. 2002b. Interferon-β activates multiple signaling cascades in primary human microglia. J. Neurochem. 81: 1361-1371.

Knobbe, C.B. and G. Reifenberger. 2003. Genetic alterations and aberrant expression of genes related to the phosphatidyl-inositol-3-kinase/protein kinase B (Akt) signal transduction pathway in glioblastomas. Brain Pathol. 13: 507-518.

Kobayashi, T. and K. Iijima, T. Mitamura, K. Toriizuka, J.C. Cyong, and H. Nagasawa. 1996. Effects of lycopene, a carotenoid, on intrathymic T cell differential and peripheral CD4/CD8 ratio in a high mammary tumor strain of SHN retired mice. Anti-Cancer Drugs 7: 195-198.

Kohlmeir, L. and J.D. Kark, E. Gomez-Gracia, B.C. Martin, S.E. Steck, A.F.M. Kardinaal, J. Ringstad, M. Thamm, V. Masaev, R. Riemersma, J.M. Martin-Moreno, J.K. Huttunen, and F.J. Kok. 1997. Lycopene and myocardial infarction risk in the EURAMIC study. Amer. J. Epidemiol. 146: 618-626.

Konishi, H. and H. Matsuzaki, M. Tanaka, Y. Takemura, S. Kuroda, Y. Ono, and U. Kikkawa. 1997. Activation of protein kinase B (Akt/RAC-protein kinase) by cellular stress and its association with heat shock protein Hsp27. FEBS Lett. 410: 493-498.

Kotake-Nara, E. and M. Kushiro, H. Zhang, T. Sugawara, K. Miyashita, and A. Nagao. 2001. Carotenoids affect proliferation of human prostate cancer cells. J. Nutr. 131: 3303-3306.

Kristenson, M. and B. Zieden, Z. Kucinskiene, L.S. Elinder, B. Bergdahl, B. Elwing, A. Abaravicius, L. Razinkoviene, H. Calkauskas, and A. Olsson. 1997. Antioxidant state and mortality from coronary heart disease in Lithuanian and Swedish men: concomitant cross sectional study of men aged 50. BMJ 314: 629-633.

Lauder, A. and A. Castellanos, and K. Weston. 2001. c-Myb transcription is activated by protein kinase B (PKB) following interleukin 2 stimulation of T cell and is required for PKB-mediated protein from apoptosis. Mol. Cell. Biol. 21: 5797-5805.

Lee, R.J. and C. Albanese, R.J. Stenger, G. Watanabe, G. Inghirami, G.K Haines 3rd, M. Webster, W.J. Muller, J.S. Brugge, R.J. Davis, and R.G. Pestell. 1999. pp60(v-src) Induction of cyclin D1 requires collaborative interactions between the extracellular signal-regulated kinase, p38, and Jun kinase pathways. A role for cAMP response element-binding protein and activating transcription factor-2 in pp60(v-src) signaling in breast cancer cells. J. Biol. Chem. 274: 7341-7350.

Lee, R.J. and C. Albanese, M. Fu, M. D'Amico, B. Lin, G. Watanabe, G.K. Haines 3rd, P.M. Siegel, M.C. Hung, Y. Yarden, J.M. Horowitz, W.J. Muller, and

R.G. Pestell. 2000. Cyclin D1 is required for transformation by activated Neu and is induced through an E2F-dependent signaling pathway. Mol. Cell. Biol. 20: 672-683.

Lee, D.K. and R.N. Grantham, J.D. Mannion, and A.L. Trachte. 2006. Carotenoids enhance phosphorylation of Akt and suppress tissue factor activity in human endothelial cells. J. Nutr. Biochem. 17: 780-786.

Lee, H.Y. and K.H. Chun, B. Liu, S.A. Wiehle, R.J. Cristiano, W.K. Hong, P. Cohen, and J.M. Kurie. 2002. Insulin-like growth factor binding protein-3 inhibits the growth of non-small cell lung cancer. Cancer Res. 62: 3530-3537.

Levi, M. and H. Ten Cate. 1999. Disseminated intravascular coagulation. New England J. Med. 341: 586-592.

Levy, J. and E. Bosin, B. Feldmen, Y. Giat, A. Miinster, M. Danilenko, and Y. Sharoni. 1995. Lycopene is a more potent inhibitor of human cancer cell proliferation than either α-carotene or β-carotene. Nutr. Cancer 24: 257-266.

Lin, S.Y. and W. Xia, J.C. Wang, K.Y. Kwong, B. Spohn, Y. Wen, R.G. Pestell, and M.C. Hung. 2000. β-Catenin, a novel prognostic marker for breast cancer: its roles in cyclin D1 expression and cancer progression. Proc. Natl. Acad. Sci. USA 97: 4262-4266.

Liu, B. and H.Y. Lee, S.A. Weinzimer, D.R. Powell, J.L. Clifford, J.M. Kurie, and P. Cohen. 2000. Direct functional interactions between insulin-like growth factor-binding protein-3 and retinoid X receptor-α regulate transcriptional signaling and apoptosis. J. Biol. Chem. 275: 33607-33613.

Liu, C. and F. Lian, D.E. Smith, R.M. Russell, and X.-D. Wang. 2003. Lycopene supplementation inhibits lung squamous metaplasia and induces apoptosis via up-regulating insulin-like growth factor-binding protein 3 in cigarette smoke-exposed ferrets. Cancer Res. 63: 3138-3144.

Liu, C. and R.M. Russell, and X.-D. Wang. 2006. Lycopene supplementation prevents smoke-induced changes in p53, p53 phosphorylation, cell proliferation, and apoptosis in the gastric mucosa of ferrets. J. Nutr. 136: 106-111.

Ma, Y.Y. and S.J. Wei, Y.C. Lin, J.C. Lung, T.C. Chang, J. Whang-Peng, J.M. Liu, D.M. Yang, W.K. Yang, and C.Y. Shen. 2000. PIK3CA as an oncogene in cervical cancer. Oncogene 19: 2739-2744.

Matsumura, I. and T. Kitamura, H. Wakao, H. Tanaka, K. Hashimoto, C. Albanese, J. Downward, R.G. Pestell, and Y. Kanakura. 1999. Transcriptional regulation of the cyclin D1 promoter by STAT5: its involvement in cytokine-dependent growth of hematopoietic cells. EMBO J. 18: 1367-1377.

Medema, R.H. and G.J. Kops, J.L. Bos, and B.M. Burgering. 2000. AFX-like Forkhead transcription factors mediate cell-cycle regulation by Ras and PKB through p27kip1. Nature 404: 782-787.

Mucci, L.A. and R. Tamimi, P. Lagiou, A. Trichopoulou, V. Benetou, E. Spanos, and D. Trichopoulos. 2001. Are dietary influences on the risk of prostate cancer mediated through the insulin-like growth factor system?. B.J.U. Int. 87: 814-820.

Muise-Helmericks, R.C. and H.L. Grimes, A. Bellacosa, S.E. Malstrom, P.N. Tsichlis, and N. Rosen. 1998. Cyclin D expression is controlled post-transcriptionally via a phosphatidylinositol 3-kinase/Akt-dependent pathway. J. Biol. Chem. 273: 29864-29872.

Muller, K. and K.L. Carpenter, I.R. Challis, J.N. Skepper, and M.J. Arends. 2002. Carotenoids induce apoptosis in the T-lymphoblast cell line Jurkat E6.1. Free Rad. Res. 36: 791-802.

Muslin, A.J. and J.W. Tanner, P.M. Allen, and A.S. Shaw. 1996. Interaction of 14-3-3 with signaling proteins is mediated the recognition of phosphoserine. Cell 84: 889-897.

Nagasawa, H. and T. Mitamura, S. Sakamoto, and K. Yamamoto. 1995. Effects of lycopene on spontaneous mammary tumour development in SHN virgin mice. Anticancer Res. 15: 1173-1178.

Nahum, A. and K. Hirsch, M. Danilenko, C.K. Watts, O.W. Prall, J. Levy, and Y. Sharoni. 2001. Lycopene inhibition of cell cycle progression in breast and endometrial cancer cells is associated with reduction in cyclin D levels and retention of p27(Kip1) in the cyclin E-cdk2 complexes. Oncogene 20: 3428-3436.

Nassif, N.T. and G.P. Lobo, X. Wu, C.J. Henderson, C.D. Morrison, C. Eng, B. Jalaludin, and E. Segelov. 2004. PTEN mutations are common in sporadic microsatellite stable colorectal cancer. Oncogene 23: 617-628.

Nelsen, C.J. and D.G. Rickheim, M.M. Tucker, T.J. McKenzie, L.K. Hansen, R.G. Pestell, and J.H. Albrecht. 2003. Amino acids regulate hepatocyte proliferation through modulation of cyclin D1 expression. J. Biol. Chem. 278: 25853-25858.

Nguyen, M.L. and S.J. Schwartz. 1999. Lycopene: chemical and biological properties. Food Tech. 53: 38-45.

Nicholson, K.M. and N.G. Anderson. 2002. The protein kinase B/Akt signaling pathway in human malignancy. Cell Signal 14: 381-395.

Obermüller-Jevic, U.C. and E. Olano-Martin, A.M. Corbacho, J.P. Eiserich, A. van der Vliet, G. Valacchi, C.E. Cross, and L. Packer. 2003. Lycopene inhibits the growth of normal human prostate epithelial cells in vitro. J. Nutr. 133: 3356-3360.

Oh, Y. and H.L. Muller, G. Lamson, and R.G. Rosenfeld. 1993. Insulin-like growth factor (IGF)-independent action of IGF-binding protein-3 in Hs578T human breast cancer cells. Cell surface binding and growth inhibition. J. Biol. Chem. 268: 14964-14971.

Osaki, M. and S. Kase, K. Adachi, A. Takeda, K. Hashimoto, and H. Ito. 2004. Inhibition of the PI3K-Akt signaling pathway enhances the sensitivity of Fas-mediated apoptosis in human gastric carcinoma cell line, MKN-45. J. Cancer Res. Clin. Oncol. 130: 8-14.

Ottilie, S. and J.-L. Diaz, W. Horne, J. Chang, Y. Wang, G. Wilson, S. Chang, S. Weeks, L. Fritz, and T. Oltersdorf. 1997. Dimerization properties of human Bad: Identification of a BH-3 domain and analysis of its binding to mutant Bcl-2 and Bcl-Xl proteins. J. Biol. Chem. 272: 30866-30872.

Ozes, O.N. and L.D. Mayo, J.A. Gustin, S.R. Pfeffer, L.M. Pfeffer, and D.B. Donner. 1999. NF-kappaB activation by tumour necrosis factor requires the Akt serine-threonine kinase. Nature 401: 82-85.

Palozza, P. and A. Sheriff, S. Serini, A. Boninsegna, N. Maggiano, F.O. Ranelletti, G. Calviello, and A. Cittadini. 2005. Lycopene induces apoptosis in immortalized fibroblasts exposed to tobacco smoke condensate through arresting cell cycle and down-regulating cyclin D1, pAKT and pBad. Apoptosis 10: 1445-1456.

Pestell, R.G. and C. Albanese, A.T. Reutens, J.E. Segall, R.J. Lee, and A. Arnold. 1999. The cyclins and cyclin-dependent kinase inhibitors in hormonal regulation of proliferation and differentiation. Endocr. Rev. 20: 501-534.

Pinot, F. and A. el Yaagoubi, P. Christie, A.T. Dinh-Xuan, and B.S. Polla. 1997. Induction of stress proteins by tobacco smoke in human monocytes: modulation by antioxidants. Cell Stress Chaperones 2: 156-161.

Pradeep, A. and C. Sharma, P. Sathyanarayana, C. Albanese, J.V. Fleming, T.C. Wang, M.M. Wolfe, K.M. Baker, R.G. Pestell, and B. Rana. 2004. Gastrin-mediated activation of cyclin D1 transcription involves β-catenin and CREB pathways in gastric cancer cells. Oncogene 23: 3689-3699.

Prakash, P. and R.M. Russell, and N.I. Krinsky. 2001. In vitro inhibition of proliferation of estrogen-dependent and estrogen-independent human breast cancer cells treated with carotenoids or retinoids. J. Nutr. 131: 1574-1580.

Pugazhenthi, S. and A. Nesterova, C. Sable, K.A. Heidenreich, L.M. Boxer, L.E. Heasley, and J.E. Reusch. 2000. Akt/protein kinase B up-regulates Bcl-2 expression through cAMP-response element-binding protein. J. Biol. Chem. 275: 10761-10766.

Rao, A.V. and N. Fleshner, and S. Agarwal. 1999. Serum and tissue lycopene and biomarkers of oxidation in prostate cancer patients: a case control study. Nutr. Cancer 33: 159-164.

Rao, A.V. and S. Agarwal. 2000. Role of antioxidant lycopene in cancer and hearth disease. J. Am. Coll. Nutr. 19: 563-569.

Rechler, M. 1997. Growth inhibition by insuline-like growth factor (IGF) binding protein-3: what's IGF got to do with it?. Endocrinology 138: 2645-2647.

Reusch, J.E. and D.J. Klemm. 2002. Inhibition of cAMP-response element-binding protein activity decreases protein kinase B/Akt expression in 3T3-L1 adipocytes and induces apoptosis. J. Biol. Chem. 277: 1426-1432.

Rissanen, T. and S. Voutilainen, K. Nyyssönen, R. Salonen, and J.T. Salonen. 2000. Low plasma lycopene concentration is associated with increased intima-media thickness of the carotid artery wall. Arterioscler. Thromb. Vasc. Biol. 20: 2677-2681.

Rissanen, T. and S. Voutilainen, K. Nyyssönen, T.A. Lakka, R. Salonen, G.A. Kaplan, and J.T. Salonen. 2001. Low serum lycopene is associated with excess risk of acute coronary events and stroke: The Kuopio Ischaemic Risk Factor Study. Brit. J. Nutr. 85: 749-754.

Roy, H.K. and B.F. Olusola, D.L. Clemens, W.J. Karolski, A. Ratashak, H.T. Lynch, T.C. Smyrk. 2002. AKT proto-oncogene overexpression is an early event during sporadic colon carcinogenesis. Carcinogenesis 23: 201-205.

Samuels, Y. and Z. Wang, A. Bardelli, N. Silliman, J. Ptak, S. Szabo, H. Yan, A. Gazdar, S.M. Powell, G.J. Riggins, J.K. Willson, S. Markowitz, K.W. Kinzler,

B. Vogelstein, and V.E. Velculescu. 2004. High frequency of mutations of the PIK3CA gene in human cancers. Science 304: 554.

Sato, S. and N. Fujita, and T. Tsuruo. 2000. Modulation of Akt kinase activity by binding to Hsp. Proc. Natl. Acad. Sci. 97: 10832-10837.

Sesso, H.D. and J.E. Buring, E.P. Norkus, and J.M. Gaziano. 2004. Plasma lycopene, other carotenoids, and retinol and the risk of cardiovascular disease in women. Amer. J. Clin. Nutr. 79: 47-53.

Shayasteh, L. and Y. Lu, W.L. Kuo, R. Baldocchi, T. Godfrey, C. Collins, D. Pinkel, B. Powell, G.B. Mills, and J.W. Gray. 1999. PIK3CA is implicated as an oncogene in ovarian cancer. Nat. Genet. 21: 99-102.

Sherr, C.J. 2003. The Pezcoller Lecture: cancer cell cycles revisited. Cancer Res. 60: 689-3695.

Shiojima, I. and K. Walsh. 2006. Regulation of cardiac growth and coronary angiogenesis by the Akt/PKB signaling pathway. Genes Dev. 20: 3347-3365.

Shtutman, M. and J. Zhurinsky, I. Simcha, C. Albanese, M. D'Amico, R. Pestell, and A. Ben-Ze'ev. 1999. The cyclin D1 gene is a target of the β-catenin/LEF-1 pathway. Proc. Natl. Acad. Sci. USA 96: 5522–5527.

Siler, U. and L. Barella, V. Spitzer, J. Schnorr, M. Lein, R. Goralczyk, and K. Wertz. 2004. Lycopene and vitamin E interfere with autocrine/paracrine loops in the Dunning prostate cancer model. FASEB J. 18: 1019-1021.

Simpson, L. and R. Parsons. 2001. PTEN: life as a tumor suppressor. Exp. Cell. Res. 264: 29-41.

Song, D.H. and B. Rana, J.R. Wolfe, G. Crimmins, C. Choi, C. Albanese, T.C. Wang, R.G. Pestell, and M.M. Wolfe. 2003. Gastrin-induced gastric adenocarcinoma growth is mediated through cyclin D1. Am. J. Physiol. Gastrointest. Liver Physiol. 285: G217-G222.

Stahl, W. and H. Sies. 1996. Lycopene: a biologically important carotenoid for humans?. Arch. Biochem. Biophys. 336: 1-9.

Stephens, L. and K. Anderson, D. Stokoe, H. Erdjument-Bromage, G.F. Painter, A.B. Holmes, P.R. Gaffney, C.B. Reese, F. McCormick, P. Tempst, J. Coadwell, and P.T. Hawkins. 1998. Protein kinase B kinases that mediate phosphatidylinositol 3,4,5-trisphosphate-dependent activation of protein kinase B. Science 279: 710-714.

Stokoe, D. and L.R. Stephens, T. Copeland, P.R. Gaffney, C.B. Reese, G.F. Painter, A.B. Holmes, F. McCormick, and P.T. Hawkins. 1997. Dual role of phosphatidylinositol-3,4,5-trisphosphate in the activation of protein kinase B. Science 277: 567-570.

Street, D.A. and G.W. Comstock, R.M. Salkeld, W. Schuep, and M.J. Klag. 1994. Serum antioxidants and myocardial infarction. Are low levels of carotenoids and alpha-tocopherol risk factors for myocardial infarction? Circulation 90: 1154-1161.

Suzui, M. and M. Masuda, J.T. Lim, C. Albanese, R.G. Pestell, and I.B. Weinstein. 2002. Growth inhibition of human hepatoma cells by acyclic retinoid is associated with induction of p21(CIP1) and inhibition of expression of cyclin D1. Cancer Res. 62: 3997-4006.

Takahashi, A. and Y. Kureishi, J. Yang, Z. Luo, K. Guo, D. Mukhopadhyay, Y. Ivashchenko, D. Branellec, and K. Walsh. 2006. Carotenoids enhance phosphorylation of Akt and suppress tissue factor activity in human endothelial cells. J. Nutr. Biochem. 17: 780-786.

Talapatra, S. and C.B. Thompson. 2001. Growth factor signaling in cell survival: implications for cancer treatment. J. Pharm. Exp. Ther. 298: 873-878.

Tanno, S. and N. Yanagawa, A. Habiro, K. Koizumi Y., Nakano, M. Osanai, Y. Mizukami, T. Okumura, J.R. Testa, and Y. Kohgo. 2004. Serine/threonine kinase AKT is frequently activated in human bile duct cancer and is associated with increased radioresistance. Cancer Res. 64: 3486-3490.

Van Eewyck, J. and F.G. Davis, and P.E. Bowen. 1991. Dietary and serum carotenoids and cervical intraepithelial neoplasia. Int. J. Cancer 48: 34-38.

Vivanco, I. and C.L. Sawyers. 2002. The phosphatidylinositol 3-Kinase AKT pathway in human cancer. Nat. Rev. Cancer 2: 489-501.

Wang, C. and N. Pattabiraman, J.N. Zhou, M. Fu, T. Sakamaki, C. Albanese, Z. Li, K. Wu, J. Hulit, P. Neumeister, P.M. Novikoff, M. Brownlee, P.E. Scherer, J.G. Jones, K.D. Whitney, L.A. Donehower, E.L. Harris, T. Rohan, D.C. Johns and R.G. Pestell. 2003. Cyclin D1 repression of peroxisome proliferator-activated receptor gamma expression and transactivation. Mol Cell Biol. 23: 6159-6173.

Wang, C. and Z. Li, M. Fu, T. Bouras, and R.G. Pestell. 2004. Signal transduction mediated by cyclin D1: from mitogens to cell proliferation: a molecular target with therapeutic potential. Cancer Treat. Res. 119: 217-237.

Wang, J.M. and J.R. Chao, W. Chen, M.L. Kuo, J.J. Yen, and H.F. Yang-Yen. 1999. The anti-apoptotic gene mcl-1 is up-regulated by the phosphadylinositol 3-kinase/Akt signaling pathway through a transcription factor complex containing CREB. Mol. Cell. Biol. 19: 6195-6206.

Watanabe, G. and A. Howe, R.J. Lee, C. Albanese, I.W. Shu, A.N. Karnezis, L. Zon, J. Kyriakis, K. Rundell, and R.G. Pestell. 1996. Induction of cyclin D1 by simian virus 40 small tumor antigen. Proc. Natl. Acad. Sci. USA 93: 12861-12866.

Woenckhaus, J. and K. Steger, E. Werner, I. Fenic, U. Gamerdinger, T. Dreyer, and U. Stahl. 2002. Genomic gain of PIK3CA and increased expression of p110alpha are associated with progression of dysplasia into invasive squamous cell carcinoma. J. Pathol. 198: 335-342.

Wymann, M.P. and M. Zvelebil, and M. Laffargue. 2003. Phosphoinositide 3-kinase signaling—which way to target?. Trends Pharmacol. Sci. 24: 366-376.

Yaffe, M.B. and K. Rittinger, S. Volinia, P.R. Caron, A. Aitken, H. Leffers, S.J. Gamblin, S.J. Smerdon, and L.C. Cantley. 1997. The structural basis for 14-3-3: Phosphopeptide binding specificity. Cell 91: 961-971.

Yang, E. and J. Zha, J. Jockel, L.H. Boise, C.B. Thompson, and S.J. Korsmeyer. 1995. Bad, a heterodimeric partner for Bcl-Xl and Bcl-2, displaces Bax and promotes cell death. Cell. 80: 285-291.

Yu, H. and T. Rohan. 2000. Role of the insulin-like growth factor family in cancer development and progression. J. Natl. Cancer Inst. (Bethesda) 92: 1472-1489.

Zha, J. and H. Harada, E. Yang, J. Jockel, and S.J. Korsmeyer. 1996. Serine phosphorylation of death agonist BAD is response to survival factor results in binding to 14-3-3 not BCL-XL. Cell 87: 619-628.

Zha, J. and H. Harada, K. Osipov, J. Jockel, G. Waksman and S.J. Korsmeyer. 1997. BH3 domain of BAD is required for heterodimerization with BCL-XL and pro-apoptotic activity. J Biol Chem. 26: 24101-24104.

Zha, J. and H. Harada, K. Osipov, J. Jockel, G. Waksmann, and S.J. Korsmeyer. 1997. BH3 domain of Bad is required for heterodimerization with Bcl-Xl and pro-apoptotic activity. J. Biol. Chem. 272: 24101-24104.

Zhang, J.M. and Q. Wei, X. Zhao, and B.M. Paterson. 1999. Coupling of the cell cycle and myogenesis through the cyclin D1-dependent interaction of MyoD with cdk4. EMBO J. 18: 926-933.

Zhang, L.-X. and R.V. Cooney, and J.S. Bertram. 1991. Carotenoids enhance gap junctional communication and inhibit lipid peroxidation in C3H/10T1/2 cells: relationship to their cancer chemopreventive action. Carcinogenesis 12: 2109-2114.

Zhang, L.-X. and R.V. Cooney, and J.S. Bertram. 1992. Carotenoids up-regulate connexin43 gene expression independent of their provitamin A or antioxidant properties. Cancer Res. 52: 5707-5712.

Zhou, B.P. and Y. Liao, W. Xia, B. Spohn, M.H. Lee, and M.C. Hung. 2001. Cytoplasmic localization of p21Cip1/WAF1 by Akt-induced phosphorylation in HER-2/neu-overexpressing cells. Nat. Cell. Biol. 3: 245-252.

Lycopene and Chylomicrons

[1]Kathleen M. Botham and [2]Elena Bravo
[1]Department of Veterinary Basic Sciences, The Royal Veterinary College
Royal College St., London NW1 0TU, UK
[2]Department of Haematology, Oncology and Molecular Medicine
Viale Regina Elena 299, 00161 Rome, Italy

ABSTRACT

Lycopene, like other lipophilic dietary components, is absorbed from the intestine in chylomicrons. These large, triacylglycerol-rich lipoproteins are metabolized in the blood by removal of some of the triacylglycerol, leaving smaller chylomicron remnants particles that deliver the remaining lipids to the liver. Lycopene may then re-enter the circulation and is transported in low density lipoprotein. The proportion of dietary lycopene absorbed in chylomicrons is relatively low and is influenced by a number of factors. It is increased by the consumption of tomato products rather than the raw fruit, as well as by the presence of fat in the diet, and the *cis* isomer is more easily incorporated into chylomicrons than the *all-trans* form. Conversely, absorption is reduced by non-absorbable fat substitutes such as sucrose polyester, and in older as compared to younger subjects. In addition, other carotenoids in the diet may compete with lycopene for absorption in chylomicrons, although this does not appear to influence its bioavailability in the medium term. Some studies have suggested that lycopene may protect against atherosclerosis, possibly because its antioxidant properties protect low density lipoprotein from oxidation. However, the presence of lycopene in chylomicron remnants, which are also known to be atherogenic, has been found to increase their induction of macrophage foam cell formation and decrease their uptake by the liver, thus potentially promoting atherosclerotic lesion development. The type of

A list of abbreviations is given before the references.

lipoprotein carrier of lycopene, therefore, may be important for its effects on atherosclerosis, with its presence being beneficial in low density lipoprotein, but deleterious in chylomicron remnants. Since chylomicron remnants are normally rapidly cleared from the blood, the beneficial effect may predominate in healthy subjects. However, remnant particles accumulate in the blood in some conditions, including obesity and diabetes, and the potentially detrimental effects may be more important in these circumstances. Despite the potential protective effects of dietary lycopene against cardiovascular and other diseases, the mechanisms regulating its absorption in chylomicrons are not completely understood, and more research is essential to establish the optimum conditions for the dietary intake of lycopene in order to obtain the maximum benefit for health.

INTRODUCTION

Atherosclerosis, the principal cause of heart attacks and stroke, is a multifactorial disease, with dietary as well as environmental and genetic factors playing an important role in its initiation and progression (Moreno and Mitjavila 2003, Yusuf et al. 2001). It has been known for many years that the type of fat in the diet modulates the risk of atherosclerosis development, so that it is decreased by consumption of diets rich in monounsaturated and polyunsaturated fats and increased by intake of saturated fats (Keys 1970, Harris 1996, Assman et al. 1997, Moreno and Mitjavila 2003). More recently, however, it has become clear that the high intake of fruit and vegetables associated with, for example, the Mediterranean diet also protects against heart disease (Kok and Kromhout 2004), and there is good evidence to suggest that micronutrients such as plant carotenoids, including lycopene, contribute to this beneficial effect (Sesso 2006, Voutilainen et al. 2006).

Atherosclerosis begins with dysfunction of the vascular endothelium, which is followed by the appearance of fatty streaks, the first visible lesions in the artery wall. Fatty streaks form when macrophages take up cholesterol and other lipids from the plasma lipoproteins, eventually becoming so engorged that they become foam cells (Kadar and Glasz 2001). Low density lipoprotein (LDL) is known to play a major role in these processes, but prior oxidation of the particles is required for many of its effects (Albertini et al. 2002). There is evidence to suggest that lycopene may protect against atherosclerotic lesion development , with a number of studies reporting an inverse relationship between serum/tissue lycopene levels and intimal wall thickness or lesions in the carotid artery and aorta (Arap and Steck 2000, Rao et al. 2006, Sesso 2006). Lycopene is a powerful antioxidant, the most efficient of the common plant carotenoids because of the long chromophore of its polyene chain (Di Mascio et al. 1989, Harker and Hirschberg 1997). Thus, its beneficial action may be due to the protection of LDL from oxidation, although it is possible that other mechanisms such as inhibition of cholesterol synthesis and/or enhancement of LDL degradation may also be involved (Arap and Steck 2000, Rao et al. 2006), as well as indirect effects via inflammation and immune function (Heber and Lu 2002).

Chylomicrons are large lipoproteins of 75-500 nm diameter or more that are formed in intestinal cells and secreted into lymph. They consist of about 98% lipid, of which about 90% is triacylglycerol, and 2% protein, including apolipoprotein (apo) B48, which is integral to the particles, and apoAs and Cs, which are bound to the surface (Redgrave 1983) (Fig. 1).

Dietary lipids, including fats, cholesterol and lipid soluble micronutrients such as lycopene, are absorbed in the intestine and secreted into lymph in chylomicrons. The large, triacylglycerol-rich lipoproteins then enter the blood via the thoracic duct and are lipolysed by lipoprotein

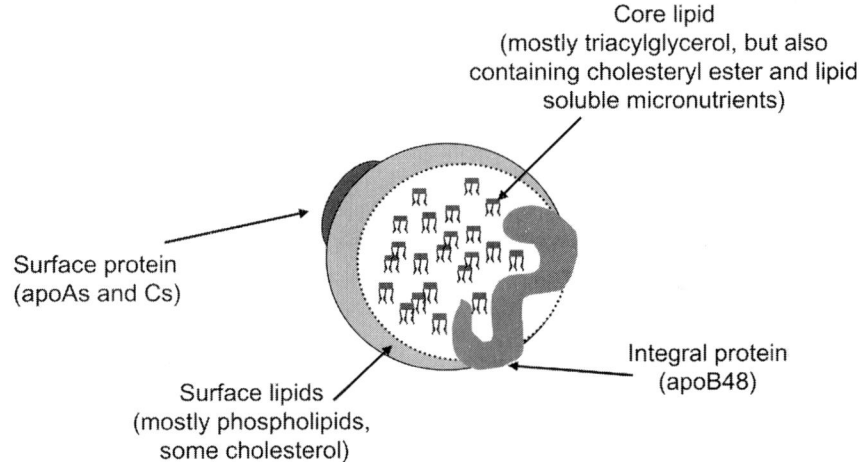

Fig. 1 Chylomicrons. Chylomicrons are large lipoproteins containing 98% lipid and 2% protein. The surface layer of lipids consists of mostly phospholipid with some cholesterol, while triacylglycerol (about 90% of the lipid component), cholesteryl ester and lipid-soluble micronutrients are found in the core. Apolipoprotein (apo) B48 is an integral protein, and apoAs and Cs are bound to the surface.

lipase in extra hepatic capillary beds, a process that removes some of the triacylglycerol and forms smaller chylomicron remnant particles, which retain all of the cholesterol and minor lipid components and subsequently deliver them to the liver for processing (Redgrave 1983) (Fig. 2).

For many years the potential atherogenicity of chylomicron remnants was neglected, since it was thought that they were too large to enter the artery wall. Studies by Mamo and colleagus, however, have demonstrated unequivocally that the remnant particles penetrate the artery wall as efficiently as LDL and are retained within the sub-endothelial space (Mamo and Wheeler 1994, Proctor et al. 2002), and there is now compelling evidence to indicate that they are indeed atherogenic (Yu and Cooper 2001, Willhem and Cooper 2003, Botham et al. 2005). The accumulation of chylomicron remnants in the blood in dyslipidemias in humans or in animal models such as apoE$^{-/-}$ mice is associated with premature atherosclerosis development (Boren et al. 2000, Botham et al. 2005), and particles resembling chylomicron remnants have been isolated from atherosclerotic plaque (Ghung et al. 1994, Pal et al. 2003). Furthermore, experiments *in vitro* have demonstrated that these lipoproteins cause endothelial dysfunction by inhibiting endothelium-dependent vascular relaxation and also cause the extensive lipid accumulation in macrophages associated with foam cell formation. In sharp contrast to LDL, however, prior oxidation of the particles is not required for these effects (Grieve et al. 1998, Evans et al. 2004, Batt et al. 2004, Botham et al. 2005). Lipids from the diet, therefore, may influence atherogenic events in the artery wall during their transport from the gut to the liver in chylomicron remnants.

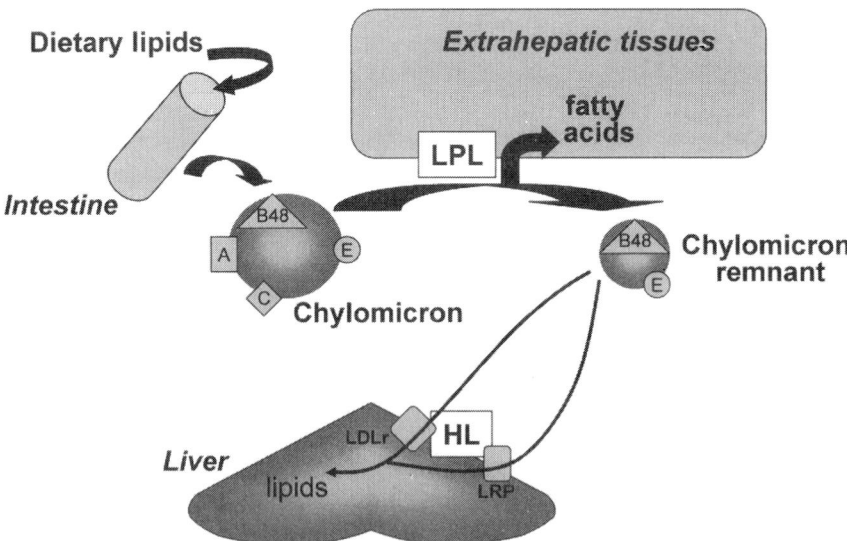

Fig. 2 Absorption and transport of dietary lipids from the intestine to the liver. Chylomicrons containing apoA, B48, C and E are lipolysed by lipoprotein lipase (LPL), resulting in the loss of apoA and C and the formation of chylomicron remnants that are taken up by the liver via the low density lipoprotein receptor (LDLr) or the LDL r-related protein (LRP) after modification by hepatic lipase (HL).

Work in our laboratory has shown that the lipid composition of chylomicron remnants can affect both the rate of their clearance from the blood by the liver (Bravo et al. 1995, Lambert et al. 1995, 2001) and their induction of macrophage foam cell formation (De Pascale et al. 2006). It is clear, therefore, that chylomicrons play a role in the effects of lycopene in cardiovascular disease, first because they are the vehicle for its absorption, and so have a part in determining its bioavailability, and second because of the potential influence of the carotenoid on the direct interaction of chylomicron remnants with the artery wall and/or their uptake by the liver. This chapter reviews the factors influencing the absorption of lycopene in chylomicrons and discusses the evidence that it modifies the effects of the remnant particles on their interactions with macrophages and hepatocytes.

ABSORPTION OF LYCOPENE IN CHYLOMICRONS

Since mammals are unable to synthesize carotenoids, most of the dietary intake, in particular in Western countries, derives from tomato sources. Dietary supplementation with lycopene as tomato juice, spaghetti sauce or tomato oleoresin results in a significant increase in serum lycopene levels and diminished amounts of serum thiobarbituric acid-reactive substances, indicating that lycopene is absorbed from tomato products and may act as an antioxidant *in vivo* (Rao and Agarwal 1998).

The absorption of lycopene and other carotenoids requires their release from the vegetable matrix, incorporation into mixed micelles and absorption by enterocytes before they are incorporated into chylomicrons for secretion into lymph. Chylomicron remnants formed from chylomicrons are taken up by the liver, and the lycopene may be re-secreted into blood in very low density lipoprotein (VLDL) which is then converted to LDL (Fig. 3).

Absorption of lycopene by the mucosa of the small intestine occurs mainly in the duodenum. Mucosal uptake requires its incorporation into mixed micelles that mediate gradient-driven diffusion into the enterocyte, and absorption seems to be limited by micellar incorporation (Borel et al. 1996). The incorporation of lycopene into chylomicrons has been measured directly in animal models by collecting the lymph after cannulation of the mesenteric duct, and in human subjects by isolating the triglyceride-rich lipoprotein (TRL) fraction from blood samples by ultracentrifugation. In rats given lipid emulsions containing lycopene, Clark et al. (1998, 2000) found that the time course for the appearance of lycopene and triacylglycerol in chylomicrons was similar and reached a steady state by 6 hr, although the efficiency of incorporation of lycopene

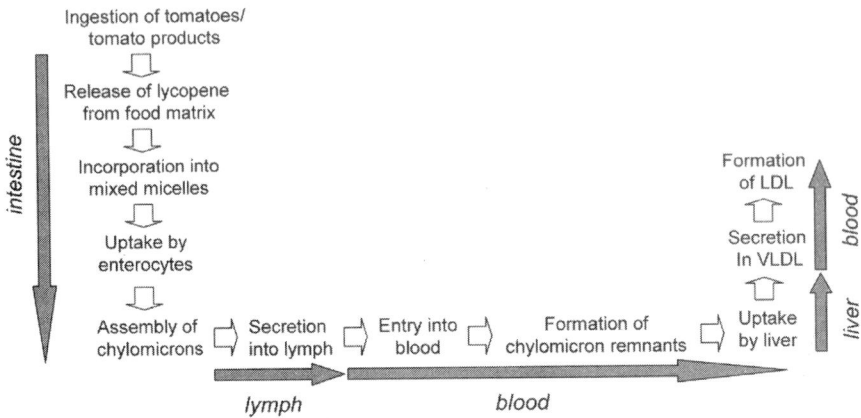

Fig. 3 Stages in the absorption of lycopene in chylomicrons and its transfer to LDL in the blood via the liver. After ingestion of tomatoes or tomato products, lycopene is released from the vegetable matrix, incorporated into mixed micelles, absorbed by enterocytes and incorporated into chylomicrons in the intestine. Chylomicrons are secreted into lymph, enter the blood and are converted into chylomicron remnants that are taken up by the liver. Lycopene may then be re-secreted into blood in very low density lipoprotein (VLDL), which is metabolized to form low density lipoprotein (LDL).

into the lipoproteins was low (2.5-6%). Lycopene can also be detected in chylomicrons in humans after a meal containing tomatoes or tomato products (Tyssandier et al. 2003, Gustin et al. 2004) , and in a physiological pharmacokinetic model developed by Diwadkar-Navsariwala et al. (2003), a peak in serum lycopene after consumption of a tomato beverage could be related to its flow into this lipoprotein fraction. However, no significant differences in the level of lycopene found in chylomicrons was detected after oral doses ranging from 10 to 120 mg in two studies (Diwadkar-Navsariwala et al. 2003, Gustin et al. 2004), and the authors suggested this may be either because the absorption process is saturable at low concentrations, or because of the large variation in chylomicron lycopene levels between subjects in the same group.

Estimates of the proportion of lycopene consumed in the diet that is absorbed in chylomicrons vary enormously. Tyssandier et al. (2003) found that, after a meal containing tomato puree providing 10 mg lycopene, only 2-3% of the dose was transferred to the micellar phase in the duodenum and, perhaps not surprisingly given this observation, they were unable to detect any significant increase in lycopene in chylomicrons. In contrast, in compartmental modelling studies, absorption after ingestion of a tomato beverage was estimated as 33.9% at a dose of 10 mg, although this fell to 5.3% at a higher dose of 120 mg (Diwadkar-Navsariwala et al. 2003).

Studies in this area have also been consistently hampered by very large inter-subject variations (Tyssandier et al. 2003, Gustin et al. 2004). It is likely that these differences are caused by a number of inter-related factors that influence the amount of lycopene from the diet transferred to chylomicrons, including the isomer present in the gut (*cis/trans*), the type of food consumed (raw, cooked or processed tomatoes or other fruit sources), the amount and type of fat or other cartenoids in the diet, and the age of the subject (Table 1).

Cis/trans Isomers and Raw, Cooked or Processed Tomatoes

Lycopene is an acyclic carotenoid with 11 conjugated double bonds, and its configuration in raw tomatoes is about 95% *all-trans*, although processed tomato products contain between 1.7% and 10.1% *cis* isomers. The *cis* isomers (primarily 5, followed by 9, 13, or 15) are also found in plants and in plasma (Holloway et al. 2000). However, the *all-trans* isomer accounts for about 65% of total lycopene in chylomicrons and only about 45%, or lower, in serum (Stahl et al. 1992, Clinton et al. 1996). There is some evidence to suggest that this change may be due to enhanced incorporation of *cis* as compared to *all-trans* lycopene into chylomicrons. In experiments with lymph-cannulated ferrets, Boileau et al. (1999) found that the percentage of *cis*-lycopene isomers increased from < 10% in the dose to 77% in the mesenteric lymph, and this was significantly greater than the proportion in any of the tissues analysed, including the intestinal mucosa (47-58% *cis*), suggesting that *cis*-lycopene is more readily removed from the intestinal membrane for incorporation into chylomicrons. In the *all-trans* configuration, lycopene is a longer molecule that is less likely to crystallize than the *cis* form (Britton 1995) and appears to be less soluble in bile acid micelles, so that its ability to move across plasma membranes and be incorporated into chylomicrons is decreased (Boileau et al. 1999). *cis*-Lycopene, therefore, is more bioavailable than the *all-trans* form, and this may be due to better extraction of these isomers from food and incorporation into chylomicrons during digestion. However, it is also possible that some tissue-activated isomerization of *all*-trans dietary lycopene occurs after absorption.

A number of studies have shown that lycopene is absorbed more easily from processed tomatoes than from the raw fruit (Stahl and Sies 1992, Porrini et al. 1998, Arap and Steck 2000). Concentrations in the chylomicron fraction of human serum of five normolipidemic volunteers have been found to be three times as high after a test meal containing

Table 1 Factors affecting the absorption of lycopene in chylomicrons directly and as assessed by plasma levels.

Factor	Effect	
	Plasma levels	Absorption in chylomicrons
Cis/trans isomers	Ratio cis: all-trans higher than in food (Stahl et al. 1992, Gartner et al. 1997).	Ratio cis: all-trans higher in chylomicrons than in food (Gartner et al. 1997, Boileau et al. 1999).
Food type (raw/cooked/processed)	Increased by consumption of cooked or processed as compared to raw tomatoes (Stahl and Sies 1992, Porrini et al. 1998, Arap and Steck 2000), and by tomato puree as compared to cooked tomatoes (Holloway et al. 2000).	Increased after consumption of tomato paste as compared to raw tomatoes (Gartner et al. 1997).
Amount and type of fat in diet	Higher after consumption of tomatoes cooked with, as compared to without, olive oil (Fielding et al. 2005)	Increased after consumption of salad containing tomatoes with increased fat content of salad dressing (Brown et al. 2004); lower absorption in chylomicrons after giving lycopene with corn oil, as compared to olive oil, in rats (Clark et al. 2000)
Presence of other cartenoids in diet	Increased when taken combined with β-carotene (Johnson et al. 1997); amount in LDL reduced by β-carotene (Graziano et al. 1995); unaffected (Tyssandier et al. 2002)	Increase after consumption of tomato puree smaller when lutein was added (Tyssandier et al. 2002); unaffected by β-carotene (van den Berg and van Vliet 1998)
Age	Lower in younger subjects (Brady et al. 1996)	Triacylglycerol-adjusted lycopene levels in chylomicrons lower (−40%) in older as compared to younger subjects (Cardinault et al. 2003)
Non-absorbable fat substitutes	Reduced by dietary sucrose polyester (Weststrate and van het Hof 1995, van Koonvitsky et al. 1997, Schlagheck et al. 1997)	Decreased due to reduced incorporation into micelles (van het Hof et al. 2000)
Cholestyramine	Decreased by about 30% (Elinder et al. 1995)	Not measured directly, but likely to be reduced as cholestyramine decreased intestinal absorption of lipids
Smoking	Basal levels unaffected in smokers as compared to non-smokers (Arap and Steck 2000), but decreased immediately after smoking; postprandial concentrations always lower than fasting levels in smokers (Rao and Agarwal 1998)	No information
Hormonal status	Peaks at mid-luteal phase of the menstrual cycle; inversely related to androgen status (Erdman 2005)	No information

tomato paste as compared to raw tomatoes, a ratio similar to that of *cis* as compared to *all-trans* isomers, even though the *cis* : *all-trans* ratio of the lycopene in the two food sources was not different (Gartner et al. 1997). This raised the hypothesis that heating increases lycopene bioavailability by increasing the *trans-to-cis* isomerization (Stahl et al. 1992). However, findings on the isomerization response to heating are controversial. Boiling tomatoes for up to 3 hr, either in water or in oil, produced only a slight increase in *cis*-isomers in tomato paste (Schierle et al. 1997), and Nguyen and Schwartz (1998) have reported that lycopene is stable in the tomato matrix and resistant to heat-induced geometrical conversion during typical food processing of tomatoes and related products, except in extreme conditions not regularly employed in the food industry or during food preparation. In contrast, Shi and Le Maguer (2000) have found that thermal processing (bleaching, retorting, and freezing processes) generally causes some loss of lycopene in tomato-based foods, and that heat induces isomerization of the *all-trans* to *cis*-forms. The small increases in *cis*-lycopene content that occur during food processing, however, cannot account for the large proportion of *cis*-isomers in the plasma and different tissues (Boileau et al. 2002), and other physiological processes must be responsible for the large differences in the percentage of *cis* isomers between foods and plasma and tissues.

As well as influencing lycopene absorption via *cis* : *trans* isomerization, food processing may improve its bioavailability by breaking down cell walls, which weakens the bonding forces between lycopene and the tissue matrix, thus making lycopene more accessible (Shi and Le Maguer, 2000). Holloway et al. (2000) have reported that, in healthy volunteers, consumption of a diet containing tomato puree for 2 wk caused a significant increase in lycopene levels in plasma, while no increase was seen when cooked tomatoes were used, showing that the lycopene within intact cells is less bioavailable than that from processed tissue. In addition, the observed increase in lycopene serum concentration after heating has been attributed to the extraction of lycopene into the lipophilic phase during boiling, facilitating its intestinal absorption (Stahl et al. 1992).

Amount and Type of Fat

The amount and type of dietary fats in food is of considerable importance in carotenoid absorption (Parker 1996). Plasma lycopene concentrations have been shown to be higher after consumption of tomatoes cooked with, as compared to without, olive oil (Fielding et al. 2005). Brown et al. (2004) compared the effects of three different salad dressings containing variable amount of fat on the appearance of lycopene and other carotenoids from raw tomatoes and salad vegetables in the plasma chylomicron fraction. Essentially no absorption was found when salads were eaten with fat-free dressing, and substantially greater amounts were absorbed when full fat as compared to reduced fat dressing was used. Thus, the amount of fat in the diet is limiting in terms of the incorporation of lycopene into chylomicrons. Similarly, data from human studies in India have suggested that a minimum of 5-10 g of fat in a meal is required for the absorption of carotenoids (Reddy 1995). In addition, some evidence suggests that the type of dietary fat also plays a role. The recovery of lycopene in rat mesenteric lymph was found to be increased on average from 2.55% to up to 6% when lycopene was infused into the duodenum with corn or olive oil, respectively (Clark et al. 2000). Thus, the extraction of lycopene into lipophilic phase favours the intestinal absorption of extracted carotenoid (Stahl et al. 1992), but polyunsaturated fatty acids decrease lycopene absorption in comparison with more saturated fats (Clark et al. 2000).

Other Carotenoids

Lycopene and other carotenoids are not well absorbed, and interactions, mostly competitive, between carotenoids during absorption and post-absorptive metabolism have been demonstrated to influence their bioavailability in animal and human feeding or supplementation experiments, as well as *in vitro* investigations of intestinal beta-carotene cleavage (van den Berg 1999). Studies on these interactions are difficult and often confusing, because the results are influenced by the species used, the dose, the study design (single-dose, short- or long-term supplementation) as well as the response measured (plasma concentration or postprandial triacylglycerol-rich lipoprotein levels).

Graziano et al. (1995) have reported that dietary supplementation with 100 mg β-carotene per day for 6 d reduced the amount of lycopene in LDL. Conversely, however, Johnson et al. (1997) found that ingestion of a single combined dose of lycopene and beta-carotene improves lycopene absorption in men, while another study (van den Berg and van Vliet 1998), which measured carotenoid concentrations in the TRL fraction after a 15 mg dose of β-carotene alone or combined with 15 mg lycopene, suggested that β-carotene has no effect on lycopene absorption. The difference in dose may partially explain these apparently contradictory results, as at high doses the solubility of lycopene may be limited (Borel et al. 1996). However, a better explanation may come from the study of Tyssandier et al. (2002), who demonstrated that, in the post-prandial phase, lycopene competes with lutein, and possibly with β-carotene, for incorporation into the chylomicron fraction, but in the medium term this competition has no adverse effect on plasma concentration of carotenoids, including lycopene, suggesting that, at least in the short term, carotenoids do not interfere with each other in terms of bioavailability and may indeed provide some synergic effects.

Age and Other Factors

There is some evidence to suggest that aging affects the absorption of lycopene. In a study of a population-based sample of 400 human subjects, Brady et al. (1996) found that lower serum lycopene concentrations were associated with younger age, and other work showing that the lycopene found in chylomicrons (after adjusting for triacylglycerol levels) was 40% lower in younger (20-35 yr) as compared to older (60-75 yr) subjects suggests that this effect is caused by changes in lycopene absorption (Cardinault et al. 2003).

Other factors that have been shown to influence lycopene absorption include the consumption of non-absorbable fat substitutes, such as sucrose polyester (van het Hof et al. 2000), hormonal status (Erdman 2005), the hypocholesterolemic drug cholestyramine, and smoking (Arap and Steck 2000). The effect of the sucrose polyester is to decrease absorption in chylomicrons, since lycopene is incorporated into the non-absorbable lipid rather than the micelles formed from dietary fats (van het Hof et al. 2000). Cholestyramine reduces serum cholesterol levels by reducing lipid absorption in the intestine, and this is likely to be the explanation for its effects in reducing serum lycopene concentrations (Elinder et al. 1995). It is not yet clear, however, whether the effects of hormonal status and smoking are caused by changes in absorption or by other mechanisms such as increased use of lycopene in the body as, for example, an antioxidant to protect against free radical damage (Arap and Steck 2000).

Studies *in vitro*

The *in vivo* studies described above provide important information about the overall process of the absorption of lycopene, but the elucidation of the mechanisms involved in absorption at the cellular level requires an *in vitro* model. The human intestinal cell line, CaCo-2, is able to secrete chylomicrons when incubated in the presence of oleic acid and taurocholate, and During et al. (2002) and During and Harrison (2004) have developed this model for the study of the uptake of carotenoids and their subsequent secretion in chylomicrons by enterocytes. Much of the data obtained using this model relates to the absorption of β-carotene

(During and Harrison 2004), but it has been shown that differentiated CaCo-2 cells take up lycopene at the apical side and secrete it at the basolateral side, although the extent of absorption of lycopene (2.5%) was lower than that of other carotenoids (e.g., β-carotene, 11%) (During et al. 2002). In the same study, both cellular uptake and secretion of lycopene were found to be reduced in the presence of β-carotene (lycopene to β-carotene, molar ratio 1:5). Moreover, despite the relatively large difference in the secretion of the different carotenoids tested, the proportion taken up by the cells was similar, ranging between 15 and 18%, suggesting that the individual structures of the different compounds may be an important factor in their incorporation into chylomicrons.

LYCOPENE AND CHYLOMICRON REMNANTS

After its absorption from the intestine in chylomicrons, lycopene is carried to the liver in chylomicron remnants. Since it is now clear that the remnant particles enter and are retained in the artery wall (Proctor et al. 2002), dietary lycopene could potentially affect the development of atherosclerotic lesions during this transport phase.

Foam cell formation occurs when macrophages that have invaded the artery wall take up lipid from lipoproteins that have become trapped in sub-endothelial space and store it intracellularly (Wilhelm and Cooper 2003). Extensive studies have established that LDL has a major role in macrophage foam cell formation; however, modification of the particles either chemically or by oxidation, a process that has been demonstrated to occur within the artery wall, is necessary before lipid accumulation is induced (Albertini et al. 2002). The importance of prior oxidation for the effects of LDL on foam cell formation and other initiating events in atherosclerosis (Albertini et al. 2002) has led to the development of the hypothesis that antioxidants may have a protective effect against heart disease, and this view is supported by epidemiological evidence indicating that diets rich in fruits and vegetables that contain relatively high levels of natural antioxidants such as carotenoids, including lycopene, and vitamin E reduce the risk of atherosclerosis development (Trichopoulou and Vasilopoulou 2000, Rao 2002). Despite this, large-scale trials of dietary supplementation with β-carotene and vitamin E have failed to show beneficial effects for heart disease (Clarke and Armitage 2002).

Studies in our laboratory and others have shown that chylomicron remnants cause macrophage foam cell formation without prior oxidation (Yu and Mamo 2000, Batt et al. 2004, Botham et al. 2005). Our experiments have demonstrated that chylomicron remnant-like particles (CRLPs), which resemble physiological chylomicron remnants in their size, density and lipid composition and contain human apoE, are taken up by human macrophages and cause intracellular lipid accumulation comparable to that observed with a similar concentration of cholesterol in oxidized LDL (Batt et al. 2004). Since lycopene is a powerful singlet oxygen quencher (Di Mascio et al. 1989), we hypothesized that its possible protective effect against the development of atherosclerosis (Arap and Steck 2000, Rao et al. 2006, Sesso 2006) may be due to inhibition of the effects of chylomicron remnants on macrophage foam cell formation caused by its antioxidant properties. When lycopene was incorporated into CRLPs (lycCRLPs), we found that they were indeed protected from oxidation. To our surprise, however, Oil red O staining suggested that lipid accumulation in human macrophages was markedly increased in the presence of lycCRLPs as compared to CRLPs (Fig. 4).

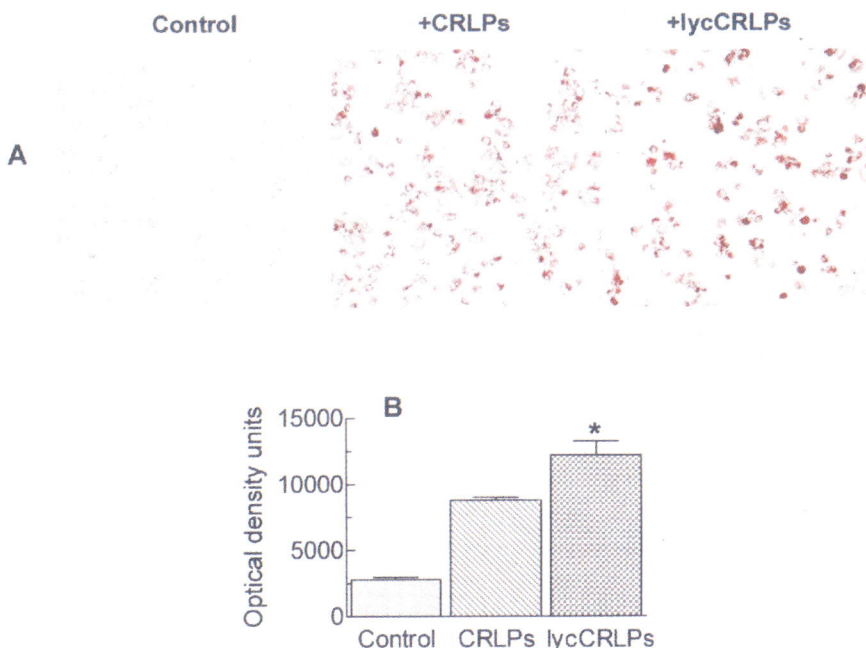

Fig. 4 The presence of lycopene in chylomicron remnants enhances their induction of macrophage foam cell formation. Macrophages derived from the human monocyte cell line THP-1 were incubated with or without (control) chylomicron remnant-like particles (CRLPs) or CRLPs containing lycopene (lycCRLPs) for 48 hr, then stained with Oil Red O. A. Images captured by light microscopy; B. Quantification of Oil Red O staining by optical density volume analysis. Data shown are the mean from three separate experiments and error bars show the SEM. *P < 0.05 vs CRLPs.

This result was confirmed by quantitative lipid analysis, which showed that triacylglycerol and cholesterol levels in the cells were raised by 100% and 62% respectively (Fig. 5).

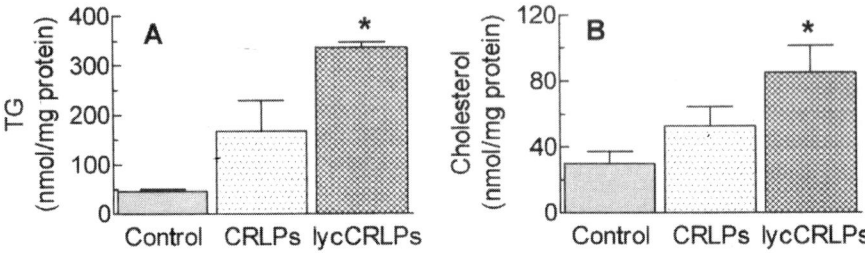

Fig. 5 The triacylglycerol and cholesterol content of macrophages exposed to chylomicron remnants is increased by lycopene. Macrophages derived from the human monocyte cell line, THP-1, were incubated with or without (control) chylomicron remnant-like particles (CRLPs) or CRLPs containing lycopene (lycCRLPs) for 48 hr and the triacylglycerol (TG) (A) and cholesterol (B) content of the cells was determined. Data shown are the mean from three separate experiments and error bars show the SEM. *P < 0.05 vs CRLPs (Moore et al. 2003).

To investigate whether this effect is caused by the protection of the remnant particles from oxidation or is a specific effect of lycopene, similar experiments were carried out with CRLPs containing the lipophilic antioxidant drug probucol, which is structurally unrelated to lycopene (Niguchi and Niki 2000). As found with lycCRLPs, the CRLPs containing probucol were protected from oxidation, and they also caused an increase in lipid accumulation of a similar magnitude (+120%) to that observed with the lycopene-containing particles (Moore et al. 2004). Further experiments established that the effect of the antioxidants in raising intracellular lipid levels is due to an increased rate of uptake of the particles (Moore et al. 2004). Thus, although the presence of lycopene in chylomicron remnants protects them from oxidation, contrary to expectations, this appears to enhance, rather than inhibit, foam cell formation.

A number of studies on dietary supplementation with tomato products or lycopene have found little effect on the oxidative modification of LDL or its resistance to oxidation, despite significant increases in the lycopene content of the lipoprotein (Kaliora et al. 2006). Aviram and Furman (1998), on the other hand, have suggested that oxidative state of LDL is reduced by an increase in its content of lycopene; in another study, serum LDL oxidation was reported to be decreased as serum lycopene levels increased (Agarwal and Rao 1998). No information is available on the effects of LDL containing lycopene on macrophage lipid accumulation, but experiments in which antioxidants such as vitamin E or probucol were contained in modified LDL have indicated that foam cell formation is suppressed

(Yamamoto et al. 1988, Suzukawa et al. 1994). Thus, the type of lipoprotein carrier of lycopene and other dietary antioxidants appears to be crucial for their effects on foam cell formation and, consequently, atherosclerosis development. When they are carried in LDL, the protection from oxidation they confer inhibits the process, but it has the opposite effect during their transport post-prandially in chylomicron remnants, so that in this case lipid accumulation is promoted. Since chylomicron remnants are normally rapidly cleared from the circulation, the beneficial effect of their transport in LDL may predominate in healthy individuals, but in conditions where the clearance of remnants from the circulation is delayed, as occurs in obesity and diabetes mellitus (Botham et al. 2005), the potentially deleterious effects of their presence in chylomicron remnants may become more important.

The unexpected finding that lycopene enhances foam cell formation when carried in chylomicron remnants suggests that it may promote atherosclerotic lesion development in the vasculature in the post-prandial phase, but more rapid clearance of the particles by the liver could be advantageous, as the time available for interaction with the artery wall would be reduced. However, in studies using CRLPs and the human hepatoma cell line HepG2, we found that, in contrast to the marked increase in uptake of CRLPs by macrophages found when lycopene was present in the particles (Moore et al. 2003), the uptake of lycCRLPs by HepG2 cells was decreased by up to 40% in comparison to that of CRLPs (Fig. 6).

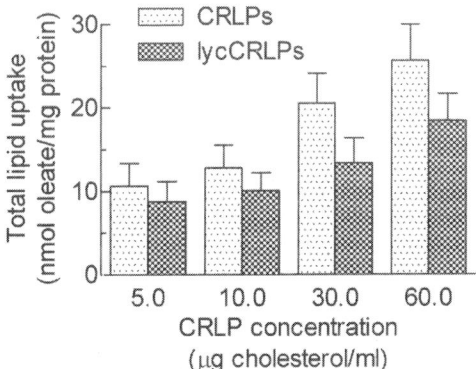

Fig. 6 The incorporation of lycopene into chylomicron remnants inhibits their uptake by hepatocytes. [³H]Triolein-labelled chylomicron remnant-like particles (CRLPs) or CRLPs containing lycopene (lycCRLPs) were incubated with HepG2 cells for 4 hr and the total uptake of radioactivity by the cells was measured. Each point represents the mean from four experiments and error bars show the SEM. Significance limits (ANOVA repeated measures) CRLPs vs lycCRLPs, P < 0.01 (Bejta et al. 2007).

In order to investigate the mechanism of this effect, further experiments were carried out using selective inhibitors of the receptor-mediated processes involved in chylomicron remnant uptake by the liver (Bejta et al. 2007). The hepatic uptake of chylomicron remnants is mediated by the LDL receptor (LDLr) and the LDLr-related protein (LRP) (Rohlmann et al. 1998, Zeng et al. 1998, Yu and Cooper 2001). It is believed that the particles are taken up after direct binding to the LDLr, or alternatively after initial binding to heparan sulphate proteoglycans (HSPG) on the hepatocyte surface followed by internalization by the LRP (Ji et al. 1993, Herz et al. 1995, Zeng et al. 1998, Mahley and Ji 1999, Yu and Cooper 2001). Addition of excess LDL to the cells to block the internalization of remnants via the LDLr markedly reduced CRLP uptake, while the uptake of lycCRLPs was not significantly changed (Fig. 7.)

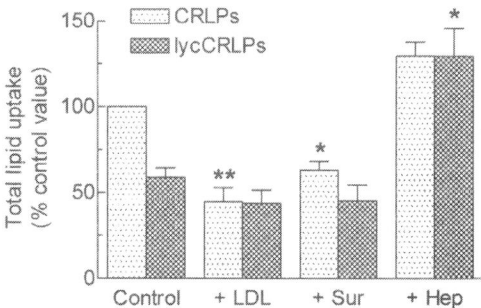

Fig. 7 The low density lipoprotein receptor (LDLr), LDLr-related protein (LRP) and heparan sulphate proteoglycans (HSPG) are involved in the inhibition of hepatic chylomicron remnant uptake by lycopene. HepG2 cells were pre-incubated for 1 hr with or without (control) the following: low density lipoprotein (LDL) (100 μg protein/ml); suramin (500 μg/ml) (Sur) or heparinase (Hep) (2 U/ml). [^{3}H]Triolein-labelled chylomicron remnant-like particles (CRLPs) or CRLPs containing lycopene (lycCRLPs) were then added, the incubation was continued for a further 4 hr and the uptake of radioactivity by the cells was determined. Data are expressed as a percentage of the value obtained with CRLPs in the absence of inhibitors (control) and are the mean $^+$ SEM from three experiments. Error bars show the SEM. *P < 0.05, **P < 0.01 vs CRLP control.

Similarly, the presence of the polysulphated drug suramin, a known inhibitor of the LRP (Vassiliou 1997), inhibited CRLP, but not lycCRLP, uptake (Fig. 7). These findings suggest that differential internalization via both the LDLr and LRP may be partly responsible for the decreased uptake of CRLPs containing lycopene by HepG2 cells.

It has been suggested that HSPG, which are abundant on the surface of hepatocytes, are involved in chylomicron remnant uptake by the liver either alone or via a "secretion capture" pathway in which remnants bind initially to HSPG, become enriched in apoE, and are then transported to the LRP for internalization by endocytosis (Mahley and Ji 1999, Ji et al. 1997). In our experiments, lycCRLP uptake was increased by about 2-fold by treatment with heparinase (Fig. 7), raising it to the level found with CRLPs, which was not significantly changed. Similar results were obtained after pre-treatment of the cells with trypsin, which would also be expected to remove surface HSPG (Bejta et al. 2007). Thus, surface HSPG appears to be involved in the decreased uptake of the lycopene-containing remnants. One possible explanation for this is that the incorporation of lycopene into the particles alters the distribution of apoE on their surface, causing their binding to HSPG to be inhibited, while still allowing interaction with the LDLr and the LRP. This could also provide an explanation for the strikingly different effect of incorporation of lycopene into CRLPs in the liver and macrophages (where uptake is increased) (Moore et al. 2003), as surface HSPG levels are much lower in the latter cell type than in hepatocytes (Deng et al. 1997).

Delayed clearance of chylomicron remnants containing lycopene by the liver coupled with increased uptake by macrophages is likely to increase the atherogenicity of the particles. This again highlights the differential effects of lycopene when it is carried in chylomicron remnants as compared to LDL and may provide part of the explanation for the failure to demonstrate protective effects of dietary supplementation with lipophilic antioxidants against cardiovascular disease in large-scale clinical trials (Clarke and Armitage 2002).

SUMMARY AND CONCLUSIONS

It has been established that lycopene, as might be expected for a lipophilic compound, is absorbed in chylomicrons. However, the amount incorporated depends on a number of factors. Levels of the *cis* as compared to the *all-trans* isomer are increased in chylomicrons as compared to food, possibly because of its increased solubility in bile acid micelles (Boileau et al. 1999); there is more incorporation of lycopene from tomato products than from raw tomatoes (because the breakdown of the food matrix makes it more accessible) (Shi and Le Maguer, 2000, Holloway et al. 2000) and when a source of fat is ingested simultaneously (Brown et al. 2004). In addition, the lycopene content of chylomicrons is reduced in older as compared to younger subjects (Cardinault et al. 2003), by non-absorbable fat substitutes such as sucrose polyester (van het Hof et al. 2000), and probably by the hypo-cholesterolemic drug cholestyramine (Elinder et al. 1995). The effects of the presence of other carotenoids in the diet on lycopene absorption in chylomicrons are not entirely clear; it has been reported to be unaffected by β-carotene (van den Berg and van Vliet 1998) and increased by lutein (Tyssandier et al. 2002); however, this latter change does not appear to influence serum lycopene concentrations in the medium term.

Chylomicron remnants are formed from chylomicrons after their entry into the blood and carry lipophilic compounds such as lycopene to the liver. Surprisingly, our studies have shown that the presence of lycopene in the particles increases their induction of foam cell formation and decreases their uptake by the liver, effects that are likely to increase the atherogenicity of the particles (Moore et al. 2003, Bejta et al. 2007). However, other studies have suggested that lycopene may protect against atherosclerotic lesion development (Arap and Steck 2000, Rao et al. 2006, Sesso 2006), possibly by protecting LDL from oxidation (Aviram and Furman 1998, Agarwal and Rao 1998). Thus, the effects of lycopene on atherogenesis may depend on the type of lipoprotein carrier, so that its presence is beneficial in LDL, but detrimental in chylomicron remnants. Normally, because of the rapid clearance of remnants from the circulation, the beneficial effect is likely to be more prominent, but the potentially deleterious effect of carriage in chylomicron remnants may be of more concern when the particles accumulate in the blood in conditions such as obesity and diabetes (Botham et al. 2005).

Many studies have investigated the influence of tomatoes and other dietary lycopene sources on plasma concentrations of the carotenoid, but relatively few have specifically examined absorption in chylomicrons, and a great deal more research is essential, as a more complete understanding of the mechanisms regulating the process will help to define the optimum conditions for the intake of lycopene in order to maximize its health benefits.

ABBREVIATIONS

Apo: apolipoprotein; CRLPs: chylomicron remnant-like particles; HSPG: heparan sulphate proteoglycans; LDL: low density lipoprotein; LDLr: low density lipoprotein receptor; LRP: low density lipoprotein receptor-related protein; lycCRLPs: chylomicron remnant-like particles containing lycopene; TRL: triglyceride-rich lipoproteins; VLDL: very low density lipoprotein

References

Agarwal, S. and A.V. Rao. 1998. Tomato lycopene and low density lipoprotein oxidation: a human dietary intervention study. Lipids 33: 981-984.

Albertini, R. and R. Moratti, and G. DeLuca. 2002. Oxidation of low-density lipoprotein in atherosclerosis from basic biochemistry to clinical studies. Curr. Mol. Med. 2: 579-592.

Arap, L. and S. Steck. 2000. Lycopene and cardiovascular disease. Amer. J. Clin. Nutr. 71 (suppl): 1691S-1695S.

Assmann, G. and G. de Backer, S. Bagnara, J. Betteridge, G. Crepaldi, A. Fernandez-Cruz, J. Godtfredsen, B. Jacotot, R. Paoletti, S. Renaud, G. Ricci, E. Rocha, E. Trautwein, G.C. Urbinati, G. Varela, and C. Williams. 1997. Olive oil and the Mediterranean diet: implications for health in Europe. Br. J. Nurs. 6: 675-677.

Aviram, M. and B. Furhman. 1998. LDL oxidation by arterial wall macrophages depends on the oxidative status in the lipoprotein and in cells: role of prooxidants vs. antioxidants. Mol. Cell. Biochem. 188: 149-159.

Batt, K.V. and M. Avella, E.H. Moore, B. Jackson, K.E. Suckling, and K.M. Botham. 2004. Differential effects of low density lipoprotein and chylomicron remnants on lipid accumulation in human macrophages. Exp. Biol. Med. 229: 528-537.

Bejta, F. and M. Napolitano, K.M. Botham, and E. Bravo. 2007. Incorporation of lycopene into chylomicron remnant-like particles inhibits their uptake by HepG2 cells. Life Sci. 80: 1699-1705.

Boileau, T.W. and A.C. Boileau, and J.W. Erdman. 2002. Bioavailability of *all-trans* and *cis*-isomers of lycopene. Exp. Biol. Med. (Maywood). 227: 914-919.

Boileau, A.C. and N.R. Merchen , K. Wasson, C.A. Atkinson, and J.W. Erdman. 1999. *Cis*-lycopene is more bioavailable that *trans*-lycopene *in vitro* and *in vivo* in lymph-cannulated ferrets. J. Nutr. 129: 1176-1181.

Borel, P. and P. Grolier, M. Armand, A. Partier, H. Lafont, D. Lairon, and V. Azais-Braesco. 1996. Carotenoids in biological emulsions: solubility, surface-to-core distribution, and release from lipid droplets. J. Lipid Res. 37: 250-261.

Boren, J. and M. Gustafsson, K. Skalen, C. Flood, and T.L. Innerarity. 2000. Role of extracellular retention of low density lipoproteins in atherosclerosis. Curr. Opin. Lipidol. 11: 451-456.

Botham, K.M. and E. Bravo, J. Elliott, and C.P.D. Wheeler-Jones. 2005. Direct interaction of dietary lipids carried in chylomicron remnants with cells of the artery wall: implications for atherosclerosis development. Curr. Pharmaceut. Design 11: 3681-3695.

Brady, W.E. and J.A. Mares-Perlman, P. Bowen, and M. Stacewicz-Sapuntzakis. 1996. Human serum carotenoid concentrations are related to physiologic and lifestyle factors. J. Nutr. 126: 129-137.

Bravo, E. and G. Ortu, A. Cantafora, M.S. Lambert, M. Avella, P.A. Mayes, and K.M. Botham. 1995. Comparison of the hepatic uptake and processing of cholesterol from chylomicrons of different fatty acid composition in the rat in vivo. Biochim. Biophys. Acta 1258: 328-336.

Britton, G. 1995. Structure and properties of carotenoids in relation to function. FASEB J. 9: 1551-1558.

Brown, M.J. and M.G. Ferruzzi, M.L. Nguyen, D.A. Cooper, A.L. Eldridge, S.J. Schwartz, and W.S. White. 2004. Carotenoid bioavailability is higher from salads ingested with full-fat than with fat-reduced salad dressings as measured with electrochemical detection. Amer. J. Clin. Nutr. 80: 396-403.

Cardinault, N. and V. Tyssandier, P. Grolier, B.M. Winklhofer-Roob, J. Ribalta, C. Bouteloup-Demange, E. Rock, and P. Borel. 2003. Comparison of the postprandial chylomicron carotenoid responses in young and older subjects. Eur. J. Nutr. 42: 315-323.

Clark, R.M. and L. Yao, L. She, and H.C. Furr. 1998. A comparison of lycopene and canthaxanthin absorption; using the rat to study the absorption of non pro-vitamin A carotenoids. Lipids 33: 159-163.

Clark, R.M. and L. Yao, L. She, and H.C. Furr. 2000. A comparison of lycopene and astaxanthin absorption from corn oil and olive oil emulsions. Lipids 35: 803-806.

Clarke, R. and J. Armitage. 2002. Antioxidant vitamins and risk of cardiovascular disease. Review of large scale randomised trials. Cardiovasc. Drugs Ther. 16: 411-415.

Clinton, S.K. and C. Emenhiser, S.J. Schwartz, D.G. Bostwick, A.W. Williams, B.J. Moore, and J.W. Erdman. 1996. cis-trans lycopene isomers, carotenoids, and retinol in the human prostate. Cancer Epidemiol. Biomarkers Prev. 5: 823-833.

Deng, J. and V. Rudick, M. Rudick, and L. Dory. 1997. Investigation of plasma membrane-associated apolipoprotein E in primary macrophages. J. Lipid Res. 38: 217-227.

De Pascale, C. and M. Avella, J.S. Perona, V. Ruiz-Gutierrez, C.P.D. Wheeler-Jones, and K.M. Botham. 2006. The fatty acid composition of chylomicron remnant-like particles influences their uptake and induction of lipid accumulation in macrophages. FEBS J. 273: 5632-5640.

Di Mascio, P. and S. Kaiser, and H. Sies. 1989. Lycopene as the most efficient biological carotenoid single oxygen quencher. Arch. Bochem. Biophys. 274: 532-538.

Diwadkar-Navsariwala, V. and J.A. Novotny, D.M. Gustin, J.A. Sosman, K.A. Rodvold, J.A. Crowell, M. Stacewicz-Sapuntzakis, and P.E. Bowen. 2003. A physiological pharmacokinetic model describing the disposition of lycopene in healthy men. J. Lipid. Res. 44: 1927-1939.

During, A. and E.H. Harrison. 2004. Intestinal absorption and metabolism of carotenoids: insights from cell culture. Arch. Biochem. Biophys. 430: 77-88.

During, A. and M.M. Hussain, D.W. Morel, and E.H. Harrison. 2002. Carotenoid uptake and secretion by CaCo-2 cells: beta-carotene isomer selectivity and carotenoid interactions. J. Lipid Res. 43: 1086-1095.

Elinder, E.S. and K. Hadell, J. Johansson, J. Molgaard, I. Holme, A.G. Olsson, and G. Walldius. 1995. Probucol treatment decreases serum concentrations of diet-derived antioxidants. Arteioscler. Thromb. Vasc. Biol. 15: 1057-1063.

Erdman, J.W. 2005. How do nutritional and hormonal status modify the bioavailability, uptake and distribution of different isomers of lycopene. J. Nutr. 135: 2046S-2047S.

Evans, M. and Y. Berhane, K.M. Botham, J. Elliott, and C.P.D. Wheeler-Jones. 2004. Chylomicron remnant-like particles modify endothelial cell production of vasoactive mediators. Biochem. Soc. Trans. 32: 110-112.

Fielding, J.M. and K.G. Rowley, P. Cooper, and K. O'Dea. 2005. Increases in plasma lycopene concentration after consumption of tomatoes cooked with olive oil. Asia Pac. J. Clin. Nutr. 14: 131-136.

Gartner, C. and W. Stahl, and H. Sies. 1997. Lycopene is more bioavailable from tomato paste than from fresh tomatoes. Amer. J. Clin. Nutr. 66: 116-122.

Ghung, B.H. and G. Talis, V. Yalamoori, G.M. Anantharamaiah, and J.P. Segrest. 1994. Liposome-like particles isolated from human atherosclerotic plaques are structurally similar to surface remnants of triglyceride-rich lipoproteins. Arterioscler. Thromb. 14: 622-635.

Graziano, J.M. and E.J. Johnson, R.M. Russell, J.E. Manson, M.J. Stampfer, P.M. Ridkcr, B. Frei, C.H. Hennekens, and N.I. Krinsky. 1995. Discrimination in absorption or transport of β-carotene isomers after oral supplementation with either *all-trans-* or 9-*cis*-β-carotene. Amer. J. Clin. Nutr. 61: 1248-1252.

Grieve, D.J. and M.A. Avella, K.M. Botham, and J.E. Elliott. 1998. Effects of chylomicrons and chylomicron remnants on endothelium-dependent relaxation of rat aorta. Eur. J. Pharmacol. 348: 181-190.

Gustin, D.M. and K.A. Rodvold, J.A. Sosman, V. Diwadkar-Navsariwala, M. Stacewicz-Sapuntzakis, M. Viana, J.A. Crowell, J. Murray, P. Tiller, and P.E. Bowen. 2004. Single-dose pharmacokinetic study of lycopene delivered in a well-defined food-based lycopene delivery system (tomato paste-oil mixture) in healthy adult male subjects. Cancer Epidemiol. Biomarkers Prev. 13: 850-860.

Harker, M. and J. Hirschberg. 1997. Biosynthesis of ketocarotenoids in transgenic cyanobacteria expressing the algal gene for beta-C-4-oxygenase. FEBS Lett. 404: 129-134.

Harris, W.S. 1996. Dietary fish oils and blood lipids. Curr. Opin. Lipidol. 7: 3-7.

Heber, D. and Q.Y. Lu. 2002. Overview of mechanisms of action of lycopene. Exp. Biol. Med. (Maywood) 227: 920-923.

Herz, J. and S-Q. Qiu, H. Oesterle, V. DeSilva, S. Shafi, and R.J. Havel. 1995. Initial hepatic removal of chylomicron remnants is unaffected but endocytosis is delayed in mice lacking the low density lipoprotein receptor. Proc. Natl. Acad. Sci. USA 92: 4611-4615.

Holloway, D.E. and M. Yang, G. Paganga, C.A. Rice-Evans, and P.M. Bramley. 2000. Isomerization of dietary lycopene during assimilation and transport in plasma. Free Radic. Res. 32: 93-102.

Ji, Z-S. and S. Fazio, Y.L. Lee, and R.W. Mahley. 1993. Secretion-capture role for apolipoprotein E in remnant lipoprotein metabolism involving cell surface heparan sulfate proteoglycans. J. Biol. Chem. 269: 2674-2772.

Ji, Z-S. and H.L. Dichek, R.D. Miranda, and R.W. Mahley. 1997. Heparan sulfate proteoglycans participate in hepatic lipase and apolipoprotein E-mediated binding and uptake of plasma lipoproteins, including high density lipoproteins. J. Biol. Chem. 272: 31285-31292.

Johnson, E.J. and J. Qin, N.I. Krinsky, and R.M. Russell. 1997. Ingestion by men of a combined dose of β-carotene and lycopene does not affect the absorption of β-carotene but improves that of lycopene. J. Nutr. 127: 1833-1837.

Kadar, A. and T. Glasz. 2001. Development of atherosclerosis and plaque biology. Cardiovasc. Surg. 9: 109-121.

Kaliora, A.C. and G.V. Dedoussis, and H. Schmidt. 2006. Dietary antioxidants in preventing atherosclerosis. Atherosclerosis 187: 1-17.

Keys, A. 1970. Coronary heart disease in seven countries. Circulation 41 (Suppl 1): 1-198.

Kok, F.J. and D. Kromhout. 2004. Atherosclerosis-epidemiological studies on the health effects of a Mediterranean diet. Eur. J. Nutr. 43 (suppl 1): I/2-5.

Koonsvitsky, B.P. and D.A. Berry, M.B. Jones, P.Y.T. Lin, D.A. Cooper, D.Y. Jones, and J.E. Jackson. 1997. Olestra affects serum concentrations of α-tocopherol and carotenoids but not vitamin D or vitamin K status in free-living subjects. J. Nutr. 127: 1636S-1645S.

Lambert, M.S. and K.M. Botham, and P.A. Mayes. 1995. Variations in composition of dietary fats affect hepatic uptake and metabolism of chylomicron remnants. Biochem. J. 310: 845-852.

Lambert, M.S. and M.A. Avella, Y. Berhane, E. Shervill, and K.M. Botham. 2001. The fatty acid composition of chylomicron remnants influences their binding and internalisation by isolated hepatocytes. Eur. J. Biochem. 268: 3983-3992.

Mahley, R.W. and Z-S. Ji. 1999. Remnant lipoprotein metabolism: key pathway involving cell surface heparan sulfate proteoglycans and apolipoprotein E. J. Lipid Res. 40: 1-16.

Mamo, J.C. and J.R. Wheeler. 1994. Chylomicrons or their remnants penetrate the rabbit thoracic aorta as efficiently as do smaller macromolecules, including low-density lipoprotein, high-density lipoprotein, and albumin. Coron. Artery Dis. 5: 695-705.

Moore, E.H. and M. Napolitano, M. Avella, F. Bejta, K.E. Suckling , E. Bravo, and K.M. Botham. 2004. Protection of chylomicron remnants from oxidation by incorporation of probucol into the particles enhances their uptake by human macrophages and increases lipid accumulation in the cells. Eur. J. Biochem. 271: 2417-2427.

Moore, E.H. and M. Napolitano, A. Prosperi, M. Avella, K.E. Suckling, E. Bravo, and K.M. Botham. 2003. Incorporation of lycopene into chylomicron remnant-like particles enhances their induction of lipid accumulation in macrophages. Biochem. Biophys. Res. Commun. 312: 1216-1219.

Moreno, J.J. and M.T. Mitjavila. 2003. The degree of unsaturation of dietary fatty acids and the development of atherosclerosis. J. Nutr. Biochem. 14: 182-195.

Niguchi, N. and E. Niki. 2000. Phenolic antioxidants: a rationale for design and evaluation of novel antioxidant drugs for atherosclerosis. Free Rad. Biol. Med. 28: 1538-1546.

Nguyen, M.L. and S.J. Schwartz. 1998. Lycopene stability during food processing. Proc. Soc. Exp. Biol. Med. 218: 101-105.

Pal, S. and K. Semorine, G.F. Watts, and J. Mamo. 2003. Identification of lipoproteins of intestinal origin in human atherosclerotic plaque. Clin. Chem. Lab. Med. 41: 792-795.

Parker, R.S. 1996. Absorption, metabolism, and transport of carotenoids. FASEB J. 10: 542-551.

Porrini, M. and P. Riso, and G. Testolin. 1998. Absorption of lycopene from single or daily portions of raw and processed tomato. Br. J. Nutr. 80: 353-361.

Proctor, S.D. and D.F. Vine, and J.C.L. Mamo. 2002. Arterial retention of apolipoprotein B48- and B100-containing lipoproteins in atherogenesis. Curr. Opin. Lipidol. 13: 461-470.

Rao, A.V. 2002. Lycopene, tomatoes and the prevention of coronary heart disease. Exp. Biol. Med. 227: 908-913.

Rao, A.V. and S. Agarwal. 1998. Bioavailability and in vivo antioxidant properties of lycopene from tomato products and their possible role in the prevention of cancer. Nutr. Cancer 31: 199-203.

Rao, A.V. and M.R. Ray, and L.G. Rao. 2006. Lycopene. Adv. Food Nutr. Res. 51: 99-164.

Redgrave, T.G. 1983. Formation and metabolism of chylomicrons. Int. Rev. Physiol. 28: 103-130.

Reddy, V. 1995. Vitamin A status and dark green leafy vegetables. Lancet 346: 1634-1636.

Rohlmann, A. and M. Gotthardt, R.E. Hammer, and J. Herz. 1998. Inducible inactivation of hepatic LRP gene by cre-mediated recombination confirms role of LRP in clearance of chylomicron remnants. J. Clin. Invest. 101: 689-695.

Schierle, J. and W. Bretzel, I. Buhler, N. Faccin, D. Hess, K. Steiner, and W. Schuep. 1997. Content and isomeric ratios of lycopene in food and human blood plasma. Food Chem. 59: 459-465.

Schlagheck, T.G. and K.A. Riccardi, N.L. Zoric, S.A. Torri, L.D. Dugan, and J.C. Peters. 1997. Olestra dose response on fat-soluble and water-soluble nutrients in humans. J. Nutr. 127: 1646S-1665S.

Sesso, H.D. 2006. Carotenoids and cardiovascular disease: what research gaps remain? Curr. Opin. Lipidol. 17: 11-16.

Shi, J. and M. Le Maguer. 2000. Lycopene in tomatoes: chemical and physical properties affected by food processing. Crit. Rev. Food Sci. Nutr. 40: 1-42.

Stahl, W. and H. Sies. 1992. Uptake of lycopene and its geometrical isomers is greater from heat-processed than from unprocessed tomato juice in humans. J. Nutr. 1992122: 2161-2166.

Stahl, W. and W. Schwarz , A.R. Sundquist, and H. Sies. 1992. cis-trans isomers of lycopene and beta-carotene in human serum and tissues. Arch. Biochem. Biophys. 294: 173-177.

Suzukawa, M. and M. Abbey, P. Clifton, and P.J. Nestel. 1994. Effects of supplementing with vitamin E on the uptake of low density lipoprotein and the stimulation of cholesteryl ester formation in macrophages. Atherosclerosis 110: 77-86.

Trichopoulou, A. and E. Vasilopoulou. 2000. Mediterranean diet and longevity. Br. J. Nutr. 84: 205-209.

Tyssandier, V. and N. Cardinault, C. Caris-Veyrat, M.J. Amiot, P. Grolier, C. Bouteloup, V. Azais-Braesco, and P. Borel. 2002. Vegetable-borne lutein, lycopene, and beta-carotene compete for incorporation into chylomicrons, with no adverse effect on the medium-term (3-wk) plasma status of carotenoids in humans. Amer. J. Clin. Nutr. 75: 526-534.

Tyssandier, V. and E. Reboul, J.F. Dumas, C. Bouteloup-Demange, M. Armand, J. Marcand, M. Sallas, and P. Borel. 2003. Processing of vegetable-borne carotenoids in the human stomach and duodenum. Amer. J. Physiol. Gastrointest. Liver Physiol. 284: G913-G923.

van den Berg, H. 1999. Carotenoid interactions. Nutr. Rev. 57: 1-10.

van den Berg, H. and T. van Vliet. 1998. Effect of simultaneous, single oral doses of beta-carotene with lutein or lycopene on the beta-carotene and retinyl ester responses in the triacylglycerol-rich lipoprotein fraction of men. Amer. J. Clin. Nutr. 68: 82-89.

van het Hof, K.H. and C.E. West, J.A. Weststrate, and J.G.A.J. Hautvast. 2000. Dietary factors that affect the bioavailability of carotenoids. J. Nutr. 130: 503-506.

Vassiliou, G. 1997. Pharmacological concentrations of suramin inhibit the binding of α-2 macroglobulin to its cell surface receptor. Eur. J. Biochem. 250: 320-325.

Voutilainen, S. and T. Nurmi, J. Mursu, and T.H. Rissanen. 2006. Carotenoids and cardiovascular health. Amer. J. Clin. Nutr. 83: 1265-1271.

Weststrate, J.A. and K.H. van het Hof. 1995. Sucrose polyester and plasma carotenoid concentrations in healthy subjects. Amer. J. Clin. Nutr. 62: 591-597.

Wilhelm, M.G. and A.D. Cooper. 2003. Induction of atherosclerosis by human chylomicron remnants: a hypothesis. J. Atheroscler. Thromb. 10: 132-139.

Yamamoto, A. and H. Hara, S. Takaichi, S. Wakasugi, and M. Tomikawa. 1988. Effect of probucol on macrophages, leading to regression of xanthomas and atheromatous vascular lesions. Amer. J. Cardiol. 62: 31B-36B.

Yu, K.C. and A.D. Cooper. 2001. Postprandial lipoproteins and atherosclerosis. Front. Biosci. 6: D332-D354.

Yu, K.C. and J.C.L. Mamo. 2000. Chylomicron remnant-induced foam cell formation and cytotoxicity: a possible mechanism for cell death in atherosclerosis. Clin. Sci. 98: 183-192.

Yusuf, S. and S. Reddy, S. Ounpuu, and S. Anand. 2001. Global burden of cardiovascular diseases. Circulation 104: 2746-2753.

Zeng, B.J. and B.C. Mortimer, I.J. Martins, U. Seydel, and T.G. Redgrave. 1998. Chylomicron remnant uptake is regulated by the expression and function of heparan sulfate proteoglycan in hepatocytes. J. Lipid Res. 39: 845-860.

2.4

Lycopene and Chromosomal Aberrations

Lusânia Maria Greggi Antunes and Maria de Lourdes Pires Bianchi
Departamento Análises Clínicas, Toxicológicas e Bromatológicas
Faculdade de Ciências Farmacêuticas de Ribeirão Preto – USP
Av. do Café s/n, 14040-903, Ribeirão Preto - São Paulo, Brasil

ABSTRACT

Dietary constituents that are inhibitors of mutagenesis and carcinogenesis are of particular importance because they may be useful in preventing human cancer by their non-toxic effects. Many endogenous substances, usually present in foods, possess some inhibitory activity towards natural or man-made mutagenic substances that often increase cancer incidence.

In vivo investigations of the mutagenic potential of chemical or dietary compounds in mammals, including the chromosomal aberrations test, are still scarce. Studies *in vitro* and *in vivo* have revealed that dietary components could be associated with the formation of fragile chromosomal sites, chromosome breaks, micronuclei induction and DNA hypomethylation. There are few studies in literature on the mutagenicity of carotenoids.

The high antioxidant capacity of carotenoids may have specific functions and they could be effective chemopreventive agents. It is becoming increasingly evident that carotenoids, including lycopene, may also have bioactivities capable of modifying gene expression and repairing DNA. Since little information is available on the mutagenic and antimutagenic effects of lycopene, the purpose of this chapter is to retrieve from the literature papers that discuss the effects of lycopene on chromosomal aberrations or micronucleus analysis and summarize the possible mechanisms of lycopene in preventing chromosomal damage.

A list of abbreviations is given before the references.

INTRODUCTION

There is considerable evidence that the high incidence of cardiovascular diseases, cancer, and other degenerative diseases is linked to oxidative stress. Several investigations have shown a positive association between high consumption of fruits and vegetables and reduced risk of cancer, suggesting that dietary components offer protection against carcinogenesis. These foods contain relatively high amounts of components with inherent antioxidant properties, including vitamins E and C, folic acid, carotenoids and plant polyphenols and flavonoids. There is a growing interest in identifying factors in the human diet that could help contain the damage caused by oxidative species. Chemoprevention by dietary antioxidants has emerged as a cost-effective approach in preventing mutagenesis and carcinogenesis. Chemoprevention is also a promising alternative to reduce the inevitable human exposure to environmental and dietary carcinogens (Pool-Zobel et al. 1997, Rauscher et al. 1998, Velmurugan et al. 2004a).

Dietary constituents that are inhibitors of mutagenesis and carcinogenesis are of particular importance because they may be useful in preventing human cancer by their non-toxic effects. Many endogenous substances, usually present in foods, possess some inhibitory activity towards natural or man-made mutagenic substances that often increase cancer incidence. Recently, much attention has been given to the antimutagenic effects of natural dietary compounds. Researchers identifying and characterizing dietary antimutagenic agents have been focusing on fruits and vegetables and their compounds, assuming that these natural compounds are very likely to offer chemoprotection against the damage induced by environmental mutagens and pharmacologically active substances (Rauscher et al. 1998).

Antioxidant micronutrients are important components of the endogenous antioxidant defense system for their ability to quench oxidation reactions, inhibit free radical propagation and prevent undesirable DNA damage. Among such natural compounds are vitamins, polyphenols and a number of carotenoids, whose antimutagenic activities is well established in various biological assays. Investigations on the beneficial properties of plants that contain considerable amounts of carotenoids as well as isolated carotenoids have shown that their intake can improve human health and that they are active against environmental and man-made mutagens and carcinogens.

Carotenoids are widely distributed in fruits and vegetables. They are frequently consumed by humans and are regularly detected in human serum and tissues, with α- and β-carotene, β-cryptoxanthin, lutein, and lycopene being the most common. There is epidemiological support for the protective roles of carotenoids in humans and experimental evidence of the protective effects against experimentally induced carcinogenesis in the majority of the animals studied (Rauscher et al. 1998, Riso et al. 1999).

The high antioxidant capacity of carotenoids may have specific functions and they could be effective chemopreventive agents. It is becoming increasingly evident that carotenoids, including lycopene, may also have bioactivities capable of modifying gene expression and repair DNA. Since little information is available on the mutagenic and antimutagenic effects of lycopene, the purpose of this chapter is to retrieve from the literature papers that discuss the effects of lycopene on chromosomal aberrations or micronucleus analysis and summarize the possible mechanisms of lycopene in preventing chromosomal damage.

LYCOPENE

There is growing evidence that supports the hypothesis that dietary carotenoids found in fruits and vegetables can play an important role in human health, providing important substances that strengthen the defense system against several diseases. Research has focused mainly on the action of single carotenoids that may offer chemoprotection against toxic, mutagenic and carcinogenic chemicals. Among other carotenoids, evidence indicates that lycopene, a predominant carotenoid in serum, may play a protective role in human health. Because lycopene has no provitamin A activity, part of its potentially beneficial properties has been attributed to its ability to scavenge free radicals and to physically quench singlet molecular oxygen (Matos et al. 2001).

Lycopene provides the red coloration to tomatoes and it is the main carotenoid present in tomatoes and processed tomato products such as juice, sauce, paste, soup and ketchup, which contain the highest concentrations of bioavailable lycopene. In the Western diet, there are other sources of fruits and vegetables besides tomatoes with high concentrations of lycopene, such as watermelon, guava, papaya and pink grapefruit. Epidemiological studies have revealed an inverse relationship between regular high consumption of fruits and vegetables containing lycopene and the incidence of cardiovascular disease and some types of cancer, especially prostate, esophagus, oral cavity, and colon cancers (Stahl and Sies 1996).

In the body, lycopene is deposited in the liver, lungs, prostate gland, colon and skin. Its concentration in body tissues tends to be higher than that of all other carotenoids. Lycopene possesses the most powerful ability to quench the singlet oxygen among all carotenoids (Di Mascio et al. 1989).

The hydrocarbon carotenoid lycopene is highly conjugated and is an effective scavenger of reactive oxygen species. Thus, consumption of tomato should be increased because of its therapeutically active carotenoid lycopene, which is a very effective antioxidant agent. However, the amount of lycopene and other carotenoids needed to achieve relevant antioxidant activity in cells and tissues has not been determined. Since lycopene has been shown to attenuate oxidative stress, it has now become the focus of attention. Lycopene is expected to be a potential chemopreventive agent against mutagenic DNA damage caused by reactive oxygen and probably other radical species.

The antioxidant effects of lycopene have led to interest in tomato as a functional or nutraceutical food with potential antitumor properties. Lycopene is also examined against the possible tumor-promoting effects of reactive oxygen species. The identification of compounds with chemopreventive potential requires their screening in animal models using short-term tests to evaluate mutagenic and antimutagenic effects. The assessment of chromosomal aberrations has gained wide acceptance since chromosomal aberrations are early biomarkers of carcinogen-induced DNA damage (Velmurugan et al. 2004a). Therefore, we evaluated the effects of lycopene on DNA-induced damage by quantifying micronuclei and chromosomal aberrations as cytogenetic endpoints of chemoprevention.

Mutagenicity of Lycopene

The field of mutation research is mainly based on investigations and original observations that mutations are caused by DNA replication errors during cell proliferation or by exposure to specific physical agents, chemical mutagens, or viruses that are present in our environment as natural or anthropogenic products. These mutations can occur deliberately under cellular control during processes such as meiosis or mitosis and can be subdivided into germ line mutations, which can be passed on to descendants, and somatic mutations, which cannot. These events can produce significant critical alterations to the genome and are a cause of cancer, neuronal diseases and developmental defects.

Numerous studies have been performed on the effect of *in vivo* exposure to chemical carcinogens on gene mutations and risk assessment in human populations, but it is only recently that the effect of dietary imbalance has been taken into consideration. According to Fenech (2002), the focus on diet as a key factor in determining genomic stability is more important than previously imagined because we now know that it impacts all relevant pathways, namely exposure to dietary carcinogens, activation and detoxification of carcinogens, DNA repair, DNA synthesis, and apoptosis. Studies *in vitro* and *in vivo* have revealed that micronutrient deficiency could be associated with the formation of fragile chromosomal sites, chromosome breaks, micronuclei induction and DNA hypomethylation. Some of these deficiencies also cause mitochondrial decay with oxidant leakage and cellular aging and are associated with late onset diseases such as cancer (Ames 2006).

Previous studies have suggested that DNA single-strand breaks and double-strand breaks are the principal lesions in the process of DNA damage that results in chromosomal aberrations (Fig. 1). Cytogenetic biomarkers, such as the frequency of chromosomal aberrations and micronuclei, have been applied as intermediate endpoints to evaluate the DNA-damaging effects resulting from a wide range of exposures in the diet and dietary supplementation. These biomarkers are the most frequently used endpoint in human population studies, and the most sensitive to measure exposure to mutagenic agents and their role as early predictors of cancer risk (Bonassi et al. 2005).

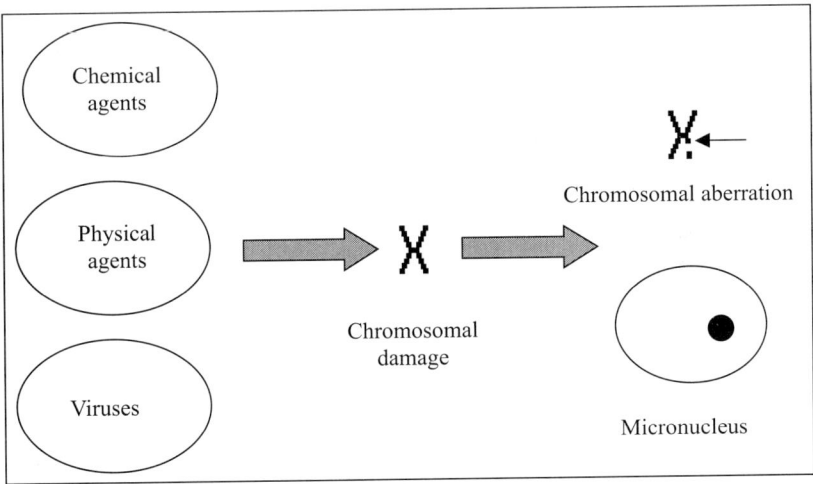

Fig. 1 Exposure to chemical, physical and biological agents inducing chromosomal damage and subsequently chromosomal aberrations (break) or micronuclei. Micronucleus represented in an anucleated cell.

Since mutations are important early factors in carcinogenesis, short-term genetic tests have been successfully used for the detection of mutagens or antimutagens, and carcinogens or anticarcinogens as well. Finally, the evaluation of the potential carcinogenic risk posed by a chemical agent or a natural compound from plants requires determination of its mutagenicity (Rauscher et al. 1998). Investigations on the *in vivo* mutagenic potential of chemical compounds in mammals, including chromosomal aberrations, are still scarce. There are few studies in the literature on the mutagenicity or antimutagenicity of the carotenoid lycopene.

Chromosomal aberrations

Direct and indirect evidence suggests that DNA is the main target of mutagenic agents responsible for the induction of chromosomal aberrations. These alterations are disruptions in the normal chromosome of a cell. Chromosomal aberrations in lymphocytes and bone marrow cells are thought to represent a surrogate endpoint for more specific chromosome alterations in target tissues of carcinogenesis. The clastogenic agents, for example, those that break chromosomes, can be classified as S-dependent agents that produce chromatid-type aberrations, or as S-independent agents that are able to induce chromosome-type and various others types of damage.

The mutagenicity of lycopene has been investigated in *in vivo* tests. Bone marrow cells have a high mitogenic activity, are highly susceptible and provide great sensitivity for detecting whether or not a tested chemical or natural product can induce chromosomal aberrations at a frequency significantly higher than that found in unexposed control animals. The number of treatment doses should allow an analysis of the presence or absence of a dose response. The use of several doses is important, with three being considered a minimum. The most appropriate way to predict effects in humans is to use a route of exposure that most resembles that anticipated or known to be the route of human exposure. The selection of the most appropriate dosage range is important. The conclusions that can be drawn from the *in vivo* cytogenetic bone marrow cells are that a compound is clearly clastogenic or that it is non-clastogenic within the restrictions of the protocol (Preston et al. 1987).

Previous studies from our laboratory have assessed the cytotoxicity of lycopene in bone marrow cells of male Wistar rats. The results of the cytotoxic assay with different doses of lycopene 0.5, 1.0, 1.5, 2.0, 4.0, 6.0, 8.0, 10.0, and 12.0 mg/kg body weight administered by gavage in an acute treatment, and the untreated control group are presented in Fig. 2. In the treatments with the medium doses, 2, 4 and 6 mg/kg bw, the cells showed mitotic indices similar to the untreated control group. In doses above 8 mg/kg bw, there was a considerable decrease in the mitotic index when compared with the untreated control. This was more evident for the lycopene dose of 12 mg/kg bw, with a mitotic index of 0.9% when compared with 1.6% of the negative control (Sendão et al. 2006). The

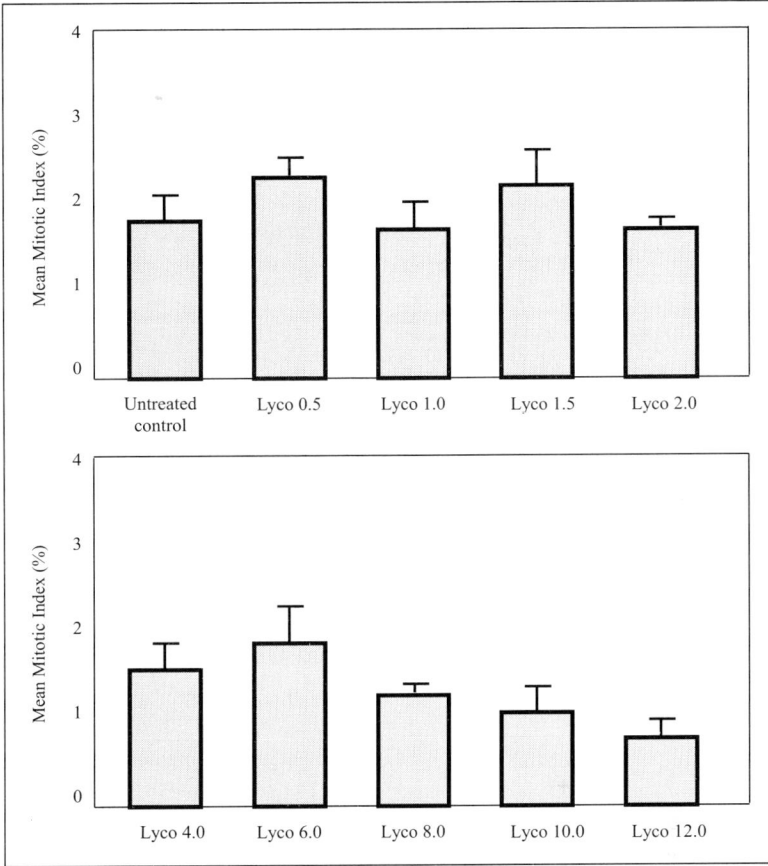

Fig. 2 Mitotic index after lycopene treatment. Histogram showing the reduction in mitotic indexes in bone marrow cells of male Wistar rats treated by gavage with different doses of lycopene (Lyco 0.5, 1.0, 1.5, 2.0, 4.0, 6.0, 8.0, 10.0 and 12.0 mg/kg bw), and respective untreated group, euthanized 24 hr after the treatment with lycopene or saline 0.9%. One thousand cells were analyzed per animal, for a total of 6,000 cells per treatment (Sendão et al. 2006).

mitotic index is an important endpoint in the cytogenetic tests. It can be used to determine the maximum dose that will be investigated in the chromosomal aberrations tests. If there is a significant cytotoxicity, or slowed cell cycle progression, a reduction in the mitotic index can be observed.

Another objective of our laboratory was to evaluate the mutagenic effects of lycopene in rat bone marrow cells. The experiments were performed with 6- to 7-wk-old healthy male Wistar rats, weighing

100-110 g. The animals were maintained under laboratory conditions on a 12 hr light/dark cycle at $23 \pm 2°C$, were housed in polycarbonate cages with steel wire tops, and had free access to standard rat chow and fresh water. The study was approved by the Animal Ethics Committee of Universidade de São Paulo, Campus de Ribeirão Preto. Based on the analyses of the preliminary experiments that focus on the mitotic index and considering that lycopene was cytotoxic at the dose of 12 mg/kg bw, the doses selected were 2, 4, and 6 mg/kg bw for the acute treatment and one-fourth of each dose for the subacute treatment, i.e., 0.5, 1.0, and 1.5 mg/kg bw, respectively.

The results showed that acute and subacute treatments with the chosen lycopene doses did not induce any increase in total chromosomal aberrations or in the number of metaphases with aberrations in rat bone marrow cells when compared with the untreated controls. Thus, lycopene was not clastogenic in the test conditions (Sendão et al. 2006). We could find only one paper in literature on lycopene-induced chromosomal aberrations *in vivo* to compare to our data. Velmurugan et al. (2004b) demonstrated that the administration of lycopene at a dose of 1.25 mg/kg bw for 5 d showed no significant change in the frequency of chromosomal aberrations when compared with the untreated control group. Few other studies have investigated lycopene mutagenicity and only a small number have assessed tomato products or lycopene in the bone marrow micronucleus test.

Micronucleus assay

Micronuclei are nuclear remnants that are formed during mitosis when an acentric chromosome fragment detaches from a chromosome after a clastogenic event or when a whole lagging chromosome does not migrate in the daughter nuclei. These inclusions may be found in any kind of cell, both somatic and germinal. Scoring of micronuclei can be exploited in *in vivo* and *in vitro* cytogenetic assays and can be performed relatively easily and on different cell types: erythrocytes, lymphocytes, fibroblasts and exfoliated epithelial cells. Since mammalian erythrocytes are anucleated, micronuclei can be detected easily in circulating erythrocytes, making this tissue particularly attractive for conducting mutagenicity tests (Schmid 1975).

Cross-sectional studies using chromosomal aberrations and micronuclei in bone marrow cells have a distinct advantage that detailed and accurate information can be collected and it is possible to process samples for chromosomal aberrations and micronuclei within a relatively short period of time after collection. Most micronucleus studies have evaluated chemical mutagens but the number of studies on diet, dietary supplementation or pharmaceutical compounds is increasing. In addition, quantitation of bone marrow micronuclei has gained wide acceptance as a reliable and early index of carcinogen-induced DNA damage and chemoprevention.

The bone marrow micronucleus test was carried out in our laboratory to evaluate the mutagenicity of subacute treatment with lycopene for three consecutive days at doses of 0.5 and 1.0 mg/kg bw. Healthy male Wistar rats were randomized into experimental and untreated control groups. All animals receiving the vehicle (corn oil) and the lycopene-treated groups appeared normal after dosing and remained healthy until the 24 hr harvest time. No mortality occurred in any of the exposed or untreated control groups. Both femurs were removed immediately after euthanasia. Bone marrow cells from control and experimental animals were processed for analysis of lycopene-induced micronucleated polychromatic erythrocytes (MnPCE) in the bone marrow in male Wistar rats. The bone marrow micronucleus test was carried out according to Schmid (1975). The micronucleus frequency, expressed as percentage of micronucleated cells, was determined by analyzing the number of MnPCEs from at least 6,000 PCEs per treatment group. The staining procedure permitted the PCEs, transparent purple, and normochromatic erythrocytes (NCEs), transparent orange, to be differentiated by color. To avoid false negative results and as a measure of toxicity on bone marrow, the PCE:NCE ratio was scored in 400 cells per treatment.

In our investigation, lycopene did not induce statistically significant increases in micronucleated PCEs at any dose level examined (Table 1). This carotenoid was not significantly cytotoxic to the bone marrow, i.e., no statistically significant decreases in the PCE:NCE ratios were observed at any dose level. In agreement with our results, the absence of mutagenic effects of tomatoes, tomato products or lycopene in the micronuclei bone marrow test had already been demonstrated. The administration of a mixture of four carotenoids, β-cryptoxanthin, canthaxanthin, lutein and lycopene at the dose of 180 mg/kg bw did not induce an increase in the frequency of MnPCE in mice bone marrow cells. The amount of the mean MnPCE/1000 PCEs analyzed, based on three animals per group, was 2.50 in the carotenoids group and it was the same value found in the untreated controls (Rauscher et al. 1998).

The treatment with male Swiss albino mice receiving the intragastric administration of tomato paste at a concentration of 0.5, 1.0, and 2.0 g/kg bw for 5 d was also investigated in the micronuclei assay. Lycopene is a major carotenoid present in tomato and tomato products. The score of MnPCE in animals treated with tomato paste alone, at all concentrations tested, was 7.4, 6.0 and 7.2, respectively. It was not significantly different

Table 1 Incidences of polychromatic erythrocytes (PCE), micronucleated PCE (MnPCE) and the ratio of PCE to normochromatic erythrocytes (PCE:NCE) in the bone marrow cells of animals treated with different doses of lycopene 48 hr, 24 hr and 30 min by gavage, before saline or cisplatin (cDDP) intraperitoneally (i.p.), and respective controls. The animals were euthanized 24 hr after the i.p. treatments.

Treatment group	Dose (mg/kg bw)	Number of PCE scored	Mean (%) MnPCE (± SD)	Ratio PCE:NCE (mean ± SD)
Untreated control	–	6,000	0.03 ± 0.01	1.33 ± 0.09
Corn oil	–	6,000	0.05 ± 0.02	1.16 ± 0.02
Lyco I	0.5	6,000	0.02 ± 0.01	1.26 ± 0.13
Lyco II	1.0	6,000	0.04 ± 0.01	1.37 ± 0.01
cDDP + corn oil	0.5	6,000	0.57 ± 0.40[a]	1.38 ± 0.44
Lyco I + cDDP	0.5 + 0.5	6,000	0.18 ± 0.11[b**]	1.52 ± 0.14
Lyco II + cDDP	1.0 + 0.5	6,000	0.40 ± 0.12[b*]	1.43 ± 0.04

Significant difference: [*]$P < 0.05$; [**]$P < 0.001$ (ANOVA, Tukey test).
[a]Different from the untreated control and vehicle groups.
[b]Different from the cDDP group.

from the 7.0 MnPCE/2500 PCEs analyzed that was observed in the untreated control group (Velmurugan et al. 2004a). These authors also investigated the mutagenicity of the carotenoid lycopene isolated in the micronuclei in male Wistar rat bone marrow cells. After receiving intragastric administration of lycopene at the dose of 1.25 mg/kg bw for 5 d, the femurs of the animals were removed and used for analysis of MnPCE. The administration of lycopene showed no significant change in the incidence of MnPCE compared with the untreated control group. The absence of mutagenicity of lycopene was confirmed when this carotenoid was administered to animals in association with 100 mg/kg bw of S-allylcysteine and no increase in the frequency of MnPCE was observed (Velmurugan et al. 2004b). Recently, the relationship between oral concentrations of carotenoids and a putative biomarker of oral cancer, the micronucleus index in human cells, was studied. Among non-smokers, the frequency of micronuclei was negatively correlated with total serum lycopene as well as individually for the serum *trans* and *cis* isomers of lycopene. Among non-smokers, no other serum or oral mucosa cell nutrient levels were significantly correlated with micronucleus frequency (Gabriel et al. 2006).

The absence of mutagenic effects of lycopene in rodent bone marrow cells is of great importance, since this carotenoid is widely distributed in tomatoes and tomato products. These results led to other experimental investigations to determine whether lycopene could be an antimutagenic agent when tested in association with chemical agents.

Possible Mechanisms of Lycopene Antimutagenicity

The identification of antimutagenic agents, especially those found in fruits and vegetables and their compounds, is a promising approach to ensure the integrity and stability of the genome against attack from chemical or physical mutagens. There are a number of studies that support the assumption that natural plant products could inhibit or reduce cancer risk in humans, forming the first level of DNA protection from the mutagenic influences of endogenous and exogenous agents. The common mechanism of the protective action of exogenous antimutagens is that they act through antimutagenic components, antioxidants and/or repair systems.

Recently, much attention has been focused on the antimutagenic effects of natural compounds from food plants by using cytogenetic biomarkers, such as chromosomal aberrations or micronuclei. Among known dietary compounds, carotenoids from tomatoes and carrots with potential antioxidant properties have been especially investigated in the field of antimutagenesis and anticarcinogenesis. It is necessary to elucidate the cellular and molecular mechanisms by which carotenoids reduce cancer risks.

Many compounds that are well-known mutagenic agents in mammalian system or substances used as cancer chemotherapeutic drugs that are able to induce chromosomal aberrations and micronuclei have been tested in short-term animal tests in association with possible antimutagenic agents, to reduce or inhibit their mutagenic or clastogenic activity. The antimutagenic properties of four carotenoids including lycopene against benzo[a]pyrene, a known mutagen, and the antitumor drug cyclophosphamide were investigated by mice bone marrow cells micronucleus assay. The incidence of MnPCEs induced by benzo[a]pyrene and cyclophosphamide in the bone marrow of mice was significantly reduced by pretreatment with carotenoids. β-Cryptoxanthin was the most potent carotenoid, but lutein and lycopene also were effective. The administration of 180 mg/kg bw of lycopene reduced cyclophosphamide-induced MnPCE by 29.6% and benzo[a]pyrene-induced micronuclei by 25.1%. This indicates that different events caused by these mutagens, including the metabolic activation of reactive species leading to DNA damage, and the process of cancer initiation, could be inhibited by carotenoids (Rauscher et al. 1998).

Pretreatment with tomato paste for five consecutive days at doses of 0.5, 1.0, and 2.0 g/kg bw significantly decreased the incidence of N'-methyl-N'-nitro-N-Nitrosoguanidine(MNNG)-induced MnPCE in mice bone marrow cells and significantly decreased lipid peroxidation and increased glutathione-dependent antioxidant and detoxifying enzymes. The increased frequency of bone marrow micronuclei found was ascribed to

MNNG-induced oxidative stress. Oxidative stress arising because of an overproduction of reactive oxygen species and breakdown of antioxidant defenses has been documented to induce chromosomal damage and formation of micronuclei (Velmurugan et al. 2004a). Furthermore, pretreatment with lycopene, the major carotenoid of tomato paste, at the concentration of 1.25 mg/kg bw also significantly reduced the frequency of MNNG-induced bone marrow micronuclei and chromosomal aberrations in male Wistar rats. This antimutagenic effect of lycopene reduced MnPCE more intensely when combined with S-allylcysteine, an organosulfur constituent of garlic. The data showed that S-allylcysteine and lycopene can interact synergistically to protect bone marrow cells against MNNG mutagenicity (Velmurugan et al. 2004b).

These results on lycopene antimutagenicity are in agreement with our results obtained from chromosomal aberrations or micronucleus analysis in rat bone marrow cells. Acute and subacute treatments with the chosen doses of lycopene reduced the chromosomal damage induced by the chemotherapeutic drug cisplatin by 40.5, 47.8 and 58.3%, respectively in the acute treatment (2, 4, and 6 mg/kg bw) and in the subacute treatment (0.5, 1, and 1.5 mg/kg bw) by 66.2, 33.7 and 36.8%, respectively (Sendão et al. 2006). The administration of 0.5 mg/kg of lycopene for four consecutive days before the single dose of the antitumor agent seemed most effective in protecting bone marrow cells against cisplatin mutagenicity. Typical micrographs showed the results obtained from the bone marrow chromosome of the animals of the groups—untreated control (Fig. 3A); lycopene 1.0 mg/kg bw by gavage (Fig. 3B); and intraperitoneal (i.p.) injection with the antitumor drug cisplatin (Fig. 3C-D)—that illustrate the acentric fragments and chromatid breaks induced by cisplatin.

For the subsequent micronucleus assay in rats, lycopene doses of 0.5 and 1.0 mg/kg bw were chosen. Simultaneously, untreated control (distilled water), vehicle control (corn oil), and positive control cisplatin (5.0 mg/kg bw) were also maintained for comparative analysis of the data. In our antimutagenicity assay, the results clearly demonstrate that increased frequency of MnPCEs was obtained after treatment with the antitumor drug alone. Cisplatin induced statistically significant increases in micronucleated PCEs compared with that of the untreated or vehicle controls, with a mean and standard deviation of 0.57 ± 0.40% compared with 0.03 ± 0.01% observed in the untreated control, demonstrating the expected mutagenic effect and sensitivity of the experimental system. Typical micrographs of the results obtained from the bone marrow of the animals of the three groups—untreated control, lycopene 1.0 mg/kg bw by gavage, and i.p. injected with cisplatin—are shown in Figs. 4 and 5, which illustrate PCE and NCE (Fig. 4A and B) and MnPCEs (Fig. 5C and D).

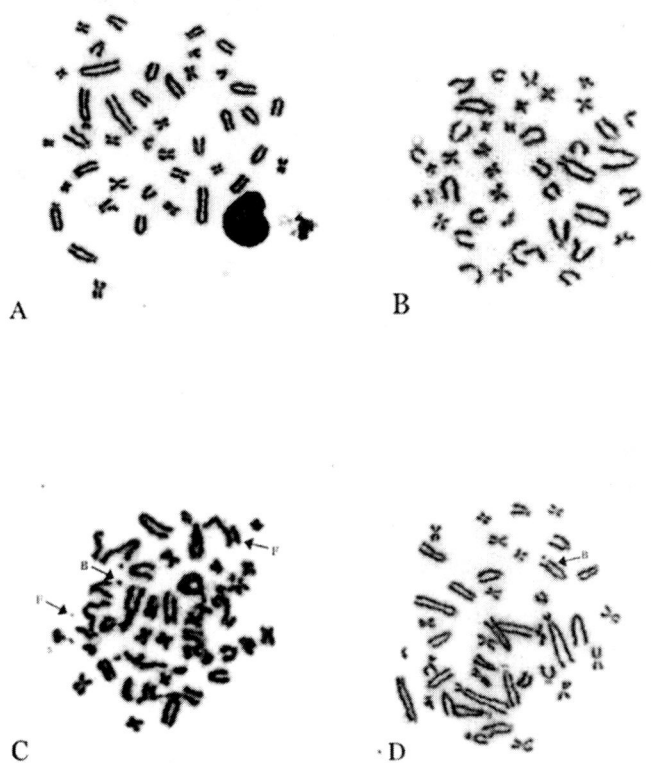

Fig. 3 Chromosomal aberrations. Photomicrographs showing chromosomal aberrations in the bone marrow cells of *Rattus norvegicus* with the following treatments: (A) untreated control with normal metaphase; (B) treatment with 1.0 mg/kg bw of lycopene by gavage with normal metaphase; (C and D) treatment with 5.0 mg/kg bw of the antitumor drug cisplatin i.p. F = fragment, B = chromatid break.

Animals that received both doses of lycopene administration associated with cisplatin displayed a significant reduction in the incidence of MnPCE, as compared with the cisplatin group alone. The mean frequencies of MnPCE in the groups lycopene 0.5 mg/kg bw + cisplatin and 1.0 mg/kg bw + cisplatin were recorded to be 0.18 ± 0.11 and 0.40 ± 0.12 respectively, compared with cisplatin. The administration of 0.5 mg/kg of lycopene also seemed most effective in protecting bone marrow cells against cisplatin-induced micronuclei. The ratio of PCE:NCE was not affected by cisplatin or lycopene plus cisplatin treatments (Table 1). The antimutagenic effects of lycopene were similar to those observed in our previous studies (Sendão et al. 2006).

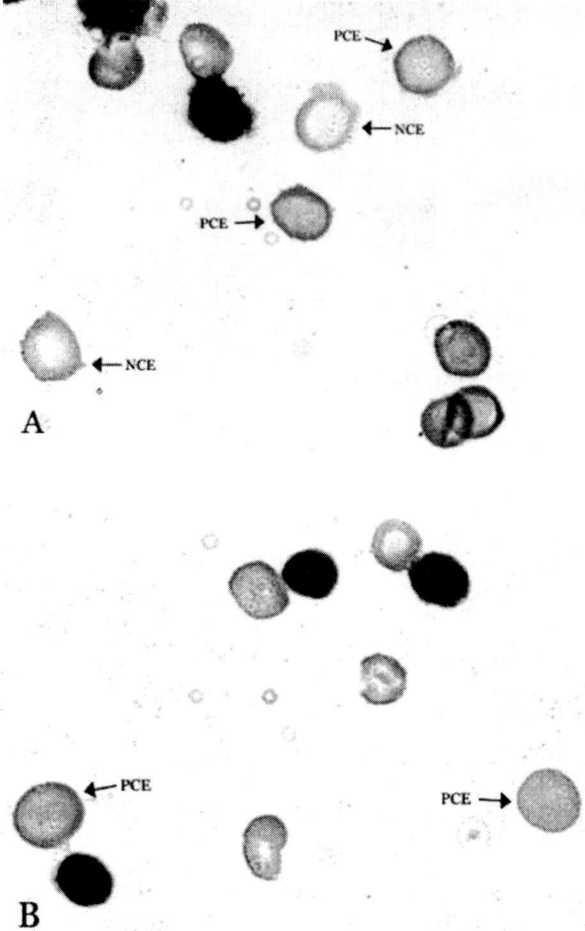

Fig. 4 Polychromatic (PCE) and normochromatic erythrocytes (NCE). Photomicrographs showing micronuclei in the bone marrow cells of *Rattus norvegicus* with the following treatments: (A) untreated control; (B) treated with 1.0 mg/kg bw of lycopene by gavage.

The protective effects of lycopene on rodent bone marrow cells could be explained by various mechanisms. The most likely mechanism of lycopene activity in these antimutagenicity studies is that this carotenoid acting as an antioxidant inhibited the damages that would be induced by reactive oxygen species released by mutagenic agents. The antioxidant properties of lycopene, like other carotenoids, are most likely ascribed to the ability of scavenging peroxyl radicals and its chemical and physical quenching of the singlet oxygen (Heinrich et al. 2003).

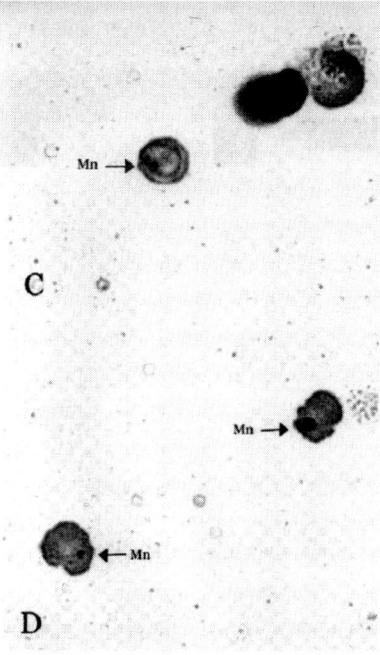

Fig. 5 Polychromatic erythrocytes with micronucleus (Mn). Photomicrographs showing micronuclei in the bone marrow cells of *Rattus norvegicus* (C and D) treated with 5.0 mg/kg bw of the antitumor drug cisplatin.

Previous studies have shown that the consumption of tomato or carrot juices significantly reduced the oxidative damage and strand breaks in DNA (Pool-Zobel et al. 1997). These effects were partly due to' the carotenoids lycopene and β-carotene found in the juices. There is some evidence that carotenoid antioxidants could influence the recovery of cells from oxidative DNA damage.

Consumption of 60 g of tomato purée containing 16.5 mg lycopene and 0.6 mg β-carotene for 21 d significantly increased both lycopene plasma concentrations and DNA resistance to oxidative stress after exposing the lymphocytes to H_2O_2 (Riso et al. 1999). The same authors observed that an intake of 25 g tomato purée containing 7 mg lycopene for 14 consecutive days also resulted in increased carotenoid concentrations and the resistance of DNA strand breaks in human lymphocytes exposed to oxidative stress. It has been suggested that tomato consumption should be increased because of its antioxidant substances such as lycopene. However, the amount necessary to achieve antioxidant activity and bioavailability of lycopene and the other carotenoids to cells and tissues have not been determined (Porrini and Riso 2000).

Pretreatment for 5 d with 10 mg/kg bw of lycopene markedly decreased liver cell necrosis induced by ferric nitrilotriacetate (Fe-NTA) administration in male Wistar rats. Fe-NTA is an effective inducer of lipid peroxidation and DNA strand breaks. Lycopene significantly reduces hepatic lipid peroxidation caused by Fe-NTA by 90% and provides effective protection against Fe-NTA-induced DNA-base oxidation (Matos et al. 2001). The antioxidant properties of lycopene are suggested to have an important role in the protective effects of this carotenoid on cells and tissues, although other possible mechanisms besides antioxidant activity have been suggested regarding carotenoids, such as influence on DNA repair mechanisms and modulation of gene expression. There are evidences in human lymphocyte assays that analyze oxidative lesions in DNA or DNA damage measured in single cell gel electrophoresis, that lycopene could apparently exert two distinct effects: as an antioxidant by preventing reactive oxygen species from attacking the DNA and as a modulator of DNA repair mechanisms (Astley and Elliot 2005).

In conclusion, the results presented in this chapter should help elucidate that lycopene is not mutagenic and that it has antimutagenic effects in cytogenetic short-test *in vivo*. However, more investigations are needed to understand the activity of lycopene and other carotenoid antioxidants in foods, so that the beneficial and protective properties of these carotenoids on human health can be characterized before suggesting its use in future chemoprevention trials.

ACKNOWLEDGEMENTS

The authors are grateful to their co-workers at the Faculdade de Ciências Farmacêuticas de Ribeirão Preto–USP, Mrs. Joana D'Arc C. Darin, MsC, Milena Cristina Sendão, MsC, Eliziani Mieko Konta, and Dr. Regislaine Valéria Burim, for constant support and valuable technical assistance, and Mr. Leandro Greggi Penha for assistance with Photoshop.

ABBREVIATIONS

cDDP: cisplatin; Fe-NTA: ferric nitrilotriacetate; i.p.: intraperitoneal; MNNG: N'-methyl-N'-nitro-N-Nitrosoguanidine; MnPCE: polychromatic erythrocyte with micronucleus; NCE: normochromatic erythrocyte; PCE: polychromatic erythrocyte

References

Ames, B.N. 2006. Low micronutrient intake may accelerate the degenerative diseases of aging through allocation of scarce micronutrients by triage. Proc. Natl. Acad. Sci. USA 103: 17589-17594.

Astley, S.B. and R.M. Elliot. 2005. How strong is the evidence that lycopene supplementation can modify biomarkers of oxidative damage and DNA repair in human lymphocytes? J. Nutr. 135: 2071S-2073S.

Bonassi, S. and D. Ugolini, M. Kirsch-Volders, U. Stromberg, R. Vermeulen, and J.D. Tucker. 2005. Human population studies with cytogenetic biomarkers: review of the literature and future prospectives. Environ. Mol. Mutagen. 45: 258-270.

Di Mascio, P. and S. Kaiser, and H. Sies. 1989. Lycopene as the most efficient biological carotenoid singlet oxygen quencher. Arch. Biochem. Biophys. 274: 532-538.

Fenech, M. 2002. Micronutrients and genomic stability: a new paradigm for recommended dietary allowances (RDAs). Food Chem. Toxicol. 40: 1113-1117.

Gabriel, H.E. and Z. Liu, J.W. Crott, S.-W. Choi, B.C. Song, J.B. Mason, and E.J. Johnson. 2006. A comparison of carotenoids, retinoids, and tocopherols in the serum and buccal mucosa of chronic cigarette smokers versus nonsmokers. Cancer Epidemiol. Biomarkers Prev. 15: 993-999.

Heinrich, U. and C. Gartner, M. Wiebusch, O. Eichler, H. Sies, H. Tronnier, and W. Stahl. 2003. Supplementation with beta-carotene or a similar amount of mixed carotenoids protects humans from UV-induced erythema. J. Nutr. 133: 98-101.

Matos, H.R. and V.L. Capelozzi, O.F. Gomes, P. Di Mascio, and M.H.G. Medeiros. 2001. Lycopene inhibits DNA damage and liver necrosis in rats treated with ferric nitrilotriacetate. Arch. Biochem. Biophys. 396: 171-177.

Pool-Zobel, B.L. and A. Bub, H. Muller, I. Wollowsky, and G. Rechkemmer. 1997. Consumption of vegetables reduces genetic damage in humans: first results of a human intervention trial with carotenoid-rich foods. Carcinogenesis 18: 1847-1850.

Porrini, M. and P. Riso. 2000. Lymphocyte lycopene concentration and DNA protection from oxidative damage is increased in women after a short period of tomato consumption. J. Nutr. 130: 189-192.

Preston, R.J. and B.J. Dean, S. Galloway, H. Holden, A.F. McFee, and M. Shelby. 1987. Analysis of chromosome aberrations in bone marrow cells. Mutat. Res. 189: 157-165.

Rauscher, R. and R. Edenharder, and K.L. Platt. 1998. In vitro antimutagenic and in vivo anticlastogenic effects of carotenoids and solvent extracts from fruits and vegetables rich in carotenoids. Mutat. Res. 413: 129-142.

Riso, P. and A. Pinder, A. Santangelo, and M. Porrini. 1999. Does tomato consumption effectively increase the resistance of lymphocyte DNA to oxidative damage? Amer. J. Clin. Nutr. 69: 712-718.

Schmid, W. 1975. The micronucleus test. Mutat. Res. 31: 9-15.

Sendão, M.C. and E.B. Behling, R.A. Santos, L.M.G. Antunes, and M.L.P. Bianchi. 2006. Comparative effects of acute and subacute lycopene administration on chromosomal aberrations induced by cisplatin in male rats. Food Chem. Toxicol. 44: 1334-1339.

Stahl, W. and H. Sies. 1996. Lycopene: a biologically important carotenoid for humans? Arch. Biochem. Biophys. 336: 1-9.

Velmurugan, B. and V. Bhuvaneswari, S.K. Abraham, and S. Nagini. 2004a. Protective effect of tomato against N-Methyl-N'-Nitro-N-Nitrosoguanidine-induced in vivo clastogenicity and oxidative stress. Nutrition 20: 812-816.

Velmurugan, B. and S.T. Santhiya, and S. Nagini. 2004b. Protective effect of S-allylcysteine and lycopene in combination against N-Methyl-N'-Nitro-N-Nitrosoguanidine-induced genotoxicity. Pol. J. Pharmacol. 56: 241-245.

2.5

Lycopene and Lycopene-enriched Prostasomes

Anuj Goyal, Mridula Chopra and Alan Cooper
Urology Research Group, Department of Pharmacy and Biomedical Sciences
University of Portsmouth, Portsmouth, United Kingdom

ABSTRACT

There is an increasing focus on chemoprevention in diseases with rising incidence, long latency or slow progression, such as prostate cancer and hyperplasia.

Lycopene is a potent antioxidant. Epidemiological studies show that increased lycopene consumption reduces the risk of prostate cancer and it also holds some promise in treating male infertility, chiefly by countering oxidative damage to DNA. Lycopene also has activities independent of antioxidant function. It can improve gap-junction communication, induce apoptosis, influence angiogenesis and suppress cell proliferation through influences on cytokines.

Prostasomes are microparticles secreted by prostatic epithelial cells into semen and are also liberated from malignant prostatic tissue. They are multilamellar membranous lipoprotein vesicles. Proteomics reveal over 130 associated proteins and their lipid composition displays abundance of cholesterol and sphingomyelin. Prostasomes are facilitators of male fertility; however, their role in prostate cancer remains unresolved.

Lycopene is found in abundance in the prostate gland. From their structure, prostasomes are candidate vehicles for transporting lipophilic compounds such

A list of abbreviations is given before the references.

as lycopene, offering a relative degree of shielding to this bioactive molecule against untimely degradation in transit. In return, lycopene offers protection to these lipoperoxidation-prone moieties, thereby also allowing them to reach their destination unscathed. We have experimentally examined this mutually beneficial association. The simplest model demonstrates prostasomes incorporating exogenous lycopene in a cell-free system. Further studies show lycopene to be taken up and subsequently expelled in microparticles by cells, suggesting that dietary lycopene may enter prostatic cells, to be exported as prostasomal lycopene. Human dietary supplementation studies with lycopene also support this hypothesis.

Lycopene remains a promising chemopreventive agent in human prostate cancer and has potential in the management of benign prostatic hyperplasia and male infertility. While exact mechanisms remain unclear, we present evidence that prostasomes facilitate trafficking of lycopene into semen or the stroma of prostate tissues, where it may impinge on fertility and prostatic pathologies.

INTRODUCTION

We continue to witness an ever-increasing focus on chemopreventive measures in medicine. This is of particular importance in diseases with rising incidence rates and relatively long latency periods or slow progression. Such an example is seen in prostate pathologies such as cancer and hyperplasia, which therefore act as excellent models to study chemopreventive strategies.

Approximately 40 different carotenoids are present in the human diet, of which 25 are found in human serum and tissues (Khachik et al. 1992). The most abundant carotenoids in human blood plasma are lycopene and β-carotene (Stahl and Sies 1996). In some studies lycopene has been shown to significantly exceed the levels of all other carotenoids (Khachik et al. 1992). The other predominant carotenoids of note are α-carotene, leutin and β-cryptoxanthin. Lycopene also features as one of the most prominent carotenoids in most human tissues. It accounts for up to 70% of the total carotenoids in the adrenal glands and testicular tissues and is present to a lesser extent (10 to 40%) in other tissues (Kaplan et al. 1990). Lycopene accumulates in the human prostate gland to the extent of being one of the most abundant carotenoids in this organ and on average accounts for approximately a third of the total carotenoid load in the prostate gland (albeit showing a large inter-individual variation ranging from 2 to 61%) (Clinton et al. 1996). The various mechanisms responsible for this differential tissue specificity in lycopene patterns have not been fully identified. Because of lycopene's lipophilic nature, following absorption from the gut it is transported via lipoproteins in the human circulation mainly incorporated into low density lipoproteins (LDL) with smaller amounts in high density and very low density lipoproteins (HDL and VLDL) (Erdman et al. 1993). There are no other described binding proteins or transport proteins for lycopene in human blood plasma. To date there are also no known membrane receptors described for specific lycopene uptake into cells. The differential distribution in human tissues may be partly explained by the LDL receptor density and the propensity of LDL uptake by the various human organs or tissues. However, this may be further influenced by selective tissue-specific uptake mechanisms that may co-exist.

Lycopene is extremely hydrophobic and therefore once assimilated tends to be located in the inner hydrocarbon sections of cellular membranes. *In vivo,* like all other carotenoids, it associates with the hydrophobic regions of proteins and, as already discussed, the lipid elements of lipoproteins. Such associations enable it to be transported to and function effectively in aqueous environments. However, the free radical-scavenging abilities of lycopene tend to be more pronounced in lipophilic environments per se.

Various health benefits of carotenoids have been reported and are mainly ascribed to their antioxidant capabilities (Mayne 1996, Tapiero et al. 2004). Lycopene has been shown to be one of the most potent antioxidants among all the carotenoids (Di Mascio et al. 1989, Miller et al. 1996). It is obtained mainly from tomatoes and tomato-based products in the human diet. Evidence, mainly epidemiological, exists that increased lycopene consumption reduces the risk of developing prostate cancer. It is thought to counter oxidative and potentially mutagenic damage to DNA. It also modulates many other activities independent of its antioxidant function. These include improving gap-junctional communication and inhibition of cytokine signalling and expression. It can also induce apoptosis, influence angiogenesis and suppress proliferation of both normal and malignant prostatic cell lines in culture. Lycopene also shows some promise in the treatment of male infertility, where it is implicated as part of the non-enzymatic defence system in semen, combating reactive oxygen species (ROS) detrimental to the health and function of spermatozoa.

Prostasomes are submicron-sized organelles secreted by acinar prostatic epithelial cells into semen and they are also liberated from malignant (including metastatic) prostatic tissue. They appear as multilamellar membranous lipoprotein vesicles surrounding less organized material. Proteomic studies have revealed in excess of 130 associated proteins (Utleg et al. 2003). Prostasomes display a characteristic lipid composition with an abundance of cholesterol and sphingomyelin. They are postulated to play a key role in aiding the reproductive physiology underlying male fertility and this is discussed further below. Their role in prostate cancer remains unresolved.

In view of lycopene abundance in the human prostate gland (Clinton et al. 1996), and the lipid abundance in prostasomes it is highly likely that these organelles act as vehicles for lycopene transport once it is sequestered into the prostatic tissues, thereby enabling delivery of this lipophilic bioactive molecule to sites where it may impinge on fertility and prostatic pathologies. In this chapter we systematically describe the methodology used to explore this hypothesis and present the ensuing evidence in support of such a process. From expanding on the physico-chemical findings already documented in the literature, focus is then shifted to cellular mechanisms by which dietary lycopene may enter prostatic epithelial cells and get packaged and exported, incorporated into prostasomes. The simplest system identifies the ability of purified prostasomes to load up and retain lycopene in a cell-free experimental model. The argument is advanced forward through supporting results from cellular processing experiments and dietary supplementation studies using volunteers. Spectroscopic techniques such as Raman microspectroscopy (Goyal 2007a) can also be used to study aspects of intracellular lycopene distribution and trafficking.

Lycopene remains a promising chemopreventive agent in human prostate cancer and also shows some potential in treating benign prostatic hyperplasia and male infertility. The exact mechanisms of action remain under investigation; however, this research sheds light on the putative role of prostasomes in trafficking lycopene into semen and into the milieu surrounding prostatic cells. In return, these exosomes with specific important biological function are protected from lipid peroxidation in environments where they are most vulnerable to being rendered ineffective by free radical damage.

PROSTASOMES

Discovery and Origins

Prostasomes were first described in the late 1970s by Ronquist and co-workers at the University of Uppsala in Sweden. The discovery of these microscopic vesicles in seminal plasma indicated an origin in the prostate gland and thus the term prostasomes was coined (Ronquist 1978, Stegmayr and Ronquist 1982, Ronquist and Brody 1985).

It is now well established that prostasomes are produced by the glandular epithelial cells of prostatic tissue (Sahlen et al. 2002). They are subsequently secreted by this prostatic ductal epithelium into the lumen where they form part of the seminal ejaculate. Prostasomes have also been shown to be produced by malignant prostatic tissue that has metastasized to distant sites (Sahlen et al. 2004). Analogous structures are also shed by prostate cell lines into culture media (Nilsson et al. 1999). These are referred to as exosomes through the course of this chapter to differentiate their origin from seminal derived prostasomes. Both form part of a wider area of increasing interest in which the cellular products are often called microparticles and are studied in the circulation as well as in tissues.

Structural Properties and Composition

Prostasomes are membrane-bound submicron-sized particles ranging from 50 to 400 nm in diameter, with an average diameter of approximately 150 nm (Ronquist and Brody, 1985). They are multilammelar vesicles and their corpuscular nature is well displayed using transmission electron microscopy as shown in Fig. 1. This reveals two sub-types: a small electron-dense variety and the larger electron-light ones (Stewart et al. 2004).

Although very small, these organelles contain and carry many different biomolecules. Proteomic analysis has demonstrated in excess of 130 such proteins (Utleg et al. 2003). These fall into varying functional categories

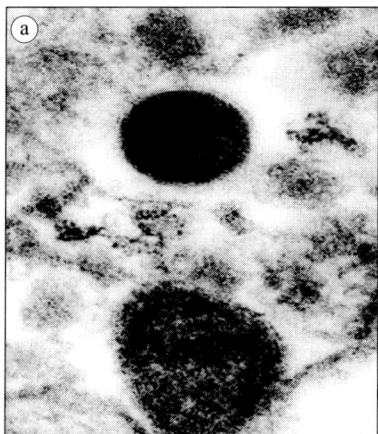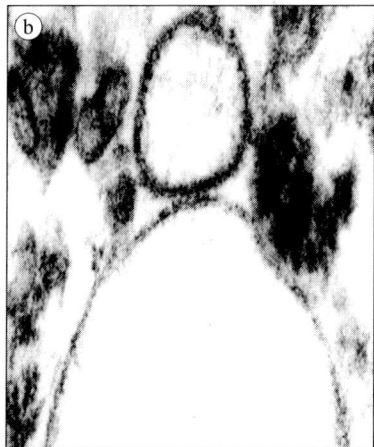

Fig. 1 Transmission electron micrographs of prostasomes demonstrating the (a) smaller dark prostasomes, average diameter ≈ 150 nm, and (b) larger light variety, average diameter ≈ 300 nm. Reproduced with permission from Blackwell Publishing (Source: BJU Int. 94(7): 987).

including enzymes, transport proteins, chaperone proteins, GTP proteins and signalling transduction proteins. Prostasomes also display an unusual lipid profile in stark contrast to that seen in sperm or even somatic cell membranes. The most striking difference is the rich abundance of cholesterol. They demonstrate a very high cholesterol to phospholipid ratio of 2:1 (Arvidson et al. 1989); in comparison the cell membranes of spermatozoa have a cholesterol to phospholipid ratio of 0.83 (Poulos and White 1973, Mack et al. 1986).

The predominance of cholesterol is an intriguing feature because of the well-described connection between cholesterol and human reproduction. Modulation of cholesterol in sperm membranes is known to play a key role in capacitation (Benoff, 1993, Martinez and Morros, 1996).

Sphingomyelin is also found in abundance and constitutes the main phospholipid (50%) in prostasomes. They are also rich in saturated glycerophospholipid (Arienti et al. 1998). The fatty acid pattern of the phospholipids in prostasomes is significantly different to that seen in sperm cellular membranes with the saturated variety of fatty acids predominating. In comparison, in all mammalian sperm membranes, phosphotidylcholine and phosphatidylethanolamine are the main phospholipids and polyunsaturated fatty acids are very common. All these are rare in prostasomes.

Biological Functions

Several physiological functions have been ascribed to prostasomes. Their main role seems to lie in the promotion of sperm survival and function, thereby improving chances of successful fertilization. These widely varying known functions include the following:

- Involvement in the liquefaction of semen (Lilja and Laurell 1984).
- Aiding sperm motility (Stegmayr and Ronquist 1982). Motility is one of the key factors for the successful fertilizing ability of spermatozoa. Buffer-washed sperm cells usually exhibit no motility. Prostasomes can restore their motility in a dose-dependent fashion (Fabiani et al. 1994).
- Activation of spermatozoa by transfer of selected proteins and lipids to sperm cells (Arienti et al. 2004).
- Immunosuppressive activity, which is thought to offer the alloantigenic spermatozoa protection by neutralizing the immune defences of the female genital tract (Kelly et al. 1991).
- Replenishing spermatozoa with inhibitors of complement system activation (Rooney et al. 1993).

- Provision of an antioxidant buffering capacity in seminal plasma to protect spermatozoa against ROS. This is thought to be mediated through reducing the actual production of ROS (Saez et al. 1998).
- Prostasomes also possess significant antibacterial activity against several bacterial strains (Carlsson et al. 2000). This is thought to be mediated via the actions of the proteolytic enzymes on their surface.

Of the above multiple and varying functions that prostasomes are known to possess, it is their interaction with spermatozoa to enhance their health and function that has been most studied and best validated. This is discussed further below.

Interactions with Sperm and Sperm Modifications

Prostasomes are known to interact and even fuse with spermatozoa (Arienti et al. 1997a). This activity takes place despite both entities exhibiting a net-negative charge on their surface (Ronquist et al. 1990). Furthermore, their bonding has been confirmed to be of a hydrophobic nature (Ronquist et al. 1990). Via this manner of interaction, prostasomes are able to transfer lipids (Carlini et al. 1997) and membrane-bound proteins (Arienti et al. 1997b, 1997c) and even participate in regulating the influx of ions such as calcium into spermatozoa (Palmerini et al. 1999). Transfer of lipid can be inferred by observing relocation of fat-soluble fluorescent probes such as Octadecylrhodamine R_{18} (used widely to study membrane fusion) and associated changes in membrane fluidity (Carlini et al. 1997). Using laser scanning confocal microscopy our group has also been able to demonstrate transfer of lipid-like markers such as PKH-26 from prostasomes to endothelial cells (Human Umbilical Vein Endothelial Cells (HUVEC) (Delves et al. 2005), benign (PNT2) and malignant (PC3) prostatic epithelial cells (Goyal et al. 2005) in culture.

LYCOPENE ADSORPTION AND RETENTION BY SEMINAL PROSTASOMES

Prostate gland secretions are rich in prostasomes, as described above. Prostasomes have been shown to play an important role in promoting fertility. They are involved in modulating and regulating the microenvironment of the spermatozoa in order to optimize their function (Arienti et al. 1997a). Specifically, they aid the motility of spermatozoa (Stegmayr and Ronquist 1982, Fabiani et al. 1994, Arienti et al. 1999a) and also protect them from free-radical damage (Saez et al. 1998). They adhere to and fuse with spermatozoa; their actual physical contact with spermatozoa has been shown to involve exchange of important lipids and membrane-bound proteins, which helps in stabilizing the sperm cellular membrane and facilitating important intracellular reactions.

Lycopene forms an integral part of the non-enzymatic defence system in human semen, helping protect spermatozoa against damage induced by free radicals. A clear correlation exists between low lycopene levels in seminal plasma and the increased incidence of male immuno-infertility (Palan and Naz, 1996).

Our research programme explored the potential role prostasomes may play in the delivery of this important antioxidant into semen. Studies in our laboratory showed lycopene to accumulate quite rapidly in purified seminal prostasomes exposed to lycopene *in vitro*, with uptake being approximately 80% of the peak uptake by 1 hr (Goyal et al. 2006a). This effect was noted with differing lycopene concentrations, including low concentrations representative of typical physiological levels of the antioxidant found in human serum following consumption of diets rich in fruits and vegetables (Yeum et al. 1996). Subsequent analysis using high performance liquid chromatography (HPLC) for quantification of lycopene extracted from the lycopene-enriched prostasomes showed it to be in the order of micrograms per milligram of prostasomal protein. A further rise in prostasomal lycopene content was then seen following longer incubation periods (6 hr), with the levels remaining at least at this level with increasing incubation times. In comparison to control preparations of solubilized lycopene subjected to similar conditions (incubation at $37°C$ and atmospheric oxygen), a rapidly progressive degradation with increasing incubation time was witnessed (Fig. 2). Studies involving lycopene concentrations representative of physiological levels (0.876 µM) showed a greater percentage loss of measurable activity when compared to the higher concentration of lycopene (Fig. 3), which presumably saturated the oxidizing capacity of the system. Therefore, in contrast to lycopene's steady degradation in free solution, there appears to be an element of significant protection of this susceptible molecule within the prostasomal microenvironment for up to 24 hr, allowing it to reach its site of action relatively unscathed.

One of the ways of counteracting oxidative stress in seminal plasma is by increasing its radical scavenging capacity. Physiologically, lycopene carried within prostasomes may represent the most efficient means by which this important antioxidant finds its way into semen. The relative protection offered by prostasomes would help maintain lycopene's antioxidant potency as it is delivered into the microenvironment of the oxidation-prone spermatozoa. Prostasomes are known to possess antioxidant potential, which may in part be due to the synergy from the

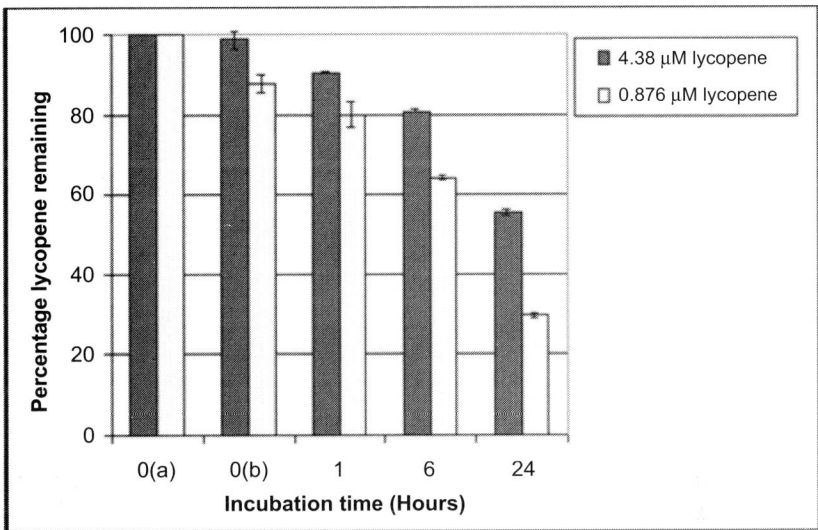

Fig. 2 Degradation of two concentrations of solubilized lycopene with length of incubation. The results are displayed as means ± SD. 0(a) represents stock lycopene dissolved directly in solvent (no extraction required). 0(b) represents lycopene diluted with PBS and extracted immediately. Further lycopene-PBS preparations were then incubated at 37°C in the dark for 1, 6 and 24 hr and the lycopene extracted at the end of the stated incubation period. Reproduced with permission from Blackwell Publishing (Source: Int. J. Androl. 29(5): 531).

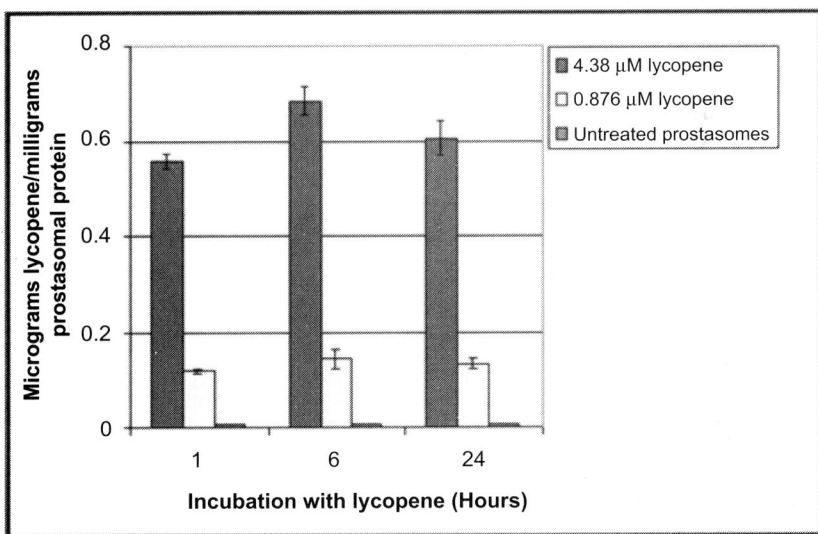

Fig. 3 The lycopene content of prostasomes normalized to prostasomal protein following various incubation periods. The results are shown as means ± SD. $p < 0.05$ between the 1 and 6 hr incubations at both lycopene concentrations used. Between the 6 and 24 hr incubation $p = 0.054$ (at the higher lycopene concentration) and $p = 0.502$ (at the lower lycopene concentration). No lycopene was detectable in the untreated prostasomes; these have been assigned a nominal "close to zero" value to facilitate pictorial representation. Reproduced with permission from Blackwell Publishing (Source: Int. J. Androl. 29(5): 531).

lycopene contained within their lipid-rich membranes. Lycopene could thus have an important role in enhancing the biological function and antioxidant potential of prostasomes.

This work provided the first line of proof for the ability of seminal prostasomes to assimilate and retain with reduced degradation lipophilic antioxidant agents such as lycopene. Complementary studies employing *in vitro* cell-based experimental systems provide a useful means of exploring this notion further, shedding light on the mechanisms by which lycopene may be trafficking through prostatic tissue, piggybacking on prostasomes.

LYCOPENE EXPORT VIA EXOSOMES LIBERATED FROM PROSTATIC EPITHELIAL CELLS

Lycopene's lipophilic nature ensures its strong affinity for membranes or organelles rich in lipids. It is therefore reasonable to hypothesize that once this antioxidant is sequestered into prostate cells it may be packaged, transported and exported via lipid-rich moieties such as prostasomes. Cell culture–based models provide one such means of studying biochemical and molecular events at the sub-cellular level. Such studies help build upon the findings already described above that showed that fully pre-formed and purified seminal prostasomes possess the ability to passively accumulate lycopene from an *in vitro* cell-free milieu (Goyal et al. 2006a). This uptake occurs in a non-metabolic and non-physiological manner.

In vitro prostatic epithelial cells exposed to lycopene are known to accumulate the molecule (Hwang and Bowen 2004). Using similar cell culture experimental models in our laboratory, both benign (virally immortalized PNT2) and malignant (PC3) human prostatic cell lines in culture exposed to lycopene have also been shown to accumulate the antioxidant and subsequently package it into exosomes (the putative *in vitro* analogues of prostasomes) for export (Goyal et al. 2006b). Quantification of the exosomal lycopene content using HPLC showed it to be in the order of nanograms per milligram of exosomal protein.

Although the main research focus when studying lycopene tends to be on its chemopreventive properties, it is of interest to note that the phenomenon is not strictly restricted to normal prostatic epithelial cells and that malignant cells also possess the ability to sequester and subsequently export lycopene-loaded exosomes. Export destinations may therefore vary from delivery into seminal plasma to interstitial spaces in prostatic tumours, including metastases. It is therefore speculated that lycopene may have a role to play in the treatment of already established disease in addition to its preventive properties. Indeed, there are several such small-scale studies that present lycopene's ability to induce regression and limit progression in established prostate cancer (Kucuk et al. 2001, Mohanty et al. 2005) and also on its metastatic component (Ansari and Gupta 2004).

Packaging into exosomes for export also appears to offer a degree of protection to this potent but fragile antioxidant. Although losses in the order of 40 to 55% were seen in the PC3 and PNT2 exosomal lycopene content (Fig. 4) due to metabolism and degradation, they are nowhere as profound as the rapid degradative loss of lycopene freely solubilized in culture medium (Fig. 5). Furthermore, the *in vitro* models demonstrated that the accumulated substance continues to be exported from the cells via the exosomes for some time (at least 6 hr) after cessation of lycopene exposure. Organizing delivery in this manner to allow gradual release potentially offers a level of continuous protection.

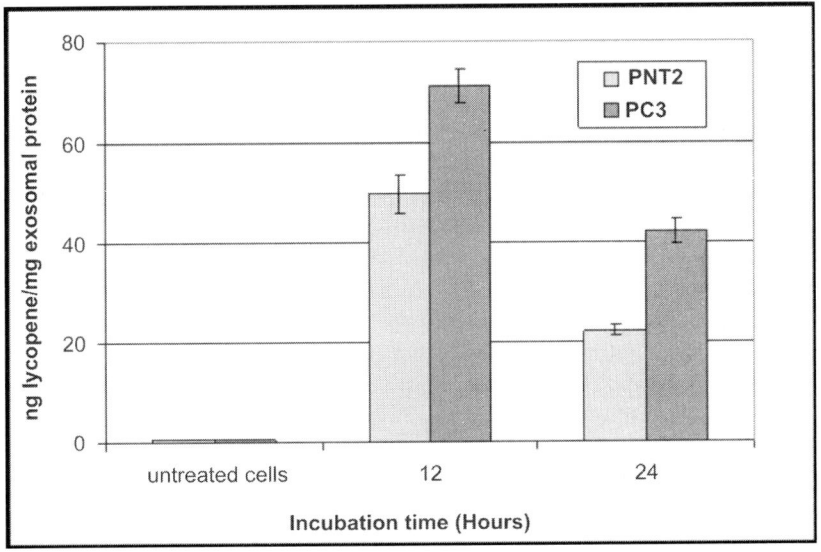

Fig. 4 Lycopene content following extraction from exosomes normalized to protein content, derived from cells exposed to lycopene for varying incubation periods. Exosomes from untreated cells were also collected following incubation in lycopene-free medium for 24 hr. No lycopene was detectable in these exosomes; they have been assigned a nominal "close to zero" value to facilitate pictorial representation. All results are shown as mean ± SD. Reproduced with permission from Blackwell Publishing (Source: BJU Int. 98(4): 910).

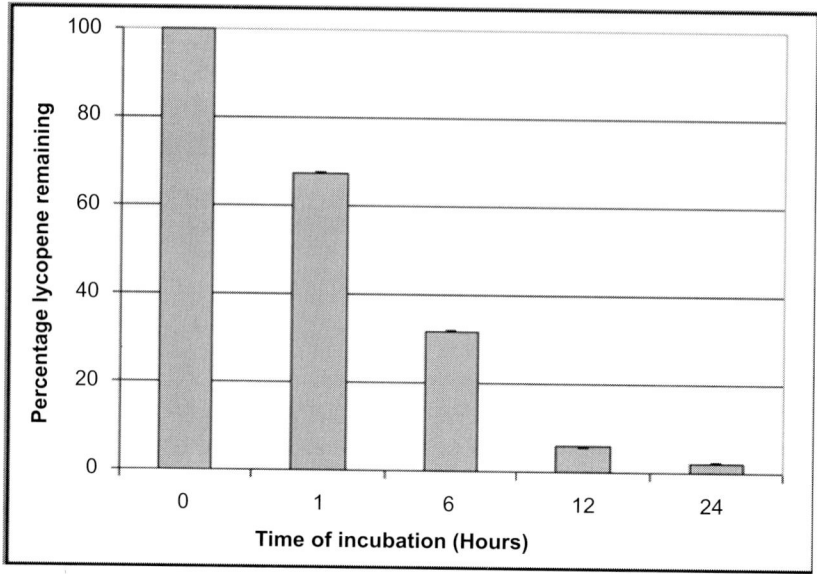

Fig. 5 Degradation of 3.58 µM lycopene solubilized in cell culture medium over time. A rapid decline in measurable lycopene is seen as a result of its accelerated degradation in free solution, when incubated under conventional cell culture conditions. Approximately only 2% of the original lycopene is recoverable at 24 hr. Reproduced with permission from Blackwell Publishing (Source: BJU Int. 98(4): 909).

The exact mechanism of cellular uptake of lycopene is not known. No specific membrane receptors have been identified for lycopene uptake into prostatic cells. It is hypothesized that uptake could be via LDL receptors on the surface of these cells as lycopene is transported in blood via lipoproteins (mainly of the low-density variety); moieties similar in nature to the lipid-rich prostasomes. However, LDL-mediated uptake seems unlikely in a cell culture-based experimental system. We suspect the lipophilic lycopene molecules are likely to seek out and equilibrate in the lipid-rich segments of cell membranes, possibly by diffusion. Thereafter, they would be internalized through the constant and rapid recycling pathways that all cellular membranes are known to undergo (Steinman et al. 1983).

Cell culture-based experimental models as described above move the argument forward by demonstrating uptake into live cells capable of processing and packaging the molecule into concurrently synthesized prostasomal machinery for later release. They also shed some light on the transport of lycopene at the sub-cellular level, expanding upon the mechanisms of how this important but labile molecule may be effectively delivered to actual or potential sites of action in reproductive physiology or urological pathologies. The accumulation of lycopene in the prostate gland may represent one of the precursory steps in the route by which it is delivered into semen, i.e., contained within exosomes found in prostatic glandular secretions. Once in seminal plasma, lycopene is known to impinge on fertility by forming an important part of the non-enzymatic antioxidant defence system against free radical damage to human spermatozoa. The mean levels of lycopene in the seminal plasma of infertile men have been shown to be significantly lower than those seen in the semen of fertile men (Palan and Naz 1996). Furthermore, oral lycopene supplementation has been shown to improve male fertility (Gupta and Kumar 2002).

Prostasomes are known to possess intrinsic antioxidant potential (Saez 2002), which should be enhanced by lycopene adsorbed and contained in them. Lycopene may thus act to potentiate the significant role the prostasome plays in promoting sperm health and function. Elucidating delivery mechanisms of such dietary compounds helps in designing further trials to investigate the best manner in which nutritional intervention could be used to facilitate improved and enhanced delivery to the sites of action.

EFFECTS OF DIETARY LYCOPENE SUPPLEMENTATION ON HUMAN SEMINAL PLASMA

From the series of *in vitro* studies presented thus far it can be inferred that lycopene sequestered in prostatic cells can be subsequently packaged and exported via prostasomes. Physiologically such lycopene-loaded exosomes represent a significant means by which this important chemopreventive agent may be delivered into tissues and semen. This forms the conceptual basis for a functional antioxidant delivery system in the male genital organs, the male reproductive tract and distant metastatic tissues potentially capable of impinging on regulation of prostatic volume, fertility and the progression of prostatic disease. The next useful step is to try and establish whether this model truly represents the physiological processes *in vivo*. This is partly addressed in a small feasibility study designed to determine the effects on seminal lycopene levels in male volunteers subjected to supplementation with a dietary source rich in lycopene (Goyal et al. 2007a). Acquisition and analysis of seminal prostasomes from the semen of such subjects offers one means of understanding whether lycopene is truly delivered into semen via prostasomes *in vivo*.

Free radicals are continuously generated in the body as part of normal metabolic activities and disease processes. They have been implicated in the aetiological basis of several chronic diseases and also play a particularly important role in the pathogenesis of male infertility (Sharma and Agarwal 1996, Agarwal and Saleh 2002). Elevated levels of ROS in semen are found in up to 40% of infertile males (Iwasaki and Gagnon 1992). Protection against this free radical damage is provided by the enzymatic and non-enzymatic antioxidant buffering capacity in seminal plasma (Sikka et al. 1995, Sharma and Agarwal 1996). Several organic molecules act in a concerted effort to scavenge these detrimental free radical species.

One of the ways of increasing the radical scavenging capacity of seminal plasma is by increasing its antioxidant content. It is proposed this may be achieved by boosting the levels of the naturally occurring antioxidants, which afford a key defence mechanism against ROS in seminal plasma. Several of these are diet-derived nutrient antioxidants and attempts have been made at boosting their levels by means of dietary supplementation in the hope they would improve seminal parameters indicative of male fertility (Vezina et al. 1996, Keskes-Ammar et al. 2003). Our research programme next investigated whether the seminal plasma levels of lycopene rise following such a supplementation period with a natural source of the molecule.

Although some epidemiological evidence is available, a direct scientific understanding of how lycopene improves impaired seminal parameters in humans is lacking. There is, however, some strong supporting evidence of lycopene's ability to protect sperm from toxicity in animal models (Atessahin et al. 2006). This study reported that treatment of rats with intraperitoneal cisplatin resulted in decreased sperm concentration, sperm motility and an increase in abnormal sperm morphology compared to the control group. However, rats also given lycopene following cisplatin administration showed a significant normalization of all the above parameters. A reduction was also seen in the levels of reactive products of oxidation (malondialdehyde) produced in the testes of the lycopene-treated group. It was concluded that adjuvant lycopene administration plays a protective role against cisplatin-induced spermiotoxicity.

Significant parallel rises in the blood and seminal plasma levels of lycopene following oral supplementation with a natural source of this molecule are demonstrable, as illustrated in Fig. 6 (Goyal et al. 2007b). Although strongly correlated, in comparison to blood there appeared to be a narrower window for increasing lycopene levels in semen following dietary supplementation. This overall smaller percentage increment is suggestive of a set upper threshold resulting from saturation in seminal plasma. Diet was supplemented in this study by means of tomatoes processed in the form of soup (equal to 22.8 mg lycopene per day) because

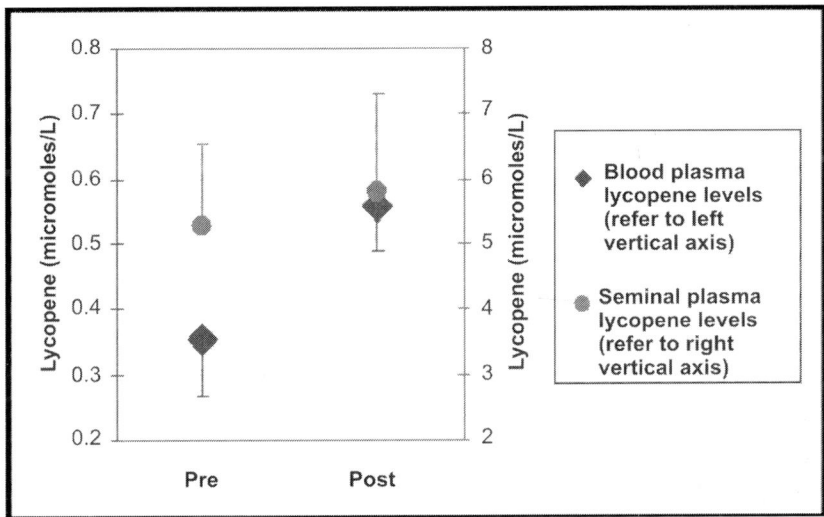

Fig. 6 Increases in blood and seminal plasma lycopene levels following a 2 wk dietary supplementation period with lycopene. The chart illustrates increases in the mean levels of lycopene in the blood (P = 0.004) (left vertical axis) and seminal plasma (P = 0.003) (right vertical axis) following a 2 wk dietary supplementation period with lycopene. The vertical bars depict the SD. Reproduced with permission from Blackwell Publishing (Source: BJU Int. 99: 1458).

in this state the lycopene has been proven to be more bioavailable (Gartner et al. 1997).

Although this study demonstrated an unequivocal rise in seminal lycopene levels following dietary supplementation, antioxidant buffering capacity assays did not demonstrate a net objective increment in the radical-scavenging capacity of the lycopene-enriched seminal plasma (Goyal et al. 2007b). Lycopene is not a ubiquitous element in human diets and therefore its levels are often undetectable in semen (Lewis et al. 1997), not to mention quantifying its antioxidant capacity in an *in vitro* assay. It has also been suggested that the lipophilicity of antioxidants remains a problematic factor in such assays, most of which are based on aqueous model systems. It is possible the lycopene may be encrypted in the hydrophobic core of membranous constituents in seminal plasma and therefore not able to exert its scavenging effect in an artificial largely aqueous environment, as was the case in this study. We have already presented laboratory evidence on how lycopene accumulates in lipid-rich seminal prostasomes, which affords it relative protection from degradation (Goyal et al. 2006a). Further work using prostatic epithelial cells in culture also confirmed reduced lycopene degradation through packaging of the substance into analogous microparticles, prior to subsequent secretion from the cell (Goyal et al. 2006b). As already discussed, these multilamellar lipid-rich moieties capable of mopping up intracellular or free lycopene are found in copious quantities in seminal plasma. In view of their inherent lipophilic nature, reliably assaying antioxidants incorporated into their membranous architecture might prove difficult. In addition, attempts at isolating these prostasomal fractions from single ejaculate specimens and quantifying their lycopene content also poses difficulties in view of the small sample volumes involved and the exceedingly meagre amounts of material acquired from each for subsequent analysis.

To get around this impracticality, pooled samples from supplemented individuals could be used, but there are clearly other shortcomings with this type of methodology. There can be large inter-individual variations in seminal plasma lycopene levels and a widely variable response to dietary supplementation. Alternatively, techniques such as Raman microspectroscopy (Goyal 2007a) could be employed to image such feeble quantities in order to ascertain the presence of lycopene in specific or individual seminal fractions.

Similar studies have looked at the effects of supplementation with other antioxidants known to be of importance in reproduction, health and functioning of spermatozoa. One such study demonstrated a significant rise in the mean seminal plasma vitamin E levels following oral alpha-tocopherol supplementation in infertile men. This rise, however, was not shown to have a demonstrable effect on overall sperm quality (Moilanen et al. 1993).

In general, antioxidant function and mechanisms of action in semen remain an under-investigated area of andrology. Confirming the presence of lycopene in human semen and demonstrating that its levels can be significantly increased following short-term dietary intervention with a natural source of the molecule is an important step forward. Furthermore, the rise in seminal lycopene levels appears to positively correlate with the increase in blood levels. Further studies are warranted to establish whether this would also be the case in infertile males with the possibility of associated improvements in their seminal parameters. It is only through continuing research efforts and analysis that we will be able to develop a better understanding of the role nutritional agents such as lycopene may play in infertility. Thus far, lycopene has shown great promise as a chemopreventive agent in diverse urological conditions (Wertz et al. 2004). It is possible such nutritional interventions may one day play a significant role in the enhancement of normal physiological processes as well as the management of certain urological pathologies.

CONCLUSIONS AND FUTURE DIRECTIONS

Reactive oxygen species and the oxidative damage they induce in biomolecules have been implicated in the causation and progression of many chronic diseases, including cancers, cardiovascular morbidity, diabetes and even infertility. Consequently, dietary antioxidants with a potential to counter such molecules and their associated oxidative damage are being increasingly investigated as strategic chemopreventive agents. Being natural products, they provoke fewer concerns about significant and adverse side-effects.

Lycopene is best known for its potency as a natural dietary antioxidant. The main evidence for its beneficial properties comes from epidemiological human health studies. It has established a reputation for providing significant health benefits and playing a notable preventive role as an antioxidant against free radical damage in cancer, ageing, and cardiovascular disease.

Lycopene appears to play an important role in a number of urological conditions including chemoprevention of prostate cancer and benign prostatic hyperplasia. Lycopene also appears to have an impact on male fertility with oral lycopene therapy holding some promise in the treatment of male immuno-infertility. Its presence in human semen contributes to the non-enzymatic defence system that helps combat reactive oxygen species that are potentially detrimental to the health and function of spermatozoa.

Lycopene's potent antioxidant ability is thought to help reduce oxidative damage to biological macromolecules such as DNA, thereby abrogating the potential for gene mutations, a well-established tumourigenic trigger. Although the exact underlying mechanisms have not been fully elucidated, it has also been shown to act through a variety of alternative modes of action independent of its antioxidant function to ward off the effects of free radical-induced damage to cellular components.

Prostasomes are submicron-sized organelles of prostatic origin. They are abundant in semen and are also secreted by metastatic prostatic tissues. They appear as a lipoprotein membrane surrounding less organized material. Proteomic studies reveal that they carry over 130 proteins belonging to various sub-classes. They also exhibit a peculiar lipid composition with an abundance of cholesterol and sphingomyelin. These lipid-rich vesicles are postulated to play a key role in aiding reproductive mechanisms underlying male fertility. Close interaction, adhesion and even fusion between prostasomes and human spermatozoa have been incontrovertibly proven, thereby allowing the sperm cells to acquire new membrane-bound lipids, proteins and also cytosolic ion increments from prostasomes. Similar logic applies to somatic cells wherever disrupted architecture brings them into contact with prostasomes.

In view of their lipid-rich nature it is highly likely that prostasomes act as suitable delivery vehicles for the highly lipophilic lycopene molecule once it is sequestered into the prostatic tissues. In this manner they would be able to deliver this important molecule to potential sites of action where it may impinge on fertility and prostatic pathologies.

Integration of biochemical and cell biology studies allows systematic investigation and an in-depth understanding of lycopene trafficking at the cellular level. The studies described above identify important synergistic associations between lycopene's actions through prostasomes. The key findings show that *in vitro* uptake of lycopene by seminal prostasomes appears to confer a degree of protection against the untimely degradation of this molecule, potentially allowing it to reach its site of action more efficaciously. *In vitro* models employing prostatic cell lines of benign (PNT2) and malignant (PC3) origins can also be used to investigate the mode of lycopene transport and delivery via exosomes further. This demonstrated that cellular exosomes expressed by these cells, when exposed to lycopene, contain lycopene in quantities measurable by HPLC. Furthermore, the cells continued to liberate lycopene-loaded exosomes even after the exposure had been removed. Export destinations from cells may therefore vary from delivery into seminal plasma to interstitial spaces in prostatic metastases where lycopene is potentially capable of limiting progression (and perhaps promoting regression) in established disease.

Supplementation with a lycopene-enriched diet results in significant increases in blood and seminal plasma lycopene levels in healthy human subjects. This is, however, not associated with a parallel measurable increase in the antioxidant capacity of the enriched seminal plasma. It is possible the assays employed cannot detect what perhaps translates into a very small increment in antioxidant buffering capacity. Lycopene encryption in lipid-rich prostasomes may further add to this problem.

The benefits of the ability to deliver increased lycopene into semen need to be further assessed by means of supplemental clinical studies in immuno-infertile males and evaluating whether dietary lycopene supplementation improves their seminal parameters against age-matched healthy subjects. A specific analysis of the prostasomal content for the presence of lycopene following dietary supplementation in such individuals would also be of added value to elucidate whether lycopene is truly delivered into semen via prostasomes. If this is the case such efforts may have a place in helping design therapeutic interventions in fertility disorders underpinned by reactive oxygen species.

Biophotonic techniques such as Raman microspectroscopy can be used to determine carotenoid levels in lycopene-treated single living cells in real time. This remains a promising tool to study the distribution of lycopene in live prostatic cells.

By aggressively pursuing fundamental studies concerning the biology of lycopene, including absorption, metabolism, excretion and biological functions in experimental models and humans, one can begin to decipher the complexities underlying lycopene's health benefits. Nutritional interventions involving dietary lycopene supplementation may one day prove to be a useful adjunct to the standard treatment in prostatic pathologies and immuno-infertility.

ACKNOWLEDGEMENTS

We would like to thank Dr. David Yeung and Dr. Tristan Robinson of the Research and Development Division at Heinz Inc. for their support and educational grant.

ABBREVIATIONS

HDL: High-density lipoproteins; HPLC: high performance liquid chromatography; HUVEC: Human Umbilical Vein Endothelial Cell; LDL: low-density lipoproteins; ROS: reactive oxygen species; SD: standard deviation; VLDL: very low-density lipoproteins; WHO: World Health Organization

References

Agarwal, A. and R.A. Saleh. 2002. Role of oxidants in male infertility: rationale, significance, and treatment. Urol. Clin. North Amer. 29: 817-827.

Al-Delaimy, W.K. and A.L. van Kappel, P. Ferrari, et al. 2004. Plasma levels of six carotenoids in nine European countries: report from the European Prospective Investigation into Cancer and Nutrition (EPIC). Publ. Health Nutr. 7: 713-722.

Allen, C.M. and A.M. Smith, S.K. Clinton, and S.J. Schwartz. 2002. Tomato consumption increases lycopene isomer concentrations in breast milk and plasma of lactating women. J. Amer. Diet Assoc. 102: 1257-1262.

Ames, B.N. and M.K. Shigenaga, and T.M. Hagen. 1993. Oxidants, antioxidants and the degenerative diseases of aging. Proc. Natl. Acad. Sci. 90: 7915-7922.

Ansari, M.S. and N.P. Gupta. 2004. Lycopene: a novel drug therapy in hormone refractory metastatic prostate cancer. Urol. Oncol. 22: 415-420.

Arienti, G. and E. Carlini, and C.A. Palmerini. 1997a. Fusion of human sperm to prostasomes at acidic pH. J. Membr. Biol. 155: 89-94.

Arienti, G. and E. Carlini, R. Verdacchi, and C.A. Palmerini. 1997b. Transfer of aminopeptidase activity from prostasomes to sperm. Biochim. Biophys. Acta 1336: 269-274.

Arienti, G. and E. Carlini, R. Verdacchi, E.V. Cosmi, and C.A. Palmerini. 1997c. Prostasome to sperm transfer of CD13/aminopeptidase N (EC 3.4.11.2). Biochim. Biophys. Acta 1336: 533-538.

Arienti, G. and E. Carlini, A. Polci, E.V. Cosmi, and C.A. Palmerini. 1998. Fatty acid pattern of human prostasome lipid. Arch. Biochem. Biophys. 358: 391-395.

Arienti, G. and E. Carlini, A. Nicolucci, E.V. Cosmi, F. Santi, and C.A. Palmerini. 1999a. The motility of human spermatozoa as influenced by prostasomes at various pH levels. Biol. Cell. 91: 51-54.

Arienti, G. and C. Saccardi, E. Carlini, R. Verdacchi, and C.A. Palmerini. 1999b. Distribution of lipid and protein in human semen fractions. Clin. Chim. Acta 289: 111-120.

Arienti, G. and E. Carlini, C. Saccardi, and C.A. Palmerini. 2004. Role of human prostasomes in the activation of spermatozoa. J. Cell. Mol. Med. 8: 77-84.

Arvidson, G. and G. Ronquist, G. Wikander, and A.C. Ojteg. 1989. Human prostasome membranes exhibit very high cholesterol/phospholipid ratios yielding high molecular ordering. Biochim. Biophys. Acta 984: 167-173.

Atessahin, A. and I. Karahan, G. Turk, S. Gur, S. Yilmaz, and A.O. Ceribasi. 2006. Protective role of lycopene on cisplatin-induced changes in sperm characteristics, testicular damage and oxidative stress in rats. Reprod. Toxicol. 21: 42-47.

Babiker, A.A. and B. Nilsson, G. Ronquist, L. Carlsson, and K.N. Ekdahl. 2005. Transfer of functional prostasomal CD59 of metastatic prostatic cancer cell origin protects cells against complement attack. Prostate 62: 105-114.

Benoff, S. 1993. Preliminaries to fertilization: The role of cholesterol during capacitation of human spermatozoa. Hum. Reprod. 8: 2001-2008.

Carlini, E. and C.A. Palmerini, E.V. Cosmi, and G. Arienti. 1997. Fusion of sperm with prostasomes: effects on membrane fluidity. Arch. Biochem. Biophys. 343: 6-12.

Carlsson, L. and C. Pahlson, M. Bergquist, G. Ronquist, and M. Stridsberg. 2000. Antibacterial activity of human prostasomes. Prostate 44: 279-286.

Clinton, S.K. and C. Emenhiser, S.J. Schwartz, D.G. Bostwick, A.W. Williams, B.J. Moore, and J.W. Erdman Jr. 1996. Cis-trans lycopene isomer, carotenoids and retinol in the human prostate. Cancer Epidemiology, Biomarkers and Prevention 5: 823-833.

Clinton, S.K. 1998. Lycopene: chemistry, biology, and implications for human health and disease. Nutr. Rev. 56: 35-51.

Delves, G.H. and A.B. Stewart, B.A. Lwaleed, and A.J. Cooper. 2005. In vitro inhibition of angiogenesis by prostasomes. Prostate Cancer Prostatic Dis. 8: 174-178.

Di Mascio, P. and S. Kaiser, and H. Sies. 1989. Lycopene as the most efficient biological carotenoid singlet oxygen quencher. Arch. Biochem. Biophys. 274: 532-538.

Erdman, J.W., Jr. and T.L. Bierer, and E.T. Gugger. 1993. Absorption and transport of carotenoids. pp. 76-85. *In*: L.M. Canfield, N.I. Krinsky and J.A. Olson [eds.] Carotenoids in Human Health, Vol 691. New York Academy of Sciences, New York.

Fabiani, R. and L. Johansson, O. Lundkvist, U. Ulmsten, and G. Ronquist. 1994. Promotive effect by prostasomes on normal human spermatozoa exhibiting no forward motility due to buffer washings. Eur. J. Obstet. Gynecol. Reprod. Biol. 57: 181-188.

Gartner, C. and W. Stahl, and H. Sies. 1997. Lycopene is more bioavailable from tomato paste than from fresh tomatoes. Amer. J. Clin. Nutr. 66: 116-122.

Giovannucci, E. 2002. A review of epidemiological studies of tomatoes, lycopene and prostate cancer. Exp. Biol. Med. 227: 852-859.

Giovannucci, E. and A. Ascherio, E.B. Rimm, M.J. Stampfer, G.A. Colditz, and W.C. Willett. 1995. Intake of carotenoids and retinol in relation to risk of prostate cancer. J. Natl. Cancer Inst. 87: 1767-1776.

Goyal, A. 2007a. Lycopene trafficking through prostate tissues and its synergistic association with prostasomes. MD Thesis, University of Portsmouth, UK.

Goyal, A. and M. Chopra, G. Delves, and A. Cooper. 2005. Do prostasomes play a role in lycopene delivery? Proc. Nutr. Soc. 64 OCA/B: 94A.

Goyal, A. and G.H. Delves, M. Chopra, B.A. Lwaleed, and A.J. Cooper. 2006a. Can lycopene be delivered into semen via prostasomes? *In vitro* incorporation and retention studies. Int. J. Androl. 29: 528-533.

Goyal, A. and G.H. Delves, M. Chopra, B.A. Lwaleed, and A.J. Cooper. 2006b. Prostate cells exposed to lycopene *in vitro* liberate lycopene-enriched exosomes. BJU Int. 98: 907-911.

Goyal, A. and M. Chopra, B. Lwaleed, B. Birch, and A.J. Cooper. 2007b. The effects of dietary lycopene supplementation on human seminal plasma. BJU Int. 99: 1456-1460.

Gupta, N.P. and R. Kumar. 2002. Lycopene therapy in idiopathic male infertility — a preliminary report. Int. Urol. Nephrol. 34: 369-372.

Hwang, E.S. and P.E. Bowen. 2004. Cell cycle arrest and induction of apoptosis by lycopene in LNCaP human prostate cancer cells. J. Med. Food 7: 284-289.

Iwasaki, A. and C. Gagnon. 1992. Formation of reactive oxygen species in spermatozoa of infertile patients. Fertil. Steril. 57: 409-416.

Kaplan, L.A. and J.M. Lau, and E.A. Stein. 1990. Carotenoid composition, concentrations and relationship in various human organs. Clin. Physiol. Biochem. 8: 1-10.

Kelly, R.W. and P. Holland, G. Skibinski, C. Harrison, L. McMillan, T. Hargreave, and K. James. 1991. Extracellular organelles (prostasomes) are immuno-suppressive components of human semen. Clin. Exp. Immunol. 86: 550-556.

Keskes-Ammar, L. and N. Feki-Chakroun, T. Rebai, Z. Sahnoun, H. Ghozzi, S. Hammami, K. Zghal, H. Fki, J. Damak, and A. Bahloul. 2003. Sperm oxidative stress and the effect of an oral vitamin E and selenium supplement on semen quality in infertile men. Arch. Androl. 49: 83-94.

Khachik, F. and G.R. Beecher, M.B. Goli, W.R. Lusby, and J.C. Smith. 1992. Separation and identification of carotenoids and their oxidation products in the extracts of human plasma. Anal. Chem. 64: 2111-2122.

Khachik, F. and L. Carvalho, P.S. Bernstein, G.J. Muir, D.-Y. Zhao, and N.B. Katz. 2002. Chemistry, distribution and metabolism of tomato carotenoids and their impact on human health. Exp. Biol. Med. 227: 845-851.

Kucuk, O. and F.H. Sarkar, W. Sakr, Z. Djuric, M.N. Pollak, F. Khachik, Y.-W. Li, M. Banerjee, D. Grignon, J.S. Bertram, J.D. Crissman, E.J. Pontes, and D.P. Wood Jr. 2001. Phase II randomised clinical trial of lycopene supplementation before radical prostatectomy. Cancer Epidemiology, Biomarkers and Prevention 10: 861-868.

Lewis, S.E.M. and E.S.L. Sterling, I.S. Young, and W. Thompson. 1997. Comparison of individual antioxidants of sperm and seminal plasma in fertile and infertile men. Fertil. Steril. 67: 142-147.

Lilja, H. and C.B. Laurell. 1984. Liquefaction of coagulated semen. Scand. J. Clin. Lab. Invest. 44: 447-452.

Mack, S.R. and J. Everingham, and L.J. Zaneveld. 1986. Isolation and partial characterization of the plasma membrane from human spermatozoa. J. Exp. Zool. 240: 127-136.

Martinez, P. and A. Morros. 1996. Membrane lipid dynamics during human sperm capacitation. Front. Biosci. 1: d103-d117.

Mayne, S.T. 1996. β-carotene, carotenoids and disease prevention in humans. FASEB J. 10: 690-701.

Miller, N.J. and J. Sampson, L.P. Candeias, P.M. Bramley, and C.A. Rice-Evans. 1996. Antioxidant activities of carotenes and xanthophylls. FEBS Lett. 384: 240-242.

Mohanty, N.K. and S. Saxena, U.P. Singh, N.K. Goyal, and R.P. Arora. 2005. Lycopene as a chemoprevetative agent in the treatment of high-grade prostate intraepithelial neoplasia. Urol. Oncol. 23: 383-385.

Moilanen, J. and O. Hovatta, and L. Lindroth. 1993. Vitamin E levels in seminal plasma can be elevated by oral administration of vitamin E in infertile men. Int. J. Androl. 16: 165-166.

Nilsson, B.O. and L. Lennartsson, L. Carlsson, S. Nilsson, and G. Ronquist. 1999. Expression of prostasome-like granules by the prostate cancer cell lines PC3, Du145 and LnCaP grown in monolayer. Ups. J. Med. Sci. 104: 199-206.

Nilsson, B.O. and L. Carlsson, A. Larsson, and G. Ronquist. 2001. Autoantibodies to prostasomes as new markers for prostate cancer. Ups. J. Med. Sci. 106: 43-49.

Palan, P. and R. Naz. 1996. Changes in various antioxidant levels in human seminal plasma related to immunoinfertility. Arch. Androl. 36: 139-143.

Palmerini, C.A. and E. Carlini, A. Nicolucci, and G. Arienti. 1999. Increase of human spermatozoa intracellular Ca^{2+} concentration after fusion with prostasomes. Cell Calcium 25: 291-296.

Palmerini, C.A. and C. Saccardi, E. Carlini, R. Fabiani, and G. Arienti. 2003. Fusion of prostasomes to human spermatozoa stimulates the acrosome reaction. Fertil. Steril. 80: 1181-1184.

Poulos, A. and White. 1973. The phospholipid composition of human spermatozoa and seminal plasma. J. Reprod. Fertil. 35: 265-272.

Ronquist, G. and I. Brody. 1985. The prostasome: its secretion and function in man. Biochim. Biophys. Acta 822: 203-218.

Ronquist, G. and B.O. Nilsson. [Eds.]. 2002. Prostasomes: Proceedings from a symposium held at the Wenner-Gren Centre, Stockholm, in June 2001. Wenner-Gren International Series, Vol. 81, Portland Press, London.

Ronquist, G. and B.O. Nilsson. 2004. The Janus-faced nature of prostasomes: their pluripotency favours the normal reproductive process and malignant prostate growth. Prostate Cancer Prostatic Dis. 7: 21-31.

Ronquist, G. and I. Brody, A. Gottfries, and B. Stegmayr. 1978. Andrologia. An Mg^{2+} and Ca^{2+}-stimulated adenosine triphosphatase in human prostatic fluid: part I 1978; 10(4): 261-272 and part II, 10: 427-433.

Ronquist, G. and B.O. Nilsson, and S. Hjerten. 1990. Interaction between prostasomes and spermatozoa from human semen. Arch. Androl. 24: 147-157.

Rooney, I. and J. Atkinson, E. Krul, G. Schonfeld, K. Polakoski, J. Saffitz, and P. Morgan. 1993. Physiologic relevance of the Membrane Attack Complex Inhibitory Protein CD59 in human seminal plasma: CD59 is present on extracellular organelles (Prostasomes), binds cell membranes and inhibits complement-mediated lysis. J. Exp. Med. 177: 1409-1420.

Saez, F. 2002. Antioxidant properties of prostasomes. pp. 101-108. *In:* G. Ronquist and B.O. Nilsson. [Eds.] Prostasomes: Proceedings from a symposium held at the Wenner-Gren Centre, Stockholm, in June 2001. Wenner-Gren International Series, Vol. 81. Portland Press, London.

Saez, F. and C. Motta, D. Boucher, and G. Grizard. 1998. Antioxidant capacity of prostasomes in human semen. Mol. Hum. Reprod. 4: 667-672.

Sahlen, G.E. and L. Egevad, A. Ahlander, B.J. Norlen, G. Ronquist, and B.O. Nilsson. 2002. Ultrastructure of the secretion of prostasomes from benign and malignant epithelial cells in the prostate. Prostate 53: 192-199.

Sahlen, G.E. and A. Ahlander, A. Frost, G. Ronquist, B.J. Norlen, and B.O. Nilsson. 2004. Prostasomes are secreted from poorly differentiated cells of prostate cancer metastases. Prostate 1; 61: 291-297.

Sharma, R.K. and A. Agarwal. 1996. Role of reactive oxygen species in male infertility. Urology 48: 835-850.

Sikka, S.C. and M. Rajasekaran, and W.J. Hellstrom. 1995. Role of oxidative stress and antioxidants in male infertility. J. Androl. 16: 464-481.

Stahl, W. and H. Sies. 1992. Uptake of lycopene and its geometrical isomers is greater from heat-processed than from unprocessed tomato juice in humans. J. Nutr. 122: 2161-2166.

Stahl, W. and H. Sies. 1996. Lycopene a biologically important carotenoid for humans? Arch. Biochem. Biophys. 336: 1-9.

Stahl, W. and W. Schwarz, A.R. Sundquist, and H. Sies. 1992. Cis-trans isomers of lycopene and beta-carotene in human serum and tissues. Arch. Biochem. Biophys. 294: 173-177.

Steinman, R.M. and I.S. Mellman, W.A. Muller, and Z.A. Cohn. 1983. Endocytosis and the recycling of plasma membrane. J. Cell. Biol. 96: 1-27.

Stegmayr, B. and G. Ronquist. 1982. Promotive effect on human sperm progressive motility by prostasomes. Urol. Res. 10: 253-257.

Stewart, A.B. and W. Anderson, G. Delves, B.A. Lwaleed, B. Birch, and A. Cooper. 2004. Prostasomes: a role in prostatic disease? BJU Int. 94: 985-989.

Tapiero, H. and D.M. Townsend, and T.D. Tew. 2004. The role of carotenoids in the prevention of human pathologies. Biomed. Pharmacother. 58: 100-110.

Utleg, A. and E. Yi, T. Xie, P. Shannon, J. White, D. Goodlett, L. Hood, and B. Lin. 2003. Proteomic analysis of human prostasomes. Prostate 56: 150-161.

Vezina, D. and F. Mauffette, K.D. Roberts, and G. Bleau. 1996. Selenium-vitamin E supplementation in infertile men. Effects on semen parameters and micronutrient levels and distribution. Biol. Trace Elem. Res. 53: 65-83.

Wertz, K. and U. Siler, and R. Goralczyk. 2004. Lycopene: modes of action to promote prostate health. Arch. Biochem. Biophys. 430: 127-134.

WHO. 1999. WHO Laboratory Manual for the Examination of Human Semen and Sperm Cervical Mucus Interaction, 4th Edition. Cambridge University Press, Cambridge, UK.

Yeum, K.-J. and S. Booth, J. Sadowski, C. Liu, G. Tang, N. Krinsky, and R. Russell. 1996. Human plasma carotenoid response to the ingestion of controlled diets high in fruits and vegetables. Amer. J. Clin. Nutr. 64: 594-602.

2.6

Topically Applied Lycopene and Antioxidant Capacity

Marco Andreassi and Lucio Andreassi
[1]Centre of Cosmetic Science and Technology, University of Siena, Italy
[2]Department of Clinical Medicine and Immunological Sciences
Section of Dermatology, University of Siena, Italy

ABSTRACT

Various antioxidant substances have a protective effect on the skin when applied in suitable formulations. Natural lycopene (extracted from tomato), incorporated in an oil/water emulsion at a concentration of 0.03% and applied to the skin 30 min before ultraviolet exposure, decreased the erythema induced by UV radiation. Lycopene was more effective when associated with α-tocopherol palmitate and ascorbic acid, at concentrations of 0.5% and 1%, respectively. Since lycopene does not screen UV radiation, its protective ability is due to antioxidant activity. The potentiation of its effectiveness by vitamins E and C can probably be attributed to an interaction between the three substances in the cascade of events leading to the antiradical effect. As proposed for β-carotene, lycopene may act by repairing the tocopherol radical, after which it is repaired by the ascorbic acid. The efficiency of the system could be related to the different affinities (lipophilic or hydrophilic) of the components of this triple association. Because of its characteristics, lycopene is particularly suitable for topical application in sunscreens or in any preparation designed to prevent or reduce the harmful effects of oxidative stimuli.

A list of abbreviations is given before the references.

INTRODUCTION

Many cutaneous inflammatory conditions are completely or partly mediated by phenomena related to oxidative stress. Ultraviolet radiation and the related damage are the most common expression of these events, although various other physical, chemical and biological stimuli can damage the skin via the action of free radicals and reactive oxygen species (ROS). The skin is particularly exposed to oxidative damage since, in addition to being pervaded by oxygen deriving from the blood, it exchanges this element with the external environment. Moreover, the oxygen in skin cells can be activated by light and this makes the integumentary system particularly vulnerable to oxidative damage (Bickers and Athar 2006).

Stimuli able to activate oxidative stress in the skin include inflammatory agents that attract neutrophils and monocytes, cells specialized in the production and release of ROS, via the enzyme NADPH oxidase (El-Benna et al. 2005). This process is particularly important since the released ROS can send signals for the recruitment of other neutrophils, establishing an extremely dangerous cascade mechanism. The activated neutrophils mainly produce superoxide anion and hydrogen peroxide; together with specialized peptides and proteases, they represent an efficient antimicrobial arsenal, which, however, can also attack and destroy some structural components of the skin, such as collagen (Quinn et al. 2006).

In addition, free radicals mediate the reactions that occur in vascular pathologies and, in particular, in cutaneous reperfusion following ischemic events (Jokuszies et al. 2006). An increase of the level of hypoxanthine leads to activation of the enzyme xanthine oxidase and catalyzes the production of superoxide radical and other radical forms including nitric oxide, the latter produced mainly by endothelial cells (McCord 1985).

Another substance with oxidative activity in the skin is ozone (Weber et al. 2001, Valacchi et al. 2005); it can be present at ground level as a product of photochemical smog and is toxic at concentrations of 0.1-0.5 ppm, thus representing a serious threat to urban air quality (Mustafa 1990). Ozone reacts quickly with most biomolecules and has a high redox potential; thus, it is one of the most powerful oxidizers, able to induce oxidation and peroxidation of biomolecules both directly and via secondary reactions. Known mainly for its pulmonary toxicity, ozone is also highly toxic to the skin, as it compromises the integrity of the horny layer via oxidative damage to lipids and proteins (Pryor 1994).

Nevertheless, sun exposure is by far the most important factor in the formation of free radicals in the skin. UV radiation, particularly between 290 and 400 nm, is the main cause of oxidative stress (Ravanat et al. 2001). Its ability to penetrate the skin varies according to the wavelength: radiation around 300 nm (UVB) loses its energy in the epidermis, while radiation around 350 nm (UVA) penetrates more deeply. Therefore, while the principal site of action of UVB is the epidermis, UVA acts mainly in the dermis (Kawanishi et al. 2001).

UV radiation can generate reactive forms of oxygen *in vivo*, which are responsible for a large amount and variety of tissue damage (Black 1987). Skin exposed to UV radiation frequently presents lipid peroxidation and the formation of sunburn cells, a marker of UVB-damaged epidermis that could be modulated by substances with antioxidant activity. In addition, and very importantly, exposure of the skin to UV radiation leads to compromise of the antioxidative defenses. Under the effect of UV radiation, the activities of superoxide dismutase, catalase and glutathione peroxidase decrease, while the levels of antioxidant vitamins are markedly reduced; these events create further oxidative stress (Shindo et al. 1993). Finally, UV radiation can activate photosensitizing substances, endogenous ones such as porphyrin and flavin (Dalle Carbonare and Pathak 1992) as well as exogenous ones such as some drugs (Fuchs 1992), which are responsible for the formation of highly reactive ROS, such as singlet oxygen.

It is reasonable to believe that the well-known biological effects following sun exposure, especially erythema and hyperpigmentation, are the direct consequence of oxidative stress induced by UV radiation. A similar cause-effect relationship cannot easily be demonstrated for delayed biological damage. However, if we accept the principle that photo-ageing and skin tumors are also related to photoexposure, we must accept that their pathogenesis is due to oxidative stress (Gasparro et al. 1998).

Numerous substances have been proposed to counteract the harmful effects of oxidative stress, especially that due to UV radiation. Particularly interesting are the carotenoids, a class of about 600 compounds (chemically defined as tetraterpenoids) that are abundant in many common foods. Carotenoids are efficient singlet oxygen quenchers but are also involved in other antioxidant reactions. The ability to neutralize ROS and free radicals is not equal for all carotenoids, as it varies according to their reduction potential, which is maximal for lycopene (Edge et al. 1997). Therefore, lycopene appears to be the most suitable carotenoid for use in topical formulations.

BIOAVAILABILITY AND BIOCHEMISTRY OF LYCOPENE IN THE SKIN

The action of any topically applied active principle depends on several factors: first, stability and solubility in the vehicle; second, ability to penetrate into the epidermis; and finally, the eventual biotransformation it might undergo in the tissue. The destiny of lycopene included in topical preparations is still not completely known, as the available data are few and incomplete. Therefore, to outline the possible events this antioxidant is involved in when applied on the skin, we must refer to data from studies conducted on other tissues and to knowledge about other antioxidant substances chemically similar to lycopene.

Regarding the stability, it is important to remember that lycopene can be obtained synthetically or extracted from natural sources. Synthetic crystalline lycopene is sensitive to light and oxygen and thus is stable only if preserved in a refrigerated environment and in a container saturated with inert gas, conditions that are very difficult to create (McClain and Bausch 2003). In contrast, lycopene extracted from natural sources, especially tomato, is more resistant to degradation than the synthetic form. The stability of natural lycopene can probably be attributed to the presence of other constituents, such as tocopherols, ascorbic acid and phenolic antioxidants (Takeoka et al. 2001).

Lycopene is a lipophilic compound completely insoluble in water. This property is particularly important with regard to the selection of the vehicle to use in topical formulations. A greasy or oily vehicle would seem to be the best choice, as it has been demonstrated that when lipophilic substances (like those with a terpene structure) are incorporated in an oily vehicle they reach higher concentrations in the horny layer than can be obtained with oil/water emulsions (Cal 2006). However, in current products, lycopene is often associated with other antioxidants, such as ascorbic acid and α-tocopherol, that increase the antioxidant effect. Therefore, the most appropriate vehicle appears to be an oil/water emulsion. In the preparation of such an emulsion, it is essential to solubilize the lycopene in a suitable solvent that is non-toxic to skin and subsequently to incorporate the lipid phase in a stable formulation.

After high-dose oral administration, lycopene is distributed in various organs and tissues; it also reaches the skin, at the concentration 0.42 nmol/g, i.e., within the range (0.2-0.8 nmol/g) recorded for most of the other biological structures investigated (Rao et al. 2006). We do not know how much lycopene reaches the cutaneous structures after topical application, although we can hypothesize that the concentration must be much higher. In fact, many active principles reach much higher concentrations in the skin after topical application than after oral administration. Moreover, some topically applied antioxidant substances permeate the epidermis and reach the maximum concentration in the horny layer, where they remain for several days, even after the skin surface has been washed (Burke 2004). This has been shown for vitamin E, which is present in higher concentration in the deep part of the horny layer than in the superficial part. Therefore, the horny layer clearly acts as a reserve for some antioxidants, assuring a constant supply to the underlying structures, according to a pharmacokinetic model that may also be valid for lycopene.

In foods, over 90% of the lycopene is in the all-trans isomeric form, while in human tissues it occurs mainly as *cis* isomers (Boileau et al. 2002). This suggests that the cis isomers are absorbed better than the *trans* ones, probably because of their small molecular length, greater micellar solubility and lower tendency to aggregate. Studies carried out on the ferret, an animal that absorbs carotenoids without inducing modifications, have shown that after the administration of a certain quantity of lycopene the proportion of *cis* isomers in the gastric and intestinal juices is 6-18%, whereas it is 77% in the mesenteric lymph (Ferriera et al. 2000). These studies support the hypothesis that the *cis* isomers of lycopene are much more bioavailable than the *trans* forms, and this modification could also occur in the skin after transepidermal absorption.

The metabolism of lycopene in humans has been studied by analyzing tissue extracts separated by HPLC and comparing the resulting data with the composition of the carotenoid in tomato (Khachik et al. 2002). In some tissues, including skin, lycopene largely occurs in the form of 1,2-epoxide and 5,6-epoxide, the latter able to be transformed into 2,6-cyclolycopene-1,5-epoxide and 2,6-cyclolycopene-1,5-diol, compounds that could be the main products of oxidative metabolism of lycopene as the result of tissue reactions mediated by enzymes. However, this hypothesis is still not completely confirmed since the same compounds also occur, albeit at low concentrations, in tomatoes and their alimentary by-products as a result of a chemical oxidative process.

ASSESSMENT OF ANTIOXIDANT CAPACITY OF TOPICALLY APPLIED FORMULATIONS

The skin is an excellent model for clinical evaluations of antioxidant products: it is very accessible and reliable experimental techniques allow us to induce graded, quantifiable skin reactions. The most common method for *in vivo* investigation of the protective ability of topical formulations containing antioxidants is the induction of graded erythematous reactions in healthy volunteers. The efficacy of the tested substances is shown by the intensity of the reaction in the treated areas in comparison with the response in the untreated areas.

Exposure of the skin to UV radiation causes biological effects that can be classified as immediate and late. Erythema is the most important immediate effect and can occur in individuals of all races with an intensity usually inversely proportional to the degree of skin pigmentation. Erythema can be induced by radiation in the entire UV spectrum, but the reaction varies according to the spectral characteristics of the radiation band. Short wavelength radiation is more erythemogenic than long wavelength radiation, with a peak around 300 nm; beyond that wavelength, the erythemogenic power decreases until it becomes about 1000 times as low for UV radiation of 320 nm (de Gruijl 1995). Therefore, the best solar simulators are those with xenon lamps consisting of tungsten electrodes in a quartz glass bulb filled with xenon at 20-40 atm pressure. At these pressures, the emission spectrum of the xenon is continuous, with a flux that includes all the wavelengths rather than specific spectral lines.

Erythema is due to increased blood flow in the dermal vessels induced by vasoactive substances released in response to UV stimulation. Its intensity, once evaluated visually, is now measured instrumentally, especially when photobiological procedures requiring quantification are involved. The techniques to objectively assess erythema are laser Doppler fluximetry, based on measurement of the velocity of blood flow by means of the Doppler effect (Choi and Bennett 2003), and colorimetry based on measurement of the radiation reflected by a surface when illuminated with a suitable source. The best instrument to assess erythema is the tri-stimulus colorimeter, as it provides a direct measurement of the chromaticity coordinates. Within the CIE*L*a*b color space, variations of the positive hemivector of the *a axis express the different intensities of red and thus indicate the intensity of the erythema (Park et al. 2002).

This experimental device has been used to study various antioxidants, including melatonin, vitamin E, vitamin C, caffeic acid, ferulic acid, phytic acid and lycopene. Melatonin shows the greatest protective activity at the concentration of 0.5% when applied before exposure to UV radiation, while it is inactive when applied after UV radiation exposure (Bangha et al. 1996, Dreher et al. 1999). The mechanism of action has been explained as radical-scavenging activity and neutralization of the free radicals produced by UV radiation exposure (Fischer and Elsner 2001). Caffeic acid and ferulic acid, dissolved in a saturated aqueous solution and applied on the skin for 3 hr by means of a Hill Top chamber immediately after UV radiation exposure, reduced the erythematous response (Saija et al. 2000). Topically applied vitamin E reduced the UV radiation-induced erythema as well as the number of sunburn cells and chronic damage from UVB, including carcinogenic effects (Eberlein-Konig and Ring 2005). The most interesting effect of α-tocopherol, however, is that, when used in association with vitamin C, it potentiates the topical protective ability of other antioxidants, including melatonin and lycopene (Dreher et al. 1998, Andreassi et al. 2004a).

In summary, the availability of a reliable and reproducible method to assess topical antioxidants is extremely useful to compare the efficacy of different active principles, to identify the best concentration of each substance, and to determine the utility of salification or esterification. Finally, although some antioxidants are unstable, it might be possible to use them in the form of controlled-release microparticles; however, their efficacy would have to be demonstrated clinically.

PERSONAL EXPERIENCE

From the above data, it is clear that the *in vivo* assessment of topical antioxidant products requires a reliable method and specific instruments. A suitable UV source and an instrument to measure the degree of erythema are indispensable. To define the magnitude of the photodamage, it is also advisable to have an evaporimeter able to evaluate the status of the barrier function by means of transepidermal water loss (TEWL). Finally, the intensity of the UV source must be checked periodically with a radiometer.

According to our experience, the best UV source is the Multiport 601 solar simulator. It is provided with a 150 W xenon lamp with a continuous emission spectrum between 290 and 400 nm and has six liquid light guide outputs that directly convey the radiation band on to the skin. The window of each output is regulated so as to emit graded doses with a constant increase. An exposure of 120 sec is sufficient to evoke the minimal erythemal dose (MED) in the range of doses emitted by the photostimulator in all subjects with phototype II/III.

The best instrument to measure the degree of the erythematous response is the Minolta Chroma Meter CR-200/300 tri-stimulus colorimeter, which is easy to manage and provides high reproducibility of results. Using the CIE L*a*b* color space (and in particular the positive hemivector of "*a"), the Chroma Meter CR-200/300 can provide values in a fairly wide range, to detect differences not appreciable upon simple visual observation. To perform a measurement, the instrument's measuring head is positioned orthogonally over the area to be assessed without pressing the skin so as to avoid compression ischemia. With these precautions, measurement of the erythema has proved to be reliable and reproducible (Andreassi and Flori 1995).

The integrity or damage of the barrier function is assessed by measuring the TEWL: the Tewameter® TM 210 instrument (Courage & Khazaka) provides reproducible results expressed in $g/m^2/hr$ (Miteva et al. 2006).

The methodology is as follows. Rectangular areas corresponding to the section of the solar simulator's output head are delimited on the volar surface of the forearm. The product is applied inside these areas (3 mg/cm^2), taking care to obtain a uniform distribution by means of a glass rod. Different areas can be used to test different formulations, with a control area for each product treated with vehicle alone. When assessing active principles that have not yet been sufficiently tested, it is good practice to prepare each formulation daily. After 30 min, each treated area is irradiated via contact of the solar simulator's output head with the skin surface for 2 min. As mentioned previously, this time period is adequate to induce MED in all subjects with intermediate pigmentation. The intensity of the erythema is assessed before the beginning of the test and then 24 and 48 hr after photostimulation (Fig. 1). The recorded values must be analyzed statistically.

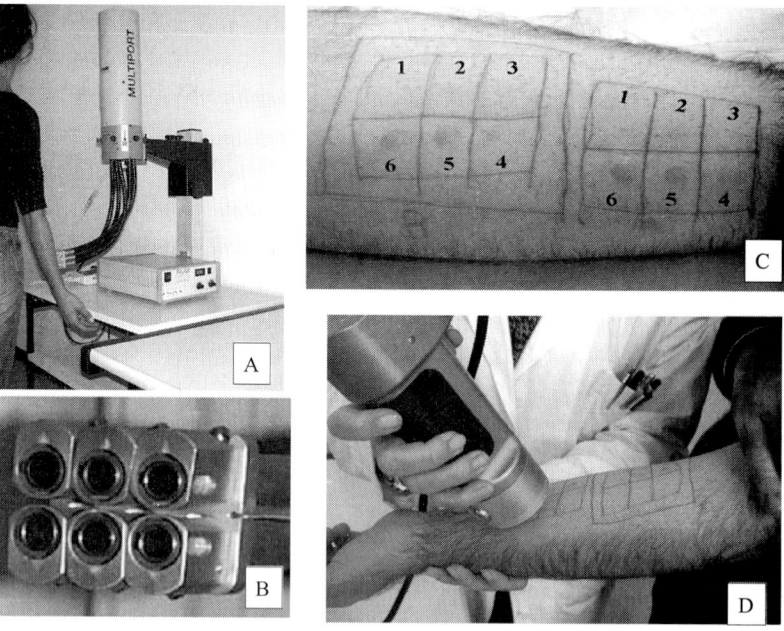

Fig. 1 The main phases of the procedure to assess topically applied antioxidant products. (A) Photostimulation of the volar surface of a forearm in a healthy volunteer. (B) Solar simulator output head provided with six windows that emit graded UV radiation doses. (C) Example of erythematous reactions after the application of formulations containing antioxidants. (D) Measurement of the intensity of the erythema with a tri-stimulus colorimeter.

With this method, we conducted a series of trials using a preparation containing lycopene, two preparations containing α-tocopherol and ascorbic acid respectively, and a third formulation including a mixture of the three active principles (Andreassi et al. 2004a, b). The gel-emulsion used as vehicle was employed as the control. All samples were prepared daily to avoid the need of a preservative and to prevent loss of activity of the active principles.

Lycopene was used at the final concentration of 0.03% in the form of a dry extract of *Lycopersicon esculentum*, a powder titered at 6% lycopene supplied by Polichimica Srl, Bologna, Italy. To assess whether the active principle acted as a UV radiation filter, we determined the absorption spectrum of an ethyl ether solution of the *L. esculentum* extract. As seen in Fig. 2, the absorption spectrum has a peak at around 340 nm, while in the 280 and 320 nm range the absorption is virtually absent; hence, the lycopene extract is not able to filter erythemogenic UV radiation.

The basic formulation was prepared by combining the aqueous phase with the lipophilic phase in steel containers and stirring the mixture until complete formation of the gel-emulsion. The product containing lycopene was prepared by dissolving the dry extract of *L. esculentum* in the lipophilic phase, adding water and stirring until complete formation of the gel-emulsion. Before choosing the final lycopene concentration, we carried out preliminary trials to determine the concentration that proved active

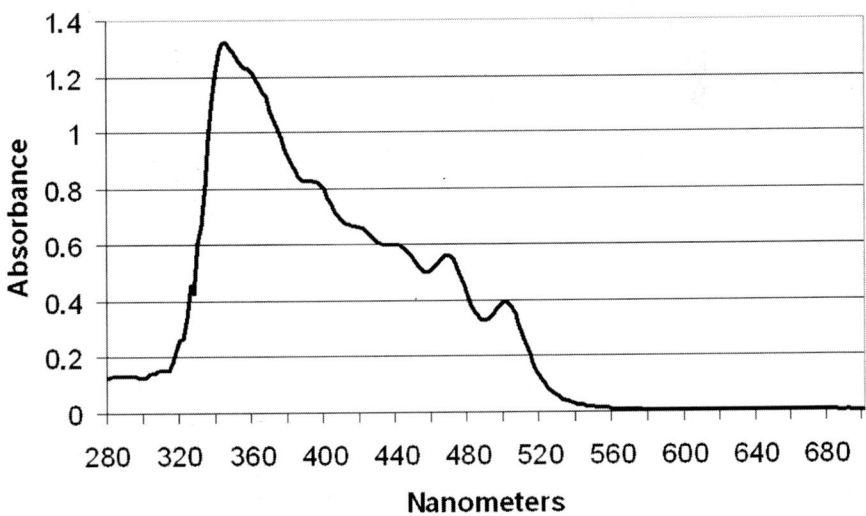

Fig. 2 UVB-VIS absorption spectrum measured using an ethyl ether solution of the dry extract of *Lycopersicon esculentum*.

without altering the rheological characteristics of the product. The trials were performed by adding graded concentrations of extract to the basic formulation.

Vitamins E and C were used at the respective concentrations of 0.5% and 1%. The formulations were prepared according to the same procedure, dissolving vitamin E palmitate in the lipophilic phase and ascorbic acid in the aqueous phase. These concentrations were chosen because they had proved active in preliminary investigations.

Upon visual examination, the intensity of the erythematous response to photostimulation progressively increased starting from the third UV radiation dose. The erythema was more intense in the control areas than in those treated with the products containing active principles. Instrumental measurements confirmed the results of the visual examination and provided numerical values to define the characteristics of the erythematous response with greater precision (Table 1). Use of all the products with active principles led to a less intense erythematous response than in the control areas, although the difference was significant only in the case of the formulation containing lycopene. The greatest differences between the lycopene product and the control were observed after 24 hr, especially in the areas that received the fourth and fifth doses of photostimulation. Figures 3 and 4 graphically summarize the results. The histograms express the intensity of the erythematous reaction and clearly illustrate the differences between the products. Table 2 reports the evaporimetry values after application of the products and photostimulation. The increase of TEWL was proportional to the dose, indicating that the damage to the barrier function was lower in the areas treated with the active principles than in the control areas, although the differences were not statistically significant.

CONCLUSIONS

Many inflammatory processes in the skin involve oxidative events related to the release of free radicals and ROS. Exposure to solar radiation, particularly the erythemogenic UV component, is the most frequent cause of skin damage mediated by oxidative stress. Topically applied antioxidant substances are useful to counteract these pathological events.

An antioxidant substance must possess certain requisites to be effective for topical use. Above all, it must not be locally or systemically toxic; it must reach high concentrations in the cutaneous structures, particularly the external ones; it must protect the skin from oxidizing stimuli. In addition to these fundamental qualities, potential topical antioxidant substances must have physicochemical characteristics that allow their

Table 1 Erythematous reactions expressed in a* values (CIE L*a*b*)[1] after application of the products and subsequent photostimulation in 10 healthy volunteers.

T[2]	Product	Standard irradiation with Multiport 601:Output No.[3]					
		1	2	3	4	5	6
24 hr	Control	7.05 ± 1.29	7.17 ± 1.20	7.83 ± 1.25	11.19 ± 1.14	12.54 ± 1.14	12.86 ± 1.02
	Vit C/E (1.0/0.5%)	6.91 ± 1.24	7.07 ± 1.12	7.31 ± 1.17	10.08 ± 1.63	11.80 ± 1.51	12.81 ± 1.52
	Lycopene 0.03%	6.90 ± 1.23	7.07 ± 1.13	7.23 ± 1.16	9.48^S ± 0.65	11.22^S ± 0.84	12.46 ± 0.95
	Lycopene/Vit C/E(0.03/1.0/0.5%)	6.88 ± 1.29	7.05 ± 1.16	7.15 ± 1.21	8.55^S ± 1.36	9.85^S ± 1.89	11.59 ± 2.05
48 hr	Control	6.91 ± 1.34	7.06 ± 1.24	7.69 ± 1.11	10.84 ± 1.19	12.11 ± 1.17	13.17 ± 1.06
	Vit C/E (1.0/0.5%)	6.94 ± 1.23	7.06 ± 1.38	7.29 ± 1.13	9.96 ± 1.56	11.36 ± 1.77	12.38 ± 1.84
	Lycopene 0.03%	6.90 ± 1.27	7.07 ± 1.10	7.24 ± 1.15	9.46^S ± 0.60	11.21 ± 0.86	12.45 ± 0.98
	Lycopene/Vit C/E(0.03/1.0/0.5%)	6.87 ± 1.26	6.97 ± 1.16	7.07 ± 1.26	8.56^S ± 1.53	9.77^S ± 1.95	11.40^S ± 2.25

(1) Values are means ± SD.
(2) Interval between photostimulation and evaluation of the response.
(3) Outputs numbered in increasing order according to the amount of UV delivered to the skin.
Values marked with '**S**' are significantly different vs. control.

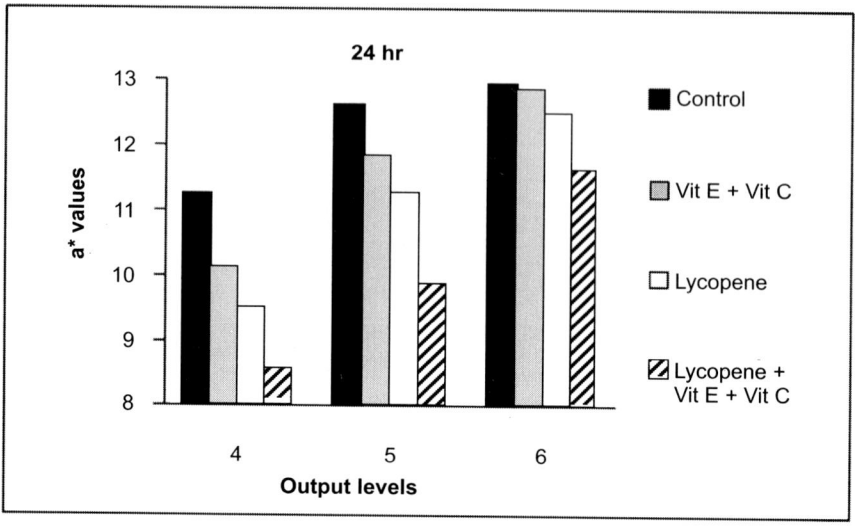

Fig. 3 Instrumental assessment of the intensity of erythema 24 hr after photostimulation. The histograms express the mean values (± SD) of *a in the CIE L*a*b* system.

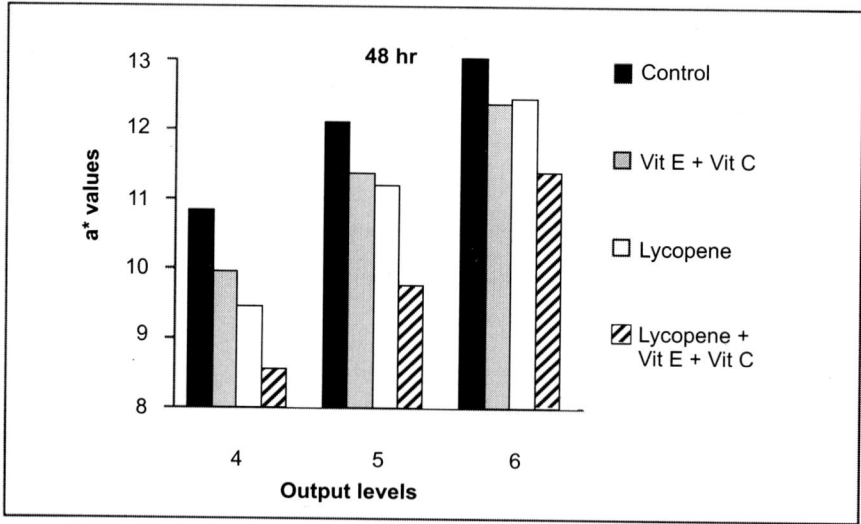

Fig. 4 Instrumental assessment of the intensity of erythema 48 hr after photostimulation. The histograms express the mean values (± SD) of *a in the CIE L*a*b* system.

incorporation in skin care products; they must be stable once included in the final product; they must have organoleptic properties that ensure good compliance.

Table 2 TEWL values expressed in g/m^2 per hr[1] after application of the products and subsequent photostimulation in 10 healthy volunteers.

T[2]	Product	Standard irradiation with Multiport 601:Output No.[3]					
		1	2	3	4	5	6
24 hr	Control	3.9 ± 1.3	3.8 ± 1.1	3.8 ± 1.4	4.7 ± 1.2	5.0 ± 1.2	5.3 ± 1.0
	Vit E/C (0.5/1%)	3.9 ± 1.1	3.8 ± 1.4	3.8 ± 1.5	4.3 ± 1.4	4.6 ± 1.5	4.8 ± 1.4
	Lycopene 0.03%	3.9 ± 1.3	3.9 ± 1.2	4.1 ± 1.2	4.3 ± 1.3	4.4 ± 1.3	4.6 ± 1.2
	Lycopene/Vit C/E(0.03/1.0/0.5%)	3.8 ± 1.2	3.7 ± 1.1	4.2 ± 1.2	4.1 ± 1.3	4.5 ± 1.3	4.8 ± 1.2
48 hr	Control	3.8 ± 1.3	3.8 ± 1.2	3.8 ± 1.3	4.3 ± 1.2	4.8 ± 1.2	4.9 ± 1.2
	Vit E/C (0.5/1%)	3.8 ± 1.3	3.8 ± 1.4	3.8 ± 1.3	4.1 ± 1.3	4.3 ± 1.4	4.4 ± 1.6
	Lycopene 0.03%	3.8 ± 1.2	3.8 ± 1.1	3.8 ± 1.2	4.0 ± 1.2	4.1 ± 1.1	4.1 ± 1.3
	Lycopene/Vit C/E(0.03/1.0/0.5%)	3.9 ± 1.2	3.8 ± 1.2	4.0 ± 1.4	3.7 ± 1.3	4.4 ± 1.1	4.4 ± 1.3

(1) Values are means ± SD.
(2) Interval between photostimulation and evaluation of the response.
(3) Outputs numbered in increasing order according to the amount of UV delivered to the skin.

One of the most common active principles with these characteristics is vitamin E, which has been shown to significantly inhibit the erythematous response to UV stimulation. This inhibition can be attributed to an antioxidant mechanism, since the product does not have a screening effect (Dreher and Maibach 2001). When applied before UV exposure, this vitamin reduces erythema, oedema, the formation of sunburn cells, lipid peroxidation and the formation of DNA adducts.

Ascorbic acid also has an antioxidant effect, although the effect is weaker than that of α-tocopherol, probably due to its instability and ease of oxidation in aqueous vehicles (Darr et al. 1992, Austria et al. 1997). However, its high solubility in water makes the presence of ascorbic acid particularly useful for the activity of other antioxidants whose oxido-reductive power is higher but requires rapid reconstitution in relation to the position occupied in the cell structures (Kobayashi et al. 1996).

Our investigation on lycopene and preparations containing vitamins E and C showed that the lycopene-based product had greater protective activity than a formulation containing a mixture of the two vitamins. Lycopene protected the skin more efficiently when used in association with both vitamins than when used alone. The protective activity of lycopene can be attributed to an antioxidant mechanism since the substance does not filter erythemogenic radiation, as demonstrated by our investigations of the absorption spectrum of the extract of *Lycopersicon esculentum* used to prepare the studied products.

The antioxidant activity of lycopene is due to its high oxido-reductive capacity, the highest in a series of carotenoids including astaxanthin, β-apo-8'-carotenal, canthaxanthin, lutein, zeaxanthin, β-carotene and lycopene (Black and Lambert 2001, Edge et al. 1997). Because of this peculiarity, lycopene appears to be the most suitable carotenoid for use as an antioxidant. Its higher efficacy when used in association with α- tocopherol and ascorbic acid could be due to a synergistic interaction with the other antioxidants (Packer et al. 1979).

Studies on the redox potential of β-carotene and vitamins E and C suggest that β-carotene participates along with the two vitamins to effectively repair the superoxide radical: α-tocopherol intercepts the superoxide radical to form the tocopherol radical cation, which in turn is repaired by β-carotene to produce the carotenoid radical cation. The latter has high pro-oxidizing power but the final intervention of ascorbic acid, which is hydrosoluble and easily metabolized, neutralizes the carotenoid radical cation and restores the efficiency of the system (Black 2004). The model proposed for β-carotene can be extended to lycopene and, according to the lipophilic or hydrophilic affinity of the components of the system, seems particularly adaptable to the anatomical and functional characteristics of the skin.

Whatever the mechanism of action, our study, based on a reliable and reproducible method, has shown that topically applied lycopene provides strong protection against UV radiation and that its activity is reinforced when it is associated with α-tocopherol and ascorbic acid. Because of these characteristics, lycopene is particularly suitable for topical application in sunscreens or in any preparation designed to prevent or reduce the harmful effects of oxidative stimuli.

ABBREVIATIONS

HPLC: high performance liquid chromatography; MED: minimal erythemal dose; ROS: reactive oxygen species; TEWL: transepidermal water loss; UVA: ultraviolet A; UVB: ultraviolet B

References

Andreassi, L. and L. Flori. 1995. Practical applications of cutaneous colorimetry. Clin. Dermatol. 13: 369-373.

Andreassi, M. and A. Ettorre, E. Stanghellini, A. Di Stefano, and L. Andreassi. 2004a. In vivo and in vitro evaluation of lycopene as antioxidant substance. J. Invest. Dermatol. 122: A137.

Andreassi, M. and E. Stanghellini, A. Ettorre, A. Di Stefano, and L. Andreassi. 2004b. Antioxidant activity of topically applied lycopene. J. Eur. Acad. Dermatol. Venereol. 18: 52-55.

Austria, R. and A. Semenzato, and A. Bettero. 1997. Stability of vitamin C derivatives in solution and topical formulations. J. Pharm. Biomed. Anal. 15: 795-801.

Bangha, E. and P. Elsner, and G.S. Kistler. 1996. Suppression of UV-induced erythema by topical treatment with melatonin (N-acetyl-5-methoxytryptamine). A dose response study. Arch. Dermatol. Res. 288: 522-526.

Bickers, D.R. and M. Athar. 2006. Oxidative stress in the pathogenesis of skin disease. J. Invest. Dermatol. 126: 2565-2575.

Black, H.S. 1987. Potential involvement of free radical reactions in ultraviolet light-mediated cutaneous damage. Photochem. Photobiol. 46: 213-221.

Black, H.S. 2004. Pro-carcinogenic activity of beta-carotene, a putative systemic photoprotectant. Photochem. Photobiol. Sci. 3: 753-758.

Black, H.S. and C.R. Lambert. 2001. Radical reactions of carotenoids and potential influence on UV carcinogenesis. Curr. Probl. Dermatol. 29: 140-156.

Boileau, T.W. and A.C. Boileau, and J.W. Erdman Jr. 2002. Bioavailability of all-*trans* and *cis*-isomers of lycopene. Exp. Biol. Med. 227: 914-919.

Burke, K.E. 2004. Photodamage of the skin: protection and reversal with topical antioxidants. J. Cosmet. Dermatol. 3: 149-155.

Cal, K. 2006. Skin penetration of terpenes from essential oils and topical vehicles. Planta. Med. 72: 311-316.

Choi, C.M. and R.G. Bennett. 2003. Laser Dopplers to determine cutaneous blood flow. Dermatol. Surg. 29: 272-280.

Dalle Carbonare, M. and M.A. Pathak. 1992. Skin photosensitizing agents and the role of reactive oxygen species in photoaging. J. Photochem. Photobiol. 14: 105-124.

Darr, D. and S. Combs, S. Dunston, T. Manning, and S. Pinnell. 1992. Topical vitamin C protects porcine skin from ultraviolet radiation-induced damage. Br. J. Dermatol. 127: 247-253.

de Gruijl, F.R. 1995. Action spectrum for photocarcinogenesis. Recent Results Cancer Res. 139: 21-30.

Dreher, F. and H. Maibach. 2001. Protective effects of topical antioxidants in humans. Curr. Probl. Dermatol. 29: 157-164.

Dreher, F. and B. Gabard, D.A. Schwindt, and H.I. Maibach. 1998. Topical melatonin in combination with vitamins E and C protects skin from ultraviolet-induced erythema: a human study in vivo. Br. J. Dermatol. 139: 332-339.

Dreher, F. and N. Denig, B. Gabard, D.A. Schwindt, and H.I. Maibach. 1999. Effect of topical antioxidants on UV-induced erythema formation when administered after exposure. Dermatology 198: 52-55.

Eberlein-Konig, B. and J. Ring. 2005. Relevance of vitamins C and E in cutaneous photoprotection. J. Cosmet. Dermatol. 4: 4-9.

Edge, R. and D.J. McGarvey, and T.G. Truscott. 1997. The carotenoids as antioxidants – a review. J. Photochem. Photobiol. B. 41: 189-200.

El-Benna, J. and P.M. Dang, M.A. Gougerot-Pocidalo, and C. Elbim. 2005. Phagocyte NADPH oxidase: a multicomponent enzyme essential for host defenses. Arch. Immunol. Ther. Exp. 53: 199-206.

Ferreira, A.L. and K.J. Yeum, C. Liu, D. Smith, N.I. Krinsky, X.D. Wang, and R.M. Russell. 2000. Tissue distribution of lycopene in ferrets and rats after lycopene supplementation. J. Nutr. 130: 1256-1260.

Fischer, T.W. and P. Elsner. 2001. The antioxidative potential of melatonin in the skin. Curr. Probl. Dermatol. 29: 165-174.

Fuchs, J. 1992. Oxidative Injury in Dermatopathology. Springer-Verlag, Berlin.

Gasparro, F.P. and M. Mitchnic, and J.F. Nash. 1998. A review of sunscreen safety and efficacy. Photochem. Photobiol. 68: 243-256.

Jokuszies, A. and A. Niederbichler, M. Meyer-Marcotty, L.U. Lahoda, K. Reimers, and P.M. Vogt. 2006. Influence of transendothelial mechanisms on microcirculation: consequences for reperfusion injury after free flap transfer. Previous, current, and future aspects. J. Reconstr. Microsurg. 22: 513-518.

Kawanishi, S. and Y. Hiraku, and S. Oikawa. 2001. Mechanism of guanine-specific DNA damage by oxidative stress and its role in carcinogenesis and aging. Mutat. Res. 488: 65-76.

Khachik, F. and L. Carvalho, P.S. Bernstein, G.J. Muir, D.Y. Zhao, and N.B. Katz. 2002. Chemistry, distribution, and metabolism of tomato carotenoids and their impact on human health. Exp. Biol. Med. 227: 845-851.

Kobayashi, S. and M. Takehana, S. Itoh, and E. Ogata. 1996. Protective effect of magnesium-L-ascorbyl-2 phosphate against skin damage induced by UVB irradiation. Photochem. Photobiol. 64: 224-228.

McClain, R.M. and J. Bausch. 2003. Summary of safety studies conducted with synthetic lycopene. Regul. Toxicol. Pharmacol. 37: 274-285.

McCord, J.M. 1985. Oxygen-derived free radicals in postischemic tissue injury. N. Engl. J. Med. 312: 159-163.

Miteva, M. and S. Richter, P. Elsner, and J.W. Fluhr. 2006. Approaches for optimizing the calibration standard of Tewameter TM 300. Exp. Dermatol. 15: 904-912.

Mustafa, M.G. 1990. Biochemical basis of ozone toxicity. Free Radic. Biol. Med. 9: 245-265.

Packer, J.E. and T.F. Slater, and R.L. Willson. 1979. Direct observation of a free radical interaction between vitamin E and vitamin C. Nature 278: 737-738.

Park, S.B. and C.H. Huh, Y.B. Choe, and J.I. Youn. 2002 Time course of ultraviolet-induced skin reactions evaluated by two different reflectance spectrophotometers: DermaSpectrophotometer and Minolta spectrophotometer CM-2002 Photodermatol. Photoimmunol. Photomed. 18: 23-28.

Pryor, W.A. 1994. Mechanisms of radical formation from reactions of ozone with target molecules in the lung. Free Radic. Biol. Med. 17: 451-465.

Quinn, M.T. and M.C. Ammons, and F.R. Deleo. 2006. The expanding role of NADPH oxidases in health and disease: no longer just agents of death and destruction. Clin. Sci. 111: 1-20.

Rao, A.V. and M.R. Ray, and L.G. Rao. 2006. Lycopene. Adv. Food Nutr. Res. 51: 99-164.

Ravanat, J.L. and T. Douki, and J. Cadet. 2001. Direct and indirect effects of UV radiation on DNA and its components. J. Photochem. Photobiol. 63: 88-102.

Saija, A. and A. Tomaino, D. Trombetta, A. De Pasquale, N. Uccella, T. Barbuzzi, D. Paolino, and F. Bonina. 2000. In vitro and in vivo evaluation of caffeic and ferulic acids as topical photoprotective agents. Int. J. Pharm. 199: 39-47.

Shindo, Y. and E. Witt, and L. Packer. 1993. Antioxidant defense mechanisms in murine epidermis and dermis and their responses to ultraviolet light. J. Invest. Dermatol. 100: 260-265.

Takeoka, G.R. and L. Dao, S. Flessa, D.M. Gillespie, W.T. Jewell, B. Huebner, D. Bertow, and S.E Ebeler. 2001. Processing effects on lycopene content and antioxidant activity of tomatoes. J. Agric. Food Chem. 49: 3713-3717.

Valacchi, G. and V. Fortino, and V. Bocci. 2006. The dual action of ozone on the skin. Br. J. Dermatol. 153: 1096-1100.

Weber, S.U. and N. Han, and L. Packer. 2001. Ozone: an emerging oxidative stressor to skin. Curr. Probl. Dermatol. 29: 52-61.

2.7

Lycopene and Cardiovascular Diseases

Martha Verghese, Rajitha Sunkara, Louis Shackelford and Lloyd T. Walker

Nutritional Biochemistry, P.O. Box 1628, Alabama A&M University
Department of Food and Animal Sciences, Normal, AL 35762, USA

ABSTRACT

Lycopene is a fat-soluble pigment (carotenoid) without provitamin A activity and it is found in selected plant foods such as tomato, watermelon, and pink grapefruit. Diets rich in carotenoids have potential in reducing chronic diseases. Considerable evidence suggests that lycopene may play a significant role in reducing the risk of cardiovascular diseases (CVD) in humans. The factors that contribute to the onset of CVD include hypertension, hypercholesterolemia, hyperlipidemia, obesity, and insulin resistance as well as genetic, environmental and lifestyle factors such as physical inactivity, smoking, and tobacco consumption. This chapter focuses on the mechanisms of action of lycopene in reducing the risk of CVD. Several *in vitro* assays have shown that lycopene is a potent antioxidant, and its antioxidant mechanisms are likely to provide health benefits against development of CVD. The most accepted theory about the mechanisms of lycopene in controlling various diseases is its singlet oxygen quenching properties. Lycopene has high numbers of conjugated double bonds and is the most potent singlet oxygen quencher among the natural carotenoids. This property of lycopene draws attention in the field of chronic diseases and carotenoids research. However, some non-antioxidant mechanisms or effects may play a role in the protection offered by lycopene against CVD, such as inhibition of 3-hydroxy-3-methyl glutaryl coenzyme A (HMG-CoA) reductase, reduction in intima-media thickness of the carotid artery, preventing platelet aggregation and thrombus

A list of abbreviations is given before the references.

formation, and suppression of endothelial cell tissue factor activity. The data supporting the protective effects of lycopene are mainly derived from epidemiological studies on normal and high-risk populations. Lycopene seems to play a role in the antioxidant as well as non-antioxidant mechanisms related to reduction in CVD.

INTRODUCTION

Cardiovascular diseases (CVD), diseases that affect the blood vessels and heart, are the leading cause of death in the United States and in industrialized countries and by 2010 are expected to be the leading cause of death in developing countries (WHO 2000). They account for 36.3% of deaths in the United States, and 47.22% of those who die are 65 years and older (Rosamond et al. 2007). The magnitude and economic costs of this problem are profound, as atherosclerosis claims more lives than all types of cancers combined. Although atherosclerosis is currently a problem of the developed world, the World Health Organization predicts that global economic prosperity could lead to an epidemic of atherosclerosis as developing countries acquire Western habits. Factors contributing to CVD are listed in Table 1.

Atherosclerosis accounts for over 50% of CVD and is characterized by the accumulation of cholesterol deposits in macrophages in large and medium-sized arteries. It includes both coronary artery disease and cerebrovascular disease, which are associated with the loss of heart and brain function as a result of reduced blood flow and are termed heart attack and stroke. The factors that contribute to the onset of the disease include hypertension, hypercholesterolemia, hyperlipidemia, obesity, and insulin resistance as well as genetic, environmental and life style factors such as physical inactivity, smoking, and tobacco consumption. An increase in the incidence of CVD may be due to reduced physical activity as well as the consumption of foods with high energy density. Among all the environmental risk factors, nutrition plays a role in decreasing abnormalities related to cardiac health. Antioxidant nutrients and phytochemicals present in fruits and vegetables play an important role in prevention (Halliwell et al. 1995).

The major risk factors in the underlying pathology of atherosclerosis are discussed below.

1. *Age.* Although not a modifiable risk factor, age is among the most important risk factors for predicting incidence of CVD.

Table 1 Factors affecting cardiovascular diseases.

Biological factors
High blood pressure
High blood cholesterol
High lipids
Hypoalphalipoproteinemia
Hyperhomocysteinemia
Hypertriglyceridemia
Elevated concentrations of lipoproteins (LDL, VLDL)
Obesity, overweight (especially abdominal obesity)
Insulin resistance
High fibrinogen

Genetic factors
Heredity
Family history

Lifestyle factors
Physical inactivity/sedentary lifestyle
Smoking
Tobacco
Stress

Others
Age
Sex
Inflammation, increased CRP expression

Dietary components
Fruits and vegetables
High fiber intake
Antioxidant vitamins: A, C, E
Carotenoids
Polyphenols
Omega 3 fatty acids

Oxidative stress
ROS
Superoxide
NO
Cytokines: IL-1β, IL-6β, IFN-γ, TNF-α

Oxidative enzymes
NADPH oxidase
Xanthine oxidases

Antioxidative enzymes
Catalase
Superoxide dismutase

Adhesion molecules
P-selectin, GPIIbIIIa, VCAM, ICAM

LDL, low density lipoproteins; VLDL, very low density lipoproteins; CRP, C-reactive protein; ROS, reactive oxygen species; IL-1β, Interleukin-1β; IL-6β, Interleukin-6β; IFN-γ, Interferon-gamma; TNF-α, tumor necrosis factor alpha; VCAM, Vasucular Cell Adhesion Molecule; ICAM, Intracellular Adhesion Molecule; GPIIbIIIa, Glycoprotein IIbIIIa.

2. *Gender.* Numerous observational studies have indicated that men exhibit a higher risk for CVD than women of the same age (Barrett-Connor and Bush 1991). There has been considerable speculation that estrogens offer a "protective" effect to women, as CVD accelerates in women after menopause. This relationship has been difficult to substantiate, as estrogen treatment has not reduced the incidence of CVD in post-menopausal women (Hulley et al. 1998). Some protection may be offered by the relatively higher concentrations of high-density lipoprotein (HDL) cholesterol in women compared to their age-matched men.

3. *Obesity.* Obesity, defined as excess body weight with abnormally high body fat, is a condition that increases the incident risk of CVD. A number of other risk factors for CVD, such as hypertension, low HDL cholesterol, and diabetes mellitus, often co-exist with obesity (Wilson et al. 1999). This relation between obesity and CVD has become a matter of significance as the prevalence of obesity in the industrialized world is increasing.

4. *Cigarette smoking.* Many studies have linked smoking to the incidence of CVD (Gaziano 1996), as cessation of smoking has been quite effective in lowering the risk of heart attack; the incidence of CVD in ex-smokers approaches that of non-smokers in just two years.

5. *Hypertension.* Hypertension is defined as a systolic blood pressure in excess of 140 mmHg or a diastolic blood pressure above 90 mmHg. The elderly are predisposed to hypertension, as up to 75% of individuals over 75 years of age suffer from this condition. There appears to be a linear relationship between blood pressure elevation and the increased incidence of CVD (MacMahon et al. 1990).

6. *Diabetes mellitus.* The risk of CVD in diabetics is 3- to 5-fold that in non-diabetics, despite controlling for other risk factors (Pyorala et al. 1987). A number of other known risk factors for CVD such as hypertension and abnormal lipids are also more common in diabetics than in the general population.

7. *Serum cholesterol.* The increased risk of CVD with increased low-density lipoprotein (LDL) cholesterol is supported by observations that cholesterol-lowering therapy diminishes the clinical manifestations of atherosclerosis, particularly inhibitors of 3-hydroxy-3-methylglutaryl coenzyme A reductase (i.e., statins) that significantly lower LDL cholesterol (Gotto and Grundy 1999). Familial hypercholesterolemia is an autosomal dominant disorder

that affects ~1 in 500 persons from the general population. In contrast to LDL cholesterol, there is an inverse relationship between HDL cholesterol and atherosclerosis (Gordon et al. 1977).

ABSORPTION OF LYCOPENE

Passive diffusion helps the uptake of lycopene by the brush border membrane of the intestinal mucosal cell. Lycopene exits the mucosal cell in chylomicrons, which are secreted via the mesenteric lymph system into the blood. Owing to the action of lipoprotein lipase on chylomicrons, lycopene has the potential to be taken up passively by various tissues (adrenals, kidney, adipose, spleen, lung, and reproductive organs). Lycopene can accumulate in the liver or can be repackaged into very low-density lipoprotein (VLDL) and sent back into the blood. Low-density lipoprotein receptors are responsible for the uptake of lycopene into the tissues from VLDL and LDL, and the tissues with the highest concentrations of lycopene (liver, adrenal, testes) tend to have high LDL receptor activity. Lycopene is a predominant carotenoid in the human liver, adrenals, adipose tissue, testes, and prostate (Dugas et al. 1998, Fuhrman et al. 1997).

LYCOPENE AND CARDIOVASCULAR DISEASE

Research has focused on identifying ways to prevent the disease through dietary changes, since diet is believed to play a major role in the development of CVD. Epidemiologic evidence indicates that persons who ingest more lycopene, or who have higher concentrations of lycopene in plasma or in adipose tissue, are at a reduced risk of certain chronic diseases, including cancer and coronary heart disease (Gerster 1997, Hoffman and Weisburger 1997). Kohlmeier et al. (1997) conducted a multi-center case-control study of antioxidant nutrients in adipose tissue and risk of myocardial infarction in Europe and found that subjects who had lower levels of three carotenoids, α-carotene, β-carotene, and lycopene, compared to their matched controls (0.33-0.82, comparing the 90th to the 10th percentile), had lower risk of myocardial infarction.

Lycopene is a natural pigment synthesized in plants and microorganisms, where it serves as an accessory light-gathering pigment and protects plants and microorganisms against the toxic effects of oxygen and light. Lycopene is an ayclic isomer of β-carotene, which is lipophilic in nature, and has no vitamin A activity. It is an open chain hydrocarbon containing 11 conjugated and two non-conjugated double bonds arranged in a linear fashion (Stahl et al. 1992). Because of the abundance of double bonds in its structure, there are many isomers of lycopene found in nature (Britton 1995). The linear all-*trans* configuration is the predominant form of lycopene, which constitutes approximately 90% of its dietary sources (Boileau et al. 2002). Clinton (1998) reported that lycopene in food is found primarily in the all-*trans* form, a variety of cis forms are commonly found in tissues and serum (Clinton 1998). The main sources of lycopene are tomato, watermelon, pink grapefruit, papaya, and apricot.

Dietary intake of tomato and tomato products containing lycopene has been shown to be associated with decreased risk of chronic diseases such as cancer and CVD. The most accepted theory of mechanism of lycopene in controlling various diseases is its singlet oxygen quenching properties. Lycopene, which has high numbers of conjugated double bonds, is the most potent singlet oxygen quencher among the natural carotenoids (DiMascio et al. 1989). This property of lycopene draws abundant attention in the field of chronic diseases and carotenoid research. Several *in vitro* assays have shown that lycopene is a potent antioxidant, and therefore its antioxidant mechanisms are likely to provide health benefits against development of CVD. However, some non-antioxidant mechanisms or effects may play a role in the protection offered by lycopene against CVD (Table 2).

ANTIOXIDANT MECHANISMS

Oxidative Stress and Coronary Heart Disease

Formation of free radicals from exogenous and endogenous sources and higher levels of molecular oxygen contributes to the progression of CVD. Reactive oxygen species (ROS) are formed during the electron transport in the mitochondria. Oxidation of LDL that carry cholesterol into the bloodstream seems to play a key role in the pathogenesis of CVD (Witztum 1994, Halliwell et al. 1995). The progression of CVD may be slowed by dietary antioxidants such as lycopene because of their ability to inhibit the oxidative processes (Heller et al. 1998, Parthasarathy 1998). Reactive oxygen species can be generated endogenously during normal cellular metabolism or provided exogenously and can play a significant role in the oxidation of LDL (Ames et al. 1995, Pincemail 1995). The oxidation of LDL is the initial step, finally leading to its uptake by the macrophages inside the arterial wall and the formation of foam cells and atherosclerotic plaque (Witztum 1994) (Fig. 1). Low-density lipoprotein is the major transport protein for cholesterol in the body and contains several lipophilic antioxidants including carotenoids and vitamin E. The stages leading to the complete oxidation of LDL include the native LDL, which is transformed to the seeded LDL, minimally modified LDL, and finally the fully oxidized LDL (Parthasarathy 1998). The characteristics of these LDL forms are shown in Fig. 1. Oxidative modifications of the native LDL molecule result in several biologically active metabolic breakdown products of oxidized fatty acids that facilitate the recognition of the

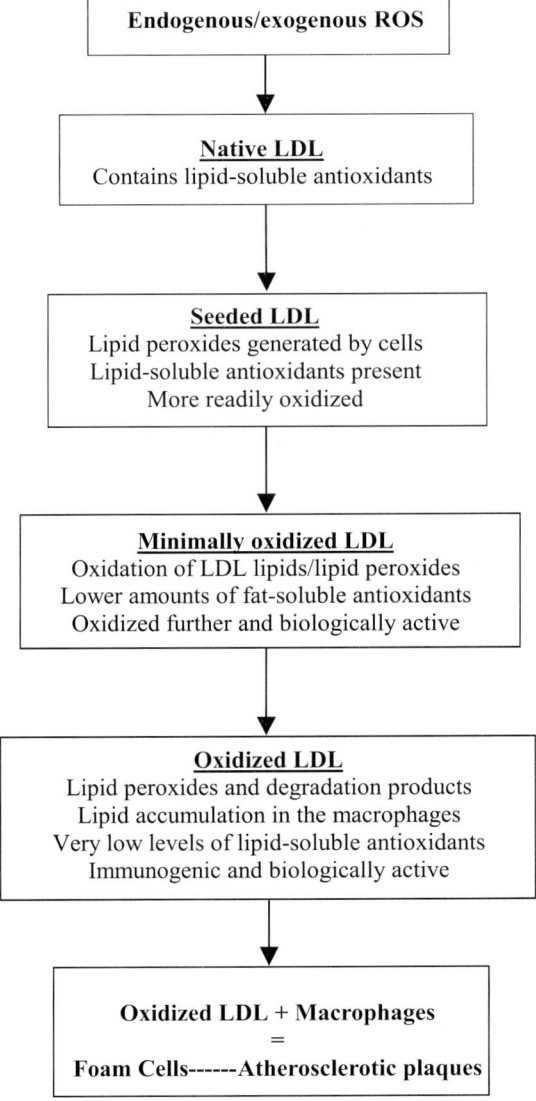

Fig. 1 Stages and characteristics of low density lipoprotein (LDL) oxidation (Parthasarathy 1998). Free radicals and reactive oxygen species (ROS) can be formed exogenously and endogenously and may contribute to the progression of CVD. Oxidation of LDL plays a role in the pathogenesis of CVD. It is the initial step, finally leading to LDL uptake by the macrophages inside the arterial wall and the formation of foam cells and atherosclerotic plaque.

modified LDL by the scavenger receptors on macrophages resulting in the formation of foam cells and plaque in the arterial walls.

Endogenous antioxidative systems that quench superoxide and ROS are catalase, superoxide dismutase, glutathione reductase and glutathione peroxidase (Halliwell 1999, Betteridge 2000). All these systems are present in blood, epithelial tissues, organs, and also in vascular smooth muscle cells (VSMC). Increased production of superoxide and ROS overwhelm the antioxidative enzyme systems, resulting in progression of CVD by increasing oxidative stress.

Oxidation of macromolecules, particularly lipoproteins, that are present or deposited in the endothelial cell activates cellular and cytokine networks that are involved in lesion formation and progression (Table 3). Oxidized LDL cause the stimulation of monocyte and platelet adhesion, vasoconstriction, proliferation of smooth muscle (Holvoet and Collen 1994, Frei 1995), and apoptosis of endothelial cells. Nuclear condensation in ROS-induced cardiac apoptosis is caused by lipid peroxidation followed by activation of caspase, protooncogene and p-53 activation. Cytokines such as Interleukin 1β (IL-1β), IL-6, IFN-γ, TNF-α and certain growth factors are found to be present at higher levels in CVD patients. Cytokines induce overexpression of inducible nitric-oxide synthase, thereby producing elevated levels of nitric oxide (NO). Formation of peroxynitrile by reaction of O_2^- with NO diminishes the level of NO in cells, which plays an important role in reducing the progression of CVD (Halliwell 1999). Increased level of cytokines was noticed in myocardial infarction and ischemia/repurfusion injuries (Kacimi et al. 1998). There are endogenous sources of ROS and O_2^- production other than mitochondria. Neutrophils through their bactericidal action release NADPH and xanthine oxidases, which in turn enhances the production of O_2^-. NADPH oxidase is also activated by angiotensin II, which mediates endothelial dysfunction associated with pathologies such as atherogenesis and heart failure (Rajagopalan et al. 1996, Warnholtz et al. 1999). Higher level of circulating xanthine oxidases are responsible for endothelial dysfunction by accumulating in vascular tissues (Houston et al. 1999).

Whatever the source, ROS and O_2^- act at both blood cell levels and within the vascular tissue to cause CVD. Formation of O_2^- induces the expression of adhesion molecules such as P-selectin, GPIIb/IIIa, vascular cell adhesion molecule (VCAM), and intracellular adhesion molecule (ICAM) in endothelial cells, platelets, leucocytes, monocytes and VSMCs. Platelets, monocytes and neutrophiles then co-adhere to the endothelial cells and VSMC, causing the release of O_2^- (Jeremy et al. 2004). This release of O_2^- reacts with NO, resulting in low NO availability, and also up-regulates the expression of adhesion molecules (Jeremy et al. 2002). The adherence also releases mitogens, cytokines, thrombogens, leucotrienes, peptide growth factors and transition metals. All these factors may contribute to the proliferation and migration of VSMC. This process occurs in the pathogenesis of angina and acute thromobosis. Oxidation in monocytes results in differentiation to macrophages, ultimately resulting in the formation of foam cell, which is an initial step in atherosclerotic plaque formation (Jeremy et al. 2004).

Need for Antioxidants

Oxidative stress plays an important role in the etiology of acute and chronic CVD. There is a need for antioxidant therapy to reduce the progression of this disease. Antioxidants are molecules capable of neutralizing free radicals by donating electrons. Endogenous and exogenous antioxidants act synergistically to quench free radicals. These consist of antioxidative enzymes that catalyze quenching of free radicals, metal-binding proteins and diet-derived antioxidants such as vitamins C and E, carotenoids and polyphenols. Endogenous antioxidants are responsible for maintenance of cell integrity and protection from oxidative stress. However, in pathological conditions and stress, endogenous antioxidative enzyme activity is depleted and there is an enhanced need for supplementation of antioxidants through the diet. Lycopene is a highly unsaturated carotenoid, containing 11 conjugated and 2 non-conjugated double bonds. Lycopene exists naturally in tomatoes in the *trans* form, which is poorly absorbed into the body. Lycopene is one of the most effective antioxidants and free radical scavengers and maintains other antioxidant substances in reduced form (Nelson et al. 2003). Research that supports the *in vivo* oxidation of LDL contributes to the hypothesis that dietary components may be important for LDL oxidative modification. Carotenoids such as lycopene are highly lipophilic and are commonly found within cell membranes and other lipid components (Clevidence and Bieri 1993). It is therefore thought that the ability of carotenoids to scavenge free radicals may be greatest in a lipophilic environment.

Antioxidative Effect

Epidemiological studies have linked an elevated plasma lycopene concentration with a lower risk for developing CVD (Petr and Erdman 2005). The protective effect of lycopene on CVD or any other disease is due mainly to its protection against oxidative stress. Molecules having unpaired or an odd number of electrons are referred to as free radicals. These free radicals are quenched by either antioxidative system enzymes or by antioxidants derived from the diet. Lycopene acts as an antioxidant through quenching free radicals by both physical and chemical means. Physical quenching brings the oxygen molecule to the ground state by transferring the excitation energy from singlet oxygen or free radical to the carotenoid. The exited carotenoid loses its energy by rotation and vibration and comes back to its original ground state and the quenching cycle starts again. Lycopene has a higher quenching capacity than other carotenoids because of its high number of double bonds (DiMascio et al. 1989, Stahl and Sies 1993). Lycopene is twice as effective as β-carotene in protecting the lymphocytes from NO_2-induced cell death and membrane damage (Bohm et al. 1995, Tinkler et al. 1994).

Chemical quenching involves the dissociation of the lycopene molecule and forms many other compounds such as apocarotenones and epoxides (Woodall et al. 1997). These by-products are unstable and being at low concentrations make it difficult to assess the process of lycopene breakdown. Lycopene was found to react more rapidly with oxidizing agents than any other carotenoids, indicating its potent role as an antioxidant.

Lycopene plays a protective role in CVD by decreasing the lipid peroxidation because of its hydrophobic nature. It offers protection against very specific types of oxidative damage and also complements other antioxidative defense mechanisms. The effect of free radicals on lycopene is limited to certain subcellular surfaces and inner hydrophobic regions of the cell membrane (Tsuchiya et al. 1993).

Lycopene and Type 2 Diabetes

Oxidative stress, which is a condition of excessive ROS, may play a role in the etiology of type 2 diabetes by impairing insulin secretion from pancreatic β-cells and inducing insulin resistance in the peripheral tissues (Oberley 1988, Paolisso and Giugliano 1996, Ceriello and Motz 2004). Lycopene may play a significant role in interrupting the chain reaction of lipid oxidation quenching peroxyl radicals (El-Agamey et al. 2004). Its potent antioxidant capacity may therefore provide protection against the development of type 2 diabetes (DiMascio et al. 1989).

Several studies (Ford et al. 1999, Amstrong et al. 1996, Polidori et al. 2000, Chuang et al. 1998, Jang et al. 2004, Coyne et al. 2005) have reported an inverse correlation between dietary or plasma carotenoids and type 2 diabetes or related symptoms, such as fasting blood glucose, glucose tolerance, and glycosylated hemoglobin. Higher dietary intake of β-cryptoxanthin but not other major carotenoids was associated with a significantly reduced risk of type 2 diabetes (Montonen et al. 2004). Wang et al. (2006) examined the association of baseline plasma lycopene and other carotenoids with the risk of type 2 diabetes in middle-aged and older women from the Women's Health Study. The study was a randomized, double-blind, placebo-controlled clinical trial of low-dose aspirin and vitamin E in the primary prevention of CVD and cancer in women. The researchers found little evidence for an association between baseline plasma lycopene and other carotenoids with the risk of type 2 diabetes after adjusting for multiple risk factors in middle-aged and older women. There is no clear evidence on whether increased plasma carotenoids will improve insulin sensitivity and glucose metabolism. A marginal inverse association for lycopene was seen after adjusting for the risk factors for diabetes.

Role of Oxidation/Free Radical Production in the Pathogenesis of Type 2 Diabetes

Oxidation and the production of free radicals have the following effects:

- They impair insulin action and glucose disposal in the peripheral tissues.
- They contribute to pancreatic β-cell dysfunction and blunt insulin secretion
- They result in chronic inflammation, which is an underlying pathway for development of insulin resistance, diabetes, and CVD (Paolisso and Giugliano 1996, Ceriello and Motz 2004, Paolisso et al. 1993, Evans et al. 2003, Robertson et al. 2003, Sakai et al. 2003).

Animal studies have shown a possible protective role of the scavenging ability of antioxidants such as lycopene against the development of type 2 diabetes (Polidori et al. 2000, Slonim et al. 1983, Murthy et al. 1992). Epidemiological studies have shown that higher consumption of fruits and vegetables, which are good sources of carotenoids, have been linked with lower risk of type 2 diabetes (Feskens et al. 1992, Ford and Mokdad 2001), suggesting a potential role of carotenoids in preventing the development of type 2 diabetes. Fruits and vegetables are, however, also rich in other nutrients that may be related to glucose metabolism; therefore, the role of carotenoids in the primary prevention of type 2 diabetes remains inconclusive.

Lycopene, Smoking and CVD

Smoking is a source of free radicals and is one of the risk factors for atherosclerosis. Free radicals cause LDL oxidation, followed by foam cell formation, ultimately resulting in atherogenesis. Smokers have 13-44% lower plasma concentrations of most carotenoids than non-smokers. No relationships between smoking and plasma lycopene concentrations were observed in studies (Peng et al. 1995, Brady et al. 1996). However, Pamuk et al. (1994) observed that serum lycopene concentration was 26% lower in African-American female smokers than in non-smokers after adjusting the confounding factors and dietary lycopene. Serum lycopene concentrations were reduced by 40% and an increased concentration of thiobarbituric acid reactive substances was seen after subjects smoked three cigarettes within 30 min (Rao and Agarwal 1998). These studies did not provide a clear mechanism on the relationship between lycopene concentrations and smoking and further research is needed. Supplementation of tomato juice for 4 wk significantly increased the conjugated diene (CD) lag time and increased CD propagation time in smokers compared to the controls (Steinberg and Chait 1998).

NON-ANTIOXIDATIVE MECHANISMS

Hypocholesterolemic Effect

Inhibition of 3-hydroxy-3-methyl glutaryl coenzyme A (HMG-CoA) reductase

Hydroxymethylglutaryl-coenzyme A (HMG-CoA) is the precursor for cholesterol synthesis. It is also an intermediate in the pathway for synthesis of ketone bodies from acetyl-CoA. The enzymes for production of ketone bodies are located in the mitochondrial matrix. HMG-CoA for cholesterol synthesis is made by enzymes in the cytosol. It is formed by condensation of acetyl-CoA and acetoacetyl-CoA, catalyzed by HMG-CoA synthase. HMG-CoA reductase catalyzes production of mevalonate from HMG-CoA. HMG-CoA reductase is an integral protein of endoplasmic reticulum membrane. The catalytic domain of this enzyme remains active following cleavage from the transmembrane portion of the enzyme. The HMG-CoA reductase reaction is rate-limiting for cholesterol synthesis. This enzyme is highly regulated and the target of pharmaceutical intervention.

Lycopene exhibits a protective role in CVD by reducing the formation of cholesterol. Incubation of human macrophage cell lines (J774A.1) with lycopene inhibited cholesterol synthesis and increased LDL receptors (Fuhramn et al. 1997). In agreement with these *in vitro* observations, dietary supplementation of lycopene (60 mg/day) to six men for a 3 mon period resulted in a 14% reduction in their plasma LDL cholesterol concentrations. Dietary supplementation of lycopene may act as a moderate hypocholesterolemic agent, secondary to their inhibitory effect on macrophage HMGCoA reductase (Parker 1988).

Reduction in intima-media thickness of the carotid artery

There is some evidence linking the reduction in the thickness of the intimal wall with a reduction in the risk of myocardial infarctions in individuals with higher adipose tissue concentrations of lycopene (Arab and Steck 2000).

The Kuopio Ischemic Heart Disease Risk Factor Study examined the relationship between serum antioxidants and intima-mediated thickness of the common carotid artery (CCA-IMT), a marker related to the risk of an acute coronary event. Increased intima-media thickness of the carotid artery has been shown to predict coronary events (Rissanen et al. 2003). The authors investigated the relationship between plasma lycopene concentrations and intima-media thickness of the common-carotid artery wall (CCA-IMT) in 520 males and females between the ages of 45 and 69. After adjusting for other cardiovascular risk factors (age, serum triglycerides, serum HDL and LDL cholesterol, plasma homocysteine, and systolic blood pressure), and intake of four nutrients (proportion of saturated fatty acids of total daily energy, vitamin C, vitamin E, and fiber), low plasma lycopene level was associated with an increase in IMT by 18% in men, as compared with men with higher plasma level of lycopene ($P = 0.003$ for difference). Lower levels of plasma lycopene were seen in men who had a coronary event compared with men who did not. In women, the difference did not remain significant after the adjustments (Rissanen et al. 2002, 2003). The authors found associations between plasma lycopene concentrations and early atherosclerosis in men, but not in women, as manifested by increased CCA-IMT.

Preventing of platelet aggregation and thrombus formation

The antiplatelet activity of lycopene may involve the following pathways:

1. Inhibition of activation of phospholipase C, and inhibition of phosphoionositide breakdown and thromboxane B_2 formation, leading to inhibition of intracellular Ca^{+2} mobilization.

2. Activation of formation of cyclic AMP/nitrate in human platelets, resulting in inhibition of platelets aggregation. The initiation of intravascular thrombosis is a factor involving platelets adherence and aggregation resulting in CVD (Schlenitz et al. 2004)

Hsiao et al. (2005) evaluated the inhibitory effect of thrombus formation. Lycopene (2-12 μmol/L) inhibited platelet aggregation and ATP-release reactions stimulated by collagen (1 μg/ml), ADP (20 μmol/L), and arachidonic acid (60 μmol/L) in human platelets (Hsiao et al. 2005). The 50% inhibitory concentration of lycopene in inhibiting platelet aggregation was 170 times as potent as α-tocopherol.

One of the hallmarks of platelets activated by agonists such as collagen is the breakdown of phosphoinositide (Broekman et al. 1980). In the presence of lycopene (6-12 µmol/L) the breakdown of phosphoinositide resulting in the formation of inositol phosphate, inositol-4,5-diphosphate, and inositol-1,4,5-triphosphate (IP_3) in the presence of collagen (1 µg/mL) decreased. This finding suggests that lycopene exerts an inhibitory effect on phosphoinositide breakdown and Ca^{+2} mobilization in human platelets stimulated by collagen. Lycopene (6-12 µmol/L) also in a dose-dependent manner inhibited thromboxane B_2 formation stimulated by collagen (1 µg/mL) in platelets (Hsiao et al. 2005). Lycopene affected Ca^{+2} release from intracellular Ca^{+2} storage sites, which is responsible for platelet aggregation. Platelet aggregation induced by agonists such as collagen, ADP and arachidonic acid was reduced in the presence of lycopene, implying that lycopene blocks a common step shared by these inducers. Stimulation of platelets by agonists led to the hydrolysis of minor plasma membrane phospholipids, phosphatidyl-inositol 4,5-biphosphate with formation of IP_3 and diacylglycerol catalyzed by phospholipase C (Kirk et al. 1981).

IP_3 induces release of Ca^{+2} from intracellular stores and diacylglycerol activates protein kinase C inducing protein phosphorylation. Lycopene inhibited phosphoinositide breakdown of collagen-activated platelets, suggesting that inhibition of platelet aggregation by lycopene is related to inhibition of phospholipase C activation.

Thromboxane A_2 is an important mediator of the release reaction and aggregation of platelets. The collagen-induced formation of thromboxane B_2, which is a stable metabolite of thromboxane A_2, was inhibited by lycopene (6-12 µmol/L) (Hsiao et al. 2005). Phosphoinositide breakdown can induce thromboxane B_2 formation by releasing free arachidonic acid by diglyceride lipase or by release of endogenous phospholipase A_2 from membrane phospholipids (McKean et al. 1981).

Lycopene may inhibit thrombus formation and platelet aggregation and may thus prevent CVD.

Suppression of endothelial cell tissue factor activity

A transmembrane glycoprotein, tissue factor (molecular weight 47 kDa) is expressed in many cells such as smooth muscle cells, monocytes, activated smooth muscle cells and fibroblasts (Mackman 2004). Tissue factor activation plays a key role in vascular thrombosis through the conversion of fibrinogen to fibrin (Levi and Cate 1999). An increase of tissue factor expression and activation of tissue factor (factor and V11a complex) leads to fibrin deposition and activation of platelets as a result of injury (Wilcox et al. 1989).

Increased oxidative stress in the vascular system is reported to enhance tissue factor production in response to activated platelets, resulting in development of unstable atherosclerotic plaque and thrombosis (Gorlach et al. 2000, Weisburger 2002). Antioxidants derived from the diet such as lycopene may help to prevent and/or delay progression of pathophysiological pathways that increase production of ROS, modifying the risk of CVD caused by oxidative damage to endothelial cells.

Lycopene, an acyclic carotenoid, has been found to suppress progression of coronary heart disease and atherosclerosis by reducing oxidative modification of LDL (Rao 2002). Lycopene exhibits higher singlet oxygen quenching capacity than β-carotene because it has more double bonds (DiMascio et al. 1989).

Incubation of endothelial cells with various carotenoids (lycopene, lutein, and β-carotene) suppressed expression of tissue factor mRNA. Lycopene and β-carotene showed a significant ($P < 0.05$) effect in suppressing tissue factor activity in endothelial cells at 2 µmol/L concentration, but lutein showed a significant effect at a higher concentration (10 µmol/L). Lycopene suppressed tissue factor activity in endothelial cells by suppression of gene expression (Lee et al. 2006) and by enhancing phosphorylation of AKT, and suppressed tissue factor activity in endothelial cells.

Inhibition of pathological up-regulation of tissue factor activity may reduce thrombosis associated with CVD (Lee et al. 2006). Another study (Martin et al. 2000) demonstrated that lycopene was more effective than β-carotene or lutein in suppressing monocyte cell adhesion to human endothelial cells, an initial step in atherosclerosis plaque formation. Lycopene was also more effective than β-carotene in inhibiting LDL oxidation (Fuhrman et al. 1997). Lycopene seems to offer greater protection against development of CVD than β-carotene and lutein by suppressing multiple aspects of atherosclerosis.

The protective effects of dietary tomato against endothelial dysfunction were studied in mice that were made hypercholesterolemic by being fed atherogenic diets. Mice fed on the atherogenic diet without tomato for 4 mon had significantly increased plasma lipid peroxide and decreased vaso-relaxing activity in the aorta induced by acetylcholine when compared with mice fed on a commercial diet. However, mice fed on the atherogenic diet containing 20% (w/w) lyophilized powder of tomato showed lower increases in plasma lipid peroxide level, and acetylcholine-induced vaso-relaxation was maintained at the same level as that in normal mice. These results indicate that tomato has a preventive effect on atherosclerosis by protecting plasma lipids from oxidation (Suganuma and Inakuma 1999), thereby reducing endothelial dysfunction.

Expression of cell surface adhesion molecules and binding of monocytes

The expression of various cell surface adhesion molecules is essential for the binding of normally non-thrombogenic leukocytes such as the monocyte to the aortic endothelial surface. It is one of the earliest detectable events in atherosclerosis (Cybulsky and Gimbrone 1991). This is followed by the transendothelial migration of these adherent leukocytes, and their accumulation in the aortic intima. The monocytes are transformed into lipid-engorged foam cells, and secretion of cytokines and growth factors are events in the initiation and progression of atherosclerotic plaques (Campbell and Campbell 1994). Cell surface adhesion molecules are important regulators of direct cell-cell interactions. The inflammatory responses in atherogenesis are directed by regulation and expression of these molecules (Walpola et al. 1995). Adhesion molecule expression in LDL receptor-knockout mice fed atherogenic diets showed a significantly reduced incidence of fatty streaks, supporting a role for adhesion molecules in atherogenesis (Nageh et al. 1997). There are a number of adhesion molecules involved in the interactions of endothelial cells (EC) and immune cells. The members of the immunoglobulin superfamily expressed on EC are intercellular adhesion molecule (ICAM)-1 and -2 (CD54 and CD102) and vascular cell adhesion molecule (VCAM)-1, (CD106). There is an increase in ICAM-1 and VCAM-1 in response to various inflammatory cytokines. E-Selectin and P-selectin (CD62E and CD62P) on EC also play an early role in adhesion between these two cell types (Cartwright et al. 1995, Jang et al. 1994).

Oxidative stress and expression of adhesion molecules on vascular EC are considered to be important features in the pathogenesis of atherosclerosis and other inflammatory diseases (Hennig et al. 1996, Alexander 1995). Studies suggest a molecular linkage between an antioxidant-sensitive transcriptional regulatory mechanism and expression of adhesion molecule genes that expands on the notion of oxidative stress as an important regulatory signal in the pathogenesis of atherosclerosis (Marui et al. 1993). In the inflammatory response of the arterial wall, leukocyte recruitment to the endothelium is mediated by the interaction of adhesion molecule receptors expressed on the surface of EC and immune cells.

Although LDL oxidation is currently a mechanism in the etiology of atherosclerosis, there are conflicting results in the role of dietary antioxidants in suppressing LDL oxidation, suggesting the involvement of other factors and alternate mechanisms. Carotenoids were reported to protect against LDL oxidation in some studies but not in others. In one study, β-carotene supplementation in rabbits retarded aortic lesion formation (Shaish et al. 1995). In another study, β-carotene supplementation in combination with vitamins E and C reduced aortic valve lesion formation in LDL receptor-knockout mice (Crawford et al. 1998), which may indicate that carotenoids play a significant role in the final pathological outcome. Martin et al. (2000) suggested that carotenoids may act via alternate mechanisms within the endothelial cell to modulate adhesion molecule expression and thus reduce subsequent leukocyte binding. They examined the *in vitro* effect of the five most prevalent plasma carotenoids on expression of key adhesion molecules involved in the atherosclerosis process and determined the subsequent binding of U937 monocytic cells when carotenoids are incorporated into human aortic endothelial cells (HAEC). They incubated HAEC cultures for 24 hr with each of the five most prevalent carotenoids (α-carotene, β-carotene, β-cryptoxanthin, lutein, and lycopene) in human plasma, at a concentration of 1 μmol/L. Monolayers were then stimulated with IL-1β (5 ng/ml) for 6 hr and subsequent determination of cell surface expression of adhesion molecules was measured by an enzyme-linked immunosorbent assay. Pre-incubation of HAEC with β-carotene, lutein and lycopene significantly reduced VCAM-1 expression by 29, 28, and 13%, respectively. Pre-incubation with β-carotene and lutein significantly reduced E-selectin expression by 38 and 34%. Pre-treatment with β-carotene, lutein and lycopene significantly reduced the expression of ICAM-1 by 11, 14, and 18%, respectively. Lycopene was the only carotenoid that attenuated both IL-1 β-stimulated and spontaneous HAEC adhesion to U937 monocytic cells by 20 and 25%, respectively. The results indicate that, among the test carotenoids, lycopene appears to be the most effective in both HAEC adhesion to monocytes and expression of adhesion molecules on the cell surface reducing immune and endothelial cell interaction (Martin et al. 2000).

Recent evidence suggests that *in vivo* oxidation of LDL, primarily by free oxygen radicals, may be involved in atherogenesis. Antioxidants such as lycopene may influence atherogenesis by interfering with the oxidation process. The influence of antioxidant levels in the diet on plasma concentrations, together with their biochemical properties, raises the possibility that increased consumption of antioxidants such as lycopene could prevent or delay the atherogenesis process (Tribble 1999).

Reduction of incidence of apoptosis

Lycopene reduces the incidence of apoptosis, which impacts tissue regeneration and repair after cardiovascular injury. A lower percentage of cells were apoptotic after being treated with lycopene (15%) and a significant decrease was seen after treatment with tomato juice (9%)

compared to the control (22%) in myocardial ischemia repurfusion injury (Das et al. 2005).

EPIDEMIOLOGICAL STUDIES

The antioxidant activity of lycopene has been seen in several epidemiological studies. Oxidation of LDL, which carry cholesterol into the bloodstream, may play an important role in the development of atherosclerosis. Low serum lycopene and high lipid peroxidation were observed in volunteers after consumption of lycopene-free diets for 2 wk (Rao and Agarwal 1998). Some studies have not reported any relationships between lycopene concentrations and the reduction in risk for myocardial infarction (Street et al. 1994). However, other studies reported inverse associations with lycopene concentrations and the risk for myocardial infarctions. An increased intake resulting in higher tissue levels of lycopene is associated with a decreased risk in myocardial infarctions. Higher concentrations of lycopene in adipose tissue of men was associated with reductions in the risk of myocardial infarction (Kohlmeier et al. 1997). In the same study, stratification with smoking status resulted in a strong effect of lycopene in non-smokers compared to smokers in reducing the risk of myocardial infarction. A large prospective women's health study did not report a significant CVD risk reduction (except for stroke) with an increase in plasma lycopene concentrations (Sesso et al. 2003). No associations between lycopene and the risk of stroke were observed in a 8 yr follow-up cohort male health professional study (Ascherio et al. 1999). The European Multicenter Case-Control Study on Antioxidants, Myocardial Infarction and Breast Cancer Study (EURAMIC) examined the correlation between adipose lycopene concentration and CVD risk. Adipose samples, considered a long-term storage depot for carotenoids, were analyzed for carotenoid content. They found that men with the highest concentrations of lycopene in adipose tissue had a 48% reduction in the risk of developing CVD compared with men with the lowest lycopene concentrations. A higher lycopene concentration was found to be independently protective against CVD. In a part of the EURAMIC study conducted in the Malaga center (Gomez-Aracena et al. 1997), there was a 60% lower risk of myocardial infarction in the participants with the highest lycopene concentrations compared with those in the lowest fifth of lycopene concentrations in adipose tissue.

The odds ratio for lycopene's protective effect was 0.52 when contrasting the 10th and 90th percentiles of adipose lycopene concentrations (Kohlmeier et al. 1997). In the same EURAMIC study, researchers examined the association between lycopene concentration in fat tissue and the risk of myocardial infarction in 10 countries. The model was adjusted for CVD risk factors and adipose tissue concentrations of α- and β-carotene. Research showed that smokers with low levels of circulating carotenoids were at an increased risk for subsequent myocardial infarction (Handelman et al. 1996). Lower blood lycopene levels were also found to be associated with increased risk for and death from coronary artery disease in a population study comparing Lithuanian and Swedish cohorts with different rates of death from coronary artery disease (Kristenson et al. 1997).

Antioxidant nutrients are believed to slow the progression of atherosclerosis because of their ability to inhibit damaging oxidative processes (Rimm et al. 1993, Heller et al. 1998, Parthasarathy 1998). Several controlled clinical trials and epidemiological studies have shown the protective effect of vitamin E, a potent antioxidant (Heller et al. 1995, Hodis et al. 1995, Paolisso et al. 1995). Supplementation with 400 IU/d of vitamin E for four and a half years did not result in beneficial effects on cardiovascular events in high-risk patients in the Heart Outcomes Prevention Evaluation (HOPE) Study (HOPE 2000). However, consumption of tomatoes and tomato products containing lycopene reduced CVD risk in individuals at risk of the disease (Kohlmeier et al. 1997). The relationship between antioxidant status and acute myocardial infarction was evaluated in a multi-center case-control study in which subjects were recruited from 10 European countries to maximize the variability in exposure (Kohlmeier et al. 1997). Adipose tissue antioxidant levels, which are better indicators of long-term exposure than blood antioxidant levels, were used as markers of antioxidant status. Biopsy specimens of adipose tissue were taken directly after the infarction and were analyzed for various carotenoids. Lycopene, and not β-carotene, levels were found to be protective after adjustment for a range of dietary variables.

Sesso et al. (2004) recently described an association between plasma lycopene and the risk of CVD in middle-aged and elderly women in their prospective, nested, case-control Women's Health Study, which was conducted in 39,876 women who were free of CVD at study baseline. The women in the upper three quartiles of lycopene concentration had a 50% risk reduction compared with those in the lowest quartile. The role of serum lycopene and the risk of acute coronary events and ischemic strokes was examined in the Kuopio Ischemic Heart Disease Risk Factor Study (Rissanen et al. 2001). The subjects were 725 middle-aged men free of coronary heart disease and stroke at the study baseline. Men with a low serum concentration of lycopene (i.e., the lowest quarter) had a > 3-fold risk of experiencing an acute coronary event or stroke compared with the other men.

Table 2 Non-antioxidative mechanisms of action of lycopene in cardiovascular disease.

Inhibition of 3-hydroxy-3-methyl glutaryl coenzyme A (HMG-CoA) reductase
Reduction in intima-media thickness of the carotid artery
Prevention of platelet aggregation and thrombus formation
Suppression of endothelial cell tissue factor activity
Expression of cell surface adhesion molecules and binding of monocytes

Table 3 Formation of foam cells, lipid-engorged cells made up of macrophages laden with cholesterol.

1. Low density lipoproteins (LDL) trapped in the matrix of the endothelial cells (EC) of the artery wall can undergo proatherogenic oxidative changes
2. Mildly oxidized LDL affects gene expression (of EC)

Leading to:

3. Up-regulation of expression of chemotactic factors or adhesion molecule genes
   a. Intercellular adhesion molecule, vascular cell adhesion molecule (monocyte binding molecule)
   b. Monocyte chemoattractant protein-1
   c. Macrophage colony-stimulating factor
4. Release of chemotactic factors supports recruitment of circulating monocytes and promotes them to macrophages
5. Oxidized LDL particles recognized and internalized by macrophage scavenger receptor
6. Excess LDL substrate process promotes marked macrophage cholesterol engorgement leading to formation of foam cells
7. Transformed into lipid-engorged foam cells, and secretion of cytokines and growth factors result in the initiation and progression of atherosclerotic plaques

CONCLUSIONS

A limited number of studies have been conducted to study the relationship between lycopene intake and CVD risk as compared to a number of epidemiological data supporting the relationship between CVD and vitamin E and β-carotene. The protective effects of lycopene are mainly derived from epidemiological studies on normal and high-risk populations. Results from clinical and experimental studies are inconclusive. Conclusions from these studies are made from the data obtained from dietary estimates vs. plasma and adipose tissue levels of lycopene in relation to the disease. Low-density lipoprotein oxidation represents an important early event in the development of CVD. Lycopene, a potent antioxidant, seems to play a role in the antioxidant as well as non-antioxidant mechanisms related to reduction in the risk of development of CVD.

ACKNOWLEDGEMENTS

The authors would like to thank Mr. David Asiamah for his help in typing the manuscript.

ABBREVIATIONS

CCA-IMT: common carotid artery; CVD: Cardiovascular diseases; EC: Endothelial cells; EURAMIC: European Multicenter Case-Control Study on Antioxidants, Myocardial Infarction and Breast Cancer Study; GPIIbIIIa: Glycoprotein IIbIIIa; HAEC: Human aortic endothelial cells; HMG CoA: 3-hydroxy-3-methyl gluaryl coenzyme A; HOPE: Heart Outcomes Prevention Evaluation Study; ICAM: Intercellular adhesion molecule; IFN-γ: Interferon-gamma; IL-1β: Interleukin-1β; IL-6β: Interleukin-6β; iNOS: Inducible nitric-oxide synthase; IP3: Inositol-1,4,5-triphosphate; LDL: Low density lipoprotein; MCP-1: Monocyte chemoattractant protein-1; NO: Nitric Oxide; ROS: Reactive Oxygen Species; TNF-α: Tumor Necrosis Nactor alpha; VCAM: Vascular cell adhesion molecule; VLDL: Very low-density lipoprotein; VSMC: Vascular smooth muscle cells

References

Alexander, R.W. 1995. Theodore Cooper Memorial Lecture. Hypertension and the pathogenesis of atherosclerosis. Oxidative stress and the mediation of arterial inflammatory response: a new perspective. Hypertension 25: 155-161.

Ames, B.N. and L.S. Gold, and W.C. Willet. 1995. Causes and prevention of cancer. Proc. Natl. Acad. Sci. USA 92: 5258-5265.

Armstrong, A.M., J.E. Chestnutt, M.J. Gormley and I.S. Young. 1996. The effect of dietary treatment on lipid peroxidation and antioxidant status in newly diagnosed noninsulin dependent diabetes. Free Radic. Biol. Med. 21(5): 719-726.

Arab, L. and S. Steck. 2000. Lycopene and cardiovascular disease. Amer. J. Clin. Nutr. 71: 1691S-1695S.

Ascherio, A. and E.B. Rimm, M.A. Hernan, E. Giovannucci, I. Kawachi, M.J. Stampfer, and W.C. Willett. 1999. Relation of consumption of vitamin E, vitamin C, and carotenoids to risk for stroke among men in the United States. Ann. Intern. Med. 130: 963-970.

Barrett-Connor, E. and T.L. Bush. 1991. Estrogen and coronary heart disease in women. JAMA 265: 1861-1867.

Betteridge, D.J. 2000. What is oxidative stress? Metabolism 49: 3-8.

Bohm, F. and J.H. Tinkler, and T.G. Truscott. 1995. Carotenoids protect against cell membrane damage by the nitrogen dioxide radical. Nat. Med. 1: 98-99.

Boileau, T.W. and A.C. Boileau, and J.W. Erdman. 2002. Bioavailability of the all-*trans* and *cis*-isomer of lycopene. Exp. Biol. Med. 227: 914-919

Brady, W.E. and J.A. Mares-Perlman, P. Bowen, and M. Stacewicz-Sapuntzakis. 1996. Human serum carotenoid concentrations are related to physiologic and lifestyle factors. J. Nutr. 126: 129-137.

Britton, G. 1995. Structure and properties of carotenoids in relation to function. FASEB J. 9: 15511-15558.

Broekman, M.J. and J.W. Ward, and A.J. Marcus. 1980. Phospholipid metabolism in phosphatidyl inositol, phosphatic acid and lysophospholipids. J. Clin. Invest. 66: 275-283.

Campbell, J.H. and G.R. Campbell. 1994. Cell biology of atherosclerosis. J. Hypertens. 12: S129-S130.

Cartwright, J.E. and G.S.J. Whitley, and A.P. Johnstone. 1995. The expression and release of adhesion molecules by human endothelial cells and their consequent binding of lymphocytes. Exp. Cell. Res. 217: 329-335.

Ceriello, A. and E. Motz. 2004. Is oxidative stress the pathogenic mechanism underlying insulin resistance, diabetes, and cardiovascular disease? The common soil hypothesis revisited. Arterioscler. Thromb. Vasc. Biol. 24: 816-823.

Chuang, C.Z. and P.N. Subramaniam, B.Y. LeGardeur, et al. 1998. Risk factors for coronary artery disease and levels of lipoprotein(a) and fat-soluble antioxidant vitamins in Asian Indians of USA. Indian Heart J. 50: 285-291.

Clevidence, B.A. and J.G. Bieri. 1993. Association of carotenoids with human plasma lipoproteins. Meth. Enzymol. 214: 33-46.

Clinton, S.K. 1998. Lycopene: Chemistry, biology, and implications for human health and disease. Nutr. Rev. 56: 35-51.

Colditz, G.A. and J.E. Manson, M.J. Stampfer, et al. 1992. Diet and risk of clinical diabetes in women. Amer. J. Clin. Nutr. 55: 1018-1023.

Coyne, T. and T.I. Ibiebele, P.D. Baade, et al. 2005. Diabetes mellitus and serum carotenoids: findings of a population-based study in Queensland, Australia. Amer. J. Clin. Nutr. 82: 685-693.

Crawford, R.S. and E.A. Kirk, M.E. Rosenfled, R.C. LeBoeuf, and A. Chait. 1998. Dietary antioxidants inhibit development of fatty streak lesion in the LDL receptor-deficient mouse. Arterioscler. Thromb. Vasc. Biol. 18: 1506-1513.

Cybulsky, M.I. and M.A. Gimbrone, Jr. 1991. Endothelial expression of a mononuclear leukocyte adhesion molecule during atherosclerosis. Science 251: 788-791.

Das, S. and O. Hajime, M. Nilanjana, and D.K. Dipak. 2005. Lycopene, tomatoes, and coronary heart disease. Free Radic. Res. 39(4): 449-455.

DiMascio, P. and S. Kaiser, and H. Sies. 1989. Lycopene as the most effective biological carotenoid singlet oxygen quencher. Arch. Biochem. Biophys. 274: 532-538.

Dugas, T.R. and D.W. Morel, and E.H. Harrison. 1998. Impact of LDL carotenoid and α-tocopherol content on LDL oxidation by endothelial cells in culture. J. Lipid Res. 39: 999-1007.

El-Agamey, A. and G.M. Lowe, D.J. McGarvey, et al. 2004. Carotenoid radical chemistry and antioxidant/pro-oxidant properties. Arch. Biochem. Biophys. 430: 37-48.

Evans, J.L. and I.D. Goldfine, and B.A. Maddux. 2003. Are oxidative stress-activated signaling pathways mediators of insulin resistance and beta-cell dysfunction? Diabetes 52: 1-8.

Feskens, E.J. and S.M. Virtanen, L. Rasanen, et al. 1992. Dietary factors determining diabetes and impaired glucose tolerance. A 20-year follow-up of the Finnish and Dutch cohorts of the Seven Countries Study. Diabetes Care 18: 1104-1112.

Ford, E.S. and A.H. Mokdad. 2001. Fruit and vegetable consumption and diabetes mellitus incidence among U.S. adults. Prev. Med. 32: 33-39.

Ford, E.S. and J.C. Will, B.A. Bowman, et al. 1999. Diabetes mellitus and serum carotenoids: findings from the Third National Health and Nutrition Examination Survey. Amer. J. Epidemiol. 149: 168-176.

Frie, B. 1995. Cardiovascular disease and nutrient antioxidants: role of low density lipoprotein oxidation. Crit. Rev. Food Sci. Nutr. 351: 83-98.

Fuhrman, B. and A. Elis, and M. Aviram. 1997. Hypercholesterolemic effect of lycopene and β-carotene is related to suppression of cholesterol synthesis and augmentation of LDL receptor activity in macrophage. Biochem. Biophys. Res. Commun. 233: 6588-6662.

Gaziano, J.M. 1996. Epidemiology of risk factor reduction. pp. 569-586. *In:* J. Loscalzo, M. Creagher, and V. Dzau. Vascular Medicine. Little Brown, Boston, Massachusetts.

Gerster, H. 1997. The potential role of lycopene for human health. J. Amer. Coll. Nutr. 16: 109-126

Gomez-Aracena, J. and J. Sloots, and A. Garcia-Rodriguez. 1997. Antioxidants in adipose tissue and myocardial infarction in a Mediterranean area. The EURAMIC Study in Malaga. Nutr. Metab. Cardiovasc. Dis. 7: 376-382.

Gordon, T. and W.P. Castelli, M.C. Hjortland, W.B. Kannel, and T.R. Dawber. 1977. High density lipoprotein as a protective factor against coronary heart disease. The Framingham Study. Amer. J. Med. 62: 707-714.

Gorlach, A. and R.P. Brandes, S. Bassus, N. Kronemann, C.M. Kirchmaier, and V.B. Schini-Kerth. 2000. Oxidative stress and expression of p22phox are involved in the upregulation of tissue factor in vascular smooth muscle cells in response to activated platelets. FASEB J. 14: 1518-1528.

Gotto, A.M., Jr. and S.M. Grundy. 1999. Lowering LDL cholesterol: questions from recent meta-analyses and subset analyses of clinical trial data. Issues from the Interdisciplinary Council on Reducing the Risk for Coronary Heart Disease, Ninth Council Meeting. Circulation 99: E1-E7.

Halliwell, B. 1999. Antioxidant defence mechanisms: from the beginning to the end (of the beginning). Free Radic. Res. 31: 261-272.

Halliwell, B. and M.A. Murcia, S. Chirico, and O.I. Aruoma.1995. Free radicals and antioxidants in food and in vivo: what they do and how they work. Crit. Rev. Food Sci. Nutr. 35: 7-20.

Handelman, G.J. and L. Parker, and C.E. Cross. 1996. Destruction of tocopherols, carotenoids and retinol in human plasma by cigarette smoke. Amer. J. Clin. Nutr. 63: 559-565.

Heart Outcomes Prevention Evaluation Study Investigators. 2000. Vitamin E supplementation and cardiovascular events in high-risk patients. N. Engl. J. Med. 342: 154-160.

Heller, F.R. and O. Descamps, and J.C. Hondekijn. 1998. LDL oxidation: therapeutic perspectives. Atherosclerosis 137: S25-S31.

Hennig, B. and M. Toborek, C.J. McClain, and J.N. Diana. 1996. Nutritional implications in vascular endothelial cell metabolism. J. Amer. Coll. Nutr. 15: 345-358.

Hodis, H.N. and W.J. Mack, L. LaBree, L. Cashin-Hemphill, A. Sevanian, and R. Johnson. 1995. Serial coronary angiographic evidence that antioxidant vitamin intake reduces progression of coronary artery atherosclerosis. JAMA 273: 1849-1854.

Hoffmann, I. and J.H. Weisburger. 1997. International symposium on the role of lycopene and tomato products in disease prevention. Cancer Epidemiol. Biomarkers Prev. 6: 643-645.

Holvoet, P. and D. Collen. 1994. Oxidized lipoproteins in atherosclerosis and thrombosis. FASEB J. 8: 1279-1284.

Houston, H. and A. Estevez, P. Chumley, M. Aslan, S. Marklund, D.A. Parks, and B.A. Freeman. 1999. Binding of xanthine oxidase to vascular endothelium. Kinetic characterization and oxidative impairment of nitric oxide-dependent signaling. Biol. Chem. 274: 4985-4994.

Hsiao, G. and N. Tzu, T. Fong, M. Shen, K. Lin, D. Chou, and J. Sheu. 2005. Inhibitory effects of lycopene on in vitro platelet activation and in vivo prevention of thrombus formation. J. Lab. Clin. Med. 146: 216-226.

Hulley, S. and D. Grady, T. Bush, C. Furberg, D. Herrington, B. Riggs, and E. Vittinghoff. 1998. Randomized trial of estrogen plus progestin for secondary prevention of coronary heart disease in postmenopausal women. Heart and Estrogen/Progestin Replacement Study (HERS) Research Group. JAMA 280: 605-613.

Jang, Y. and A.M. Lincoff, E.F. Plow, and E.J. Topol. 1994. Cell adhesion molecules in coronary artery disease. J. Amer. Coll. Cardiol. 24: 1591-1601.

Jang, Y. and J.H. Lee, E.Y. Cho, et al. 2004. Differences in body fat distribution and antioxidant status in Korean men with cardiovascular disease with or without diabetes. Amer. J. Clin. Nutr. 73: 68-74.

Jeremy, J.Y. and A.P. Yim, S. Wan, and G.D. Angelini. 2002. Oxidative stress, nitric oxide and vascular disease. J. Cardiovasc. Surg. 17: 324-327.

Jeremy, J.Y. and N. Shukla, S. Muzaffar, A. Handley, and G.D. Angelini. 2004. Reactive oxygen species, vascular disease and cardiovascular surgery. Curr. Vasc. Pharmacol. 2: 229-236.

Kacimi, R. and J.S. Karliner, F. Koudssi, and C.S. Long. 1998. Expression and regulation of adhesion molecules in cardiac cells by cytokines: response to acute hypoxia. Circ. Res. 82: 576-586.

Kirk, C.J. and J.A. Creba, C.P. Downes, and R.H. Michell. 1981. Hormone-stimulated metabolism of inositol lipids and its relationship to hepatic receptor function. Biochem. Soc. Trans. 9: 377-379.

Kohlmeier, L. and J.D. Kark, E. Gomez-Garcia, B.C. Martin, S.E. Steck, A.F.M. Kardinaal, J. Ringstad, M. Thamm, V. Masaev, R. Riemersma, J.M. Martin-Moreno, J.K. Huttunen, W. Stahl, W. Schwartz, A.R. Sundquist, and H. Sies. 1992. *Cis-trans* isomers of lycopene and β-carotene in human serum and tissues. Arch. Biochem. Biophys. 294: 173-177.

Kohlmeier, L. and J.D. Kark, E. Gomez-Gracia, B.C. Martin, S.E. Steck, A.F. Kardinaal, J. Ringstad, M. Thamm, V. Masaev, R. Riemersma, J.M. Martin-Moreno, J.K. Huttunen, and F.J. Kok. 1997. Lycopene and myocardial infarction risk in the EURAMIC Study. Amer. J. Epidemiol. 146: 618-626.

Kristenson, M. and B. Zieden, Z. Kucinskiene, L.S. Elinder, B. Bergdahl, and B. Elwing. 1997. Antioxidant state and mortality from coronary heart disease in Lithuanian and Swedish men: concomitant cross sectional study of men aged 50. BMJ 314: 629-633.

Lee, D.K. and R.N. Grantham, J.D. Mannion, and A.L. Trachte. 2006. Carotenoids enhance phosphorylation of Akt and suppress tissue factor activity in human endothelial cells. J. Nutr. 17: 780-786.

Levi, M. and H. Ten Cate. 1999. Disseminated intravascular coagulation. N. Engl. J. Med. 341: 586-592.

Mackman, N. 2004. Role of tissue factor in hemostasis, thrombosis, and vascular development. Arterioscler. Thromb. Vasc. Biol. 24: 1015-1022.

MacMahon, S. and R. Peto, J. Cutler, R. Collins, P. Sorlie, J. Neaton, R. Abbott, J. Godwin, A. Dyer, and J. Stamler. 1990. Blood pressure, stroke, and coronary heart disease. Part 1. Prolonged differences in blood pressure: prospective observational studies corrected for the regression dilution bias. Lancet 335: 765-774.

Martin, K.R. and D. Wu, and M. Meydan. 2000. The effect of carotenoids on the expression of cell surface adhesion molecules and binding of monocytes to human aortic endothelial cells. Atherosclerosis 150: 265-274.

Marui, N. and M.K. Offermann, R. Swerlick, C. Kunsch, C.A. Rosen, M. Ahmad, R.W. Alexander, and R.M. Medford. 1993. Vascular adhesion molecule-1 (VCAM-1) gene transcription and expression are regulated through an antioxidant-sensitive mechanism in human vascular endothelial cells. J. Clin. Invest. 92: 1866-1874.

McKean, M.L. and J.B. Smith, and W.J. Silver. 1981. Formation of lysophosphatidylcholine in human platelets in response to thrombin. J. Biol. Chem. 256: 1522-1524.

Montonen, J. and P. Knekt, R. Jarvinen, et al. 2004. Dietary antioxidant intake and risk of type 2 diabetes. Diabetes Care 27: 362-366.

Murthy, V.K. and J.C. Shipp, and C. Hanson. 1992. Delayed onset and decreased incidence of diabetes in BB rats fed free radical scavengers. Diabetes Res. Clin. Pract. 18: 11-16.

Nageh, M. and E.T. Sandberg, K.R. Marotti, A.H. Lin, E.P. Melchior, D.C. Bullard, and A.L. Beaudet. 1997. Deficiency of inflammatory cell adhesion molecules against atherosclerosis in mice. Arterioscler. Thromb. Vasc. Biol. 17: 1517-1520.

Nelson, J.L. and P.S. Bernstein, M.C. Schmidt, M.S. Von Tress, and E.W. Askew. 2003. Dietary modification and moderate antioxidant supplementation differentially affect serum carotenoids, antioxidant levels and markers of oxidative stress in older humans. J. Nutr. 133: 3117-3123.

Oberley, L.W. 1988. Free radicals and diabetes. Free Radical Biol. Med. 5: 113-124.

Pamuk E.R. and T. Byers, R.J. Coates, J.W. Vann, A.L. Sowell, E.W. Gunter, and D. Glass. 1994. Effect of smoking on serum nutrient concentrations in African-American women. Amer. J. Clin. Nutr. 59: 891-895.

Paolisso, G. and D. Giugliano. 1996. Oxidative stress and insulin action: is there a relationship? Diabetologia 39: 357-363.

Paolisso, G. and A. D'Amore, and G. Di Maro. 1993. Evidence for a relationship between free radicals and insulin action in the elderly. Metabolism 42: 659-663.

Paolisso, G. and A. Gambardella, D. Giugliano, D. Galzerano, L. Amato, and C. Volpe. 1995. Chronic intake of pharmacological doses of vitamin E might be useful in the therapy of elderly patients with coronary heart disease. Amer. J. Clin. Nutr. 61: 848-852.

Parker, R.S. 1988. Carotenoid and tocopherol composition in human adipose tissue. Amer. J. Clin. Nutr. 47: 33-36.

Parthasarathy, S. 1998. Mechanisms by which dietary antioxidants may prevent cardiovascular diseases. J. Med. Food 1: 45-51.

Peng, Y.M. and Y.S. Peng, Y. Lin, T. Moon, D.J. Roe, and C. Ritenbaugh. 1995. Concentrations and plasma-tissue-diet relationships of carotenoids, retinoids, and tocopherols in humans. Nutr. Cancer 23: 233-246.

Petr, L. and J.W. Erdman. 2005. Lycopene and risk of cardiovascular disease. pp. 204-217. In: L. Packer, U. Obermueller-Jevic, K. Kramer and H. Sies. [Eds.] Carotenoids and Retinoids: Biological Actions and Human Health. AOCS Press, Champaign, Illinois.

Pincemail, J. 1995. Free radicals and antioxidants in human disease. pp. 83-95. In: A.E. Favier, J. Cadet, B. Kalyanaraman, M. Fontecave and J.-L. Pierre. [Eds.] Analysis of Free Radicals in Biological Systems. Birkhäuser Verlag, Basel.

Polidori, M.C. and P. Mecocci, W. Stahl, et al. 2000. Plasma levels of lipophilic antioxidants in very old patients with type 2 diabetes. Diabetes Metab. Res. Rev. 16: 15-19.

Pyorala, K. and M. Laakso, and M. Uusitupa. 1987. Diabetes and atherosclerosis: an epidemiologic view. Diabetes Metab. Rev. 3: 463-524.

Rajagopalan, S. and S. Kurz, T. Munzel, M. Tarpey, B.A. Freeman, and K.K. Griendling. 1996. Angiotensin II-mediated hypertension in the rat increases vascular superoxide production via membrane NADH/NADPH oxidase activation. Contribution to alterations of vasomotor tone. J. Clin. Invest. 97: 1916-1923.

Rao, A.V. 2002. Lycopene, tomatoes, and the prevention of coronary heart disease. Exp. Biol. Med. 227: 908-913.

Rao, A.V. and S. Agarwal. 1998. Bioavailability and in vivo antioxidant properties of lycopene from tomato products and their possible role in the prevention of cancer. Nutr. Cancer 31: 199-203.

Rimm, E.B. and M.J. Stampfer, A. Ascherio, E. Giovannucci, G.A. Colditz, and W.C. Willett. 1993.Vitamin E consumption and the risk of coronary heart disease in men. N. Engl. J. Med. 328: 1450-1456.

Rissanen, T. and S. Voutilainen, and K. Nyyssönen. 2001. Low serum lycopene is associated with excess risk of acute coronary events and stroke: The Kuopio Ischaemic Risk Factor Study. Br. J. Nutr. 85: 749-754.

Rissanen, T. and S. Voutilainen, K. Nyyssonen, J. Salonon, G. Kaplan, and J. Salonen. 2003. Serum lycopene concentration and carotid atherosclerosis: the Kuopio Ischaemic Heart Disease Risk Factor Study. Amer. J. Clin. Nutr. 77: 133-138.

Rissanen, T. and S. Voutilainen, K. Nyyssonen, and J. Salonon. 2002. Lycopene, atherosclerosis, and coronary heart disease. Exp. Biol. Med. 227: 900-907.

Robertson, R.P. and J. Harmon, and P.O. Tran. 2003. Glucose toxicity in beta-cells: type 2 diabetes, good radicals gone bad, and the glutathione connection. Diabetes 52: 581-587.

Rosamond, W. and K. Flegal, G. Friday, K. Furie, A. Go, K. Greenlund, N. Haase, M. Ho, V. Howard, B. Kissela, S. Kittner, D. Lloyd-Jones, M. McDermott, J. Meigs, C. Moy, G. Nichol, C.J. O'Donnell, V. Roger, J. Rumsfeld, P. Sorlie, J. Steinberger, T. Thom, S. Wasserthiel-Smoller, and Y. Hong. 2007. Heart disease and stroke statistics—2007 update: a report from the American Heart Association Statistics Committee and Stroke Statistics Subcommittee. Circulation. Feb 6. 115: e69-171.

Sakai, K. and K. Matsumoto, and T. Nishikawa. 2003. Mitochondrial reactive oxygen species reduce insulin secretion by pancreatic beta-cells. Biochem. Biophys. Res. Commun. 300: 216-222.

Schlenitz, M.D. and J.P. Weiss, and D.K. Owens. 2004. Clopidogrel versus aspirin for secondary prophylaxis of vascular events: a cost-effectiveness analysis. Amer. J. Med. 116: 797-806.

Sesso, H.D. and S. Liu, J.M. Gaziano, and J.E. Buring. 2003. Dietary lycopene, tomato-based food products and cardiovascular disease in women. J. Nutr. 133: 2336-2341.

Sesso, H.D. and J.E. Buring, E.P. Norkus, and J.M. Gaziano. 2004. Plasma lycopene, other carotenoids, and retinol and the risk of cardiovascular disease in women. Amer. J. Clin. Nutr. 79: 47-53.

Shaish, A. and A. Daugherty, F. O'Sullivan, G. Schonfeld, and J.W. Heinecke. 1995. β-carotene inhibits atherosclerosis in hypercholesterolemic rabbits. J. Clin. Invest. 96: 2075-2082.

Slonim, A.E. and M.L. Surber, and D.L. Page. 1983. Modification of chemically induced diabetes in rats by vitamin E. Supplementation minimizes and depletion enhances development of diabetes. J. Clin. Invest. 71: 1282-1288.

Stahl, W. and H. Sies. 1992. Physical quenching of singlet oxygen and cis-trans isomerization of carotenoids. pp. 10-19. *In:* L.M. Canfield, N.I. Krinsky and J.A. Olson. [Eds.] Carotenoids in Human Health, vol. 691. New York Academy of Sciences, New York.

Steinberg, F.M. and A. Chait. 1998. Antioxidant vitamin supplementation and lipid peroxidation in smokers. Amer. J. Clin. Nutr. 68: 319-327.

Street, D.A. and A.R. Folsom, D.R. Jacobs, W. Schuep, and M.J. Klag. 1994. Serum antioxidants and myocardial infarction: are low levels of carotenoids and alpha-tocopherol risk factors for myocardial infarction? Circulation 90: 1154-1161.

Suganuma, H. and T. Inakuma. 1999. Protective effect of dietary tomato against endothelial dysfunction in hypercholesterolemic mice. Biosci. Biotechnol. Biochem. 63: 78-82.

Suzuki, K. and Y. Ito, S. Nakamura, et al. 2002. Relationship between serum carotenoids and hyperglycemia: a population-based cross-sectional study. J. Epidemiol. 12: 357-366.

Tinkler, J.H. and F. Bohm, W. Schalch, and T.G. Truscott. 1994. Dietary carotenoids protect human cells from damage. J. Photochem. Photobiol. 26: 283-285.

Tribble, D.L. 1999. Antioxidant consumption and risk of coronary heart disease: emphasis on vitamin C, vitamin E and beta carotene. A statement for health care professionals from the nutrition committee, American Heart Association. Circulation 99: 591-595.

Tsuchiya, M. and G. Scita, D. Thompson, and L. Packer. 1993. Retinoids and carotenoids are peroxyl radical scavengers. pp. 525-536. *In:* M. Livrea and L. Packer. [Eds.] Retinoids: Progress in Research and Clinical Applications. Marcel Dekker, Inc., New York.

Walpola, P.L. and A.I. Gotlieb, M.I. Cybulsky, and B.L. Langille. 1995. Expression of ICAM-1 and VCAM-1 and monocyte adherence in arteries exposed to altered shear stress. Arterioscler. Thromb. Vasc. Biol. 15: 2-10.

Wang, Lu and S. Liu, A.D. Pradhan, J.E. Manson, J.E. Buring, J. Michael Gaziano, and H.D. Sesso. 2006. Plasma lycopene, other carotenoids, and the risk of Type 2 diabetes in women. Amer. J. Epidemiol. 164: 576-585.

Warnholtz, A. and G. Nickenig, E. Schulz, R. Macharzina, J.H. Brasen, and M. Skatchkov. 1999. Increased NADH-oxidase-mediated superoxide production in the early stages of atherosclerosis: evidence for involvement of the renin-angiotensin system. Circulation 99: 2027-2033.

Weisburger, J.H. 2002. Lycopene and tomato products in human promotion. Exp. Biol. Med. 227: 924-927.

Wilcox, J.N. and K.M. Smith, S.M. Schwartz, and D. Gordon. 1989. Localization of tissue factor in the normal vessel wall and in the atherosclerotic plaque. Proc. Natl. Acad. Sci. USA 86: 2839-2843.

Wilson, P.W. and W.B. Kannel, H. Silbershatz, and R.B. D'Agostino. 1999. Clustering of metabolic factors and coronary heart disease. Arch. Intern. Med. 159: 1104-1109.

Witztum, J.L. 1994. The oxidation hypothesis of atherosclerosis. Lancet 344: 793-795.

Woodall, A.A. and S.W. Lee, R.J. Weesie, M.J. Jackson, and G. Britton. 1997. Oxidation of carotenoids by free radicals: relationship between structure and reactivity. Biochem. Biophys. Acta (Netherlands) 1336: 33-42.

World Health Organization. 2000. Report on Cardiovascular Disease. http://www.who.int/ncd/cvd/index.htm

2.8

Effects of Lycopene and Monounsaturated Fat Combination on Serum Lycopene, Lipid and Lipoprotein Concentrations

Kiran Deep Kaur Ahuja and Madeleine Joyce Ball

Locked Bag 1320, School of Human Life Sciences, University of Tasmania
Launceston, 7250, TAS, Australia

ABSTRACT

Lycopene, the predominant carotenoid in tomatoes, has a potential health benefit because of its antioxidant properties. Monounsaturated fats also have health benefits pertaining to their antioxidant effects and their beneficial effects on blood lipids. These food substances are particularly found together in the Mediterranean diet, which is associated with reduced risk of vascular diseases. It is possible that they could have synergistic benefits, if the absorption of lycopene, a fat-soluble carotenoid, increases when lycopene is ingested in the presence of fat. There is, however, limited data available to confirm this hypothesis. Although single meal studies have suggested this hypothesis may have a basis, short-term dietary intervention studies comparing the combination effects of high fat (especially rich in monounsaturated fatty acids), high lycopene diets with that of high carbohydrate, low fat, high lycopene diets have failed to confirm this. Results of these short trials (dietary periods of 10 to 16 d) demonstrate similar increases in serum lycopene concentration with high lycopene, high carbohydrate diets (15-17% energy from fat) compared to high

A list of abbreviations is given before the references.

lycopene, high fat diets rich in monounsaturated fatty acids (35-38% of energy from fat). However, the combination of high lycopene and high monounsaturated fat in the diet still provides greater beneficial effects on serum lipids and lipoprotein concentrations and on the susceptibility of low density lipoprotein to oxidation compared to the combination of high lycopene and high carbohydrate in the diet.

INTRODUCTION

Finding healthy dietary patterns that appeal to a considerable section of the community is a major public health challenge for many countries. In many countries saturated fat intake has traditionally been far too high and has contributed to the high rates of coronary heart disease (CHD). Research has shown that people living in the Mediterranean region have a low rate of a number of chronic diseases such as CHD (Keys 1970) and this may be associated with the lifestyle, especially dietary factors. The Mediterranean diets have traditionally relied heavily on plant foods (Kromhout 1989) and included unsaturated fats, such as the olive oil, which is rich in monounsaturated fatty acid, in considerable quantities, as well as regular use of lycopene-rich tomatoes.

Saturated, polyunsaturated and monounsaturated fatty acids differ in their chemical structure. Saturated fatty acids contain no double bond, monounsaturated fatty acids contain one double bond, and polyunsaturated fatty acids contain two or more double bonds (Fig. 1). Cooking fats and oils are generally classified as saturated, monounsaturated and polyunsaturated fats depending on the fatty acid present in the largest amount. Table 1 presents the fatty acid composition of some of the commonly used cooking fats (Meadow Lea Foods 1995).

There is a vast amount of data available in support of the intake of monounsaturated fats, a large component of the Mediterranean diets, and their beneficial effects on the serum lipid profile (Grundy et al. 1988, Mensink and Katan 1989, Bonanome et al. 1992, Mata et al. 1996, 1997), which is a major determinant of CHD risk (Castelli et al. 1986, Stamler et al. 1986, Roberts 1989). When they replace saturated fats, monounsaturated fats result in a reduction of the atherogenic low density lipoprotein (LDL) cholesterol in the plasma but maintenance of the levels of the protective high density lipoprotein (HDL) cholesterol component (Mensink and Katan 1992, Clarke 1997). There is also evidence of beneficial effects of monounsaturated fatty acids in reducing the susceptibility of LDL

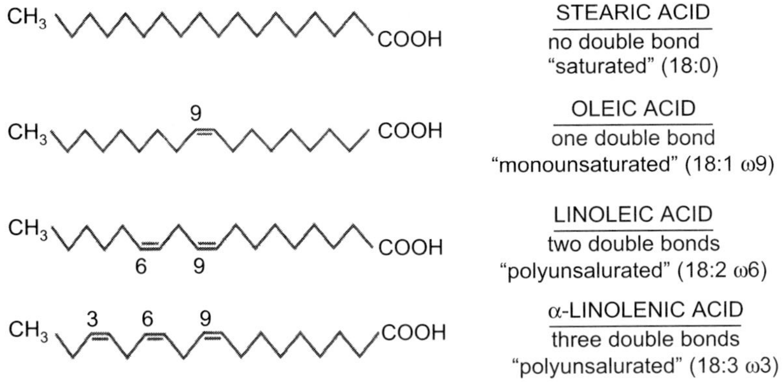

CH₃ (structure)	STEARIC ACID
	no double bond
	"saturated" (18:0)

CH₃ ——COOH   STEARIC ACID / no double bond / "saturated" (18:0)

CH₃ ——9——COOH   OLEIC ACID / one double bond / "monounsaturated" (18:1 ω9)

CH₃ ——6——9——COOH   LINOLEIC ACID / two double bonds / "polyunsaturated" (18:2 ω6)

CH₃ ——3——6——9——COOH   α-LINOLENIC ACID / three double bonds / "polyunsalurated" (18:3 ω3)

Fig. 1. Configuration of saturated and unsaturated fatty acids.

Table 1 Fatty acid composition of some commonly used cooking fats (g/100g).

| | | Fatty acids | | |
Fats and oils	Total lipids	Saturates	Polyunsaturates	Monounsaturates
Butter	83	52	3	21
Canola oil	100	8	30	62
Coconut oil	100	91	1	8
Sunola oil™	100	10	5	85
Safflower oil	100	9	77	14
Sunflower oil	100	11	66	23
Olive oil	100	14	10	76

Data modified from Meadow Lea Foods (1995).

cholesterol to oxidation (Reaven et al. 1991, Bonanome et al. 1992, Abbey et al. 1993, Mata et al. 1997) and LDL oxidation is thought to be a process associated with initiation, development and progression of atherosclerosis (Steinberg et al. 1989).

Lycopene-rich tomatoes and tomato products, another component of Mediterranean diets, have also been found to have effects that may potentially play an important role in the association between the diet in these areas and the reduced risk of CHD compared to many other countries. Epidemiological data suggests a negative association between lycopene intakes, serum/tissue concentrations of lycopene and risk factors for CHD. Serum concentrations of lycopene increase with increased intake of lycopene-rich tomatoes and tomato products (Agarwal and Rao 1998, Mayne et al. 1999, Sutherland et al. 1999, Casso et al. 2000, Upritchard et al. 2000) as well as lycopene supplements (Paetau et al. 1998, Olmedilla et al.

2002). Conversely, serum concentrations of lycopene reduce with reduced consumption of lycopene-rich foods (Rock et al. 1992, Yeum et al. 1996, Rao and Agarwal 1998, Lee et al. 2000). Dietary intervention studies with increased lycopene consumption have also been shown to prolong resistance of LDL to oxidation in healthy subjects (Agarwal and Rao 1998, Maruyama et al. 2001), type 2 diabetics (Upritchard et al. 2000) and non-smokers, but not in current smokers (Chopra et al. 2000) and renal transplant patients (Sutherland et al. 1999). Also, reduced lycopene intake increases the susceptibility of LDL to oxidation in healthy subjects (Rao and Agarwal 1998).

For many years the mainstay in attempts to reduce saturated fat intake was a low fat diet, with the caloric replacement for saturated fat being carbohydrate. High carbohydrate, low fat diets have been under investigation for more than fifty years, to assess their influence on the risk factors for CHD. Suggestions that a higher fat diet, where the fat is mainly monounsaturated fat, may be more palatable have led to advocacy for the latter and much discussion on the relative benefits. Intervention studies comparing the two types of diets have mainly shown similar reductions in serum total cholesterol and LDL cholesterol (Schaefer et al. 1995, Turley et al. 1998). Results on HDL cholesterol have shown either no change or a relative increase with the monounsaturated fat-rich, low carbohydrate diets, whereas the results with high carbohydrate, low fat diets have either been no significant change or a small reduction in HDL cholesterol (Schaefer et al. 1995, Turley et al. 1998). High monounsaturated fat diets have almost always resulted in a reduction in serum triglyceride concentrations, whereas the effects of the high carbohydrate diets have mainly depended on the type of carbohydrate included in the studies, with high refined carbohydrate diets often causing a relative increase, and high fibre diets causing little change (Schaefer et al. 1995, Turley et al. 1998). Moreover, high carbohydrate, low fat diets result in an increased *in vitro* susceptibility of LDL to oxidation compared to high monounsaturated fat diets (Ashton et al. 2001).

Research on Lycopene Monounsaturated Fat Combination

Lycopene is a fat-soluble carotenoid and it could thus be hypothesized that the absorption of lycopene might increase with higher intake of fat taken in combination with the lycopene-rich foods. A small number of studies have tested this hypothesis and have shown increased serum lycopene concentrations with high intake of lycopene-rich foods in combination with high fat diets (Stahl and Sies 1992, Fielding et al. 2005). However, these studies involved small, non-randomized, single meals or lasted 3 to

5 d and did not control for the daily diets except for the lycopene-rich meals. Another study compared the effects of the combination of lycopene with monounsaturated fat from extra-virgin olive oil to the combination of lycopene and polyunsaturated fat from sunflower oil and showed similar increases in serum lycopene concentrations but higher plasma antioxidant activity after 7 d of the lycopene and olive oil diet (Lee et al. 2000).

Although there is a considerable amount of information on the individual effects of monounsaturated fat-rich diets and lycopene-rich diets on risk factors for CHD, very little data is available on the combined effects of the two with respect to their effect on the serum lipid profile. In addition, there is paucity in the data from intervention trials comparing the combined effects of high lycopene, high monounsaturated fat diets (~35% energy from fat) and high lycopene, high carbohydrate and low fat diets (~15% energy from fat) on serum lycopene concentration and risk factors for CHD including the lipid profile and susceptibility to oxidation. To date, there are only three small intervention trials (Ahuja et al. 2003a, b, 2006) in the literature that have investigated this in humans. A collective synopsis of these three trials is presented in this chapter. Table 2 presents the population characteristics for the three studies. All three investigations were conducted in generally healthy individuals in the age range of 21-70 yr.

The number of individuals in each study was relatively small, but the results can be usefully viewed together. One of the reasons for the small size may be the long list of exclusion criteria. Reduced carotenoid absorption is associated with poor iron, zinc, and protein status, intestinal diseases, malabsorption, and liver and kidney problems (White et al. 1993, Williams et al. 1998); hence, people diagnosed with any of these conditions

Table 2 Baseline characteristics of the study population.

	Study 1	Study 2	Study 3
Number of subjects	18	13	21
Males	–	13	6
Age (yr)	40 ± 16	39 ± 11	44 ± 12
Body mass index (kg/m^2)	23.2 ± 4.1	25.5 ± 2.4	24.3 ± 3.5
Total cholesterol (mmol/L)	5.05 ± 1.26	5.54 ± 0.85	5.06 ± 0.78
LDL cholesterol (mmol/L)	3.10 ± 1.02	3.84 ± 0.88	3.09 ± 0.72
HDL cholesterol (mmol/L)	1.52 ± 0.67	1.07 ± 0.22	1.43 ± 0.38
Triglycerides (mmol/L)	0.96 ± 0.32	1.65 ± 1.06	1.20 ± 0.60
LDL : HDL	2.28 ± 0.96	3.65 ± 0.87	2.36 ± 1.00

Values presented as mean ± standard deviation; LDL, low density lipoprotein; HDL, high density lipoprotein. Data reproduced with permission: Study 1 and Study 2 from Ahuja et al. (2003a), © the Biochemical Society; Study 3 from Ahuja et al. (2006), © Elsevier.

or on medication for them were excluded from the studies. Similarly, anyone on drugs affecting cholesterol absorption was excluded, as such drugs could have affected the results (Parker 1996, Forman et al. 1998). People taking vitamin and/or carotenoid supplements were excluded as some carotenoids have been suggested to interact and affect the absorption of other carotenoids (Micozzi et al. 1992, Johnson et al. 1997). Smokers were excluded from the studies because smoking can affect carotenoid absorption (Roidt et al. 1988, Bolton-Smith et al. 1991, Albanes et al. 1997). Body mass index has been suggested to affect serum carotenoid concentrations (Roidt et al. 1988, Casso et al. 2000); therefore, people trying to gain or lose weight were also excluded from all the three studies. Women with a history of menstrual irregularity, peri-menopausal women, pregnant and nursing mothers were excluded from the studies because of the suggested association between menstrual phase and serum carotenoid concentrations (Forman et al. 1996, Heber 1996). Women on hormone replacement therapy were excluded from the first study but included in the third provided they were on hormone replacement therapy for longer than a year and did not change the management programme during the study period. These sorts of considerations are important to achieve valid and meaningful results.

STUDY DESIGNS

The study protocol for the three studies was similar and is presented in Fig. 2. They were based on a randomized crossover dietary intervention trial with a weight control protocol. The differences between the three

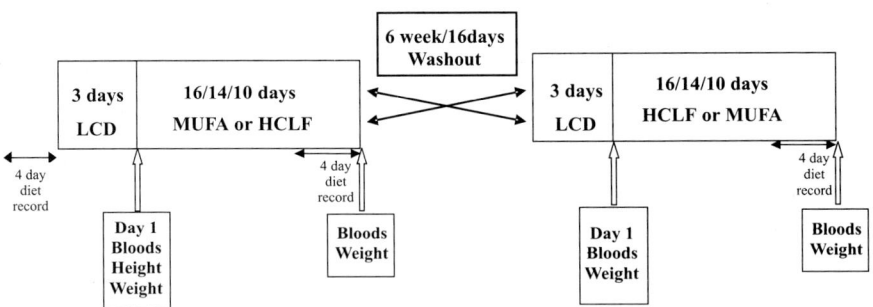

Fig. 2 Study design protocol. Figure modified from Ahuja et al. (2003), © 2003 Nature Publishing Group. LCD, low carotenoid diet especially low in lycopene; MUFA, high lycopene, high monounsaturated fat diet; HCLF, high lycopene, high carbohydrate diet. Dietary periods of 16, 14 and 10 d in Studies 1, 2 and 3 respectively. Washout period of 6 wk on Study 1 and 2 and 16 d on Study 3.

trials included duration of the dietary period, source of monounsaturated fat, source and amount of lycopene-rich products, and duration of the "washout" period between the two intervention dietary periods. Anthropometric data and blood samples were collected at the start and end of each dietary period. Ethanol intake may affect carotenoid absorption (Forman et al. 1995, Brady et al. 1996), so participants were instructed not to take more than two standard alcoholic drinks per day.

The first study was conducted in women and included dietary periods of 16 d each with a washout period of 6 wk. The second study was conducted in men and included 14 d dietary periods with a washout period of 6 wk. The third study was conducted in men and women and included dietary periods of 10 d each with a washout period of 16 d. Since serum lycopene concentrations rise in response to a high lycopene intake and plateau after about 2 wk of regular consumption of high lycopene diet (Yeum et al. 1996, Frohlich et al. 2006), the studies were of short term by design. The main aim was to investigate the serum lycopene concentration after regular high intake of lycopene-rich products in conjunction with a high monounsaturated fat diet (MUFA) or a high carbohydrate, low fat diet (HCLF). The diets may thus not have been long enough to demonstrate the maximum potential effect on lipids and lipoproteins, but that was a secondary aim of this research. Dietary periods on all the three diets were preceded with 3 d of low carotenoid diets, especially low in lycopene. This was to reduce any acute effects of the usual diets. It was important, as a meal high in carotenoids or lycopene the day/evening before the blood was taken could have increased the levels temporarily and made it difficult to evaluate the subsequent effects (Porrini et al. 1998). A washout period of 6 wk in Study 1 and 2 and of 16 d in Study 3 was included. These times were calculated to ensure that women were in the same phase of menstrual cycle on the second dietary period as they were on the first.

The first study in women provided a high lycopene intake (~24 mg) every second day with 8 mg intake on the day in between. This was to avoid overpowering of the diets with tomatoes and tomato products every day during the intervention diets (Table 3). Responses and results from the first study laid the foundation for the second study. This study was conducted in men, to investigate whether the effects might differ in men from women. In these men the dietary periods were of 14 d duration and the lycopene intake was more regular with intake (from partly different sources) of about 20 mg each day. In the third study, the same amount of lycopene was provided as in the second study but a different source of monounsaturated fat was used to ascertain whether the type of monounsaturated fat had any influence on the results. Lycopene sources in

Table 3 Lycopene intake during the intervention diets.

Product	Serving size	Study 1 Odd days (1, 3...15)	Study 1 Even days (2, 4...16)	Study 2
Guava nectar	125 g	1.5 mg	1.5 mg	
Tomato soup	300 g	15.5 mg		15.5 mg
Canned tomato	100 g	6.7 mg	6.7 mg	
Tomato paste	60 g			4.7 mg
Total		23.7 mg	8.2 mg	20.2 mg
Mean intake/day		15.8 mg		20.2 mg

For studies 1 and 2, the lycopene analysis of the products was carried out using high pressure liquid chromatography. It is expected that the lycopene content in study 3 was very similar to that in study 2 because the same amount and variety of products were used, although from different batch and canning. Data reproduced with permission from Ahuja et al. (2003a), © the Biochemical Society.

the first study included guava juice, tomato soup and tomato puree, while in the second and third studies they were tomato soup and tomato paste. An oleate-enriched variant of sunflower oil (Sunola™ oil) was used as the major source of fat in the first two studies. The monounsaturated fat component of this oil is 77.8%, compared to 19.5% in the conventional sunflower oil (Crisco oil) that was used sparingly during the high carbohydrate, low fat period. Extra-light olive oil, a rich source of monounsaturated fatty acids, was used as the main source of fat in the third study. These cooking oils rich in monounsaturated fatty acid were used rather than extra-virgin olive oil because they are more accessible, cheaper and hence more affordable to a larger population, milder in taste, and easier to use in cuisines other than just the Mediterranean. On the negative aspect, these oils contain less polyphenols, which are known to provide enhanced antioxidant activity to extra-virgin olive oil. Oleate-enriched sunflower oils do contain a higher content of vitamin E compared to olive oil. No effort was made to equalize vitamin E content between the two dietary periods in these studies, because foods rich in vitamin E could not be added as they are usually high in fat, and the use of vitamin supplements would have changed the emphasis of the studies, which was to compare diets containing the same basic foods but with different amounts of oils.

The dietary protocols for the three studies were controlled rigorously. As carotenoids can interact and influence one another's absorption and metabolism, the diets were designed to be low in carotenoid-rich foods except for the lycopene-rich tomatoes and tomato products. Fruit intake was restricted to two fruits a day (apple, banana or pear). Vegetable intake was restricted to 100 g frozen green peas and corn mix, and small amounts of mushroom, white onion, radish and iceberg lettuce. Individually packed breakfast cereal (muesli, natural for the low fat diet and toasted with Sunola™/extra-light olive oil on high fat diet) and cooking oils were provided free to the study participants. To provide the equivalent energy on the high carbohydrate diets, polyjoule—a glucose polymer (Nutricia Australia)—was provided to the participants. In all the studies, intakes were designed to be isocaloric to the participants' usual diets. Fibre content of the diets was also controlled because of its potential association with carotenoid absorption (Rock and Swendseid 1992, Erdman et al. 1993, Williams et al. 1998). Lycopene availability from food may depend on a number of factors. Season, heating and processing of tomato products may change the amount of bioavailable lycopene (Stahl and Sies 1992, Lessin et al. 1997, Schierle et al. 1997). To control these aspects, tomatoes and tomato products from the same batch of canning were used in individual studies, and participants were given cooking/heating instructions. Participants were instructed to use preferably the same cooking utensils and heat setting on microwave/stove over the two dietary periods.

The comparison of 35-38% of energy from fat versus 15-17% energy from fat has a practical and scientific basis. According to the WHO (1990) nutrient goals for populations, the lower limit for average fat intake should be 15% of energy from fat. Fat providing less than 15% of the energy signifies a bulky, low energy dense diet leading to an inadequate intake for young active adults (1990). It is also difficult to design a palatable diet for studies with less than 15% of energy from fat. Thirty-six to 38% of energy as fat is feasible, as the intake in the traditional diets of people in Mediterranean countries such as Greece and Crete was about 40% and they had a low rate of CHD (Kromhout 1989). The intakes provided a good comparison of the two extremes of fat intake in the diet on the different parameters measured in the blood.

RESULTS FROM THE STUDIES

Four-day diet records from the two dietary periods showed the anticipated results. As designed, total fat intake was significantly higher and carbohydrate intake was lower on the high lycopene, high fat diet compared to the high lycopene, high carbohydrate diet (Table 4). Similarly, monounsaturated fat intake was significantly higher on the high lycopene, high fat diet, while sugar and starch intake was significantly higher on high lycopene, high carbohydrate diets. Due to the unavailability of an Australian database on the carotenoid content of foods, these nutrients were not analysed from the 4 d diet records. However, the strict control of the diets, especially fruit and vegetable intake, was clearly visible in the individual serum carotenoid results.

Table 4 Average nutrient intake during lycopene-rich high carbohydrate, low fat (HCLF) and lycopene-rich high monounsaturated fat (MUFA) diets.

	Study 1(n = 18)		Study 2(n = 13)		Study 3(n = 18)	
	HCLF	MUFA	HCLF	MUFA	HCLF	MUFA
Energy (MJ)	7.6 ± 1.4	8.4 ± 1.3*	10.2 ± 1.2	11.4 ± 1.9*	8.3 ± 1.4	8.5 ± 1.5
Protein (% of energy)	19.8 ± 1.5	17.4 ± 1.8	18.1 ± 1.6	15.9 ± 1.2*	18.1 ± 2.2	16.7 ± 2.5
Fat (% of energy)	16.6 ± 1.9	38.2 ± 2.1*	14.2 ± 2.8	37.9 ± 2.2*	16.8 ± 4.1	35.2 ± 3.6
Monounsaturated fat (% of fat)	35.2 ± 2.3	71.9 ± 1.4*	33.4 ± 1.8	66.6 ± 2.0*	37.9 ± 4.9	56.9 ± 3.1*
Carbohydrate (% of energy)	60.0 ± 2.9	42.0 ± 2.2*	64.6 ± 3.3	42.6 ± 2.7*	64.7 ± 4.8	47.5 ± 4.3*
Sugar (g)	143.1 ± 41.2	97.7 ± 22.4*	179.9 ± 35.0	131.6 ± 31.3*	169.1 ± 33.3	112.2 ± 23.7*
Starch (g)	140.4 ± 28.1	119.2 ± 19.3*	215.7 ± 43.8	169.7 ± 31.4*	147.7 ± 37.4	127.3 ± 34.2*
Cholesterol (mg)	201.4 ± 76.7	206.7 ± 70.9	174.6 ± 42.0	206.9 ± 56.1	124.2 ± 44.7	128.3 ± 58.0
Fibre (g)	27.8 ± 5.6	24.7 ± 4.3*	32.2 ± 3.8	32.0 ± 5.6	32.4 ± 7.5	31.0 ± 7.5

Results are presented as mean ± standard deviation, analysed from 4 d weighed food diet records completed at the end of each dietary period. *Statistically significantly different, p < 0.05 when compared to the HCLF diet. Data reproduced with permission: Study 1 and Study 2 from Ahuja et al. (2003a), © the Biochemical Society; Study 3 from Ahuja et al. (2006), © Elsevier.

Serum lycopene concentrations were similar at the end of high lycopene, high monounsaturated fat diets and high lycopene, high carbohydrate/low fat diets in the individual studies. In Study 1, serum lycopene concentrations increased significantly after 16 d of high lycopene, high fat diet (mean difference 0.11 µmol/L) but were not significantly different from those in the high lycopene, high carbohydrate diet (Table 5). Study 2 showed significant increases in serum lycopene concentrations on both high lycopene, high fat and high lycopene, high carbohydrate diets by 0.55 µmol/L and 0.49 µmol/L, respectively. However, there was no significant difference between the serum lycopene concentrations at the end of the two dietary periods. As in Study 2, serum lycopene concentrations increased significantly in Study 3 by 1.52 µmol/L and 1.58 µmol/L after high lycopene, high fat and high lycopene, high carbohydrate diets, respectively, and there was no significant difference between the final results after the two diets. The change in the levels of other carotenoids indicated good compliance to the diets. Serum lutein and zeaxanthin concentrations remained similar during different study periods, with no significant differences between the ends of the diets in individual studies. Cryptoxanthin significantly reduced during each dietary period in all three studies, with no significant difference between high lycopene, high fat and high lycopene, high carbohydrate dietary periods. Beta-carotene fell significantly on both high lycopene, high fat and high lycopene, high carbohydrate diets in Study 1, while the concentrations remained stable in Studies 2 and 3. Furthermore, there was no significant difference in beta-carotene concentrations between the two dietary periods on the three studies. Serum alpha-carotene concentrations fell significantly in Studies 1 and 2 during both diets, with no significant differences between the dietary periods. In Study 3, serum alpha-carotene concentrations were not significantly altered.

Serum total and LDL cholesterol was reduced significantly on both the high fat and high carbohydrate diets with no significant difference between the ends of the two diets. The mean fall in total cholesterol after the high lycopene, high fat diet was 0.75 mmol/L in both Studies 1 and 2 and 0.3 mmol/L in Study 3. For LDL cholesterol the mean reduction was 0.62, 0.65 and 0.31 mmol/L in Studies 1, 2 and 3, respectively. For the high lycopene, high carbohydrate diets the mean reduction in total cholesterol was 0.55, 0.56 and 0.37 mmol/L, while the fall in LDL cholesterol was 0.35, 0.70 and 0.23 mmol/L after Studies 1, 2 and 3, respectively. Serum triglycerides were reduced significantly after the high lycopene, high fat diet in women (Study 1) and at the end were significantly lower, by 0.17 mmol/L, than at the end of the high lycopene, high carbohydrate diet. On the other hand, triglycerides increased significantly on the high lycopene,

Table 5 Serum carotenoids at the end of lycopene-rich high carbohydrate, low fat (HCLF) and lycopene-rich high monounsaturated fat (MUFA) diets.

Carotenoid (µmol/L)	Study 1 (n = 18)		Study 2 (n = 13)		Study 3 (n = 21)	
	HCLF	MUFA	HCLF	MUFA	HCLF	MUFA
Lycopene	0.33 ± 0.17	0.38 ± 0.18	0.95 ± 0.26	0.99 ± 0.40	3.14 ± 1.08	3.26 ± 1.13
Beta-carotene	0.37 ± 0.38	0.38 ± 0.35	0.49 ± 0.20	0.44 ± 0.21	1.41 ± 0.74	1.33 ± 0.48
Alpha-carotene	0.11 ± 0.11	0.12 ± 0.12	0.09 ± 0.04	0.08 ± 0.07	0.19 ± 0.14	0.16 ± 0.08
Lutein+zeaxanthin	0.30 ± 0.07	0.32 ± 0.09	0.27 ± 0.11	0.27 ± 0.13	0.40 ± 0.15	0.37 ± 0.12
Cryptoxanthin	0.16 ± 0.20	0.16 ± 0.18	0.15 ± 0.10	0.22 ± 0.32	0.28 ± 0.14	0.35 ± 0.25
Alpha-tocopherol	19.85 ± 7.17	22.33 ± 7.40	29.04 ± 7.07	30.20 ± 9.40	40.39 ± 9.66	39.07 ± 7.79
Gamma-tocopherol	1.10 ± 0.55	0.99 ± 0.54	1.36 ± 0.68	1.35 ± 1.14	1.36 ± 0.51	1.31 ± 0.65

Results are presented as mean ± standard deviation. Data reproduced with permission: Study 2 from Ahuja et al. (2003b), © 2003 Nature Publishing Group.

Table 6 Serum lipids and lipoproteins at the end of lycopene-rich high carbohydrate, low fat (HCLF) and lycopene-rich high monounsaturated fat (MUFA) diets.

	Study 1 (n = 18)		Study 2 (n = 13)		Study 3 (n = 21)	
	HCLF	MUFA	HCLF	MUFA	HCLF	MUFA
Total cholesterol (mmol/L)	4.34 ± 1.12	4.27 ± 1.04	4.86 ± 0.63	4.89 ± 0.77	4.72 ± 1.01	4.68 ± 0.74
LDL cholesterol (mmol/L)	2.68 ± 1.08	2.55 ± 0.83	3.13 ± 0.63	3.16 ± 0.68	2.82 ± 0.83	2.72 ± 0.65
HDL cholesterol (mmol/L)	1.22 ± 0.38	1.36 ± 0.44*	0.96 ± 0.20	1.13 ± 0.23*	1.35 ± 0.44	1.45 ± 0.40*
Triglycerides (mmol/L)	1.01 ± 0.40	0.84 ± 0.23*	1.78 ± 0.77	1.37 ± 0.50*	1.21 ± 0.83	1.13 ± 0.79*
LDL : HDL	2.36 ± 1.15	1.99 ± 0.68*	3.39 ± 0.85	2.96 ± 0.94*	2.32 ± 1.09	2.01 ± 0.73*
Lag time (min)	33.7 ± 6.0	41.1 ± 6.1*	37.2 ± 4.6	44.5 ± 6.1*	103.3 ± 25.2	102.6 ± 17.5

Results are presented as mean ± standard deviation. *Statistically significantly different, $p < 0.05$ when compared to the HCLF diet. Data reproduced with permission: Study 1 and Study 2 from Ahuja et al. (2003a), © the Biochemical Society; Study 3 from Ahuja et al. (2006), © Elsevier.

high carbohydrate diet in men (Study 2) and were significantly higher, by 0.41 mmol/L, than at the end of the high lycopene, high monounsaturated fat diet. HDL cholesterol dropped significantly after the high lycopene, high carbohydrate diets in Studies 1 and 2 and was significantly lower by 0.14 and 0.17 mmol/L, respectively, than after the high lycopene, high monounsaturated fat diet. Study 3 also showed significantly higher HDL cholesterol (by 0.10 mmol/L) after the high fat diet than after the high carbohydrate diet.

In addition to the serum carotenoids, lipids and lipoprotein analyses, *in vitro* susceptibility of isolated LDL (in Studies 1 and 2) or whole serum (in Study 3) to copper-induced oxidation was also measured and compared for the end of the diet periods. Lag time, an indicator of the protection of the lipoproteins against oxidation, was calculated as the intercept between baseline (time zero) and the tangent of the absorbance curve during the propagation phase (Fig. 3). In Studies 1 and 2, the lag time was significantly higher after the high lycopene, high monounsaturated fat diet compared to the high lycopene, high carbohydrate diet. No such difference was observed in Study 3.

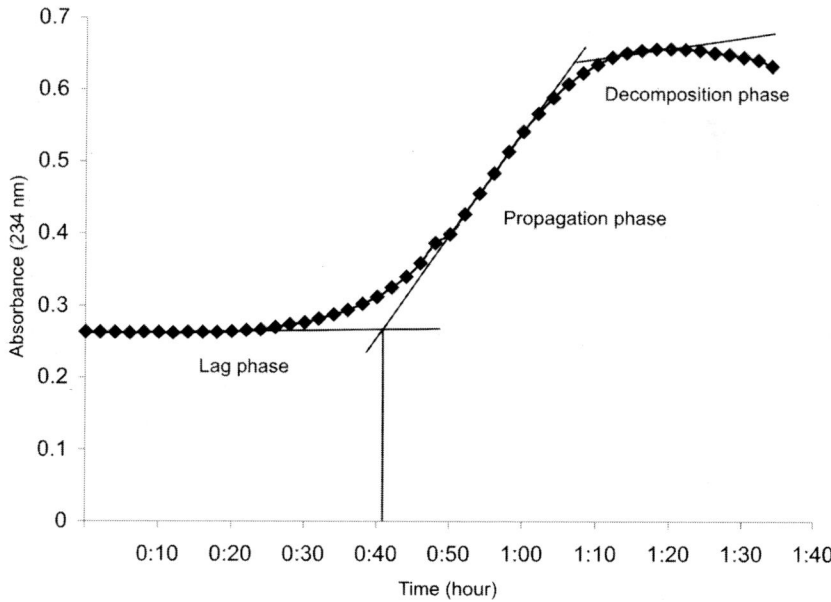

Fig. 3 Oxidation kinetics graph showing the change in absorbance with time due to changes occurring when low density lipoprotein cholesterol or serum is oxidized in a process initiated by the addition of copper.

DISCUSSION

The debate over whether monounsaturated fatty acids or high carbohydrate, low fat diets are most beneficial for their effects on the risk factors for CHD is ongoing. In addition, the influences of other dietary factors including different types of fruits and vegetables, whole grains, and wine have added to the complex debate about the optimum diet to reduce the CHD epidemic. Tomatoes and tomato products are palatable foods, and high consumption of lycopene, a fat-soluble carotenoid, has been suggested to be associated with reduced risk of CHD and some forms of cancer. The studies discussed in the present chapter were devised to investigate and compare the effects of combining tomatoes as a high lycopene source in a high monounsaturated fatty acid or a high carbohydrate, low fat diet. The results confirmed previous data and showed that increased lycopene intake at a high but reasonable concentration leads to increased serum concentration of lycopene and that this effect is quite rapid. Although no difference was observed between the serum lycopene concentrations at the end of the high lycopene, high monounsaturated fat and high lycopene, high carbohydrate diets, clear differences were noticed between the three studies. Serum lycopene was lowest after the first study, higher in the second and highest in the third. There are a number of possible reasons for these differences. Study 1 included approximately 24 mg lycopene intake on odd days of the intervention periods and only 8 mg on the even days, making an average of ~16 mg per day. Serum lycopene concentrations are affected by the immediate/acute intake of lycopene, and the analysis of serum lycopene on day 17, i.e., the day following the lower lycopene intake, may be responsible for the lower concentrations compared to Study 2, where the intake each day was kept at ~20 mg. Study 3 showed even higher serum lycopene concentrations than the previous two studies, but the results were analysed in a different laboratory and with a slightly different methodology. Although all tomato products used in the three studies came from Heinz Australia, the studies were conducted at different times and some of the products were from different batches so it is possible that their lycopene content differed. Irrespective of these differences, the results clearly show that increased serum concentrations of lycopene reflect an increased intake of lycopene.

The studies do not, however, confirm the hypothesis that higher fat intake at the same time as intake of the lycopene-containing food increases the serum lycopene concentrations. In fact, all three studies showed similar serum lycopene concentrations after diets contributing either 15% or 35% of energy from fat. These results are different from a couple of single meal studies and 5 d studies (Stahl and Sies 1992, Fielding et al. 2005) but similar to those from 7 d or 4 wk studies investigating the effects of different amounts of fat intake with other carotenoids such as alpha- and beta-carotene (Jayarajan et al. 1980, Jalal et al. 1998, Ribaya-Mercado et al. 2007). Similar to single meal and 5 d lycopene studies, another investigation showed higher serum concentrations of beta-carotene after 5 d of beta-carotene, high fat combination compared to beta-carotene, low fat combination (Dimitrov et al. 1988). The possibility cannot be excluded that serum lycopene and other carotenoid levels respond differently to lycopene intake with or without fat for the first week; and the high lycopene intake and 14% of energy from fat or 35% of energy from fat per day has no differential effect after 7 or 9 d of regular consumption.

A secondary aspect of these studies was to investigate and compare the effects of a high lycopene, monounsaturated fat-rich and a high lycopene, high carbohydrate, low fat diet on serum lipids and lipoproteins. Although the dietary periods of the three studies were short and the effects on plasma lipids may not have reached their full potential, the results point towards beneficial effects of MUFA diet, especially for its influences on triglycerides and lipoprotein risk ratios (LDL/HDL) for CHD. These results are comparable to other research trials evaluating monounsaturated fat-rich and carbohydrate-rich diets with longer dietary periods, but it is of note that they were less stringently controlled for fruit and vegetable intakes than the intervention trials detailed in this chapter.

Lycopene has been shown to reduce the susceptibility of LDL to oxidation *in vitro*. Similarly, LDL isolated from plasma after individuals have eaten monounsaturated fat diets shows significantly greater resistance to oxidation than samples taken after high carbohydrate, low fat diets. Results from the first two studies discussed in the present chapter also showed a longer lag time before changes occurred in response to the stimulation of oxidation with copper, i.e., greater resistance of LDL to oxidation after the lycopene, monounsaturated fat diet combination than after the lycopene, high carbohydrate diet combination. As dietary intake and serum concentrations of lycopene were similar between the two dietary periods, it is difficult to separate the effects of lycopene on the oxidation susceptibility of LDL from these studies. The third study failed to confirm the results of the two previous studies, possibly because of the different methodologies used. First, serum was used in the *in vitro* test system, not LDL. It is possible that the water-soluble antioxidants in the serum, which are not present in isolated LDL, have a considerable influence and mask any effects of the antioxidant capacity of monounsaturated fatty acids in serum. Higher levels of compounds with antioxidant properties such as phenol and/or vitamin E intakes may be required than were present in the MUFA diet in order to show a difference in serum oxidation. Finally, it is possible that the dietary periods of 10 d duration may not have been long enough to reveal differences in the oxidation susceptibility of serum lipoproteins.

CONCLUSION

The data discussed in this chapter clearly suggest that serum lycopene concentrations increase with increased intake of lycopene irrespective of the amount of fat present in the diet. Combinations of high lycopene and a high carbohydrate, low fat diet show similar effects on total and LDL cholesterol, but the effects on HDL, triglycerides and oxidation susceptibility of lipoproteins tend to favour lycopene-rich, high monounsaturated fat diets.

Further studies are needed to determine whether other dietary combinations optimize the antioxidant capacity of the blood and reduce CHD risk. These results need to be confirmed in longer dietary intervention trials, and such trials should ideally also include measurement of other risk factors for CHD, such as vascular compliance and inflammatory mediators.

ABBREVIATIONS

CHD: coronary heart disease; LDL: low density lipoprotein; HDL: high density lipoprotein; HCLF: high carbohydrate low fat; MUFA: monounsaturated fatty acid

References

Abbey, M. and G.B. Belling, M. Noakes, F. Hirata, and P.J. Nestel. 1993. Oxidation of low-density lipoproteins: intraindividual variability and the effect of dietary linoleate supplementation. Amer. J. Clin. Nutr. 57: 391-398.

Agarwal, S. and A.V. Rao. 1998. Tomato lycopene and low density lipoprotein oxidation: a human dietary intervention study. Lipids 33: 981-984.

Ahuja, K.D. and E.L. Ashton, and M.J. Ball. 2003a. Effect of two lipid lowering, carotenoid controlled diets on oxidative modification of low-density lipoproteins in free-living humans. Clin. Sci. 105: 355-361.

Ahuja, K.D. and E.L. Ashton, and M.J. Ball. 2003b. Effects of a high monounsaturated fat, tomato-rich diet on serum levels of lycopene. Eur. J. Clin. Nutr. 57: 832-841.

Ahuja, K.D. and J.K. Pittaway, and M.J. Ball. 2006. Effects of olive oil and tomato lycopene combination on serum lycopene, lipid profile, and lipid oxidation. Nutrition 22: 259-265.

Albanes, D. and J. Virtamo, P.R. Taylor, M. Rautalahti, P. Pietinen, and O.P. Heinonen. 1997. Effects of supplemental beta-carotene, cigarette smoking, and alcohol consumption on serum carotenoids in the Alpha-Tocopherol, Beta-Carotene Cancer Prevention Study. Amer. J. Clin. Nutr. 66: 366-372.

Ashton, E.L. and J.D. Best, and M.J. Ball. 2001. Effects of monounsaturated enriched sunflower oil on CHD risk factors including LDL size and copper-induced LDL oxidation. J. Amer. Coll. Nutr. 20: 320-326.

Bolton-Smith, C. and C.E. Casey, K.F. Gey, W.C. Smith, and H. Tunstall-Pedoe. 1991. Antioxidant vitamin intakes assessed using a food-frequency questionnaire: correlation with biochemical status in smokers and non-smokers. Br. J. Nutr. 65: 337-346.

Bonanome, A. and A. Pagnan, S. Biffanti, A. Opportuno, F. Sorgato, M. Dorella, M. Maiorino, and F. Ursini. 1992. Effect of dietary monounsaturated and polyunsaturated fatty acids on the susceptibility of plasma low density lipoproteins to oxidative modification. Arterioscler. Thromb. 12: 529-533.

Brady, W.E. and J.A. Mares Perlman, P. Bowen, and M. Stacewicz Sapuntzakis. 1996. Human serum carotenoid concentrations are related to physiologic and lifestyle factors. J. Nutr. 126: 129-137.

Casso, D. and E. White, R.E. Patterson, T. Agurs-Collins, C. Kooperberg, and P.S. Haines. 2000. Correlates of serum lycopene in older women. Nutr. Cancer 36: 163-169.

Castelli, W.P. and R.J. Garrison, P.W. Wilson, R.D. Abbott, S. Kalousdian, and W.B. Kannel. 1986. Incidence of coronary heart disease and lipoprotein cholesterol levels. The Framingham Study. JAMA 256: 2835-2838.

Chopra, M. and M.E. O'Neill, N. Keogh, G. Wortley, S. Southon, and D.I. Thurnham. 2000. Influence of increased fruit and vegetable intake on plasma and lipoprotein carotenoids and LDL oxidation in smokers and nonsmokers. Clin. Chem. 46: 1818-1829.

Clarke, R. 1997. Dietary lipids and blood cholesterol: quantitative meta-analysis of metabolic ward studies. BMJ 314: 112-117.

Dimitrov, N.V. and C. Meyer, D.E. Ullrey, W. Chenoweth, A. Michelakis, W. Malone, C. Boone, and G. Fink. 1988. Bioavailability of beta-carotene in humans. Amer. J. Clin. Nutr. 48: 298-304.

Erdman, J.W., Jr. and T.L. Bierer, and E.T. Gugger. 1993. Absorption and transport of carotenoids. Ann. NY Acad. Sci. 691: 76-85.

Fielding, J.M. and K.G. Rowley, P. Cooper, and K. O' Dea. 2005. Increases in plasma lycopene concentration after consumption of tomatoes cooked with olive oil. Asia Pac. J. Clin. Nutr. 14: 131-136.

Forman, M.R. and G.R. Beecher, E. Lanza, M.E. Reichman, B.I. Graubard, W.S. Campbell, T. Marr, L.C. Yong, J.T. Judd, and P.R. Taylor. 1995. Effect of alcohol consumption on plasma carotenoid concentrations in premenopausal women: a controlled dietary study. Amer. J. Clin. Nutr. 62: 131-135.

Forman, M.R. and G.R. Beecher, R. Muesing, E. Lanza, B. Olson, W.S. Campbell, P. McAdam, E. Raymond, J.D. Schulman, and B.I. Graubard. 1996. The fluctuation of plasma carotenoid concentrations by phase of the menstrual cycle: a controlled diet study. Amer. J. Clin. Nutr. 64: 559-565.

Forman, M.R. and E.J. Johnson, E. Lanza, B.I. Graubard, G.R. Beecher, and R. Muesing. 1998. Effect of menstrual cycle phase on the concentration of individual carotenoids in lipoproteins of premenopausal women: a controlled dietary study. Amer. J. Clin. Nutr. 67: 81-87.

Frohlich, K. and K. Kaufmann, R. Bitsch, and V. Bohm. 2006. Effects of ingestion of tomatoes, tomato juice and tomato puree on contents of lycopene isomers, tocopherols and ascorbic acid in human plasma as well as on lycopene isomer pattern. Br. J. Nutr. 95: 734-741.

Grundy, S.M. and L. Florentin, D. Nix, and M.F. Whelan. 1988. Comparison of monounsaturated fatty acids and carbohydrates for reducing raised levels of plasma cholesterol in man. Amer. J. Clin. Nutr. 47: 965-969.

Heber, D. 1996. Plasma carotenoids and the menstrual cycle. Amer. J. Clin. Nutr. 64: 640.

Jalal, F. and M.C. Nesheim, Z. Agus, D. Sanjur, and J.P. Habicht. 1998. Serum retinol concentrations in children are affected by food sources of beta-carotene, fat intake, and anthelmintic drug treatment. Amer. J. Clin. Nutr. 68: 623-629.

Jayarajan, P. and V. Reddy, and M. Mohanram. 1980. Effect of dietary fat on absorption of beta carotene from green leafy vegetables in children. Indian J. Med. Res. 71: 53-56.

Johnson, E.J. and J. Qin, N.I. Krinsky, and R.M. Russell. 1997. Ingestion by men of a combined dose of beta-carotene and lycopene does not affect the absorption of beta-carotene but improves that of lycopene. J. Nutr. 127: 1833-1837.

Keys, A. 1970. Coronary heart disease in seven countries. Circulation 41: 1-198.

Kromhout, D. 1989. Food consumption patterns in the Seven Countries Study. Seven Countries Study Research Group. Ann. Med. 21: 237-238.

Lee, A. and D.I. Thurnham, and M. Chopra. 2000. Consumption of tomato products with olive oil but not sunflower oil increases the antioxidant activity of plasma. Free Radical Biol. Med. 29: 1051-1055.

Lessin, W. and G. Catigani, and S. Schwartz. 1997. Quantification of *cis-trans* isomers of provitamin A carotenoids in fresh fruits and vegetables. J. Agric. Food Chem. 45: 3728-3732.

Maruyama, C. and K. Imamura, S. Oshima, M. Suzukawa, S. Egami, M. Tonomoto, N. Baba, M. Harada, M. Ayaori, T. Inakuma, and T. Ishikawa. 2001. Effects of tomato juice consumption on plasma and lipoprotein carotenoid concentrations and the susceptibility of low density lipoprotein to oxidative modification. J. Nutr. Sci. Vitaminol. 47: 213-221.

Mata, P. and R. Alonso, A. Lopez Farre, J.M. Ordovas, C. Lahoz, C. Garces, C. Caramelo, R. Codoceo, E. Blazquez, and M. de Oya. 1996. Effect of dietary fat saturation on LDL oxidation and monocyte adhesion to human endothelial cells in vitro. Arterioscler. Thromb. Vasc. Biol. 16: 1347-1355.

Mata, P. and O. Varela, R. Alonso, C. Lahoz, M. de Oya, and L. Badimon. 1997. Monounsaturated and polyunsaturated n-6 fatty acid-enriched diets modify LDL oxidation and decrease human coronary smooth muscle cell DNA synthesis. Arterioscler. Thromb. Vasc. Biol. 17: 2088-2095.

Mayne, S.T. and B. Cartmel, F. Silva, C.S. Kim, B.G. Fallon, K. Briskin, T. Zheng, M. Baum, G. Shor Posner, and W.J. Goodwin Jr. 1999. Plasma lycopene concentrations in humans are determined by lycopene intake, plasma cholesterol concentrations and selected demographic factors. J. Nutr. 129: 849-854.

Meadow Lea Foods (1995). The fatty acid composition of common fats and oils. Fats and oils: The facts. Australia, Meadow Lea Foods Advisory Centre: 1.7.

Mensink, R.P. and M.B. Katan. 1989. Effect of a diet enriched with monounsaturated or polyunsaturated fatty acids on levels of low-density and high-density lipoprotein cholesterol in healthy women and men. N. Engl. J. Med. 321: 436-441.

Mensink, R.P. and M.B. Katan. 1992. Effect of dietary fatty acids on serum lipids and lipoproteins. A meta-analysis of 27 trials. Arterioscler. Thromb. 12: 911-919.

Micozzi, M.S. and E.D. Brown, B.K. Edwards, J.G. Bieri, P.R. Taylor, F. Khachik, G.R. Beecher, and J.C. Smith Jr. 1992. Plasma carotenoid response to chronic intake of selected foods and beta-carotene supplements in men. Amer. J. Clin. Nutr. 55: 1120-1125.

Olmedilla, B. and F. Granado, S. Southon, A.J.A. Wright, I. Blanco, E. Gil-Martinez, H. van den Berg, D. Thurnham, B. Corridan, M. Chopra, and I. Hininger. 2002. A European multicentre, placebo-controlled supplementation study with alpha-tocopherol, carotene-rich palm oil, lutein or lycopene: analysis of serum responses. Clin. Sci. 102: 447-456.

Paetau, I. and F. Khachik, E.D. Brown, G.R. Beecher, T.R. Kramer, J. Chittams, and B.A. Clevidence. 1998. Chronic ingestion of lycopene-rich tomato juice or lycopene supplements significantly increases plasma concentrations of lycopene and related tomato carotenoids in humans. Amer. J. Clin. Nutr. 68: 1187-1195.

Parker, R.S. 1996. Absorption, metabolism, and transport of carotenoids. FASEB J. 10: 542-551.

Porrini, M. and P. Riso, and G. Testolin. 1998. Absorption of lycopene from single or daily portions of raw and processed tomato. Br. J. Nutr. 80: 353-861.

Rao, A.V. and S. Agarwal. 1998. Effect of diet and smoking on serum lycopene and lipid peroxidation. Nutr. Res. 18: 713-721.

Reaven, P. and S. Parthasarathy, B.J. Grasse, E. Miller, F. Almazan, F.H. Mattson, J.C. Khoo, D. Steinberg, and J.L. Witztum. 1991. Feasibility of using an oleate-rich diet to reduce the susceptibility of low-density lipoprotein to oxidative modification in humans. Amer. J. Clin. Nutr. 54: 701-706.

Ribaya-Mercado, J.D. and C.C. Maramag, L.W. Tengco, G.G. Dolnikowski, J.B. Blumberg, and F.S. Solon. 2007. Carotene-rich plant foods ingested with minimal dietary fat enhance the total-body vitamin A pool size in Filipino schoolchildren as assessed by stable-isotope-dilution methodology. Amer. J. Clin. Nutr. 85: 1041-1049.

Roberts, W.C. 1989. Atherosclerotic risk factors — are there ten or is there only one? Amer. J. Cardiol. 64: 552-554.

Rock, C.L. and M.E. Swendseid. 1992. Plasma beta-carotene response in humans after meals supplemented with dietary pectin. Amer. J. Clin. Nutr. 55: 96-99.

Rock, C.L. and M.E. Swendseid, R.A. Jacob, and R.W. McKee. 1992. Plasma carotenoid levels in human subjects fed a low carotenoid diet. J. Nutr. 122: 96-100.

Roidt, L. and E. White, G.E. Goodman, P.W. Wahl, G.S. Omenn, B. Rollins, and J.M. Karkeck. 1988. Association of food frequency questionnaire estimates of vitamin A intake with serum vitamin A levels. Amer. J. Epidemiol. 128: 645-654.

Schaefer, E.J. and A.H. Lichtenstein, S. Lamon-Fava, J.R. McNamara, M.M. Schaefer, H. Rasmussen, and J.M. Ordovas. 1995. Body weight and low-density lipoprotein cholesterol changes after consumption of a low-fat ad libitum diet. JAMA 274: 1450-1455.

Schierle, J. and W. Bretzel, I. Buehler, N. Faccin, D. Hess, K. Steiner, and W. Schueep. 1997. Content and isomeric ratio of lycopene in food and human blood plasma. Food Chem. 59: 459-465.

Stahl, W. and H. Sies. 1992. Uptake of lycopene and its geometrical isomers is greater from heat-processed than from unprocessed tomato juice in humans. J. Nutr. 122: 2161-2166.

Stamler, J. and D. Wentworth, and J.D. Neaton. 1986. Is relationship between serum cholesterol and risk of premature death from coronary heart disease continuous and graded? Findings in 356,222 primary screenees of the Multiple Risk Factor Intervention Trial (MRFIT). JAMA 256: 2823-2828.

Steinberg, D. and S. Parthasarathy, T.E. Carew, J.C. Khoo, and J.L. Witztum. 1989. Beyond cholesterol. Modifications of low-density lipoprotein that increase its atherogenicity. N. Engl. J. Med. 320: 915-924.

Sutherland, W.H. and R.J. Walker, S.A. De Jong, and J.E. Upritchard. 1999. Supplementation with tomato juice increases plasma lycopene but does not alter susceptibility to oxidation of low-density lipoproteins from renal transplant recipients. Clin. Nephrol. 52: 30-36.

Turley, M.L. and C.M. Skeaff, J.I. Mann, and B. Cox. 1998. The effect of a low-fat, high-carbohydrate diet on serum high density lipoprotein cholesterol and triglyceride. Eur. J. Clin. Nutr. 52: 728-732.

Upritchard, J.E. and W.H. Sutherland, and J.I. Mann. 2000. Effect of supplementation with tomato juice, vitamin E, and vitamin C on LDL oxidation and products of inflammatory activity in type 2 diabetes. Diabetes Care 23: 733-738.

White, W.S. and K.M. Peck, T.L. Bierer, E.T. Gugger, and J.W. Erdman Jr. 1993. Interactions of oral beta-carotene and canthaxanthin in ferrets. J. Nutr. 123: 1405-1413.

WHO. 1990. Diet, Nutrition and the Prevention of Chronic Diseases. WHO, Geneva.

Williams, A.W. and T.W. Boileau, and J.W. Erdman Jr. 1998. Factors influencing the uptake and absorption of carotenoids. Proc. Soc. Exp. Biol. Med. 218: 106-108.

Yeum, K.J. and S.L. Booth, J.A. Sadowski, C. Liu, G. Tang, N.I. Krinsky, and R.M. Russell. 1996. Human plasma carotenoid response to the ingestion of controlled diets high in fruits and vegetables. Amer. J. Clin. Nutr. 64: 594-602.

2.9

Lycopene: Cataract and Oxidative Stress

S.K. Gupta, Sushma Srivastava, Renu Agarwal and Shyam Sunder Agrawal

Delhi Institute of Pharmaceutical Sciences & Research
Pushp Vihar, Sector 3, MB Road, New Delhi 110 017

ABSTRACT

Nutritional health of the people has become the prime concern of health authorities in many countries. In the developing countries of Africa and South East Asia, the majority of the people suffer nutritional deficiencies resulting in various health disorders. The therapeutic efficacy of dietary supplements, especially antioxidants, is being explored for almost all diseases. Cataract, the opacification of the eye lens, is one disease in which oxidative stress plays a major role. A large number of antioxidants have been investigated for their role in prevention of development and progression of cataract. Lycopene is one potent antioxidant that has been reported to protect against the development as well as progression of cataract. This chapter presents the pathophysiology of cataract development, the role of antioxidants, and *in vivo* as well *in vitro* experiments to reveal potential benefits of lycopene in medical management of cataract. In view of the results of the above-mentioned studies, nutritional antioxidants such as lycopene are now being given prime importance in therapeutics, since they are already being used by the population and the risk of adverse events is less on long-term administration.

A list of abbreviations is given before the references.

INTRODUCTION

Blindness is currently a major concern. Approximately 40 million people are blind worldwide and in 48% of those cases blindness is caused by cataract, the degenerating opacity of the lens. Depending on the etiology, various types of cataract have been described, including age-related or senile cataract, cataract secondary to ocular or systemic condition, i.e., secondary cataract, traumatic cataract, and congenital cataract.

PATHOPHYSIOLOGY OF CATARACT

A healthy lens is transparent, as the lens fibers are arranged in a regular pattern with very few extracellular spaces. These fibers have a high concentration of low molecular weight, long-lived proteins called crystallins, which do not scatter light independently (Brown 2001). α-Crystallin is a major component of lens fiber and it belongs to the small heat shock protein family that prevents denaturation and aggregation of proteins in cells under stress (Wistow 1985). With increasing age, unfolding and cross-linking of proteins occur, forming high molecular weight aggregates. As the molecular weight increases the refractive index of the lens increases and so does the turbidity. Several factors such as exposure to light radiation, besides aging, predispose the lens to this protein denaturation as a result of exaggerated free radical formation. The normal lens has a stock of antioxidants such as vitamins C and E, carotenoids and glutathione and antioxidant enzymes such as superoxide dismutase, catalase, glutathione peroxidase and glutathione reductase providing an antioxidant defense system. In addition, proteolytic enzymes or proteases provide a second level of defense (Kyselova et al. 2004). Among all antioxidant defenses, the role of glutathione (GSH), a tripeptide thiol in the lens, has extensively been investigated. The levels of GSH decrease with age. Deficient GSH levels contribute to a faulty antioxidant defense system (Head 2001) of the eye lens. Calcium influx and protection of lens proteins against damaging effects of sugars are controlled by GSH (Highfower 1986, Ajiboye and Harding 1989, Ross et al. 1983). The protective activity of GSH and its enzyme cofactors is attributed to its role in metabolism within the lens. Glutathione prevents the formation of cataract in the following ways:

1. It maintains sulfhydryl groups on proteins in their reduced form and hence prevents disulfide cross-linkages.
2. It protects sulfhydryl groups involved in active transport and membrane permeability.

3. It prevents oxidative damage from hydrogen peroxide (Reddy and Giblin 1984).

In diabetic cataract, along with the oxidative stress, polyol formation and non-enzymatic glycation are also implicated. High levels of sugar in diabetics convert glucose to sorbitol or galactose to galactitol, which accumulate in the lens and produce an osmotic gradient. Deleterious changes in electrolyte concentrations lead to osmotic imbalance and finally cataract. In diabetic rats, accumulation of sorbitol in the lens has been reported to produce osmotic effect bringing in water and resulting in the swelling and opacification of the lens. The enzyme responsible for these chains of reactions is aldose reductase. However, its role is controversial; in case of human diabetic cataracts, the levels of aldose reductase in the lens are undetectable and Km (Michaelis Menten constant) for glucose, a natural substrate in the lens, is very high (Gupta et al. 1997). Apart from polyol pathway, some portion of the excess glucose reacts with proteins non-enzymatically and increases the rate of non-enzymatic glycation. The formation of advanced glycation end products begins with the formation of Schiff's base and amadori product. Several reports indicate the role of advanced glycation end products in accelerated cataractogenesis in hyperglycemic experimental animals and diabetic humans. Various morphological types of cataract have one or another predominant mechanism but finally these mechanisms exert their influence through oxidation of lens proteins and lipid peroxidation. (Davies and Truscott 2001, Harding 2002).

OXIDATIVE STRESS AND CATARACTOGENESIS

Oxidative stress has been implicated in a wide range of age-related diseases such as cardiovascular diseases, neurodegenerative diseases, arthritis and cataract (Ames et al. 1993, Salganik et al. 1994a, b, c, d, Salganik 2001). Several studies have supported the fundamental role of oxidative stress in cataractogenesis (Spector and Garner 1981). Oxidative stress has been implicated as a major factor leading to age-related opacification of ocular lens. The avascular and encapsulated lens is maintained in an environment of comparatively low oxygen tension, i.e., less than 30 mmHg, just sufficient to support some aerobic lens metabolism (Kwan et al. 1972). Exposure of lens to light rays of appropriate wavelength under certain conditions can lead to generation of reactive oxygen species and it has been observed that exposure to high levels of reactive oxygen species such as H_2O_2 leads to development of lens opacities in organ culture (Spector et al. 1993). Furthermore, increased

H_2O_2 levels are used as typical biochemical markers in cataract and a good correlation has been observed with oxidation of other cataractous lens components (Spector and Garner 1981).

ROLE OF ANTIOXIDANTS AND PHARMACOLOGICAL PREVENTION OF CATARACT

Surgery is the only effective remedy at present for cataract. Surgical treatment of cataract has its own drawbacks, such as psychological inhibition, postoperative complications, development of glaucoma, economic burden and need for technically trained personnel. Extensive efforts are being made to develop a potent anticataract formulation to arrest the onset and delay the progression of cataract keeping in view the different mechanisms of cataract formation. Hypoglycemic agents, aldose reductase inhibitors, inhibitors of glycation, and calpain inhibitors are being evaluated the world over to reduce the burden of blindness by pharmacological treatment of cataract. Dietary intake of vitamins and antioxidants is believed to have powerful preventive effect on cataractogenesis. Dietary antioxidants, including vitamin C, *Ocimum sanctum* and green tea leaf extract (Ayala and Soderberg 2004, Varma et al. 1995, Reddy et al. 2001, Gupta et al. 2002, 2005) have shown protective activity against cataract in experimental models. As lycopene is known to possess potent antioxidant properties, its role in cataract prevention has also been evaluated in experimental models.

LYCOPENE AND CATARACT

Lycopene is a carotenoid that is responsible for the red color of tomatoes, strawberries, guava, grapes, apricots and other fruits. It is obtained from plants and exists in a thermodynamically stable form (all *trans* configuration). The chemical structure of lycopene is shown in Fig. 1. As humans cannot produce lycopene, they should eat fruits containing lycopene to meet their nutritional requirement. Lycopene is found as an isomeric mixture, with 50% as *cis* isomers in human plasma. In human plasma, the level of lycopene is higher than that of all other dietary carotenoids, indicating its importance in the defense mechanism. It concentrates in adrenal, liver, testes, prostate and low-density fractions of serum, though there is no correlation between the intake of fruits and vegetables and the levels of lycopene in plasma and various tissues. Studies have indicated that lycopene is better absorbed from processed fruits than from raw fruits and vegetables.

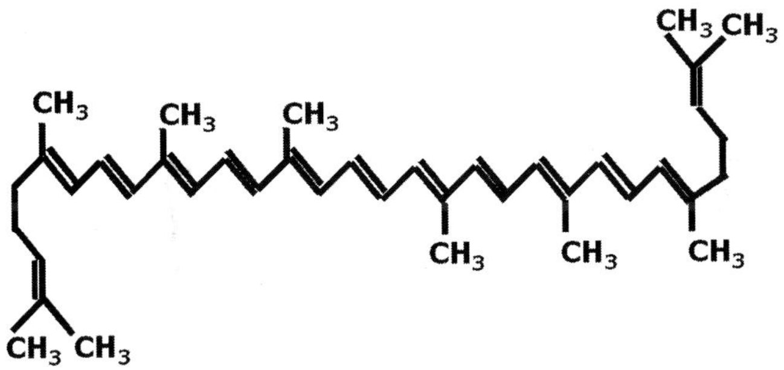

Fig. 1   Chemical structure of lycopene.

Lycopene is also a potent antioxidant and scavenges free radicals, especially those derived from oxygen. Being a singlet oxygen quencher, it is the most potent antioxidant of all carotenoids. It participates in reactions involved in the prevention of carcinogenesis and atherogenesis by protecting critical cellular biomolecules, including lipids, proteins, and DNA, and provides protection from various diseases in which oxidative stress is believed to play a significant role, such as prostate cancer, breast cancer, atherosclerosis, and associated coronary heart disease. Reports suggest that lycopene can also reduce the risk of macular degenerative disease, serum lipid oxidation, and various other cancers. Restoration of antioxidant defense system and inhibition of protein insolubilization by physiological (Gerster 1989) and nutritional (Gupta et al. 2002) antioxidants have been reported and this supports the hypothesis that oxidative stress is a major cataract-inducing factor. In view of this, the attenuation of cataract by herbal and nutritional antioxidants was evaluated by a number of workers (Gupta et al. 2002, 2003, 2005, Mohanty et al. 2002a). Among the nutritional antioxidants, carotenoids especially have been shown to delay the onset and progression of cataract (Jacques and Chylack 1991). Carotenoids such as leutin and zeaxanthin have been detected in the human lens (Yeum et al. 1999). Lycopene has also shown anticataract potential after oral feeding in rats (Pollack et al. 1996). In one of the experiments of Pollack et al. (1999) it was observed that when rats were fed with high galactose diet supplemented with lycopene (0.2, 0.4 and 0.8%) they had decreased incidence of cataract and lower grades of cataract as compared to control rats, which were given high galactose diet. In addition, levels of lens protein and reduced GSH were higher and those of aldose reductase lower in lycopene-supplemented groups than in controls. These results indicate a significant anticataract activity of lycopene in cataract induced by oxidative stress. However, in the same experiment similar anticataract activity was not observed in diabetic rats.

Subsequent portions of this chapter describe the detailed evaluation of anticataract potential of lycopene in experimental models.

EVALUATION OF ANTICATARACT POTENTIAL OF LYCOPENE IN EXPERIMENTAL MODELS

Experimental models of cataract have become an indispensable tool for evaluating the efficacy of an anticataract agent as well as in understanding the underlying mechanism of cataract formation. In these models the process of cataractogenesis can be interrupted at any stage to explain the structural and biochemical changes in the lens. Several *in vitro* and *in vivo* models are being used the world over for research in cataract for indicating crucial targets for intervention and prevention of lens damage. However, the final steps in many of these models may remain the same. The role of lycopene in prevention of onset and progression of cataract has been studied in *in vitro* and *in vivo* experimental models by various researchers including Mohanty et al. (2002a, b) and Gupta et al. (2003, 2005).

Anticataract Activity of Lycopene in vitro

In vitro models are widely used for evaluating the anticataract potential of test drugs. They are easily reproducible and can be produced within a considerably short time span.

Organ culture

Isolated animal lenses, when exposed to various stressful conditions, develop opacities. These conditions can be generated using several exogenous chemicals, which can produce osmotic, hyperglycemic or hypergalactocemic conditions resulting in cataract formation *in vitro*. The cataract can be induced by incorporating galactose, glucose, xylose and other chemicals to the culture medium. Oxidative stress can be produced *in vitro* by incorporating sodium selenite at 25 to 100 µM (Gupta et al. 2003). Photosensitized microquantities of riboflavin (4-100 µM) in culture medium can generate free radicals and cause physiological damage and opacification (Spector et al. 1995). Supplementing the plain culture medium with 50-500 µM H_2O_2 also produces cataract in isolated rodent lenses but the medium should be changed frequently as H_2O_2 is highly unstable (Cui and Lou 1993, Spector et al. 1993).

Gupta et al. (2003) evaluated the role of lycopene in cataract prevention in organ culture. Rat lenses were cultured in Dulbecco's modified Eagle's medium (DMEM) alone or in addition to 100 µM sodium selenite forming normal and control groups, while the test group comprised those cultured in the control medium supplemented with 10 µM lycopene. Being lipophilic, lycopene raised problems in terms of solubility and stability in the aqueous culture medium. Therefore, it was occasionally shaken and suspended as microcrystals and the solution appeared clear to the naked eye. The medium was changed every 6 hr. Almost all the lenses in the control group developed opacity after 24 hr incubation, while in the treatment group a highly significant number of lenses remained clear (Fig. 2).

After 24 hr incubation at 37°C, the lenses were evaluated for morphologic variations and biochemical parameters to test the antioxidant

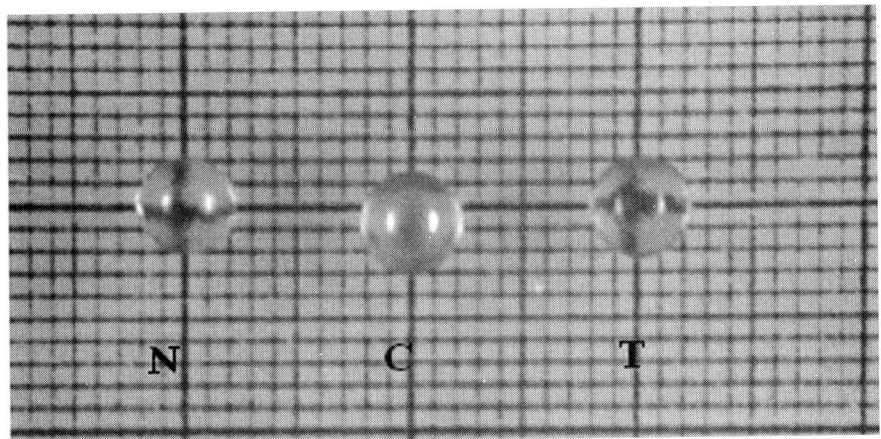

Fig. 2 *In vitro* effect of lycopene in selenite stress. N = normal group, lens incubated in DMEM alone; C = control group, lens incubated in DMEM + 100 µM sodium selenite; T = test group, lens incubated in DMEM + 100 µM sodium selenite +10 µM lycopene.

activity. A fall of 25% in GSH and 32% malondialdehyde (MDA) levels was observed in the control group in comparison to the normal group. In the treatment groups, the levels of both GSH and MDA were significantly restored. Antioxidant enzyme activity was significantly decreased in the control group as compared to normal, whereas in lycopene-supplemented group the antioxidant status was restored.

The study did not demonstrate the mechanism(s) by which lycopene enters the lens fibers. The authors hypothesized that lycopene microcrystals settle from the suspension on the epithelial membrane and being lipid soluble exert action on the surface or gain access to the epithelial cells. Thereafter, lycopene maintains lens membrane integrity by preventing the oxidation of critical sulfhydryl groups.

Cell culture

Human lens epithelial cells (HLEC) play an important role in maintaining the homeostasis and transparency of the lens, as they are the metabolic units of the lens. As they are most anteriorly placed they are the primary sites of external insult leading to cataract formation. Hence, efforts are being made to study the protective effects of potential anticataract drugs on HLEC when exposed to cataractogenic challenges. One of the studies described below assesses the anticataract potential of lycopene in cultured HLEC (Mohanty et al. 2002a).

Human lens epithelial cells obtained from donor eyes were cultured in Eagle's medium supplemented with fetal calf serum (200 ml/l) and antibiotics (100 μg streptomycin and 100 IU penicillin/ml). The primary culture took 10-12 d to attain confluency. When confluency was observed, the cells were treated with trypsin (0.5 mg/ml) and EDTA (0.2 mg/ml) for 8 min and gently dissociated. The cells were now subcultured by transferring 10^6 cells/flask. In the normal group flasks, cells were incubated with modified Eagle's medium with fetal calf serum. The control group flasks had 30 mM galactose added to the medium, whereas in test group flasks, along with galactose, 5, 10 and 20 μM lycopene was also added. Flasks of all groups were maintained in duplicate and incubated for 72 hr. During incubation, lycopene-containing medium was changed every 6 hr. At the end of the incubation period one set of flasks from each group was used to evaluate morphological changes under light and electron microscope, while the other set was used for biochemical studies.

Examination of normal HLEC under phase contrast microscope shows hexagonal cells with centrally placed, prominent nuclei and uniformly distributed cytoplasm. Electron microscopy of these cells showed demarcated cell membranes, nuclei and cytoplasm. Cells exposed to galactose for 72 hr showed prominent intracellular vacuole with an average size of 0.34 ± 0.09 μm. In groups supplemented with lycopene, the size of the vacuole (0.27± 0.08 μm) was significantly reduced (p < 0.01) (Table 1). Similar vacuole formation was observed in HLEC (Lin et al. 1991) and *in vivo* in rat lens epithelium when animals were fed with galactose (Jasmina et al. 1994).

The basal glutathione level observed in normal HLEC was 87.99 μg/mg protein. The glutathione level was found to be significantly reduced in

Table 1 Effect of lycopene on human lens epithelial cells.

	Control	Test group (10 µM lycopene)
Vacuole size	$0.34 \pm 0.090\ \mu m$	$0.27 \pm 0.08\ \mu m*$
Volume	$0.21 \pm 0.003\ \mu m^3$	$0.01 \pm 0.00\ \mu m^{3}*$
Area	$0.37 \pm 0.008\ \mu m^2$	$0.23 \pm 0.05\ \mu m^{2}*$

Electron microscopic examination of HLEC in lycopene (10 µM) group showed significantly reduced vacuole size, volume and area as compared to control (*P < 0.01). Normal HLEC group did not show vacuole formation. All values represent mean ± SD of six samples.

control group with a value of 38.91 µg/mg protein (p < 0.05). Among the lycopene-supplemented groups no change in glutathione level was observed in 5 µM group and a positive modulation with 65.30 and 67.20 µg/mg protein was observed in 10 and 20 µM groups. The fall in glutathione level in control group could be attributed to excessive utilization of NADPH for polyol synthesis and reduced availability of NADPH for glutathione reductase and overstressed antioxidant mechanism.

Estimation of MDA showed significantly increased lipid peroxidation with MDA levels of 2.6 nmol/mg protein in the control group as compared to normal group (1.4 nmol/mg protein). In lycopene-supplemented groups the lipid peroxidation was significantly reduced with MDA levels of 1.90 and 1.82 nmol/mg protein in 10 and 20 µM groups. The exact mechanism by which lycopene prevents lipid peroxidation is not clear. The poorly soluble microcrystalline solution of lycopene probably sticks to the outer surface of cells and prevents exposure to reactive oxygen species.

Lycopene (10 µM) was also found to significantly increase glutathione S-transferase and glutathione peroxidase activity by 88 and 86% respectively and decrease catalase activity by 52% as compared with the control group. Catalase plays an important role in protection against the damage caused by free radicals. Enhanced catalase activity in control group indicates overproduction of hydrogen peroxide in the presence of galactose. Reduced catalase activity in the presence of lycopene might be due to either inhibition of autooxidation of galactose itself leading to reduced generation of hydrogen peroxide or increased utilization of hydrogen peroxide by glutathione S-transferase and glutathione peroxidase, which is reflected in the increased activity of these enzymes.

The results of the studies discussed above clearly demonstrate the anticataract potential of lycopene at very low doses and these beneficial effects can be attributed to modulation of lens antioxidant mechanisms. Similar observations in relation to the protective role of lycopene against oxidative damage to epithelium cortex of human lens have been reported

(Lyle et al. 1999, Yeum et al. 1999). In view of these interesting findings, anticataract activity of lycopene was further evaluated using *in vivo* models of cataract.

Anticataract Activity of Lycopene *in vivo*

Popular animal models for the screening of anticataract agents include models in which cataract is induced by galactose, selenite, streptozotocin, alloxan and radiation as well as transgenic models (Srivastava and Joshi 2004). Sugar cataract can be induced in rats by feeding them high sugar diet including galactose (Gupta et al. 2005) or impairing the production of insulin using agents such as streptozotocin (Rodriguez 1999) and alloxan (Ahmad et al. 1985, Vats et al. 2004).

Galactose-induced cataract

Gupta et al. (2003) investigated the protective role of lycopene in a rat model with galactose-induced cataract. *Wistar* rats of either sex weighing 80-100 g were fed with 30% galactose in diet to induce cataract. The rats in the study group were administered lycopene orally at a dose of 200 µg/kg body weight. Lycopene treatment was started 1 wk before the initiation of 30% galactose diet. Cataractous changes were observed through the slit lamp periodically and were graded using slit lamp according to the method of Sippel (1966). At the end of the experimental period, i.e., 30 d, out of the total eyes, 35% had mature cataract in the test group in comparison to 100% of control group (Table 2).

In models with galactose-induced cataract, model lycopene was found to delay the onset and progression of cataract. Similar findings were reported by Pollack et al. (1996). The different stages of cataract development are as follows and are shown in Fig. 3:

Stage 0: Lenses similar to normal lenses

Stage I: Lenses showing faint peripheral opacity

Stage II: Irregular peripheral opacity with slight involvement of the lens in the center

Stage III: Faint opalescence visible with the naked eye

Stage IV: Mature nuclear cataract

Stage V: Opacity involving entire lens

Selenite-induced cataract

Attenuation of cataract using selenite model was also studied by Gupta et al. (2003) and is shown in Fig. 4. To induce cataract, sodium selenite was injected in rat pups 9 to 14 d old at 19-30 µmoles/kg body weight (Gupta

Table 2 Effect of lycopene in rat model of galactose-induced cataract.

Days on galactose	Stages of cataract	Control group	Test group(lycopene 200 µg/kg)
		% Eyes (n = 20)	% Eyes (n = 20)
Day 7	Normal	–	30%
	I	100%	70%
	II	–	–
	III	–	–
	IV	–	–
	Opacity index	**1**	**0.7****
Day 14	Normal	–	–
	I	–	25%
	II	100%	75%
	III	–	–
	IV	–	–
	Opacity index	**2**	**1.75****
Day21	Normal	–	–
	I	–	55%
	II	–	35%
	III	100%	10%
	IV	–	–
	Opacity index	**3**	**2.55*****
Day 30	Normal	–	–
	I	–	–
	II	–	20%
	III	–	45%
	IV	100%	35%
	Opacity index	**4**	**3.15*****

Lycopene treatment in test group significantly delayed the onset and progression of cataract development as indicated by the percentage of eyes showing various stages of cataract in control and lycopene-treated groups. n = number of eyes. *P < 0.05 as compared to control, **P < 0.01 as compared to control, ***P < 0.001 as compared to control. An overall grade point average was calculated to compare the rate of progression in test group with that of control. For that purpose, normal eyes were not given any points, Stage I was given one point, Stage II two points, Stage III three points and Stage IV four points. To calculate the opacity index, the number of eyes in each group divides the sum of points in the respective group.

et al. 2002, 2005, Shearer et al. 1987, 1992). Pups were housed with the mother, which was fed normal diet and water ad libitum. Pups were segregated in different groups and the sodium selenite was injected subcutaneously at the scruff of the neck. The treatment group was administered lycopene 200 µg/kg body weight intraperitoneally a few hours before the selenite administration. When the pups first opened their

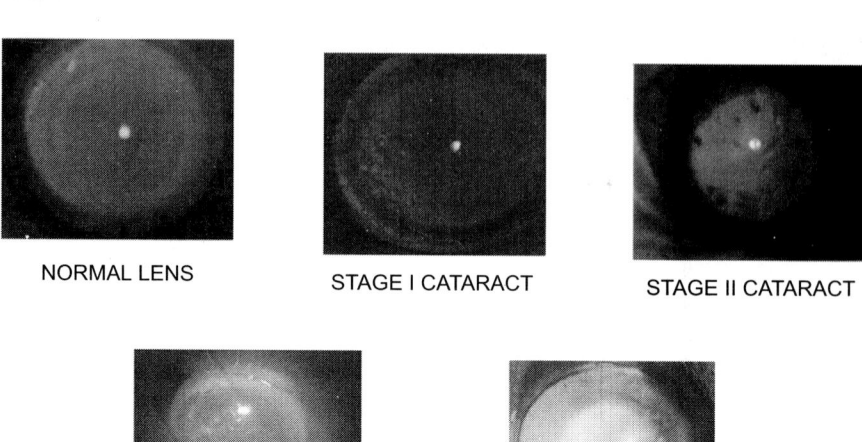

NORMAL LENS   STAGE I CATARACT   STAGE II CATARACT

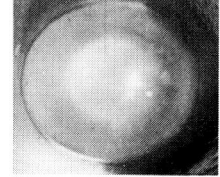

STAGE III CATARACT   STAGE IV CATARACT

Fig. 3 Stages of cataract development in galactose-induced model in rats.

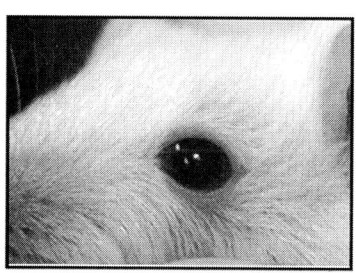
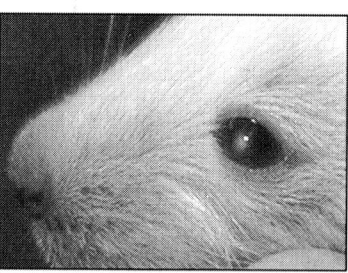

NORMAL RAT EYE   NUCLEAR CATARACT IN CONTROL GROUP

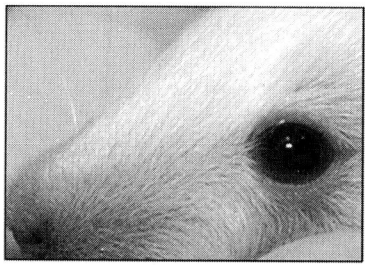

CLEAR LENS IN LYCOPENE TREATED GROUP

Fig. 4 Anticataract effect of lycopene in selenite-induced cataract in rats

eyes, incidence of bilateral cataract was compared between the group treated with selenite alone and that treated with selenite with lycopene. As compared to 83.33% of the control group, only 9.10% of eyes in the treatment group had dense nuclear opacity. In addition, the treatment group had 18.20% eyes absolutely clear, indicating that lycopene protected the lens morphology and clarity (Table 3).

Table 3 Anticataract activity of lycopene in selenite model.

Group	Clear	Pinpoint opacity	Dense nuclear opacity
	% Eyes	% Eyes	% Eyes
Control (n = 24)	–	16.67	83.33
Lycopene (200 µg/kg body weight) (n = 22)	18.20	72.70	9.10

In 9-d-old rat pups, sodium selenite subcutaneous injection at 25 µmoles/kg body weight developed 100% opacity in the control group with 83.33% eyes having dense opacity, while a single intraperitoneal administration of lycopene showed remarkable anticataract activity compared to the control group. 18.20% eyes were normal and only 9.10% developed dense cataract. n = number of eyes.

CONCLUSION

Lycopene, which has repeatedly been shown to possess potent antioxidant properties, has been studied for its therapeutic potential in a wide range of diseases for which oxidative stress is a major factor in pathogenesis, such as atherosclerosis. In addition, the experiments described above clearly demonstrate the potential of lycopene as an anticataract agent. Owing to its potent antioxidant properties, lycopene seems to provide remarkable prophylactic as well as curative activity against senile cataracts. However, research in well-controlled double-blind clinical trials is required to further substantiate the role of lycopene in cataract prevention.

ABBREVIATIONS

DMEM: Dulbecco's modified Eagle's medium; GSH: Glutathione; HLEC: Human lens epithelial cells; IU: International Units; MDA: Malondialdehyde; NADPH: Reduced nicotinamide adenosine diphosphate

References

Ahmad, S.S. and K.C. Tsou, S.I. Ahmad, M.A. Rahman, and T.H. Kirmani. 1985. Studies on cataractogenesis in humans and in rats with alloxan-induced diabetes. I. Cation transport and sodium potassium dependent ATPase. Ophthalmic Res. 17: 1-11.

Ajiboye, R. and J.J Harding. 1989. The nonenzymatic glycosylation of bovine lens proteins by glucosamine and its inhibition by aspirin, ibuprofen and glutathione. Exp. Eye Res. 49: 31-41.

Ames, B.N and M.K. Shigenaga, and T.M. Hagen. 1993. Oxidants, antioxidants, and the degenerative diseases of aging. Proc. Natl. Acad. Sci. USA 90: 7915-7922.

Ayala, M.N. and P.G. Söderberg. 2004. Vitamin E can protect against ultraviolet radiation-induced cataract in albino rats. Ophthalmic Res. 36: 264-269.

Brown, N.P. 2001. Mechanisms of cataract formation. OT. 4: 27-31.

Cui, X.L. and M.F. Lou. 1993. The effect and recovery of long term H_2O_2 exposure on lens morphology and biochemistry. Exp. Eye Res. 57: 157-167.

Davies M.J. and R.J. Truscott. 2001. Photo-oxidation of proteins and its role in cataractogenesis. J. Photochem. Photobiol. 63: 114-125.

Gerster, H. 1989. Antioxidant vitamins in cataract prevention. Z. Emahrungswiss. **28**: 56-57.

Gupta, S.K. and S. Joshi, T. Velpandian, A. Len, and J. Prakash. 1997. An update on pharmacological prospectives for prevention and development of cataract. Indian J. Pharmacol. 29: 3-10.

Gupta, S.K. and N. Halder, S. Srivastava, and D. Trivedi, S. Joshi, and S.D. Varma. 2002. Green tea (*Camellia sinensis*) protects against selenite-induced oxidative stress in experimental cataractogenesis. Ophthalmic Res. 34: 258-263.

Gupta, S.K. and D. Trivedi, S. Srivastava, S. Joshi, N. Halder, and S.D. Varma. 2003. Lycopene attenuates oxidative stress induced experimental cataract development: An *in vivo* and *in vitro* study. Nutrition 19: 794-799.

Gupta, S.K. and S. Srivastava, D. Trivedi, S. Joshi, and N. Halder. 2005. Ocimum sanctum modulates selenite-induced cataractogenic changes and prevents rat lens opacification. Curr. Eye Res. 30: 583-591.

Harding, J.J. 2002. Viewing molecular mechanisms of ageing through lens. Aging Res. Rev. 1: 465-479.

Head, K. 2001. Natural therapies for ocular disorders. Part two: cataracts and glaucoma. Alt. Med. Rev. 6: 141-164.

Highfower, K.R. 1986. Superficial membrane-SH groups inaccessible by intracellular GSH. Curr. Eye Res. 5: 421-427.

Jacques, P.F. and L.T. Chylack. 1991. Epidemiological evidence of a role for the antioxidant vitamins and carotenoids in cataract prevention. Amer. J. Clin. Nutr. 53: 352.

Jasmina, B.M. and I. Bekhor, M.H. Weiss, and B.V. Zlokovic. 1994. Galactose induced cataract formation in guinea pigs: Morphological changes and accumulation of galactitol. Invest. Ophthalmol. Vis. Sci. 35: 804-811.

Kwan, M. and J. Ninikoshi, and T.K. Hunt. 1972. *In vivo* measurements of oxygen tension in the cornea, aqueous humor and anterior lens of the open eye. Invest. Ophthalmol. 11: 108-114.

Kyselova, Z. and M. Stefek, and V. Bauer. 2004. Pharmacological prevention of diabetic cataract. J. Diabetes Complications 18: 129-140.

Lin, L.R. and V.N. Reddy, F.J. Giblin, P.F. Kador, and J.H. Kinoshita. 1991. Polyol accumulation in cultured human lens epithelial cells. Exp. Eye Res. 52: 93-100.

Lyle, B.J. and J.A. Mares-Perlman, B.E. Klein, R. Klein, M. Palta, P.E. Bowen, and J.L. Greger. 1999. Serum carotenoids and tocopherols and incidence of age related nuclear cataract. Amer. J. Clin. Nutr. 69: 272-277.

Mohanty, I. and S. Joshi, D. Trivedi, S. Srivastava, and S.K. Gupta. 2002a. Lycopene prevents sugar induced morphological changes and modulates antioxidant status of human lens epithelial cells. Br. J. Nutr. 88: 347-354.

Mohanty, I. and S. Joshi, D. Trivedi, S. Srivastava, R. Tandon, and S.K. Gupta. 2002b. Pyruvate modulates antioxidant status of cultured human lens epithelial cells under hypergalactosemic conditions. Mol. Cell. Biochem. 238: 129-135.

Pollack, A. and Z. Madar, Z. Eisner, A. Nyska, and P. Oren. 1996. Inhibitory effect of lycopene on cataract development in galactosemic rats. Metab. Pediatr. Syst. Ophthalmol. 19-20: 31-36.

Pollack, A. and P. Oren, A.H. Stark, Z. Eisner, A. Nyska, and Z. Madar. 1999. Cataract development in sand and galactosemic rats fed a natural tomato extract. J. Agric. Food Chem. 47: 5122-5126.

Reddy, V.N. and F.J. Giblin. 1984. Metabolism and function of glutathione in the lens. Ciba Found Symp. 106: 65-87.

Reddy, B. and S. Nayak, P.Y. Reddy, and K.S. Bhat. 2001. Reduced levels of rat lens antioxidant vitamins upon *in vitro* UVB irradiation G. J. Nutr. Biochem. 12: 121-124.

Rodrigues, B. 1999. Streptozotocin-induced diabetes: Induction and mechanism(s) and dose dependency. pp. 3-17. *In:* J.H. Mc Neill [Ed.]. Experimental Models of Diabetes. CRC Press LLC, Florida.

Ross, W.M. and M.O. Creighton, and J.R. Trevithick. 1983. Modelling cortical cataractogenesis: VI. Induction by glucose *in vitro* or in diabetic rats: prevention and reversal by glutathione. Exp. Eye Res. 337: 559-573.

Salganik, R.I. 2001. The benefits and hazards of antioxidants: controlling apoptosis and other protective mechanisms in cancer patients and the human population. J. Amer. Col. Nutr. 20: 464S-472S.

Salganik, R.I. and N.A. Solovyova, O.N. Grishaeva, S.I. Dikalov, V.V. Kandaurov, and L.A. Semenova. 1994a. Inherited increase of free radical production in rat: development of pathological conditions. Free Radic. Biol. Med. 16: 13-14.

Salganik, R.I. and N.A. Solovyova, O.N. Grishaeva, S.I. Dikalov, L.A. Semenova, and A.V. Popovsky. 1994b. Inherited hyperproduction of free radicals. The pathology of aging. Dokl. Russ. Akad. Nauk (Proc. Russ. Acad. Sci.) 336: 255-258.

Salganik, R.I. and N.A. Solovyova, S.I. Dikalov, O.N. Grishaeva, L.A. Semenova, and A.V. Popovsky. 1994c. Inherited enhancement of hydroxyl radical generation and lipid peroxidation in the S strain rats results in DNA rearrangements, degenerative diseases, and premature aging. Biochem. Biophys. Res. Commun. 199: 726-733.

Salganik, R.I. and I.G. Shabalina, N.A. Solovyova, N.G. Kolosova, V.N. Solovyov, and A.R. Kolpakov. 1994d. Impairment of respiratory functions in mitochondria of rats with an inherited hyperproduction of free radicals. Biochem. Biophys. Res. Commun. 205: 180-185.

Shearer, T.R. and L.L. David, and R.S. Anderson. 1987. Selenite cataract. A review. Curr. Eye Res. 6: 289-300.

Shearer, T.R. and L.L. David, R.S. Anderson and M. Azuma. 1992. Review of selenite cataract. Curr. Eye Res. 11: 357-369.

Sippel, T.O. 1966. Changes in the water, protein and glutathione contents of the lens in the course of galactose cataract development in rats. Invest. Ophthalmol. 5: 568-575.

Spector, A. and W.H. Garner. 1981. Hydrogen peroxide and human cataract. Exp. Eye Res. 33: 673-681.

Spector, A. and G.M. Wang, R.R. Wang, W.H. Garner, and H. Moll. 1993. The prevention of cataract caused by oxidative stress in cultured rat lenses. I. H_2O_2 and photochemically induced cataract. Curr. Eye Res. 12: 163-179.

Spector, A. and G.M. Wang, R-R. Wang, W-C. Li, and J.R. Kuszak. 1995. A brief photochemically induced oxidative insult causes irreversible lens damage and cataract 1. Transparency and epithelial cell layer. Exp. Eye Res. 60: 471-481.

Srivastava, S. and S. Joshi. 2004. Anticataract agents. pp. 333-346. *In:* S.K. Gupta [Ed.] Drug Screening Methods. Jaypee Brothers Medical Publishers (P) Ltd., New Delhi.

Varma, S.D. and P.S. Devamanoharan, and S.M. Morris. 1995. Prevention of cataracts by nutritional and metabolic antioxidants. Crit. Rev. Food Sci. Nutr. 35: 111-129.

Vats, V. and S.P. Yadav, N.R. Biswas, and J.K. Grover. 2004. Anti-cataract activity of *Pterocarpus marsupium* bark and *Trigonella foenum-graecum* seeds extract in alloxan diabetic rats. J. Ethnopharmacol. 93: 289-294.

Wistow, G. 1985. Domain structure and evolution in alpha-crystallins and small heat-shock proteins. FEBS Lett. 181: 1-6.

Yeum, K.J. and F.M. Shang, W.M. Schalch, R.M. Russel, and A. Taylor. 1999. Fat soluble nutrient concentrations in different layers of human cataractous lens. Curr. Eye Res. 19: 502-505.

2.10

Lycopene and Bone Tissue

[1]L.G. Rao, [1]E.S. Mackinnon and [2]A.V. Rao
Calcium Research Laboratory, Division of Endocrinology & Metabolism &
St. Michael's Hospital and Department of Medicine[1] and
Department of Nutritional Sciences[2]
University of Toronto, Ontario, Canada

ABSTRACT

Lycopene, a potent antioxidant present primarily in tomatoes and tomato products, is now proven to be important in bone health on the basis of epidemiological data, clinical studies and *in vitro* cell culture studies. Lycopene stimulates the growth and differentiation of the bone-forming osteoblasts through mechanisms not yet understood. Lycopene also inhibits the formation of osteoclasts and their ability to resorb bone by inhibiting the formation of reactive oxygen species. The evidence for the involvement of lycopene in bone health is presented through epidemiological and cross-sectional studies on men and premenopausal and postmenopausal women. Our ongoing clinical study is the first to evaluate lycopene from nutritional supplements and tomato juice for its ability to decrease the risk of osteoporosis in postmenopausal women. This review also includes the effects of oxidative stress and antioxidants in general, and the role played by lycopene in particular, on bone tissue and bone cells in culture. The final results of our study may indicate that lycopene can be used either as a dietary alternative to drug therapy or as a complement to the pharmaceuticals used by women at risk for osteoporosis.

A list of abbreviations is given before the references.

INTRODUCTION

Oxidative stress is now recognized as a major cause of several chronic diseases including cancer, cardiovascular disease, diabetes and macular degenerative disease. Lycopene, a potent carotenoid antioxidant found predominantly in tomatoes and tomato products, has been shown to counteract some of these diseases (reviewed in Giovannucci 1999, Rao and Rao 2004a, b). Although oxidative stress and antioxidants have also been associated with bone diseases, it is only recently that the important role of lycopene in bone health has been recognized. We have included in this review the reported studies on the role of oxidative stress and antioxidants in the bone-forming osteoblast and the bone-resorbing osteoclast, the two major cells involved in the pathogenesis of metabolic bone diseases.

OXIDATIVE STRESS, ANTIOXIDANTS AND BONE

The dynamic nature of bone throughout life is demonstrated by its continuous renewal through remodelling, a process of coupled events of removal of old bone by osteoclasts and formation of new bone by osteoblasts (for review, see Chan and Duque 2002, Cohen 2006, Kenny and Raisz 2002). During the remodelling process, osteoblasts and osteoclasts interact with multiple molecular agents including hormones, growth factors, and cytokines. Metabolic bone diseases result from disturbances in the remodelling process (Lerner 2006, Lindsay and Cosman 1999). As is reviewed below, the pathogenesis of the skeletal system, including osteoporosis, the most prevalent metabolic bone disease, is in part due to the actions of oxidative stress on both osteoclasts and osteoblasts; antioxidants counteract these actions.

Evidence Associating Oxidative Stress and Antioxidants with Osteoblasts

Osteoblasts can be induced to produce intracellular reactive oxygen species (ROS) (Cortizo et al. 2000, Liu et al. 1999), which can cause cell death and a decrease in alkaline phosphatase (ALP) activity (Liu et al. 1999), and these can be partly inhibited by vitamin E (Cortizo et al. 2000). Rat osteosarcoma ROS 17/2.8 cells treated with tumour necrosis factor-alpha (TNF-α) suppressed bone sialoprotein gene transcription through a tyrosine kinase dependent pathway that generates ROS (Samoto et al. 2002). Intracellular calcium (Ca^{2+}) activity in osteoblasts is modulated by hydrogen peroxide (H_2O_2), which increases Ca^{2+} release from the intracellular Ca^{2+} stores (Nam et al. 2002). Apoptosis of osteoblasts is one of the mechanisms by which ROS inflict damage to these cells (Byun et al.

2005, Chang et al. 2006, Ho et al. 2005, Park et al. 2005), possibly through the stimulation of NF-kappa-B (Bai et al. 2005).

Evidence Associating Oxidative Stress and Antioxidants with Osteoclasts

One theory suggests that ROS are in part involved in the differentiation of osteoclasts and their ability to resorb bone (Silverton 1994). The mechanisms involved are poorly understood but may involve NF-kappa-B regulation (D.J. Kim et al. 2006, H.J. Kim et al. 2006). The H_2O_2 that is produced by osteoclasts (Bax et al. 1992) and the endothelial cells (Zaidi et al. 1993) intimately associated with osteoclasts can increase osteoclastic activity and bone resorption. H_2O_2 may also be involved in the differentiation of osteoclast precursors (Steinbeck et al. 1998), osteoclast formation (Suda et al. 1993), and osteoclast motility (Bax et al. 1992). Tartrate-resistant acid phosphatase (TRAP), found on the surface of osteoclasts, reacts with H_2O_2 to degrade collagen and other proteins (Halleen et al. 1999). Using nitroblue tetrazolium (NBT), which is reduced to purple-coloured formazan by ROS, superoxide was localized both intracellularly and at the osteoclast-bone interface, suggesting the participation of superoxide in bone resorption (Key et al. 1990) and formation and activation of osteoclasts (Garrett et al. 1990). Fraser et al. (1996) suggested that H_2O_2, but not superoxide, stimulates bone resorption in mouse calvaria and that the earlier finding of stimulation by superoxide (Garrett et al. 1990, Key et al. 1994) may be due in part to conversion of this radical to H_2O_2.

Hormones known to stimulate bone resorption such as parathyroid hormone (PTH) (Datta et al. 1996) and 1,25-dihydroxyvitamin D_3 (1,25(OH)$_2$D$_3$) (Berger et al. 1999) have stimulatory effects on ROS production in osteoclasts. Additionally, hormones known to have inhibitory effect on bone resorption, such as calcitonin, can inhibit ROS production (Berger et al. 1999, Datta et al. 1995). 1,25(OH)$_2$D$_3$ has a direct non-genomic effect on the generation of superoxide anions (O_2^-), which was inhibited by estrogen (Berger et al. 1999) through its antioxidant property (Clarke et al. 2001, Wagner et al. 2001). Lean et al. (2005) later showed that H_2O_2 is the ROS responsible for signalling the bone loss of estrogen deficiency. A crucial role for ROS has recently been shown in RANKL-induced osteoclast differentiation (Lee et al. 2005, Yip et al. 2005).

Antioxidants also play a role in osteoclast activity. Osteoclasts produce the antioxidant enzyme superoxide dismutase (SOD) in the plasma membrane (Oursler et al. 1991). Production of ROS in osteoclasts was inhibited after cells were treated with antioxidant enzymes such as SOD

(Key et al. 1990) and catalase (Suda et al. 1993). It was also inhibited by estrogen (Berger et al. 1999), the superoxide scavenger deferoxamine mesylate-manganese complex (Key et al. 1994, Ries et al. 1992), pyrrolidine dithiocarbamate, and N-acetyl cysteine (Hall et al. 1995).

Evidence Associating Oxidative Stress and Antioxidants with Osteoporosis

Osteoporosis results in part from oxidative stress induced by ROS. Epidemiological evidence showed that the antioxidants vitamin C, E and beta-carotene may reduce the risk of osteoporosis (Leveille et al. 1997, Melhus et al. 1999, Morton et al. 2001, Singh 1992) and counteract the adverse effects on bone of oxidative stress produced by smoking (Melhus et al. 1999) and during strenuous exercise (Singh 1992). Vitamins A, C, and E, uric acid, the antioxidant enzymes glutathione peroxidase in plasma and SOD in plasma and erythrocyte were consistently lower in osteoporotic subjects than in control subjects. These results showed that antioxidant defences are markedly decreased in osteoporotic women (Maggio et al. 2003). Increased 8-iso-prostaglandin F alpha (8-iso-PGFα), an oxidative stress biomarker, is linked with reduced bone density (Basu et al. 2001, Sontakke and Tare 2002). The level of oxidative stress marker, lactic acid, positively correlated with the severity of osteoporosis in two men with mitochondrial deletion (mtDNA) (Varanasi et al. 1999), and osteoporosis is linked to an increase in oxidative stress in relatively young males (Polidori et al. 2001). Serum glutathione reductase correlates positively with bone densitometry values (Avitabile et al. 1991). Tryptophan metabolism is altered in osteoporosis in a manner that could contribute to oxidative stress and, thus, to the progress of the disease (Forrest et al. 2006). The cellular and molecular mechanisms involved in the role of oxidative stress in osteoporosis remain to be elucidated in spite of these reports.

In lower species, low bone density is also associated with oxidative stress. The bone-protective effect of the antioxidant melatonin in ovariectomized rats depends in part on its free radical scavenging properties (Cardinali et al. 2003). The accelerated mouse-senescence prone P/2 (SAM-P/2) that generates increased oxygen radicals is used as a mouse model to study the role of ROS in age-related disorders including osteoporosis (Hosokawa 2002, Udagawa 2002). This model is useful in studying the role of lycopene in osteoporosis.

LYCOPENE AND BONE HEALTH

The use of antioxidants polyphenols, β-carotene and lycopene from natural sources such as fruits and vegetables and nutritional supplements could be another way of inhibiting ROS. In this regard, we review the studies on bone tissue and lycopene, a potent antioxidant found predominantly in tomatoes and tomato products.

The evidence for the involvement of lycopene in bone tissue is presented through epidemiological and cross-sectional studies on men and premenopausal and postmenopausal women, as well as through its effects *in vitro* on the bone-forming osteoblasts and bone-resorbing osteoclasts.

In vitro Studies on Lycopene in Bone Cells

Effects of lycopene on osteoclasts

To date, there are only two reported studies on the effects of lycopene in osteoclasts (Ishimi et al. 1999, Rao et al. 2003). Rao et al. (2003) cultured cells from bone marrow prepared from rat femur in 16-well calcium phosphate–coated Osteologic[TM] multi-test slides (Millenium Biologix Inc). Varying concentrations of lycopene in the absence or presence of the resorbing agent parathyroid hormone (PTH-(1-34) were added at the start of culture and at each medium change every 48 hr. The effects of lycopene on mineral resorption, TRAP-positive (TRAP+) multinucleated osteoclasts formation, and NBT-staining were studied. In this study, Rao et al. (2003) showed that lycopene inhibited the TRAP+ multinucleated cell formation in both vehicle- and PTH-treated cultures, as well as the mineral resorption in osteologic slide cultures (Fig. 1). The cells that were stained with the NBT reduction product formazan were decreased by treatment with 10-5 M lycopene, indicating that lycopene inhibited the formation of ROS-secreting osteoclasts (Fig. 2). Rao et al. concluded that lycopene inhibited basal and PTH-stimulated osteoclastic mineral resorption and formation of TRAP+ multinucleated osteoclasts, as well as the ROS produced by osteoclasts. These findings are new and may be important in the pathogenesis, treatment and prevention of osteoporosis.

The effects of lycopene on osteoclast formation were also reported by Ishimi et al. (1999) in murine osteoclasts formed in co-culture with calvarial osteoblasts. Their results differed from those of Rao et al. (2003) in that they found that lycopene inhibited the PTH-induced, but not the basal, TRAP+ multinucleated cell formation. Furthermore, they could not demonstrate any effect of lycopene on bone resorption. They also did not study the effect of lycopene on ROS production.

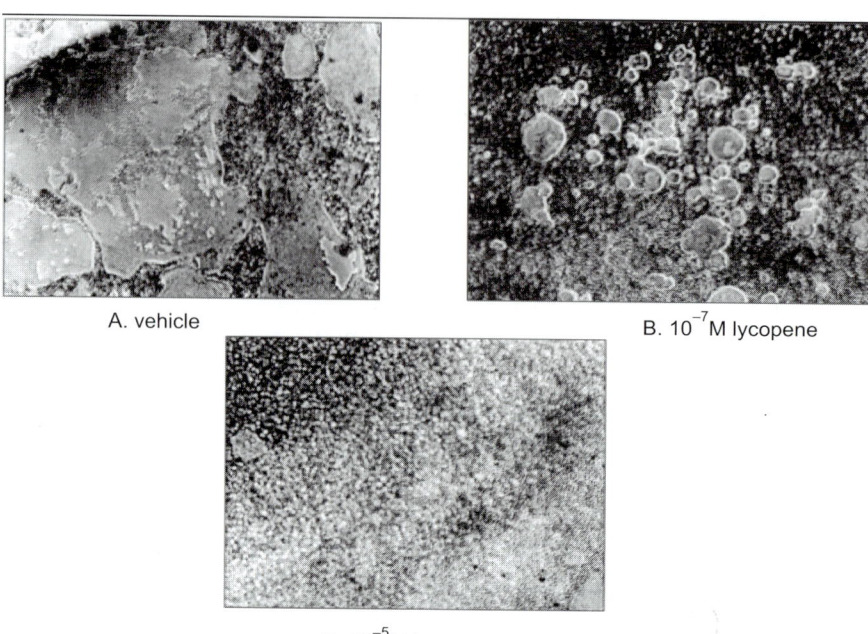

A. vehicle   B. 10^{-7} M lycopene

C. 10^{-5} M lycopene

Fig. 1 Effect of lycopene on resorption of the calcium phosphate substrate coating of Osteologic multi-test slides by osteoclasts (Rao LG et al., 2003).

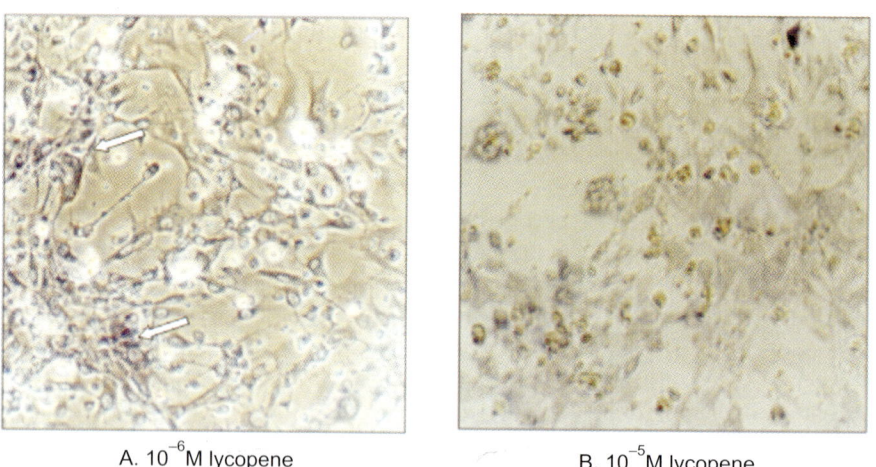

A. 10^{-6} M lycopene   B. 10^{-5} M lycopene

Fig. 2 Effect of lycopene on ROS production in osteoclasts (Rao LG et al., 2003).

Effects of lycopene on osteoblasts

The studies on the effects of lycopene on osteoblasts are limited to two reports (Kim et al. 2003, Park et al. 1997). Kim et al. (2003) showed that lycopene stimulated the proliferation of the osteoblast-like SaOS-2, as shown in Fig. 3. They also reported that lycopene had a stimulatory effect on ALP activity, a marker of osteoblastic differentiation in more mature cells in the presence, but not absence, of dexamethasone (Fig. 4).

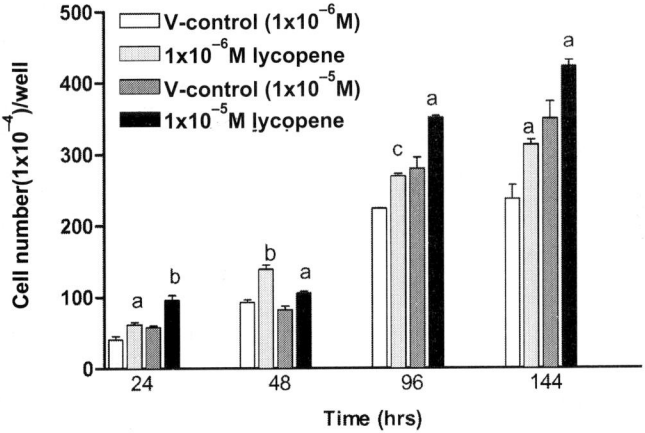

Fig. 3 Effect of lycopene on the proliferation of SaOS-2 cells. Compared with respective vehicle control of the same dilution: a = $p<0.05$; b = $p<0.001$; c = $p<0.005$. (Kim L et al., 2003).

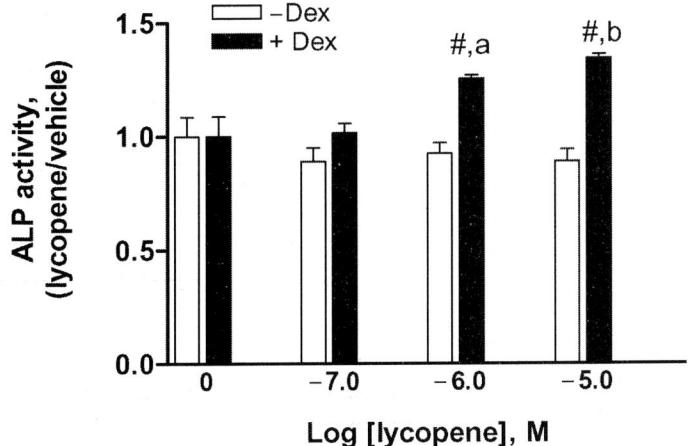

Fig. 4 Effect of lycopene on alkaline phosphotase (ALP) activity. **#** $p < 0.005$, Comparison with zero control: a < 0.05, b < 0.01 (Kim L et al., 2003).

These findings constituted the first report on the effect of lycopene on human osteoblasts. In another study by Park et al. (1997), the effect of lycopene on MC3T3 cells (the osteoblastic cells of mice) was contrary to the findings of Kim et al. (2003). Park demonstrated that lycopene had an inhibitory effect on cell proliferation. Both studies, however, reported that ALP activity was stimulated.

Studies on the role of lycopene in bone health

Very little work has been reported on the effect of lycopene on bone health. One epidemiological study of Anglo-Celtic Australian population of 68 men and 137 women showed that bone mass of total body and lumbar spine were positively related to lycopene intake in men, and to lycopene and lutein/zeaxanthin intake in premenopausal women. β-carotene, but not lycopene, had a positive association with lumbar spine bone mass in postmenopausal female volunteers (Wattanapenpaiboon et al. 2003). Lycopene therapy appears to be effective and safe in the treatment of hormone refractory metastatic prostate cancer; it not only takes care of the rising prostate antigen but also improves the bone pain associated with this disease (Ansari and Gupta 2004). Our cross-sectional clinical study is the first to evaluate the correlation between serum lycopene and the prevention of the risk of osteoporosis in postmenopausal women (Rao et al. 2007). The design and results of the study are described in detail below.

Clinical Studies on the Role of Lycopene in Postmenopausal Women at Risk of Osteoporosis

Postmenopause is associated with an overall increase in bone turnover markers (Kushida et al. 1995, Vernejoul 1998) that predict bone loss and osteoporosis in postmenopausal women (Garnero et al. 1996). We recently completed a cross-sectional study to determine the effects of lycopene on oxidative stress parameters, bone turnover markers, and antioxidant capacity in women at risk for osteoporosis (Rao et al. 2007). Thirty-three female participants were recruited who were at least one year postmenopausal, aged 50-60, and not on medications for osteoporosis, diabetes, heart disease or high blood pressure. Participants were asked to record their diet for 7 d and then provide a fasting blood sample on the eighth day. Dietary analysis of lycopene intake was performed using previously published data on lycopene content in common foods (Rao and Agarwal 1999). The following oxidative stress parameters were measured in serum: lipid peroxidation (Draper et al. 1993), reported as thiobarbituric acid reactive substances (TBARS), and protein oxidation (Hu 1994), where a high protein thiol level indicates a lower level of protein oxidation. Serum lycopene was determined using high-performance liquid chromatography (HPLC) (Stahl and Sies 1992). The following bone turnover markers were analysed by ELISA: cross-linked aminoterminal N-telopeptide (NTx) (INTER MEDICO, Ontario, Canada), a measure of bone resorption, and bone-specific alkaline phosphatase (BAP) (ESBE Scientific, Ontario, Canada), a measure of bone formation. For statistical analysis the participants were grouped into quartiles according to their serum lycopene per kilogram body weight (nM/kg) (Table 1). One-way ANOVA and the Newman-Keuls post test were used to test for significant differences between groups with respect to serum lycopene, oxidative stress parameters, and bone turnover markers.

The characteristics of participants grouped into quartiles according to serum lycopene per kilogram body weight and the summary of the values for various parameters are given in Table 1. A positive correlation between serum lycopene levels and dietary lycopene intake was found as determined from the estimated food records ($p < 0.01$) (Fig. 5). Participants with higher serum lycopene compared with those with lower serum lycopene had significantly lower protein oxidation, as indicated by increased thiols ($p < 0.05$) (Fig. 6), and significantly lower NTx values ($p < 0.005$) (Fig. 7). The positive correlation between serum lycopene and dietary lycopene intake (Fig. 5) supports our hypothesis that dietary lycopene acts as an effective antioxidant, reducing oxidative stress and bone turnover markers. Our observations suggest an important role for lycopene mediated via its antioxidant property in reducing the risk of osteoporosis (Rao et al. 2007). We are currently expanding this cross-sectional study to include premenopausal women and osteoporotic women to compare the effects of serum lycopene on bone turnover markers, oxidative stress parameters, and antioxidant capacity with those described for postmenopausal women.

A randomized clinical study is ongoing in our laboratory to determine whether intervention with lycopene in postmenopausal women will improve antioxidant capacity, while decreasing markers of bone turnover and parameters of oxidative stress. Recruited participants are the same as described above, and in addition they are asked to refrain from taking multivitamins or supplements containing antioxidants for the duration of the study. Smokers and women taking hormone replacement therapy are excluded. Following a one-month washout period during which participants are asked to refrain from consuming lycopene-containing products, fasting blood samples and dietary records are collected as baseline samples. Participants are then randomized to one of the four following supplements to be taken twice daily with meals for a period of

Table 1 Characteristics of participants grouped into quartiles according to serum lycopene per kilogram body weight.

Parameters	Mean concentrations ± SEM in groups stratified according to serum lycopene (nM/kg)			
	Group 1	Group 2	Group 3	Group 4
Number of participants	9	8	7	9
Average age (yr)	56.11 ± 1.02	56.50 ± 0.76	56.43 ± 0.87	56.33 ± 1.03
Lycopene intake (mg/day)	1.76 ± 0.76	3.68 ± 0.94	3.03 ± 1.09	7.35 ± 0.80
Serum lycopene (nM)	74.99 ± 15.09	165.60 ± 12.04	234.5 ± 24.15	502.8 ± 47.39
Serum lycopene (nM/kg)	1.13 ± 0.25	2.62 ± 0.11	4.04 ± 0.22	8.11 ± 0.63
Protein thiols (µM)	504.20 ± 16.34	501.90 ± 25.63	457.4 ± 50.10	592.2 ± 31.12
TBARS (nmol/mL serum)	7.48 ± 1.01	5.11 ± 0.30	5.04 ± 0.67	5.56 ± 0.59
NTx (nM BCE)	22.55 ± 2.28	27.12 ± 2.37	24.58 ± 1.46	17.16 ± 1.33
BAP (U/L)	23.08 ± 2.79	20.72 ± 2.62	21.77 ± 3.33	21.31 ± 2.08

Source: Rao et al. (2007), with permission of Springer Science and Business Media.

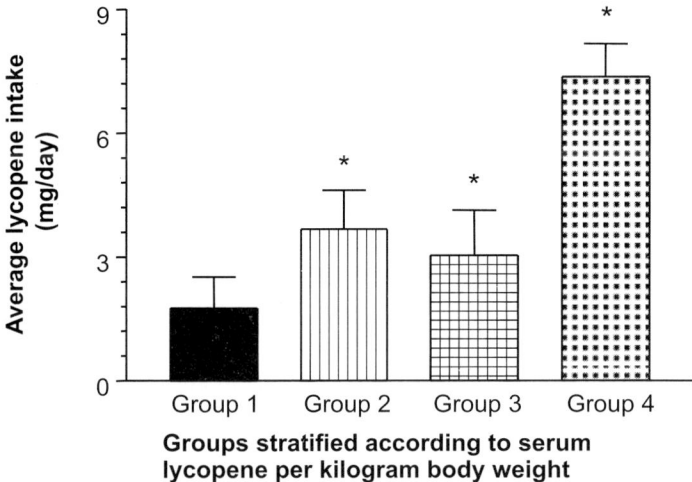

Fig. 5 Effect of lycopene intake on serum lycopene in 33 postmenopausal women participants (Rao et al. 2007), with permission of Springer Science and Business Media).

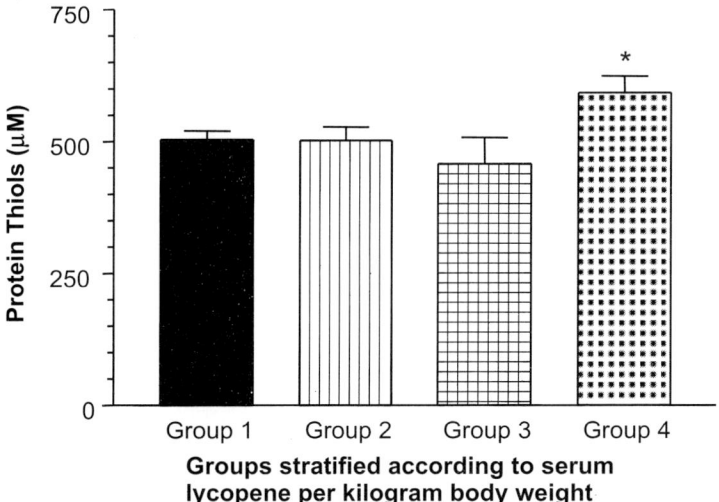

Fig. 6 Effect of serum lycopene on protein oxidation (protein thiols) in 33 postmenopausal women participants (Rao et al. 2007), with permission of Springer Science and Business Media).

4 mon: tomato juice (Heinz, Canada) (30 mg lycopene/day), lycopene-enriched tomato juice (Kagome, Japan) (70 mg lycopene/day), lycopene capsule (30 mg lycopene/day), or placebo capsule (LycoRed, Israel) (0 mg lycopene/day). Fasting blood samples and dietary records are collected after 2 and 4 mon of supplementation. Dietary records will be

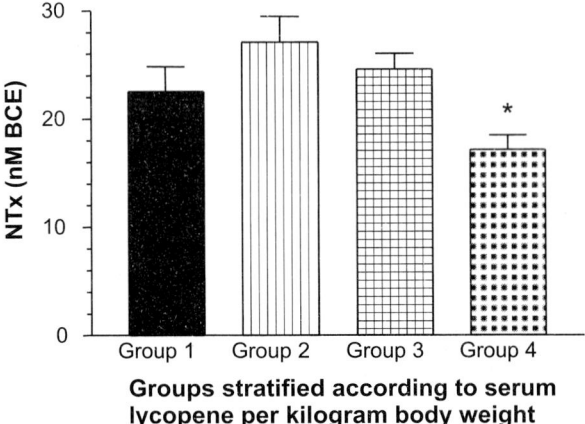

Fig. 7 Effect of serum lycopene on NTx (bone resorption) in 33 postmenopausal women (Rao et al. 2007), with permission of Springer Science and Business Media).

analyzed for intake of antioxidants and minerals known to be beneficial to bone and serum samples will be analyzed for bone turnover markers, antioxidant capacity and oxidative stress parameters. To date, 60 participants have been recruited for this study. This study is important because it will determine the beneficial effects of lycopene in the prevention and management of osteoporosis.

CONCLUSION

The direct role of lycopene in bone tissue is now evident with beneficial effects on risk prevention for osteoporosis in postmenopausal women, positive effects on bone density in men and premenopausal women, inhibition of bone resorption by osteoclasts, and stimulation of the bone-forming osteoblasts. Although it is premature to suggest that eating tomatoes and tomato products will prevent osteoporosis, it would be a healthy practice to include tomatoes, tomato products or lycopene supplements in the diet to prevent oxidative stress-related chronic diseases, including osteoporosis. The final results of our intervention study may indicate that lycopene can be used either as a dietary alternative to drug therapy or as a complement to the pharmaceuticals used by women at risk for osteoporosis.

ACKNOWLEDGEMENTS

The Fund by the Canadian Institutes of Health Research (CIHR), which matched the funds from the following industrial collaborators: Genuine

Health Inc., the H.J. Heinz Co, Millenium Biologix Inc. (Canada), Kagome Co. (Japan), and LycoRed Natural Product Industries, Ltd. (Israel). We thank the assistance of Ms. H. Shen for the HPLC analysis, and Dr. C. Derzko and Mr. M. Simms for initial participant recruitment.

ABBREVIATIONS

$1,25(OH)_2D_3$: 1,25-dihydroxyvitamin D_3; 8-iso-PGFα: 8-iso-prostaglandin F alpha; ALP: alkaline phosphatase; BAP: bone-specific alkaline phosphatase; Ca^{2+}: intracellular calcium; H_2O_2: hydrogen peroxide; HPLC: high-performance liquid chromatography; mtDNA: mitochondrial deletion; NBT: nitroblue tetrazolium; NTx: cross-linked aminoterminal N-telopeptide; O_2^-: superoxide anions; PTH: parathyroid hormone; ROS: reactive oxygen species; SAM-P/2: mouse-senescence prone P/2; SOD: superoxide dismutase; TBARS: thiobarbituric acid reactive substances; TRAP: tartrate-resistant acid phosphatase

References

Ansari, M.S. and N.P. Gupta. 2004. Lycopene: a novel drug therapy in hormone refractory metastatic prostate cancer. Urol. Oncol. 22: 415-420.

Avitabile, M. and N.E. Campagna, G.A. Magri, M. Vinci, G. Sciacca, G. Alia, and A. Ferro. 1991. Correlation between serum glutathione reductases and bone densitometry values. Bollettino-Societa Italiana Biologia Sperimentale 67: 931-937.

Bai, X.C. and D. Lu, A.L. Liu, Z.M. Zhang, X.M. Li, Z.P. Zou, W.S. Zeng, B.L. Cheng, and S.Q. Luo. 2005. Reactive oxygen species stimulates receptor activator of NF-kappaB ligand expression in osteoblast. J. Biol. Chem. 280: 17497-17506.

Basu, S. and K. Michaelsson, H. Olofsson, S. Johansson, and H. Melhus. 2001. Association between oxidative stress and bone mineral density. Biochem. Biophys. Res. Comm. 288: 275-279.

Bax, B.E. and A.S. Alam, B. Banerji, C.M. Bax, P.J. Bevis, C.R. Stevens, B.S. Moonga, D.R. Blake, and M. Zaidi. 1992. Stimulation of osteoclastic bone resorption by hydrogen peroxide. Biochem. Biophys. Res. Comm. 183: 1153-1158.

Berger, C.E. and B.R. Horrocks, and H.K. Datta. 1999. Direct non-genomic effect of steroid hormones on superoxide generation in the bone resorbing osteoclasts. Mol. Cell. Endocrinol. 149: 53-59.

Byun, C.H. and J.M. Koh, D.K. Kim, S.I. Park, K.U. Lee, and G.S. Kim. 2005. Alpha-lipoic acid inhibits TNF-alpha-induced apoptosis in human bone marrow stromal cells. J. Bone Minl. Res. 20: 1125-1135.

Cardinali, D.P. and M.G. Ladizesky, V. Boggio, R.A. Cutrera, and C. Mautalen. 2003. Melatonin effects on bone: experimental facts and clinical perspectives. J. Pineal Res. 34: s81-87.

Chan, G.K. and G. Duque. 2002. Age-related bone loss: old bone, new facts. Gerontology 48: 62-71.

Chang, C.C. and Y.S. Liao, Y.L. Lin, and R.M. Chan. 2006. Nitric oxide protects osteoblasts from oxidative stress-induced apoptotic insults via a mitochondria-dependent mechanism. J. Orth. Res. 24: 1917-1925.

Clarke, R. and F. Leonessa, J.N. Welch, and T.C. Skaar. 2001. Cellular and molecular pharmacology of antiestrogen action and resistance. Pharmacol. Rev. 53: 25-71.

Cohen, M.M. Jr. 2006. The new bone biology: pathologic, molecular, and clinical correlates. Amer. J. Med. Gen. A 140: 2646-2706.

Cortizo, A.M. and L. Bruzzone, S. Molinuevo, and S.B. Etcheverry. 2000. A possible role of oxidative stress in the vanadium-induced cytotoxicity in the MC3T3E1 osteoblast and UMR106 osteosarcoma cell lines. Toxicology 147: 89-99.

Datta, H.K. and P. Manning, H. Rathod, and C.J. McNeil. 1995. Effect of calcitonin, elevated calcium and extracellular matrices on superoxide anion production by rat osteoclasts. Exp. Physiol. 80: 713-719.

Datta, H.K. and H. Rathod, P. Manning, Y. Turnbull, and C.J. McNeil. 1996. Parathyroid hormone induces superoxide anion burst in the osteoclasts: evidence of the direct instantaneous activation of the osteoclast by the hormone. J. Endocrinol. 149: 269-275.

de Vernejoul, M.-C. 1998. Markers of bone remodelling in metabolic bone disease. Drugs & Aging 1 (suppl 1): 9-14.

Draper, H.H. and E.J. Squires, H. Mahmoodi, J. Wu, S. Agarwal, and M. Hadley. 1993. A comparative evaluation of thiobabituric acid methods for determination of malondialdehyde in biological materials. Free Radical Biol. Med. 15: 353-363.

Forrest, C.M. and G.M. Mackay, L. Oxford, N. Stoy, T.W. Stone, and L.G. Darlington. 2006. Kynurenine pathway metabolism in patients with osteoporosis after 2 yr of drug treatment. Clin. Exp. Pharmacol. Physiol. 33: 1078-1087.

Fraser, J.H. and M.H. Helfrich, H.M. Wallace, and S.H. Ralston. 1996. Hydrogen peroxide, but not superoxide, stimulates bone resorption in mouse calvariae. Bone 19: 223-226.

Garnero, P. and E. Sornay-Rendu, M-C. Chapuy, and P.D. Delmas. 1996. Increased bone turnover in late postmenopausal women is a major determinant of osteoporosis. J. Bone Min. Res. 11: 337-349.

Garrett, I.R. and B.F. Boyce, R.O.C. Oreffo, L. Bonewald, J. Pser, and G.R. Mundy. 1990. Oxygen-derived free radicals stimulate osteoclastic bone resorption in rodent bone in vitro and in vivo. J. Clin. Invest. 85: 632-639.

Giovannucci, E. 1999. Tomatoes, tomato-based products, lycopene, and cancer: review of the epidemiologic literature. J. Nat. Cancer Inst. 91: 317-331.

Hall, T.J. and M. Schaeublin, K. Fuller, and T.J. Chambers. 1995. The role of oxygen intermediates in osteoclastic bone resorption. Biochem. Biophys. Res. Comm. 207: 280-287.

Halleen, J.M. and S. Raisanen, J.J. Salo, S.V. Reddy, G.D. Roodman, T.A. Hentunen, P.P. Lehenkari, H. Kaija, P. Vihko, H.K. Vaananen, H. Zhao, M. Mulari, and J.M. Halleen. 1999. Intracellular fragmentation of bone resorption products by reactive oxygen species generated by osteoclastic tartrate-resistant acid phosphatase. J. Biol. Chem. 274: 22907-22910.

Ho, W.P. and T.L. Chen, W.T. Chiu, Y.T. Tai, and R.M. Chen. 2005. Nitric oxide induces osteoblast apoptosis through a mitochondria-dependent pathway. Ann. NY Acad. Sci. 1042: 460-470.

Hosokawa, M. 2002. A higher oxidative status accelerates senescence and aggravates age-dependent disorders in SAMP strains of mice. Mech. Ageing Dev. 123: 1553-1561.

Hu, M.L. 1994. Measurement of protein thiol groups and glutathione in plasma. Meth. Enzymol. 233: 380-385.

Ishimi, Y. and M. Ohmura, X. Wang, M. Yamaguchi, and S. Ikegami. 1999. Inhibition by carotenoids and retinoic acid of osteoclast-like cell formation induced by bone-resorbing agents in vitro. J. Clin. Biochem. Nutr. 27: 113-122.

Kenny, A.M. and L.G. Raisz. 2002. Mechanism of bone remodelling: implications for clinical practice. J. Rep. Med. 47: 63-70.

Key, L.L. and W.L. Ries, R.G. Taylor, B.D. Hays, and B.L. Pitzer. 1990. Oxygen derived free radicals in osteoclasts: the specificity and location of the nitroblue tetrazolium reaction. Bone 11: 115-119.

Key, L.L. and W.C. Wolf, C.M. Gundberg, and W.L. Ries. 1994. Superoxide and bone resorption. Bone 15: 431-436.

Kim, D.J. and J.M. Koh, O. Lee, N.J. Kim, Y.S. Lee, Y.S. Kim, J.Y. Park, K.U. Lee, and G.S. Kim. 2006. Homocysteine enhances apoptosis in human bone marrow stromal cells. Bone 39: 582-590.

Kim, H.J. and E.J. Chang, H.M. Kim, S.B. Lee, H.D. Kim, G. Su Kim, and H.H. Kim. 2006. Antioxidant alpha-lipoic acid inhibits osteoclast differentiation by reducing nuclear factor-kappaB DNA binding and prevents in vivo bone resorption induced by receptor activator of nuclear factor-kappaB ligand and tumor necrosis factor-alpha. Free Radical Biol. Med. 40: 1483-1493.

Kim, L. and A.V. Rao, and L.G. Rao. 2003. Lycopene II – Effect on osteoblasts: The caroteroid lycopene stimulates cell proliferation and alkaline phosphatase activity of SaOS-2 cells. J. Med. Food. 6: 79-86.

Kushida, K. and M. Takahashi, K. Kawana, and T. Inoue. 1995. Comparison of markers for bone formation and resorption in premenopausal and postmenopausal subjects, and osteoporosis patients. J. Clin. Endocr. Metab. 80: 2447-2450.

Lean, J.M. and C.J. Jagger, B. Kirstein, K. Fuller, and T. J. Cambers. 2005. Hydrogen peroxide is essential for estrogen-deficiency bone loss and osteoclast formation. Endocrinology 146: 728-735.

Lee, N.K. and Y.G. Choi, J.Y. Baik, S.Y. Han, D.W. Jeong, Y.S. Bae, N. Kim, and S.Y. Lee. 2005. A crucial role for reactive oxygen species in RANKL-induced osteoclast differentiation. Blood 106: 852-859.

Lerner, U.H. 2006. Bone remodeling in post-menopausal osteoporosis. J. Dent. Res. 85: 584-595.

Leveille, S.G. and A.Z. LaCroix, T.D. Koepsell, S.A. Beresford, G. VanBelle, and D.M. Buchner. 1997. Dietary vitamin C and bone mineral density in postmenopausal women in Washington State, USA. J. Epidemiol. Comm. Health 51: 479-485.

Lindsay, R. and F. Cosman. 1999. Prevention of osteoporosis. pp. 264-270. *In:* M.J. Favus [Ed.] Primer on the Metabolic Bone Diseases and Disorders of Mineral Metabolism. Lippincott Williams & Wilkins, New York.

Liu, H.-C. and R.-M. Cheng, F.-H. Lin, and H.W. Fang. 1999. Sintered beta-dicalcium phosphate particles induce intracellular reactive oxygen species in rat osteoblasts. Biomed. Eng. Appl. Basis Comm. 11: 259-264.

Maggio, D. and M. Barabani, M. Pierandrei, M.C. Polidori, M. Catani, P. Mecocci, U. Senin, R. Pacifici, and A. Cherubini. 2003. Marked decrease in plasma antioxidants in aged osteoporotic women: results of a cross-sectional study. J. Clin. Endocrinol. Metab. 88: 1523-1527.

Melhus, H. and K. Michaelsson, L. Holmberg, A. Wolk, and S. Ljunghall. 1999. Smoking, antioxidant vitamins, and the risk of hip fracture. J. Bone Min. Res. 14: 129-135.

Morton, D.J. and E.L. Barrett-Connor, and D.L. Schneider. 2001. Vitamin C supplement and bone mineral density in postmenopausal women. J. Bone Min. Res. 16: 135-140.

Nam, S.H. and S.Y. Jung, C.M. Yoo, E.H. Ahn, and C.K. Suh. 2002. H_2O_2 enhances Ca^{2+} release from osteoblast internal stores. Yonsei Med. J. 43: 229-235.

Oursler, M.J. and P. Collin-Osdoby, L. Li, E. Schmitt, and P. Osdoby. 1991. Evidence for an immunological and functional relationship between superoxide dismutase and a high molecular weight osteoclast plasma membrane glycoprotein. J. Cell. Biochem. 46: 331-344.

Park, B.G. and C.L. Yoo, H.T. Kim, C.H. Kwon, and Y.K. Kim. 2005. Role of mitogen-activated protein kinases in hydrogen peroxide-induced cell death in osteoblastic cells. Toxicol. 215: 115-125.

Park, C.K. and Y. Ishimi, M. Ohmura, M. Yamaguchi, and S. Ikegami. 1997. Vitamin A and carotenoids stimulate differentiation of mouse osteoblastic cells. J. Nutr. Sci. Vitaminol. 43: 281-296.

Polidori, M.C. and W. Stahl, O. Eichler, I. Niestroj, and H. Sies. 2001. Profiles of antioxidants in human plasma. Free Radical Biol. Med. 30: 456-462.

Rao, A.V. and S. Agarwal. 1999. Role of lycopene as antioxidant carotenoid in the prevention of chronic diseases: a review. Nutr. Res. 19: 305-323.

Rao, A.V. and L.G. Rao. 2004. Lycopene and human health. Curr. Top. Nutr. Res. 2: 127-136.

Rao, L.G. and N. Krishnadev, K. Banasikowska, and A.V. Rao. 2003. Lycopene I – Effect on osteoclasts: Lycopene inhibits basal and parathyroid hormone-stimulated osteoclast formation and mineral resorption mediated by reactive oxygen species in rat bone marrow cultures. J. Med. Food 6: 69-78.

Rao, L.G. and E.S. Mackinnon, R.G. Josse, T.M. Murray, A. Strauss, and A.V. Rao. 2007. Lycopene consumption decreases oxidative stress and bone resorption markers in postmenopausal women. Osteo. Int. 18: 109-115.

Ries, W.L. and L.L. Key, and R.M. Rodriguiz. 1992. Nitroblue tetrazolium reduction and bone resorption by osteoclasts in vitro inhibited by a manganese-based superoxide dismutase mimic. J. Bone Min. Res. 1992: 931-938.

Samoto, H. and E. Shimizu, Y. Matsuda-Honjo, R. Saito, M. Yamazaki, K. Kasai, S. Furuyama, H. Sugiya, J. Sodek, and Y. Ogata. 2002. TNF-alpha suppresses bone sialoprotein (BSP) expression in ROS17/2.8 cells. J. Cell. Biochem. 87: 313-323.

Silverton, S. 1994. Osteoclast radicals. J. Cell. Biochem. 56: 367-373.

Singh, V.N. 1992. A current perspective on nutrition and exercise. J. Nutr. 122: 760-765.

Sontakke, A.N. and R.S. Tare. 2002. A duality in the roles of reactive oxygen species with respect to bone metabolism. Clin. Chim. Acta 318: 145-148.

Stahl, W. and H. Sies. 1992. Uptake of lycopene and its geometrical isomers is greater from heat-processed than from unprocessed tomato juice in humans. J. Nutr. 122: 2161-2166.

Steinbeck, M.J. and J.K. Kim, M.J. Trudeau, P.V. Hauschka, and M.J. Karnovsky. 1998. Involvement of hydrogen peroxide in the differentiation of clonal HD-11EM cells into osteoclast-like cells. J. Cell. Physiol. 176: 574-587.

Suda, N. and I. Morita, T. Kuroda, and S. Murota. 1993. Participation of oxidative stress in the process of osteoclast differentiation. Biochim. Biophys. Acta 1157: 318-323.

Udagawa, N. 2002. Mechanisms involved in bone resorption. Biogerontology 3: 79-83.

Varanasi, S.S. and R.M. Francis, C.E. Berger, S.S. Papiha, and H.K. Datta. 1999. Mitochondrial DNA deletion associated oxidative stress and severe male osteoporosis. Osteo. Int. 10: 143-149.

Wagner, A.H. and M.R. Schroeter, and M. Hecker. 2001. 17β-estradiol inhibition of NADPH oxidase expression in human endothelial cells. FASEB J. 15: 2121-2130.

Wattanapenpaiboon, N. and W. Lukito, M.L. Wahlqvist, and B.J. Strauss. 2003. Dietary carotenoid intake as a predictor of bone mineral density. Asia Pacif. J. Clin. Nutr. 12: 467-473.

Yip, K.H. and M.H. Zheng, J.H. Steer, T.M. Giardina, R. Han, S.Z. Lo, A.J. Bakker, A.I. Cassady, D.A. Joyce, and J. Xu. 2005. Thapsigargin modulates osteoclastogenesis through the regulation of RANKL-induced signaling pathways and reactive oxygen species production. J. Bone Min. Res. 20: 1462-1471.

Zaidi, M. and A.S. Alam, B.E. Bax, V.S. Shankar, C.M. Bax, J.S. Gill, M. Pazianas, C.L. Huang, T. Sahinoglu, and B.S. Moonga. 1993. Role of the endothelial cell in osteoclast control: new perspectives. Bone 14: 97-102.

PART 3

Lycopene and Cancer

3.1

Lycopene and Its Potential Role in Prostate Cancer Prevention

[1]A. Trion, [2]F.H. Schröder and [3]W.M. van Weerden

[1,3]Department of Urology, ErasmusMC, Dr. Molewaterplein 50, P.O. Box 2040
3000 CA Rotterdam, The Netherlands
[2]Department of Urology, ERSPC, Rochussenstraat 125, 3015 EJ Rotterdam
The Netherlands

ABSTRACT

Prostate cancer is a major health problem for men in Western society. Since the development of prostate-specific antigen (PSA) screening, the number of prostate cancers detected at an early developmental stage is increasing rapidly. In these patients, local treatment with curative intent is feasible. However, screening studies indicate that approximately 30% of these men are overdiagnosed, meaning that their cancer would never have caused any symptoms during their life time. In order to minimize unnecessary treatment for such patients, active surveillance is an alternative approach. Chemoprevention may be a valuable treatment option in addition to active surveillance, as dietary factors have been shown to delay prostate cancer progression and possibly increase longevity. Lycopene is one of a group of substances that are currently being tested for their potential as a nutritional source of chemopreventive agent in prostate cancer. *In vitro* experimental studies have shown growth inhibiting effects of lycopene on prostate cancer cells. Recent preclinical studies in mice revealed that lycopene supplementation reduces growth of human prostate cancer xenografts. Short-term intervention studies in patients receiving lycopene supplementation prior to surgical intervention have already indicated an effect on circulating PSA levels. In addition, tertiary chemoprevention studies in progressive locally treated prostate cancer patients have been

A list of abbreviations is given before the references.

conducted showing reduced plasma PSA doubling times after the use of dietary supplements that also contained lycopene. Further studies are warranted to establish the role of lycopene in primary and secondary prevention of prostate cancer.

INTRODUCTION

The prostate is the largest accessory sex gland and consists of a secretory epithelium surrounded by a dense stroma that plays an essential role in the normal development, differentiation and function of the prostate epithelium. Growth and function of the prostate are primarily dependent on androgenic stimuli. In the healthy male, the major circulating androgen is testosterone, which is almost exclusively (90%) of testicular origin. The major enzymes secreted by the prostate epithelium are prostate-specific antigen (PSA), beta-microseminoprotein (beta-inhibin) and prostate acid phosphatase. The internal architecture of the human prostate is composed of three distinct histological zones: a central zone (approximately 25% of the gland), a peripheral zone (approximately 70% of the gland), and a transitional zone (approximately 5% of the gland). Most prostate carcinomas (70-80%) originate in the peripheral zone, whereas the transitional zone is most often the place of origin for benign prostate hyperplasia (BPH). Recent studies have shown that gene expression profiles differ significantly between the various zones and may underlie the distinct origination of both prostate diseases (van der Heul-Nieuwenhuijsen et al. 2006).

PROSTATE CANCER

Prostate cancer is the most prevalent malignancy among men in the Western world and has become the second leading cause of cancer-related death in males in most Western countries (Greenlee et al. 2000, Jemal et al. 2004). It is a slowly developing cancer and its incidence increases with increasing age. Although latent prostate cancer affects about 30% of men over 50 yr of age, and 60-70% of men older than 80, prostate cancers may progress into an aggressive form with clinical symptoms (Fig. 1). Clinically aggressive prostate cancer invades the seminal vesicles, the apex of the prostate and the bladder, with finally metastatic spread predominantly to lymph nodes and bone causing high morbidity and mortality.

Although the etiology of prostate cancer is not known, currently accepted risk factors for prostate cancer include age, ethnicity (Afro-Americans have more clinically detectable prostate cancer than Caucasian, Japanese or Chinese Americans (Taylor et al. 1994)) and family history (Crawford 2003).

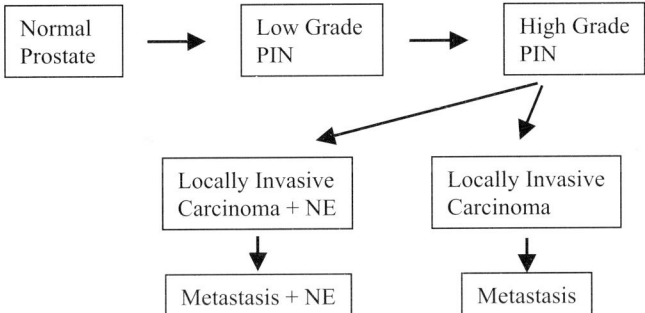

Fig. 1 A multistep model of prostate cancer progression. The development of prostate cancer can be characterized by a sequence of distinct steps. In the normal prostate of young men, histological evidence of atypical cell growth can already be observed. With increasing age, low-grade prostatic intra-epithelial neoplasia (PIN) becomes clearly evident. Low-grade PIN is thought to progress to high-grade PIN (HGPIN). HGPIN is diagnosed by prostate biopsy and may progress into prostate adenocarcinoma. Less commonly, it may have evidence of a neuroendocrine pathology. Locally invasive disease can progress to metastasize outside the prostatic capsule and spread to organs such as bone (adapted from Lamb and Zhang 2005).

The incidence of clinically detectable, aggressive prostate cancer is highest in the United States and Scandinavian countries and lowest in Japan and the Mediterranean countries (Jensen et al. 1990, Shimizu et al. 1991, Black et al. 1997). Moreover, cancer risk rises in Japanese men who move to Western countries (Haenszel and Kurihara 1968, Akazaki and Stemmerman 1973). In contrast, the presence of latent prostate cancer at autopsy is an extremely common finding, even in those countries with the lowest rates of clinically significant disease (Breslow et al. 1977). Therefore, it seems likely that environmental factors such as nutrition play an important role in the progression of latent prostate cancer to clinically manifest disease.

Detection and Treatment

Circulating PSA is currently the most important biochemical marker for the diagnosis of prostate cancer, since prostatic carcinomas almost always cause an elevation of serum PSA levels. A major drawback is its limited specificity, as benign abnormalities of the prostate such as BPH and prostatitis may also result in increased PSA blood levels. Furthermore, PSA-based diagnosis will also include clinically irrelevant tumors.

Several treatment options can be considered when prostate cancer has been detected. Conservative treatment, also referred to as watchful waiting or active surveillance, is an approach in which patients are carefully monitored using plasma PSA, digital rectal examination and prostate biopsy. If tumors progress, curative or palliative treatment options can be considered. Local treatment includes radical prostatectomy and external beam radiotherapy. Due to earlier recognition of prostate cancer, the number of patients that are undergoing radical prostatectomy or local radiation therapy is rapidly increasing and fewer patients are confronted with disseminated disease at the time of diagnosis. When prostate cancer is no longer restricted to the prostate gland, and metastatic disease is observed in lymph nodes and/or bone, hormonal therapy by either surgical or chemical castration (LHRH agonists and/or antiandrogens) is initiated aiming to reduce circulating levels of androgen. Hormonal therapy is most often used as a palliative measure. The majority of patients (60-80%) with advanced prostate cancer initially respond well to hormonal ablation therapy. However, nearly all these patients progress within 18-24 mon to a disease state that is unresponsive to further anti-hormone therapy. At this stage, chemotherapeutic therapies (taxanes, cisplatin) only temporarily affect prostate cancer growth.

It remains to be determined whether early screening for prostate cancer will ultimately affect patient survival. Recent data from screening studies suggest that up to 50% of men with PSA screen-detected prostate cancer are over-diagnosed, meaning that their cancer would never have caused any symptoms (Schroder 2005). Over-diagnosis is a serious issue as radical prostatectomy and local radiation therapy have substantial side effects and significantly affect the patients' quality of life (Parker 2003, Klotz 2005). Ideally, these treatments should be restricted to those who really need it. Active surveillance aims to individualize the management of localized prostate cancer by selecting only those men with non-indolent cancers for curative treatment (Haenszel and Kurihara 1968, Kolonel et al. 1983). We recently initiated a study (PRIAS) to validate active surveillance as a treatment option in men with localized, well-differentiated prostate cancer so as to limit overtreatment (van den Bergh et al. 2007). For the indolent cases that are advised active surveillance, development of less aggressive treatment modalities that are able to delay progression of prostate cancer is desirable. Chemoprevention may be such an alternative that delays tumor progression without causing significant side effects.

Chemoprevention

A long latency period and the putative role of environmental factors such as lifestyle and diet in the development and progression of (prostate) cancer make prostate cancer a suitable candidate for chemoprevention. The geographical differences in incidence of clinical prostate cancer suggest that an Eastern diet rich in soy and low in fat prevents the development of certain cancer types, including prostate cancer. Indeed, several studies have demonstrated that consumption of diets rich in fruits and vegetables is inversely related to the occurrence of typical Western diseases such as cancer and cardiovascular disease (Steinmetz and Potter 1996, van't Veer et al. 2000).

Chemoprevention is the use of natural or laboratory-made substances to prevent the induction of, slow down the progression of, or reverse a disease. Carotenoids are a group of nutrients that are found in brightly colored vegetables and fruits. Among the carotenoids (there are more than 600 structural variants), lycopene is a characteristic lipophilic red pigment that is mainly present in ripe tomatoes. Lycopene, a 40-carbon acyclic carotenoid with 11 linearly arranged conjugated double bonds, is a potent antioxidant and the most efficient singlet oxygen quencher among carotenoids (Di Mascio et al. 1989). Lycopene is one of a group of substances, including other carotenoids, vitamin E, selenium and soy isoflavones, that are currently being tested for their potential as a nutritional source of chemopreventive agent in prostate cancer (Table 1).

Table 1 Agents/regimens evaluated for usefulness as chemoperventive regimen.

Agent/regimen	Mechanism
Lignans (interolacton, enterodiol)	Weak estrogens
Isoflavonoids (genistein, daidzein, equol)	Weak estrogens, antiadhesion
Low fat diet	Anti-oxidant
Green tea (flavonoids, epigallocatechin, quercetin), red wine (resveratol)	
Vitamins (E, D, A)	Anti-oxidant, differentiation
Selenium	Anti-oxidant
Lycopene	Anti-oxidant
Anti-androgens	
Anti-inflammatory agents	

Tomato and tomato-based food products are the most important dietary source of lycopene. Giovannucci and colleagues (Giovannucci 1999) reviewed the epidemiologic literature on the intake of tomatoes, tomato-based products, and blood lycopene level, and the risk of various cancers. Among the 72 studies identified, 57 reported inverse associations between tomato intake and/or blood lycopene levels and the risk of cancer at a defined anatomical site. This inverse relationship was statistically significant in 35 studies. The evidence for a beneficial effect of lycopene was strongest for cancers of the prostate, lung and stomach.

Focusing on prostate cancer, several studies have reported an inverse association between plasma levels of carotenoids (including lycopene) and prostate cancer (Gann et al. 1999, Lu et al. 2001, Vogt et al. 2002, Wu et al. 2004). However, even though epidemiologic evidence strongly suggests a link between the consumption of tomatoes, tomato-based products or lycopene and the reduction in occurrence or progression of (prostate) cancer, it remains to be determined whether lycopene, the major carotenoid of tomato, is responsible for this effect.

In vitro studies have demonstrated that lycopene inhibits the proliferation of various types of human cancer cells, including endometrial, mammary and lung cancer cells (Ben-Dor et al. 2001, Nahum et al. 2001, 2006, Fornelli et al. 2006), and that it suppresses cell growth induced by insulin-like growth factor-1 (IGF-1) in human breast (MCF-7) and endometrial (ECC-1) cancer cells (Nahum et al. 2006). Under experimental conditions, lycopene (alone and in combination with other antioxidants) also inhibited the growth of the human prostate cancer cell lines Du145, PC3 and LNCaP *in vitro* (Pastori et al. 1998, Kotake-Nara et al. 2001, Tang et al. 2005). These prostate cancer cell lines accumulate lycopene, which is predominantly localized to the nuclear membranes (55%), followed by accumulation in the nuclear matrix (26%) and in microsomes (19%). LNCaP cells were able to accumulate the most lycopene: 2.5 times as much as PC-3 and 4.5 times as much as Du145. However, lycopene was found not to be a ligand for the androgen receptor (Liu et al. 2006). Studies on normal prostatic epithelial cells suggest that lycopene may inhibit cell growth and induce apoptosis in these cells (Obermuller-Jevic et al. 2003, Barber et al. 2006), thereby reducing the risk of developing BPH. Several mechanisms of action are implicated in the ability of lycopene to prevent the development and progression of prostate cancer, including reduction of oxidative DNA damage (Chen et al. 2001), initiating up-regulation of gap-junctional proteins to improve intercellular communication (Aust et al. 2003, Bertram and Vine 2005, Fornelli et al. 2006) and a reduction of local androgen signaling (for review see Wertz et al. 2004).

In vivo Models

Clinical trials usually require large cohorts, and it is difficult to control for disease stage, genetic background, diet, age and other prostate cancer risk factors. Use of experimental models would therefore greatly facilitate research on human prostate cancer. However, the presence of spontaneous prostatic adenocarcinoma is reported almost exclusively in men and dogs and has rarely been described in other (laboratory) animals. Additionally, the very poor growth of human prostate tumor tissue in culture or in immune-incompetent nude mice has been a significant drawback for prostate cancer research. As an alternative, the normal rat ventral prostate and several chemically and hormonally induced rodent models have long been used as models for prostate cancer. Recent efforts have resulted in a substantial number of human prostate cancer xenograft models (van Weerden and Romijn 2000, Marques et al. 2006) and transgenic mouse models (TRAMP, Lady, *Pten*) that are now widely applied as experimental tools for the study of prostate cancer.

Xenografts

A xenograft is a surgical graft of tissue from one species to another species. Xenografts in which human prostate cancer tissue is transplanted into an animal, usually an immune-deficient mouse, have been established. Small pieces of patient-derived tumor tissue are implanted in both shoulders of male athymic nude mice. Although subcutaneous grafting allows for easy accessibility and a high capacity, it does not resemble the prostate environment and thus may not reflect interactions that are present in the patient. Therefore, orthotopic xenografts, i.e., xenografts placed within the tissue of origin, have the advantage of providing an environment that is representative of the tissue in which the tumor originated. Human prostate xenografts are relevant models to evaluate the effects of chemopreventive therapies in patients diagnosed with prostate cancer, so-called tertiary prevention.

Transgenic mouse models

Human xenograft models are of less use to study the potential primary and secondary preventive effects of chemopreventive modalities as they typically represent established tumors of relative late stage disease. However, data obtained with tertiary prevention studies in xenografts may provide evidence for potential use in primary and secondary prevention. Transgenic mouse models may prove to be valuable in the study of the impact of chemopreventive therapies on tumor induction and early progression. In addition, they are very useful tools to study the effects of chemopreventive or therapeutic agents, and the mechanisms by which they modulate the cancer process. Several transgenic mouse models of prostate cancer have been developed. These include the TRAMP and LADY mouse models and several knockout models (for review see Kasper and Smith 2004). Targeted *Pten* knockout mice show a complete (Probasin or PSA-promotor driven) inactivation of the *Pten* tumor suppressor gene in the prostate, resulting in prostate tumor development in all mice. This conditional gene mutation model allows the study of the malignant transformation of prostate cells in the context of the appropriate, non-mutated, prostate microenvironment (Wang et al. 2003, Ma et al. 2005).

Chemoprevention in Animal Models

The effect of lycopene on prostate cancer has been studied in a number of *in vivo* models, several of which will be discussed here. Limpens et al. (2006) examined the ability of lycopene, alone and in combination with vitamin E, to reduce tumor growth in an orthotopic mouse model of human prostate cancer. They used a human prostate cancer cell line developed in Rotterdam, PC-346C, which has been described in detail elsewhere (van Weerden and Romijn 2000), for induction of prostate cancer. Two different dosages (5 or 50 mg/kg body weight daily by oral gavage) of lycopene were tested in this study. Lycopene supplementation resulted in increased levels of lycopene in the mouse liver and prostate (tumor). A non-significant, inhibitory effect of lycopene on tumor growth and PSA levels was observed (53% inhibition of tumor growth at 42 d, and an increased median survival time from 47 to 66 d). The combination of lycopene and vitamin E supplements at low dosages (5 mg/kg body weight each) had a stronger, significant, effect on tumor growth than either treatment alone: 73% inhibition of tumor growth at 42 d (Fig. 2) and an increased median survival time from 47 to 56 d. The markedly reduced tumor growth when lycopene and vitamin E are given in combination suggests a cooperative interaction (Limpens et al. 2006).

The transplantable rat MatLyLu Dunning prostate cancer model was used to study the effect of lycopene on tumor and rat prostate. Male Copenhagen rats were treated with 200 µg lycopene/g diet for 4 wk, prior to subcutaneous tumor transplantation. Tumor growth was followed for 18 d. Lycopene treatment significantly increased the necrotic area of tumors of the prostate (Siler et al. 2004, 2005). The hallmark of this lycopene effect was suppression of genes involved in steroid metabolism and signaling: a reduction in steroid 5α-reductase 1 expression as well as a reduction of a set of androgen target genes such as prostatic spermine-binding protein, prostatic steroid-binding proteins C1, C2 and C3 chain and probasin. These genes play a role in pathways that are considered to be important in prostate cancer development. Interestingly, lycopene treatment significantly inhibited local androgen signaling, IGF expression and inflammatory signals in the normal rat prostate, although this did not affect normal prostate growth.

The rat Dunning R3327-H prostate adenocarcinoma model was used to evaluate the effect of tomato and broccoli, and their combination on prostate tumor development (Canene-Adams et al. 2007). Male

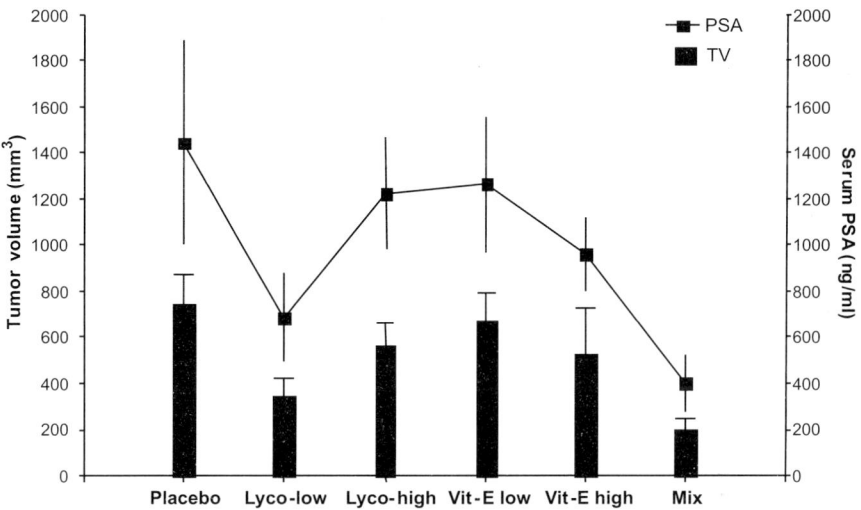

Fig. 2 Effects of lycovit/vitamin E supplementation on tumor volume and PSA secretion. Prostate tumors were induced in nude mice by an orthotopic xenograft of a human prostate cancer cell line (PC-346C). Immediately after tumor induction, mice received lycopene (LycoVit), vitamin E or a combination of the two. The graph depicts tumor volume as measured by transrectal ultrasonography of the murine prostate (TRUMP) 42 d after tumor induction, and PSA values measured at the same time point.

Copenhagen rats were fed diets containing either lycopene alone, 10% tomato or broccoli powders, or a combination of tomato and broccoli (5%:5% or 10%:10%) for 22 wk, starting 1 mon prior to receiving tumor implants subcutaneously. Lycopene reduced tumor weights by 7-18%, but this effect was not significant. In contrast, treatment with tomato, broccoli or the combination significantly reduced tumor size (by 34%, 42% and 52% respectively). These growth reductions were associated with reduced cell proliferation and increased apoptosis. The authors suggest that intake of whole foods, such as broccoli and tomatoes, could significantly affect prostate cancer development and that combining different foods may be more effective in slowing tumor growth than supplementation with a single component. Similar results were reported in the N-methyl-N-nitrosourea (NMU)-testosterone-treated rat prostate carcinoma model. Consumption of tomato powder (13 mg lycopene/kg diet), but not pure lycopene (161 mg lycopene/kg diet), inhibited prostate carcinogenesis in this model (Boileau et al. 2003). Although the authors state that this may be explained by the fact that whole tomato products contain additional compounds that add to the effect, the large difference in lycopene dose between the two groups does not allow for such a conclusion. Indeed, Limpens et al. (2006) identified in the aforementioned PC-346C xenograft study that a 10-fold higher lycopene dose was not as effective as the lower dose of lycopene. Moreover, *in vitro* studies revealed that protection of lycopene against oxidative damage followed a U-shaped curve (Lowe et al. 1999).

Administration of antioxidants (vitamin E, selenium and lycopene) as part of the diet of male *Lady* transgenic mice was shown to dramatically inhibit prostate cancer development, and increased disease-free survival (Venkateswaran et al. 2004). Treatment with antioxidants resulted in a 4-fold reduction in the incidence of prostate cancer when compared with untreated animals. Furthermore, non-quantitative immunohistochemical data suggested increased expression of proliferating cell nuclear antigen and a reduction of the p27 cell cycle inhibitor in the antioxidant-treated animals as compared to control animals. Several other studies have also demonstrated that combining lycopene treatment with another antioxidant may be more beneficial than either treatment alone (Boileau et al. 2003, Siler et al. 2005, Limpens et al. 2006, Canene-Adams et al. 2007).

The use of different model systems and administration methods of lycopene make comparison of the aforementioned *in vivo* studies difficult. Identification of the active components of whole foods remains important, however, since a well-defined, safe combination of chemopreventive agents may turn out to be beneficial in addition to conventional forms of cancer prevention and treatment.

Besides carotenoids and antioxidants, soy, a major dietary constituent of the Eastern diet, has also been associated with a reduced prostate cancer risk. Genistein, the predominant phytoestrogen in soy food, was tested for its chemopreventive effects in TRAMP mice. Indeed, lifelong supplementation of the TRAMP mice with genistein resulted in a 50% reduction in poorly differentiated cancerous lesions. Whether the chemopreventive effects of genistein were related to its anti-estrogenic and tyrosine kinase inhibitory effect is unknown (Wang et al. 2007).

Androgens are involved in the development of prostate cancer. Whether lycopene is able to modulate androgen levels was studied in 8-wk-old male F334 rats (Campbell et al. 2006). F334 rats were castrated or sham-operated and subsequently received daily oral supplementation of 0.7 mg/day phytofluene, 0.7 mg/day lycopene, or a diet supplemented with 10% tomato powder for 4 d. Sham-operated rats receiving phytofluene, lycopene or tomato powder had a 40-50% reduction in serum testosterone levels when compared to the sham-operated control group. Overall, the results from this study indicate that short-term intake of tomato carotenoids significantly alters androgen status, which may be a mechanism by which tomato intake reduces prostate cancer risk. In contrast, Venkateswaran et al. (2004) showed that dietary supplementation with antioxidants (vitamin E, selenium and lycopene) for 34 wk did not affect plasma testosterone levels in their Lady transgenic mice.

Since prostate tumors are initially androgen-dependent, chemo-prevention studies have also focused on the use of antiandrogens. The effect of the antiandrogen flutamide on the development of prostate cancer in the TRAMP model was studied by Raghow et al. (2000). Two doses of flutamide were tested: 6.6 mg/kg and 33 mg/kg. The efficacy of flutamide in inhibiting prostate tumor formation was measured by the absence of palpable tumor formation. The low-dose group did not differ significantly from the placebo group, where palpable tumors were found at 17 wk of age. In the high-dose flutamide group the development of tumors was retarded, and tumors could not be observed until 24 wk of age, which represents a lag period of 7 wk. In contrast to the placebo and low-flutamide groups, 42% of the animals in the high-dose group remained tumor free till the age of 34 wk. Therefore, flutamide is shown to suppress tumor development and decrease the incidence of prostate cancer in the TRAMP mice.

We are presently conducting a chemoprevention study in which one group receives the antiandrogen bicalutamide by oral gavage. Prostate tumors were induced in nude mice by an orthotopic xenograft of a human prostate cancer cell line (PC-346C). Immediately after tumor induction, mice were given supplements of bicalutamide at 25 mg/kg body weight/day. Tumor volume was measured by transrectal ultra-sonography of the murine prostate. Mice receiving bicalutamide had a 83% reduction in tumor size 42 d after inducing prostate cancer, and an increased survival rate as compared to the placebo group (unpublished observation).

Human Intervention Studies

Ideally, the effect of chemopreventive agents should be studied in primary prevention studies in humans. Since prostate cancer has a long latency period, a study with survival as endpoint would require at least 10-15 yr of follow-up to reach this endpoint, and a very large study population. The largest ongoing intervention study for prostate cancer prevention is the SELECT trial in the United States that has recruited 32,400 men and has a follow-up of at least 12 yr (Brawley and Parnes 2000). Clearly, such trials cannot be conducted for all potential chemopreventive combinations of interest. As an alternative, small-scale tertiary prevention studies are being developed in patients who have been diagnosed with prostate cancer, aiming to alter the course of minimal disease. Such tertiary intervention trials provide a relatively rapid, efficient, and cost-effective alternative.

Furthermore, it seems likely that if these interventions are slowing down tumor growth and tumor progression, they might also be effective in primary and secondary prevention.

Potential use of PSA as intermediate marker for treatment response

Intervention studies in patients with prostate cancer are scarce, since it is difficult to determine which are the appropriate endpoint markers to use. One of the most important biomarkers for prostate cancer is PSA. Indeed, PSA seems to be affected by lycopene treatment, as indicated from a study by Edinger and Koff (2006). In their study, the effect of tomato paste consumption on plasma PSA levels was determined in patients with BPH. They showed a significant 10.77% reduction of PSA after daily supplementation of 50 g tomato paste for 10 wk. The authors suggested that the high amount of lycopene in the tomato paste could be the cause of the decline of plasma PSA levels.

Progression of prostate cancer is usually accompanied by a rise in circulating levels of PSA. Any intervention that affects PSA levels suggests an impact on the progression of the disease, but this can only be validated when the tumor is localized and available for (biomarker) assessment. In contrast to short-term chemopreventive studies as described above, where tumor tissue becomes available for evaluation of effects, intervention therapy in progressive prostate cancer patients after local treatment is limited to PSA as the only potential biomarker. These patients with tumor progression, as based on the rise in plasma PSA levels after initial local treatment, have metastatic disease of unknown origin and thus PSA is the only indicator of the extent of the disease. However, use of PSA as potential biomarker of tumor response in chemopreventive studies requires a thorough validation of the relationship between PSA and prostate tumor mass. Studies by Thalmann et al. (1996) have shown that treatment of prostate xenograft-bearing mice with suramin, a growth factor inhibitor, clearly affected PSA levels in these animals but did not have any effect on tumor volume. Clearly, this work shows that specific treatments may affect the synergy of PSA and tumor growth and thus abrogate its use as response marker. For PSA kinetics to be used as biomarker in chemopreventive studies, experimental validation studies are essential to determine the reliability of PSA in reflecting tumor response to treatment. This approach was first suggested by Schröder et al. (2000), proposing an algorithm that includes experimental xenograft studies to validate the use of PSA as biomarker in parallel PSA-based clinical phase II and II trials (Table 2).

Table 2 Hormone-independent prostate cancer*: Proposal PSA-based phase II–III clinical evaluation (ancillary endpoints to be included)(adapted from (Schröder et al. 2000), reprinted with permission of Wiley-Liss, Inc., a subsidiary of John Wiley and Sons, Inc.)

Phase II study (PSA-based)	Experimental confirmation study (growth response)	Action on further compound development
1 Negative**	Negative	Discard drug
2 Negative	Positive	Kinetics? Dose adjustment? Discard drug?
3 Positive***	Negative	Different model? Different endpoints? Traditional phase II study
4 Positive	Positive	Phase III study ↓ Endpoints PSA response/progression

Differences (between arms)   No differences (between arms)

Continue (progression, survival, difference?)   Discontinue

*As defined by rising PSA under endocrine treatment.
**Negative study: no PSA response.
***Positive study: PSA response and response ancillary endpoints.

Several studies have been performed using this concept. Kranse et al. (2005) studied the effect of a dietary supplement containing plant estrogens, antioxidants (including carotenoids such as lycopene) and selenium on total and free PSA levels in 37 hormonally untreated men with prostate cancer and rising PSA levels. The study was designed as a randomized, placebo-controlled, double-blind crossover study in which treatment periods of 6 wk were separated by a 2 wk washout period. This dietary supplement was found to increase free PSA doubling time. In addition to PSA levels, the effect of the supplement on male sex hormone levels was studied showing a reduction in 5α-dihydrotestosterone and testosterone levels. Whether these reduced androgen levels are responsible for the reduction in PSA or the reduced PSA levels truly reflect a slowing of disease progression needs further confirmation.

Following this study, a second study was performed to evaluate the chemopreventive action of a dietary supplement containing, among other nutrients, soy isoflavone, lycopene, ascorbic acid and α-tocopherol (Schroder et al. 2005). Forty-nine patients with a history of prostrate cancer and rising PSA levels after radical prostatectomy or radiotherapy were included in this study. Again, the study was designed as a randomized, placebo-controlled, double-blind crossover study with treatment periods of 10 wk separated by a 4 wk washout period. The dietary supplement used in this study was shown to significantly delay PSA progression (PSA doubling time increased from 445 d to 1150 d). In this study, no effect of the supplement on androgen levels was observed that could be the cause of the decline in PSA.

Recently, the first data have become available from a third chemopreventive study in which the effect of lycopene (15 mg daily) and vitamin E (400 IU daily) supplementation has been investigated in men with minimal prostate cancer and rising PSA after radical prostatectomy. The aim of the study was to show an effect of a dietary supplement on PSA progression as measured by the impact of the dietary supplement on PSA doubling time. The study was performed according to the double-blind, randomized two-arm crossover study design. Eighty patients were included, of which 69 could be analyzed in the intent-to-treat analysis. Preliminary results indicate that PSA doubling times increased in the period when receiving supplement as compared to the placebo-controlled periods (unpublished observation).

Lycopene supplementation prior to surgical intervention

Studies by Bowen et al. (2002) indicated that lycopene accumulates in the prostate of patients with localized prostate adenocarcinoma. Similarly, Limpens et al. (2006) found high levels of lycopene in the mouse liver and prostate tumor after oral lycopene supplementation for 90 d. A small number of short-term intervention studies have been conducted in patients prior to surgical intervention (Chen et al. 2001, Kucuk et al. 2001). In these studies, a variety of tissue biomarkers were used such as leukocyte oxidative DNA damage, elements of the IGF pathway, biomarkers of differentiation, proliferation and apoptosis, or assessment of disease progression from BPH to high-grade prostatic intra-epithelial neoplasia (HGPIN), in addition to PSA. In the study by Kucuk et al. (2001, 2002), 26 patients were randomly assigned to receive 15 mg lycopene twice daily or no supplementation for 3 wk prior to radical prostatectomy. Treatment with lycopene decreased PSA levels with 18% during the treatment period. In contrast, the control group showed increases in PSA levels of 14%. In addition to differences in PSA levels, lycopene-treated patients had smaller tumors and less involvement of surgical margins and/or extra-prostatic tissues with cancer. Connexin 43 expression was also increased in the lycopene-treated group. Even though these data indicate a positive effect of lycopene treatment, none of the aforementioned effects reached statistical significance, probably because of small sample size, and these results should be interpreted cautiously. Chen et al. (2001) studied the effects of consumption of tomato sauce–based pasta dishes on lycopene uptake, oxidative DNA damage and PSA levels in patients already diagnosed with prostate cancer. In the 3 wk prior to scheduled radical prostatectomy, 32 patients consumed tomato sauce–based pasta dishes, which amounts to 30 mg lycopene per day. Dietary intervention significantly increased serum and prostate lycopene levels while decreasing PSA levels. In addition, oxidative damage was significantly reduced in leukocytes and prostate tissue. These data indicate a possible role for a tomato sauce constituent, possibly lycopene, in the treatment of prostate cancer (Chen et al. 2001).

High-grade prostatic intra-epithelial neoplasia (HGPIN) is a precursor of prostate cancer. The effectiveness of lycopene in preventing progression of HGPIN to prostate cancer was studied in 40 patients diagnosed with HGPIN after undergoing a transurethral resection. These patients were randomized into two groups, one of which received 4 mg lycopene twice daily for 1 yr and one control group. Both groups were followed up every 3 mon for 2 yr. Baseline serum lycopene and total PSA of both groups was obtained before therapy. During follow-up, serum lycopene and PSA levels were determined and a digital rectal examination performed. Prostate biopsy was performed as and when indicated during follow-up. Lycopene was shown to delay or prevent HGPIN from developing into prostate cancer. In addition, an inverse relationship between lycopene and PSA levels was found. Lycopene was found to be effective in the treatment of HGPIN, with no toxicity and good patient tolerance (Mohanty et al. 2005).

Ansari et al. compared the efficacy of combining lycopene supplementation with orchidectomy with orchidectomy alone in the management of advanced prostate cancer (Ansari and Gupta 2003). Fifty-four patients were randomized to orchidectomy alone or orchidectomy plus lycopene treatment. Lycopene supplementation was started on the day of orchidectomy at 2 mg twice daily. Patients were evaluated before and every 3 mon after surgical intervention for PSA levels, bone metastasis and uroflowmetry. Adding lycopene to orchidectomy produced a more reliable and consistent decrease in serum PSA levels than orchidectomy alone. In addition, lycopene treatment shrinks the primary tumor and diminishes the secondary tumors, providing better relief from bone pain and lower urinary tract symptoms, and improving survival compared with orchidectomy alone.

Androgen deprivation may also affect the progression of HGPIN to prostate cancer. Alberts et al. (2006) conducted a chemoprevention trial assessing the efficacy of flutamide in reducing the progression rate of HGPIN to prostate cancer. Patients were treated with 250 mg/d flutamide or placebo for 1 yr and repeat biopsies were performed at 12 and 24 mon after the start of treatment. In this study, no benefit of flutamide as a chemopreventive agent could be found in men with HGPIN.

Primary chemoprevention trials of prostate cancer

Prostate Cancer Prevention Trial (PCPT). Since androgens are involved in the development of prostate cancer, drugs that inhibit the production of 5α-dihydrotestosterone, the primary and most active androgen in the prostate, may reduce prostate cancer risk. The enzyme 5α-reductase metabolizes the conversion from testosterone to 5α-dihydrotestosterone. Therefore, 5α-reductase inhibitors may reduce the risk of prostate cancer. The PCPT was the first phase III clinical trial of prostate cancer chemoprevention.

In the PCPT, 18,882 men aged 55 yr or older with a normal digital rectal examination and PSA levels ≤ 3 ng/ml were randomly assigned treatment with finasteride (5α-reductase inhibitor) 5 mg per day or placebo for 7 yr. The primary endpoint was the prevalence of prostate cancer during the 7 yr of the study. Finasteride use significantly reduced the prevalence was reduced by 24.8% during the study period. However, more high-grade prostate cancers were observed in the finasteride group. Therefore, the benefits of finasteride use on prostate cancer prevalence and the reduced risk of urinary problems should be weighed against the increased risk of high-grade prostate cancer and sexual side effects observed (Thompson et al. 2003).

Reduction by Dutasteride of Prostate Cancer Events (REDUCE) Trial. This clinical trial is designed as an international, multicenter, double-blind, placebo-controlled chemoprevention study and aims to determine whether dutasteride 0.5 mg daily will decrease the risk of biopsy-detectable prostate cancer (Andriole et al. 2004). Dutasteride also inhibits the enzyme 5α-reductase. Eight thousand men at increased risk for prostate cancer will be randomized to receive dutasteride or placebo for 4 yr. Repeat biopsies will be taken at 2 and 4 yr and the rates of prostate cancer for each group will be compared. The results of this study remain to be determined.

Selenium and Vitamin E Chemoprevention Trial (SELECT). Currently ongoing is one of the largest prostate cancer chemoprevention studies to date, namely the Selenium and Vitamin E Chemoprevention Trial. The SELECT trial started July 2001 and is aimed at studying possible agents for the prevention of prostate cancer in a population of 32,400 healthy men (primary prevention) in the United States. SELECT is a phase III randomized, placebo-controlled trial of selenium (200 μg/d) and/or vitamin E (400 IU/d) supplementation, the duration of which will be a minimum of 7 yr (maximum of 12 yr). The primary endpoint is the clinical incidence of prostate cancer as determined by routine clinical management and confirmed by central pathology review. Vitamin E and selenium were chosen for their potent antioxidant properties that exert their effect primarily in cellular membranes preventing lipid peroxidation and oxidative stress (Pak et al. 2002). Patient enrollment was completed June 2004 (Lippman et al. 2005). The results of this study are eagerly awaited and will contribute to the understanding of the role of antioxidants in prostate cancer prevention.

CONCLUSION

Chemoprevention of prostate cancer is a subject of increasing interest in the management of prostate cancer, especially since the number of patients that are diagnosed with prostate cancer at a very early stage of the disease has increased rapidly. Tertiary prevention aiming to alter the course of the disease may be a very attractive treatment option for patients who have small, localized tumors and are eligible for active surveillance protocols. Epidemiological data indicate that a number of dietary substances may be of interest as chemopreventive agents, such as lycopene, the major constituent of tomatoes. Experimental data show that lycopene or tomato-based products reduce tumor growth and PSA release in both *in vitro* and *in vivo* models of prostate cancer. Also, short-term intervention studies in patients who follow diets supplemented with lycopene or tomato-based products prior to operation indicate that lycopene is taken up by the prostate and affects blood PSA levels and tissue biomarkers indicative for tumor inhibition. Finally, a small number of PSA-based randomized double-blind crossover tertiary intervention studies revealed that lycopene-containing dietary supplementation reduced PSA levels and delayed PSA progression in patients with metastatic disease. Parallel dietary studies in human prostate xenografts showed PSA response to be related to tumor growth response, suggesting that the observed PSA effects in metastatic patients might be attributed to an inhibitory effect on the tumor. The present data on lycopene as a potential chemopreventive agent are promising. Further identification of the best combination of dietary substances and the optimal supplementation regime is warranted prior to implementation of lycopene-based (tertiary) chemoprevention in the management of prostate cancer.

ABBREVIATIONS

BPH: benign prostate hyperplasia; HGPIN: high-grade prostatic intra-epithelial neoplasia; IGF-1: insulin-like growth factor-1; PIN: prostatic intra-epithelial neoplasia; PSA: prostate-specific antigen; TRAMP: transgenic adenocarcinoma of the mouse prostate; TRUMP: transrectal ultrasonography of the murine prostate

References

Akazaki, K. and G.N. Stemmerman. 1973. Comparative study of latent carcinoma of the prostate among Japanese in Japan and Hawaii. J. Nat. Cancer Inst. 50: 1137-1144.

Alberts, S.R. and P.J. Novotny, J.A. Sloan, J. Danella, D.G. Bostwick, T.J. Sebo, M.L. Blute, T.R. Fitch, R. Levitt, R. Lieberman, and C.L. Loprinzi. 2006. Flutamide in men with prostatic intraepithelial neoplasia: a randomized, placebo-controlled chemoprevention trial. Amer. J. Ther. 13: 291-297.

Andriole, G. and D. Bostwick, O. Brawley, L. Gomella, M. Marberger, D. Tindall, S. Breed, M. Somerville, R. Rittmaster, and R.S. Group. 2004. Chemoprevention of prostate cancer in men at high risk: rationale and design of the reduction by dutasteride of prostate cancer events (REDUCE) trial. J. Urol. 172: 1314-1317.

Ansari, M.S. and N.P. Gupta. 2003. A comparison of lycopene and orchidectomy vs orchidectomy alone in the management of advanced prostate cancer. BJU Int. 92: 375-378; discussion 378.

Aust, O. and N. Ale-Agha, L. Zhang, H. Wollersen, H. Sies, and W. Stahl. 2003. Lycopene oxidation product enhances gap junctional communication. Food Chem. Toxicol. 41: 1399-1407.

Barber, N.J. and X. Zhang, G. Zhu, R. Pramanik, J.A. Barber, F.L. Martin, J.D. Morris, and G.H. Muir. 2006. Lycopene inhibits DNA synthesis in primary prostate epithelial cells in vitro and its administration is associated with a reduced prostate-specific antigen velocity in a phase II clinical study. Prostate Cancer Prostatic Dis. 9: 407-413.

Ben-Dor, A. and A. Nahum, M. Danilenko, Y. Giat, W. Stahl, H.D. Martin, T. Emmerich, N. Noy, J. Levy, and Y. Sharoni. 2001. Effects of acyclo-retinoic acid and lycopene on activation of the retinoic acid receptor and proliferation of mammary cancer cells. Arch. Biochem. Biophys. 391: 295-302.

Bertram, J.S. and A.L. Vine. 2005. Cancer prevention by retinoids and carotenoids: independent action on a common target. Biochem. Biophys. Acta 1740: 170-178.

Black, R.J. and F. Bray, J. Ferlay, and D.M. Parkin. 1997. Cancer incidence and mortality in the European Union: cancer registry data and estimates of national incidence for 1990. Eur. J. Cancer 33: 1075-1107.

Boileau, T.W. and Z. Liao, S. Kim, S. Lemeshow, J.W. Erdman Jr., and S.K. Clinton. 2003. Prostate carcinogenesis in N-methyl-N-nitrosourea (NMU)-testosterone-treated rats fed tomato powder, lycopene, or energy-restricted diets. J. Natl. Cancer Inst. 95: 1578-1586.

Bowen, P. and L. Chen, M. Stacewicz-Sapuntzakis, C. Duncan, R. Sharifi, L. Ghosh, H.S. Kim, K. Christov-Tzelkov, and R. van Breemen. 2002. Tomato sauce supplementation and prostate cancer: lycopene accumulation and modulation of biomarkers of carcinogenesis. Exp. Biol. Med. (Maywood) 227: 886-893.

Brawley, O.W. and H. Parnes. 2000. Prostate cancer prevention trials in the USA. Eur. J. Cancer 36: 1312-1315.

Breslow, N. and C.W. Chan, G. Dhom, R.A. Drury, L.M. Franks, B. Gellei, Y.S. Lee, S. Lundberg, B. Sparke, N.H. Sternby, and H. Tulinius. 1977. Latent carcinoma of prostate at autopsy in seven areas. The International Agency for Research on Cancer, Lyons, France. Int. J. Cancer 20: 680-688.

Campbell, J.K. and C.K. Stroud, M.T. Nakamura, M.A. Lila, and J.W. Erdman Jr. 2006. Serum testosterone is reduced following short-term phytofluene, lycopene, or tomato powder consumption in F344 rats. J. Nutr. 136: 2813-2819.

Canene-Adams, K. and B.L. Lindshield, S. Wang, E.H. Jeffery, S.K. Clinton, and J.W. Erdman Jr. 2007. Combinations of tomato and broccoli enhance antitumor activity in dunning r3327-h prostate adenocarcinomas. Cancer Res. 67: 836-843.

Chen, L. and M. Stacewicz-Sapuntzakis, C. Duncan, R. Sharifi, L. Ghosh, R. van Breemen, D. Ashton, and P.E. Bowen. 2001. Oxidative DNA damage in prostate cancer patients consuming tomato sauce-based entrees as a whole-food intervention. J. Natl. Cancer Inst. 93: 1872-1879.

Crawford, E.D. 2003. Epidemiology of prostate cancer. Urology 62: 3-12.

Di Mascio, P. and S. Kaiser, and H. Sies. 1989. Lycopene as the most efficient biological carotenoid singlet oxygen quencher. Arch. Biochem. Biophys. 274: 532-538.

Edinger, M.S. and W.J. Koff. 2006. Effect of the consumption of tomato paste on plasma prostate-specific antigen levels in patients with benign prostate hyperplasia. Braz. J. Med. Biol. Res. 39: 1115-1119.

Fornelli, F. and A. Leone, I. Verdesca, F. Minervini, and G. Zacheo. 2006. The influence of lycopene on the proliferation of human breast cell line (MCF-7). 2007 Toxicol. In Vitro 21: 217-223.

Gann, P.H. and J. Ma, E. Giovannucci, W. Willett, F.M. Sacks, C.H. Hennekens, and M.J. Stampfer. 1999. Lower prostate cancer risk in men with elevated plasma lycopene levels: results of a prospective analysis. Cancer Res. 59: 1225-1230.

Giovannucci, E. 1999. Tomatoes, tomato-based products, lycopene, and cancer: review of the epidemiologic literature. J. Natl. Cancer Inst. 91: 317-331.

Greenlee, R.T. and T. Murray, S. Bolden, and P.A. Wingo. 2000. Cancer statistics, 2000. CA Cancer J. Clin. 50: 7-33.

Haenszel, W. and M. Kurihara. 1968. Studies of Japanese migrants. I. Mortality from cancer and other diseases among Japanese in the United States. J. Natl. Cancer Inst. 40: 43-68.

Jemal, A. and R.C. Tiwari, T. Murray, A. Ghafoor, A. Samuels, E. Ward, E.J. Feuer, M.J. Thun, and S. American Cancer. 2004. Cancer statistics, 2004. CA Cancer J. Clin. 54: 8-29.

Jensen, O.M. and J. Esteve, H. Moller, and H. Renard. 1990. Cancer in the European Community and its member states. Eur. J. Cancer 26: 1167-1256.

Kasper, S. and J.A. Smith Jr. 2004. Genetically modified mice and their use in developing therapeutic strategies for prostate cancer. J. Urol. 172: 12-19.

Klotz, L.H. 2005. Active surveillance with selective delayed intervention: walking the line between overtreatment for indolent disease and undertreatment for aggressive disease. Can. J. Urol. 12 Suppl 1: 53-57; discussion 101-102.

Kolonel, L.N. and A.M. Nomura, M.W. Hinds, T. Hirohata, J.H. Hankin, and J. Lee. 1983. Role of diet in cancer incidence in Hawaii. Cancer Res. 43: 2397s-2402s.

Kotake-Nara, E. and M. Kushiro, H. Zhang, T. Sugawara, K. Miyashita, and A. Nagao. 2001. Carotenoids affect proliferation of human prostate cancer cells. J. Nutr. 131: 3303-3306.

Kranse, R. and P.C. Dagnelie, M.C. van Kemenade, F.H. de Jong, J.H. Blom, L.B. Tijburg, J.A. Weststrate, and F.H. Schroder. 2005. Dietary intervention in prostate cancer patients: PSA response in a randomized double-blind placebo-controlled study. Int. J. Cancer 113: 835-840.

Kucuk, O. and F.H. Sarkar, Z. Djuric, W. Sakr, M.N. Pollak, F. Khachik, M. Banerjee, J.S. Bertram, and D.P. Wood Jr. 2002. Effects of lycopene supplementation in patients with localized prostate cancer. Exp. Biol. Med. (Maywood) 227: 881-885.

Kucuk, O. and F.H. Sarkar, W. Sakr, Z. Djuric, M.N. Pollak, F. Khachik, Y.W. Li, M. Banerjee, D. Grignon, J.S. Bertram, J.D. Crissman, E.J. Pontes, and D.P. Wood Jr. 2001. Phase II randomized clinical trial of lycopene supplementation before radical prostatectomy. Cancer Epidemiol. Biomarkers Prev. 10: 861-868.

Lamb, D.J. and L. Zhang. 2005. Challenges in prostate cancer research: animal models for nutritional studies of chemoprevention and disease progression. J. Nutr. 135: 3009S-3015S.

Limpens, J. and F.H. Schroder, C.M. de Ridder, C.A. Bolder, M.F. Wildhagen, U.C. Obermuller-Jevic, K. Kramer, and W.M. van Weerden. 2006. Combined lycopene and vitamin E treatment suppresses the growth of PC-346C human prostate cancer cells in nude mice. J. Nutr. 136: 1287-1293.

Lippman, S.M. and P.J. Goodman, E.A. Klein, H.L. Parnes, I.M. Thompson Jr., A.R. Kristal, R.M. Santella, J.L. Probstfield, C.M. Moinpour, D. Albanes, P.R. Taylor, L.M. Minasian, A. Hoque, S.M. Thomas, J.J. Crowley, J.M. Gaziano, J.L. Stanford, E.D. Cook, N.E. Fleshner, M.M. Lieber, P.J. Walther, F.R. Khuri, D.D. Karp, G.G. Schwartz, L.G. Ford, and C.A. Coltman Jr. 2005. Designing the Selenium and Vitamin E Cancer Prevention Trial (SELECT). J. Natl. Cancer Inst. 97: 94-102.

Liu, A. and N. Pajkovic, Y. Pang, D. Zhu, B. Calamini, A.L. Mesecar, and R.B. van Breemen. 2006. Absorption and subcellular localization of lycopene in human prostate cancer cells. Mol. Cancer Ther. 5: 2879-2885.

Lowe, G.M. and L.A. Booth, A.J. Young, and R.F. Bilton. 1999. Lycopene and beta-carotene protect against oxidative damage in HT29 cells at low concentrations but rapidly lose this capacity at higher doses. Free Radical Res. 30: 141-151.

Lu, Q.Y. and J.C. Hung, D. Heber, V.L. Go, V.E. Reuter, C. Cordon-Cardo, H.I. Scher, J.R. Marshall, and Z.F. Zhang. 2001. Inverse associations between plasma lycopene and other carotenoids and prostate cancer. Cancer Epidemiol. Biomarkers Prev. 10: 749-756.

Ma, X. and A.C. Ziel-van der Made, B. Autar, H.A. van der Korput, M. Vermeij, P. van Duijn, K.B. Cleutjens, R. de Krijger, P. Krimpenfort, A. Berns, T.H. van der Kwast, and J. Trapman. 2005. Targeted biallelic inactivation of Pten in the mouse prostate leads to prostate cancer accompanied by increased epithelial cell proliferation but not by reduced apoptosis. Cancer Res. 65: 5730-5739.

Marques, R.B. and W.M. van Weerden, S. Erkens-Schulze, C.M. de Ridder, C.H. Bangma, J. Trapman, and G. Jenster. 2006. The human PC346 xenograft and cell line panel: a model system for prostate cancer progression. Eur. Urol. 49: 245-257.

Mohanty, N.K. and S. Saxena, U.P. Singh, N.K. Goyal, and R.P. Arora. 2005. Lycopene as a chemopreventive agent in the treatment of high-grade prostate intraepithelial neoplasia. Urol. Oncol. 23: 383-385.

Nahum, A. and K. Hirsch, M. Danilenko, C.K. Watts, O.W. Prall, J. Levy, and Y. Sharoni. 2001. Lycopene inhibition of cell cycle progression in breast and endometrial cancer cells is associated with reduction in cyclin D levels and retention of p27(Kip1) in the cyclin E-cdk2 complexes. Oncogene 20: 3428-3436.

Nahum, A. and L. Zeller, M. Danilenko, O.W. Prall, C.K. Watts, R.L. Sutherland, J. Levy, and Y. Sharoni. 2006. Lycopene inhibition of IGF-induced cancer cell growth depends on the level of cyclin D1. Eur. J. Nutr. 45: 275-282.

Obermuller-Jevic, U.C. and E. Olano-Martin, A.M. Corbacho, J.P. Eiserich, A. van der Vliet, G. Valacchi, C.E. Cross, and L. Packer. 2003. Lycopene inhibits the growth of normal human prostate epithelial cells *in vitro*. J. Nutr. 133: 3356-3360.

Pak, R.W. and V.J. Lanteri, J.R. Scheuch, and I.S. Sawczuk. 2002. Review of vitamin E and selenium in the prevention of prostate cancer: implications of the selenium and vitamin E chemoprevention trial. Integr. Cancer Ther. 1: 338-344.

Parker, C. 2003. Active surveillance: an individualized approach to early prostate cancer. BJU Int. 92: 2-3.

Pastori, M. and H. Pfander, D. Boscoboinik, and A. Azzi. 1998. Lycopene in association with alpha-tocopherol inhibits at physiological concentrations proliferation of prostate carcinoma cells. Biochem. Biophys. Res. Commun. 250: 582-585.

Raghow, S. and E. Kuliyev, M. Steakley, N. Greenberg, and M.S. Steiner. 2000. Efficacious chemoprevention of primary prostate cancer by flutamide in an autochthonous transgenic model. Cancer Res. 60: 4093-4097.

Schroder, F.H. 2005. Detection of prostate cancer: the impact of the European Randomized Study of Screening for Prostate Cancer (ERSPC). Can. J. Urol. 12 Suppl 1: 2-6; discussion 92-93.

Schroder, F.H. and R. Kranse, N. Barbet, W.C. Hop, A. Kandra, and M. Lassus. 2000. Prostate-specific antigen: A surrogate endpoint for screening new agents against prostate cancer? Prostate 42: 107-115.

Schroder, F.H. and M.J. Roobol, E.R. Boeve, R. de Mutsert, S.D. Zuijdgeest-van Leeuwen, I. Kersten, M.F. Wildhagen, and A. van Helvoort. 2005. Randomized, double-blind, placebo-controlled crossover study in men with prostate cancer and rising PSA: effectiveness of a dietary supplement. Eur. Urol. 48: 922-930; discussion 930-921.

Shimizu, H. and R.K. Ross, L. Bernstein, R. Yatani, B.E. Henderson, and T.M. Mack. 1991. Cancers of the prostate and breast among Japanese and white immigrants in Los Angeles County. Br. J. Cancer 63: 963-966.

Siler, U. and L. Barella, V. Spitzer, J. Schnorr, M. Lein, R. Goralczyk, and K. Wertz. 2004. Lycopene and vitamin E interfere with autocrine/paracrine loops in the Dunning prostate cancer model. Faseb. J. 18: 1019-1021.

Siler, U. and A. Herzog, V. Spitzer, N. Seifert, A. Denelavas, P.B. Hunziker, L. Barella, W. Hunziker, M. Lein, R. Goralczyk, and K. Wertz. 2005. Lycopene effects on rat normal prostate and prostate tumor tissue. J. Nutr. 135: 2050S-2052S.

Steinmetz, K.A. and J.D. Potter. 1996. Vegetables, fruit, and cancer prevention: a review. J. Amer. Diet Assoc. 96: 1027-1039.

Tang, L. and T. Jin, X. Zeng, and J.S. Wang. 2005. Lycopene inhibits the growth of human androgen-independent prostate cancer cells in vitro and in BALB/c nude mice. J. Nutr. 135: 287-290.

Taylor, J.D. and T.M. Holmes, and G.M. Swanson. 1994. Descriptive epidemiology of prostate cancer in metropolitan Detroit. Cancer 73: 1704-1707.

Thalmann, G.N. and R.A. Sikes, S.M. Chang, D.A. Johnston, A.C. von Eschenbach, and L.W. Chung. 1996. Suramin-induced decrease in prostate-specific antigen expression with no effect on tumor growth in the LNCaP model of human prostate cancer. J. Natl. Cancer Inst. 88: 794-801.

Thompson, I.M. and P.J. Goodman, C.M. Tangen, M.S. Lucia, G.J. Miller, L.G. Ford, M.M. Lieber, R.D. Cespedes, J.N. Atkins, S.M. Lippman, S.M. Carlin, A. Ryan, C.M. Szczepanek, J.J. Crowley, and C.A. Coltman Jr. 2003. The influence of finasteride on the development of prostate cancer. N. Engl. J. Med. 349: 215-224.

van't Veer, P. and M.C. Jansen, M. Klerk, and F.J. Kok. 2000. Fruits and vegetables in the prevention of cancer and cardiovascular disease. Public Health Nutr. 3: 103-107.

van den Bergh, R.C. and S. Roemeling, M.J. Roobol, W. Roobol, F.H. Schroder, and C.H. Bangma. 2007. Prospective Validation of Active Surveillance in Prostate Cancer: The PRIAS Study. Eur. Urol., Volume 52, Issue 6, pp. 1560-1563.

van der Heul-Nieuwenhuijsen, L. and P.J. Hendriksen, T.H. van der Kwast, and G. Jenster. 2006. Gene expression profiling of the human prostate zones. BJU Int. 98: 886-897.

van Weerden, W.M. and J.C. Romijn. 2000. Use of nude mouse xenograft models in prostate cancer research. Prostate 43: 263-271.

Venkateswaran, V. and N.E. Fleshner, L.M. Sugar, and L.H. Klotz. 2004. Antioxidants block prostate cancer in lady transgenic mice. Cancer Res. 64: 5891-5896.

Vogt, T.M. and S.T. Mayne, B.I. Graubard, C.A. Swanson, A.L. Sowell, J.B. Schoenberg, G.M. Swanson, R.S. Greenberg, R.N. Hoover, R.B. Hayes, and R.G. Ziegler. 2002. Serum lycopene, other serum carotenoids, and risk of prostate cancer in US Blacks and Whites. Amer. J. Epidemiol. 155: 1023-1032.

Wang, J. and I.E. Eltoum, and C.A. Lamartiniere. 2007. Genistein chemoprevention of prostate cancer in TRAMP mice. J. Carcinog. 6: 3.

Wang, S. and J. Gao, Q. Lei, N. Rozengurt, C. Pritchard, J. Jiao, G.V. Thomas, G. Li, P. Roy-Burman, P.S. Nelson, X. Liu, and H. Wu. 2003. Prostate-specific deletion of the murine Pten tumor suppressor gene leads to metastatic prostate cancer. Cancer Cell 4: 209-221.

Wertz, K. and U. Siler, and R. Goralczyk. 2004. Lycopene: modes of action to promote prostate health. Arch. Biochem. Biophys. 430: 127-134.

Wu, K. and J.W. Erdman Jr., S.J. Schwartz, E.A. Platz, M. Leitzmann, S.K. Clinton, V. DeGroff, W.C. Willett, and E. Giovannucci. 2004. Plasma and dietary carotenoids, and the risk of prostate cancer: a nested case-control study. Cancer Epidemiol. Biomarkers Prev. 13: 260-269.

3.2

Lycopene and Urokinase Receptor Expression in Prostate Cancer Cells

Inder Sehgal
Louisiana State University, School of Veterinary Medicine
Comparative Biomedical Sciences Department, Skip Bertman Drive
Baton Rouge LA 70803

ABSTRACT

The carotenoid lycopene, found in tomatoes, has been associated with decreasing prostate cancer risk. Potential mechanisms for this risk reduction include lycopene's status as a potent antioxidant, its inhibitory effect on cell proliferation and its ability to increase intercellular gap junctional communication. Presently, in the United States, almost 200,000 men are diagnosed with the disease and approximately 30,000 succumb to its metastatic effects. Therefore, novel treatment strategies are needed for patients who currently have the disease, especially those in advanced, i.e., metastatic status. We sought to determine whether lycopene's inhibitory properties on pre-malignancy could be extended to advanced prostate cancer by assessing effects on a cell line derived through metastatic passage, the PC-3MM2. To our surprise, we found that in this cell line, lycopene has a potentially unwanted effect of up-regulating expression of the uPAR and facilitating invasion while failing to significantly inhibit proliferation or to induce detectable levels of the gap junctional protein connexin 43 expression. These results indicate that some caution should be used with regard to use of lycopene to treat potentially advanced and metastatic prostate cancers.

A list of abbreviations is given before the references.

INTRODUCTION

Lycopene, the carotenoid that gives tomatoes their red color, has generated much recent excitement as a potential chemopreventive for prostate cancer. This excitement is founded on an increasing number of epidemiological and prospective studies that associated lycopene consumption with reduced prostate cancer risk (Minorsky 2002, Barber and Barber 2002). In addition, oral intake of lycopene prior to prostatectomy has been associated with reduced incidence of high-grade prostatic intraepithelial neoplasia, increased apoptotic index, smaller tumors and reduced levels of prostate-specific antigen (Kucuk et al. 2002), suggesting a potential use for lycopene in patients with pre-existing prostate cancer in an early state of development. Although not all studies support the association of lycopene with reduced prostate cancer risk (Barber and Barber 2002, Hadley et al. 2002, Schuurman et al. 2002), much attention has been given to those studies that do show risk reduction. This attention has stimulated public interest in lycopene consumption, precipitating a proliferation of commercially available lycopene supplements (Minorsky 2002, Barber and Barber 2002). However, there is great variability in the amounts of lycopene used successfully in patient studies, suggesting a lack of consensus regarding exact doses necessary to confer specific protective effects. Table 1 shows the amounts and

Table 1 Successful reports of lycopene for prostate cancer chemoprevention.

Quantity of lycopene taken orally	Positive outcome reported	Form of lycopene	Reference
Approx. 5.0 mg daily	Reduced odds ratio for development of adenocarcinoma	Cooked tomatoes	Jian et al. 2005
Approx. 1.5 mg daily	Reduced odds ratio of development of adenocarcinoma in men with XRCC1 genotype	Tomatoes	Goodman et al. 2006
4 mg twice daily	Significantly reduced PSA, delayed prostate cancer in men with HGPIN	Supplement	Mohanty et al. 2005
30 mg/daily for 3 wk prior to radical prostatectomy	Decreased serum PSA, higher apoptotic index and reduced DNA damage in prostate cancer cells	Tomato sauce-based pasta	Bowen et al. 2002
15 mg twice daily for 3 wk preceding radical prostatectomy	Decreased PSA, smaller tumors and greater proportion of organ-confined disease in men with newly diagnosed prostate cancer	Lycopene supplement in the form of a tomato oleoresin extract (Lyc-O-Mato®)	Kucuk et al. 2002

Table summarizes oral doses of lycopene with reported benefits in preventing prostate cancer occurrence or reducing progress of pre-existing disease vs. men without lycopene supplementation.

frequency of lycopene consumed, the form in which it was taken orally, and benefit with regard to prostate cancer from recent studies that have reported positive outcomes from lycopene consumption. These data suggest that even modest quantities of lycopene (< 5.0 mg/d) can reduce the relative risk of prostate cancer in some instances, while moderate levels (30 mg/d) may reduce the size of early primary tumors if taken a few weeks prior to prostatectomy. Pharmacokinetic data indicate that plasma lycopene levels attained with supplements covering a wide range of doses (15-90 mg/d) are similar (Clark et al. 2006); therefore, it would seem that intake of quantities above 60 mg/d may be unnecessary.

Mechanistically, lycopene's antioxidant properties likely play a major role in tumor prevention; however, other reported effects could potentially be efficacious against established cancer cells. These effects would include lycopene's anti-proliferative properties and its ability to increase connexin 43 expression, which enhances gap junction intracellular communication between pre-neoplastic and normal epithelial cells (Kucuk et al. 2002, Heber and Lu 2002).

To date, most lycopene studies involve its effects on inhibiting early stages of prostate carcinogenesis — stages in which chemoprevention is effective (Minorsky 2002, Barber and Barber 2002, Kucuk et al. 2002, Hadley et al. 2002, Giovannucci 2002, Gann et al. 1999, Bowen et al. 2002). However, there are few investigations on effects of lycopene on advanced prostate cancer cells. Such investigations seem warranted, since lycopene has shown a beneficial effect in reducing the grade in patients with organ-confined cancer and because of its effects beyond protection from oxidative damage (Kucuk et al. 2002, Heber and Lu 2002). In addition, patients with existing advanced disease who hear or read the advertising may be tempted to use lycopene as a complementary treatment.

Advanced prostate cancer cells, particularly those that successfully metastasize, are different from developing cells within the primary tumor. Metastatic cells have acquired abilities to invade, intravasate, extravasate and proliferate at secondary sites such as bone and therefore, because of their altered phenotype, could respond differently to lycopene than neoplastic, non-metastatic cells.

One critical characteristic that metastatic cancer cells have acquired is the ability to dissolve basement membranes and the extracellular matrix. This degradative process is mediated largely by matrix metalloproteinases and the urokinase plasminogen activator (uPA) system consisting of uPA, its receptor uPAR and the inhibitor PAI-1 (Sidenius and Blasi, 2003). uPA activates several matrix metalloproteinases (Carmeliet et al. 1997, Davis et al. 2001) in addition to activating plasmin, which degrades several matrix proteins. uPA binds to an approximately 50-60 kDa glycoprotein receptor (uPAR), which is linked through glycosyl phosphatidylinositol to the outer leaf of the plasma membrane. Molecularly, uPAR consists of three separate domains forming a bowl that binds uPA within the central cavity (Barinka et al. 2006; Fig. 1). In prostate cancer, uPAR signaling enhances invasion and tumor cell growth (Festuccia et al. 1998, Rabbani and Xing 1998, Hoosein et al. 1991) and has become an attractive target for inhibition in cell lines (Festuccia et al. 1998) and animal models (Festuccia et al. 1998, Evans et al. 1997, Rabbani and Gladu 2002).

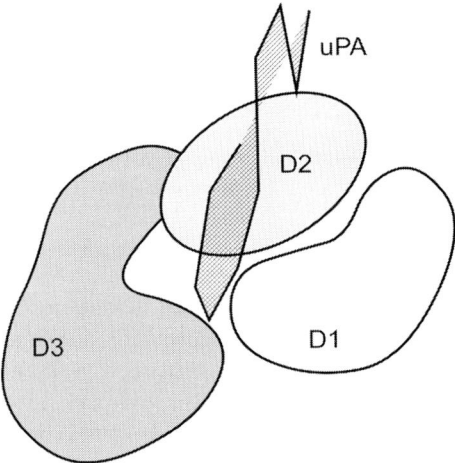

Fig. 1 Schematic representation of uPAR and associated uPA. uPAR consists of three protein domains labeled D1, D2, D3. Binding of uPA ligand occurs via a cavity formed by conformation of the three receptor domains.

We recently assessed the potential effects of lycopene on expression of uPAR and proliferation in a prostate cancer bone metastatic cell line derived from repeated metastatic selection (Forbes et al. 2003). We were surprised to find that in this PC-3MM2 cell line, lycopene has the potentially unwanted effects of up-regulating expression of the uPAR and facilitating invasion while failing to significantly inhibit proliferation or to induce detectable levels of connexin 43 expression. These results suggest caution should be used with regard to chronic use of lycopene to "treat" potentially advanced and metastatic prostate cancers.

RESULTS

In these studies, we used the human prostate cell line PC3 and breast carcinoma MCF-7, which were obtained from the American Type Culture Collection (ATCC), and the PC-3MM2 cell line, a mandibular metastasis (Pettaway et al. 1996, Delworth et al. 1995), obtained as a generous gift from Dr. Isaiah Fidler, MD Anderson Cancer Center, Houston, Texas. We have previously demonstrated that PC-3MM2 has at least two characteristics that could give it a selective advantage in metastasis—a greater level of the invasion-associated urokinase receptor (Forbes et al. 2004) and a more rapid rate of proliferation than PC-3 or PC-3M prostate cancer lines (Sehgal et al. 2003), which are two prostate cancer cell lines from which the PC-3MM2 was ultimately derived (Delworth et al. 1995). Because we had observed that uPAR levels increased with metastatic selection, we wished to determine whether lycopene could be used to reduce uPAR levels and/or invasion or proliferation in these advanced cancer cells.

Lycopene (Sigma-Aldrich Co. #L-9879, *all-trans* configuration) was added to PC-3MM2 cultures at a dose of 1.0 μM and lysates assessed for uPAR expression. The use of the 1.0 μM level was based on reported plasma and prostatic concentrations of lycopene (Pastori et al. 1998). To our surprise, lycopene treatment resulted in the increased expression of the uPAR (Fig. 2A, left panel). To examine the effects of lycopene on the other components of the uPA system, we next assessed expression of uPA and PAI-1. Our results showed that lycopene treatment had little effect on expression of uPA and PAI-1 in PC-3MM2 cell lysates (Fig. 2A, center and right panels), although activity for uPA and PAI-1 were not directly assayed. Because both uPA and PAI-1 are secreted, and uPAR can be cleaved into a soluble form, we also analyzed conditioned media for soluble uPAR, uPA and PAI-1 levels and found little to no effect of lycopene on uPA or PAI-1 but an increase in soluble uPAR (Fig. 2B). Therefore, our results indicated that lycopene selectively modulates the uPA system in this cell line by increasing cellular uPAR expression, perhaps contributing more uPAR for constitutive cleavage and release.

To determine whether this effect of lycopene stimulation on the uPAR was a general phenomenon or observed in this cell model derived from repeated metastatic passage, we checked uPAR levels in two other cancer cell lines—the PC-3 and the MCF-7. The PC-3 line was selected because it is the parental prostate cancer cell from which the PC-3MM2 was ultimately derived (Delworth et al. 1995). The MCF-7 was also used. Although these are breast cancer cells, they have previously been shown to be growth-inhibited by lycopene stimulation (Heber and Lu 2002) and we have therefore used them as a positive control. Our results demonstrated that lycopene had no effect on expression of the uPAR in the PC-3 or MCF-7 lines, although uPAR levels in MCF-7 cells were barely detectable (Fig. 3). The lack of response in these two cell lines suggests that the PC-3MM2 line has acquired a unique phenotype that can paradoxically use lycopene to enhance uPAR expression.

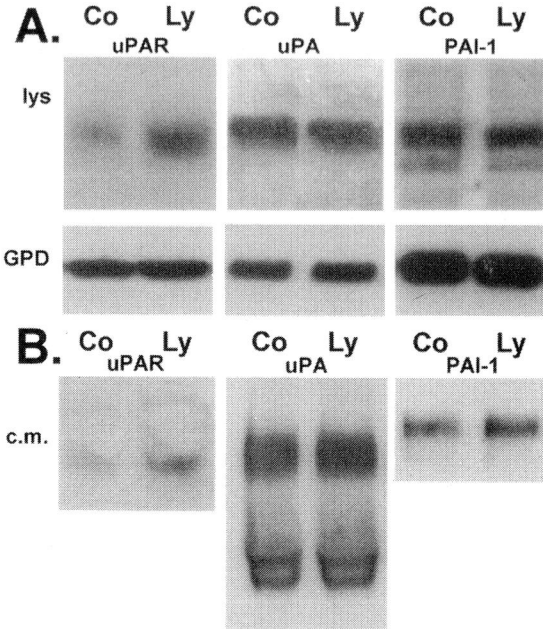

Fig. 2 Lycopene induces urokinase receptor but not uPA or PAI-1 in PC-3MM2 cells. A. Cell lysates (lys) consisting of 100 µg total protein of control (Co) and lycopene-treated (Ly) cells were immunoblotted for uPAR, uPA and PAI-1 expression. Antibodies for uPAR (R&D Systems) and connexin 43 (Sigma Chemicals) were used at dilutions of 0.5 µg/ml and 2.0 µg/ml respectively. Lycopene was used at a dose of 1.0 µM. Protein loading was verified by glyceraldehyde-phosphate dehydrogenase (GPD) expression. B. Conditioned medium (c.m.) collected and concentrated from control and lycopene-treated cells was analyzed for expression of soluble uPAR, uPA and PAI-1. Medium equal to that secreted by 100,000 cells was loaded into each well.

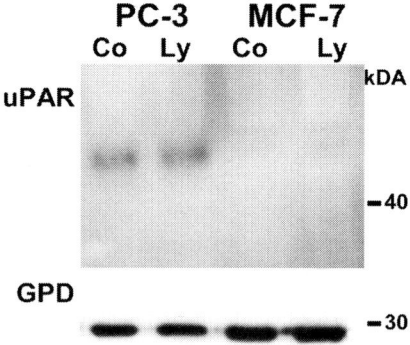

Fig. 3 Lycopene treatment does not alter urokinase receptor expression in lysates from PC-3 and MCF-7 cell lines. One hundred micrograms of total protein was immunoblotted for uPAR (upper panel) and GPD (lower panel) in control (Co) and lycopene-treated (Ly) cells.

uPAR facilitates matrix degradation through binding of uPA and subsequently localizes plasmin activation. It is possible, therefore, that increased expression of uPAR by lycopene could lead to increased invasiveness. We tested this hypothesis using a standard *in vitro* invasion assay (Hoosein et al. 1991, Greiff et al. 2002) in which Transwell polycarbonate membrane inserts with 8.0 μm pores (Falcon Corp.) were coated with 100 uL of a 1.0 mg/ml dilution of growth factor reduced Matrigel basement membrane matrix containing 5 nM uPA. Cells plated into the upper chamber were allowed to cross through the matrix towards a chemoattractant (NIH 3T3 cell conditioned medium) and attach to the under surface of the membrane over a 6 hr period. This assay revealed that, indeed, lycopene-treated cells were almost twice as invasive as untreated cells (Table 2).

Table 2 Effect of lycopene treatment on *in vitro* invasion.

	Number of cells per HPF
Control	18.2 ± 3.0
Lycopene treated	34.6 ± 6.2

PC-3MM2 cells invading over a 6 hr period are expressed as the quantity of cells per high power microscopic field (HPF). Treated cells were cultured with 1.0 μM lycopene for 24 hr prior to the assay. Data indicates the mean ± standard deviation.

Lycopene has been widely reported to inhibit cell proliferation and this effect has provided justification for its use as a prostate cancer chemopreventive (Minorsky 2002, Barber and Barber 2002, Kucuk et al. 2002, Hadley et al. 2002, Giovannucci 2002, Gann et al. 1999, Bowen et al. 2002). We tested lycopene in proliferation assays and compared its effect with effects on PC-3 and MCF-7 cell lines (Fig. 4). These proliferation assays revealed that, although lycopene inhibited growth rates of the PC-3 and MCF-7 lines over an 8 d period, it had minimal effect on growth of the PC-3MM2 cells. Therefore, it appears that the PC-3MM2 line has lost growth inhibitory response to lycopene.

The PC-3 line was previously reported to be inhibited by lycopene only with simultaneous addition of α-tocopherol (Pastori et al. 1998); however, this study used a much greater quantity of serum (10.0% vs. 1.0%) and a higher initial cell concentration and it was conducted over a shorter time period (one day). The experimental conditions we used allowed for a more sensitive assessment of lycopene-induced growth inhibition, yet there was no effect on proliferation of the PC-3MM2 line.

Lycopene has been reported to exert a stabilizing effect on pre-malignancies through a number of pathways including a critical up-regulation of cell-cell communication via the gap junction protein connexin 43 (Kucuk et al. 2002, Heber and Lu 2002). In order to determine

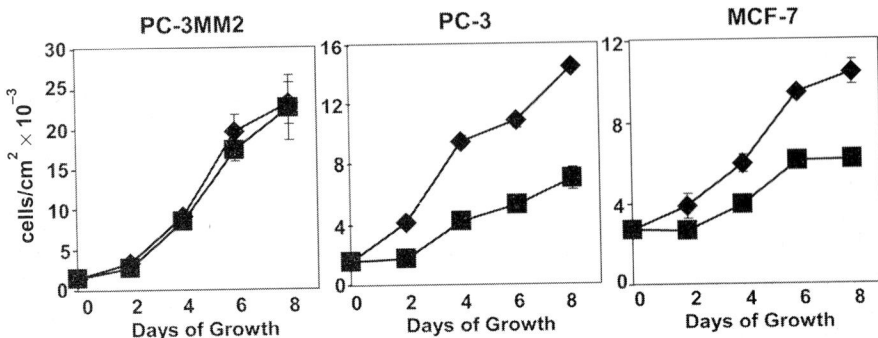

Fig. 4. Lycopene effect on proliferation of tumor cell lines. Lycopene (1.0 µM) was added to cultures of PC-3MM2, PC-3 and MCF-7 cells for an 8 d period. Growth assay medium consisted of the cell's stock media without supplements except 1.0% fetal bovine sera −/+ lycopene. Triplicate cell counts were taken every other day from treated (■) and untreated wells (◆). Error bars represent standard deviations.

whether this effect was preserved in PC-3MM2 cells, we immunoblotted cell lysates for connexin 43 expression but found no detectable quantities prior to or after lycopene treatment (Fig. 5). Interestingly, however, lycopene addition did result in increased levels in the PC-3 line. There was no change in connexin 43 levels in MCF-7 cells (data not shown). These results suggest that the effect of lycopene on connexin 43 expression in prostate cancer lines may become lost with the selections in metastatic passage.

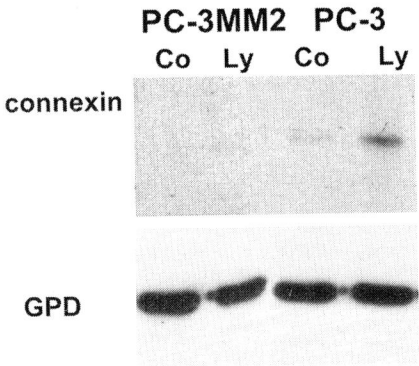

Fig. 5. Lycopene treatment does not induce detectable levels of connexin 43 expression in PC-3MM2 but does in PC-3 cells. Control (Co) and lycopene-treated (Ly) cells were immunoblotted for connexin 43 expression. Equivalence of loading was verified by GPD (lower panel).

DISCUSSION

Alternative medications that are heavily advertised as cancer preventives may be used by some patients for treatment; hence, men diagnosed with prostate cancer may consume lycopene products in an attempt to treat their disease. However, our results with the PC-3MM2 cell line do not support the thesis that lycopene use inhibits metastatic cancer cells. In fact, increased expression of uPAR in some cells by lycopene could actually enhance invasion while having little or no effect on inhibiting proliferation or cell-cell adhesion properties. Interestingly, another carotenoid supplement, beta-carotene, was once promoted as a method to decrease lung cancer in smokers, but trials with beta-carotene indicated an increased risk in some individuals (Barber and Barber 2002). It must be noted that our results were obtained exclusively with one cell line isolated experimentally from a heterogeneous population of tumor cells and our findings in culture cannot be generalized to advanced stage disease. The tumor microenvironment is a complex milieu of tumor, stroma, normal epithelium, endothelium and immunological representatives as well as growth factors, cytokines, matrix proteins and proteases. These environmental components will modulate effects of lycopene on uPAR levels and invasion *in vivo*; therefore, animal studies will be required to further assess the effects reported here. Such studies can also effectively address the potential metabolic conversion of *trans* to *cis* lycopene, tissue distribution and bioavailability.

An ideal cell culture comparison of lycopene effects would involve both cells representing advanced disease and patient-matched premalignant cells; however, with lines derived from human tumors, this is not possible. We therefore compared the effects of lycopene treatment on PC-3MM2 cells with effects on PC-3 cells and found that the PC-3 line still retained some inhibitory responses to lycopene treatment such as reduced proliferation and increased connexin 43 expression and was unaffected by lycopene treatment in expression of uPAR. The MCF-7 breast cancer line behaved similarly to the PC-3 in terms of growth inhibition, although lycopene had no effect on connexin 43 levels. These results suggest that lycopene can inhibit growth pathways in established carcinoma cells. However, some prostate carcinoma cells may develop lycopene growth resistance, perhaps through the selection pressures involved in metastasis, and also may develop advantageous uses for lycopene signals that promote cancer-stage specific matrix degradation potential.

ACKNOWLEDGEMENTS

The author wishes to thank the Society of Experimental Biology and Medicine and the Executive Director of Experimental Biology and Medicine, Dr. Felice O'Grady, for granting permission to reproduce this work.

ABBREVIATIONS

GPD: glyceraldehyde-phosphate dehydrogenase; HPF: high power microscopic field; PAI-1: plasminogen activator inhibitor-1; uPA: urokinase plasminogen activator; uPAR: urokinase plasminogen activator receptor

References

Barber, N.J. and J. Barber. 2002. Lycopene and prostate cancer. Prostate Cancer Prostatic Dis. 5: 6-12.

Bowen, P. and L. Chen, M. Stacewicz-Sapuntzakis, C. Duncan, R. Sharifi, L. Ghosh, H.S. Kim, K. Christov-Tzelkov, and R. van Breemen. 2002. Tomato sauce supplementation and prostate cancer: lycopene accumulation and modulation of biomarkers of carcinogenesis. Exp. Biol. Med. (Maywood) 227: 886-893.

Barinka, C. and G. Parry, J. Callahan, D.E. Shaw, A. Kuo, K. Bdeir, D.B. Cines, A. Mazar, and J. Lubkowski. 2006. Structural basis of interaction between urokinase-type plasminogen activator and its receptor. J. Mol. Biol. 363: 482-495.

Carmeliet, P. and L. Moons, R. Lijnen, M. Baes, V. Lemaitre, P. Tipping, A. Drew, Y. Eeckhout, S. Shapiro, F. Lupu, and D. Collen. 1997. Urokinase-generated plasmin activates matrix metalloproteinases during aneurysm formation. Nat. Genet. 17: 439-444.

Clark, P.E. and M.C. Hall, L.S. Borden Jr., A.A. Miller, J.J. Hu, W.R. Lee, D. Stindt, R. D'Agostino Jr., J. Lovato, M. Harmon, and F.M. Torti. 2006. Phase I-II prospective dose-escalating trial of lycopene in patients with biochemical relapse of prostate cancer after definitive local therapy. Urology 67: 1257-1261.

Davis, G.E. and K.A. Pintar Allen, R. Salazar, and S.A. Maxwell. 2001. Matrix metalloproteinase-1 and -9 activation by plasmin regulates a novel endothelial cell-mediated mechanism of collagen gel contraction and capillary tube regression in three-dimensional collagen matrices. J. Cell Sci. 114: 917-930.

Delworth, M.G. and K. Nishioka, C.A. Pettaway, M. Gutman, J.J. Killion, A.C. von Eschenbach, and I.J. Fidler. 1995. Systemic administration of 4-amidinoindanon-1-(2'-amidino)-hydrazone, a new inhibitor of S-adenosyl-methionine decarboxylase, produces cytostasis of human prostate cancer in athymic nude mice. Int. J. Oncol. 6: 293-299.

Evans, C.P. and F. Elfman, S. Parangi, M. Conn, G. Cunha, and M.A. Shuman. 1997. Inhibition of prostate cancer neovascularization and growth by urokinase-plasminogen activator receptor blockade. Cancer Res. 57: 3594-3599.

Festuccia, C. and V. Dolo, F. Guerra, S. Violini, P. Muzi, A. Pavan, and M. Bologna. 1998. Plasminogen activator system modulates invasive capacity and proliferation in prostatic tumor cells. Clin. Exp. Metastasis 16: 513-528.

Forbes, K. and K.M. Gillette, and I. Sehgal. 2003. The antitumor anitoxidant lycopene induces expression of the urokinase receptor in bone metastasis-derived prostate cancer cells. Exp. Biol. Med. 228: 967-971.

Forbes, K. and K. Gillette, L.A. Kelley, and I. Sehgal. 2004. Increased levels of urokinase plasminogen activator receptor in prostate cancer cells derived from repeated metastasis. World J. Urol. 22: 67-71.

Gann, P.H. and J. Ma, E. Giovannucci, W. Willett, F.M. Sacks, C.H. Hennekens, and M.J. Stampfer. 1999. Lower prostate cancer risk in men with elevated plasma lycopene levels: results of a prospective analysis. Cancer Res. 59: 1225-1230.

Giovannucci, E. 2002. A review of epidemiologic studies of tomatoes, lycopene, and prostate cancer. Exp. Biol. Med. (Maywood) 227: 852-859.

Goodman, M. and R.M. Bostick, K.C. Ward, P.D. Terry, C.H. van Gils, J.A. Taylor, and J.S. Mandel. 2006. Lycopene intake and prostate cancer risk: effect modification by plasma antioxidants and the XRCC1 genotype. Nutr. Cancer 55: 13-20.

Greiff, A. and W.M Fischer, and I. Sehgal. 2002. Paracrine communication between malignant and non-malignant prostate epithelial cells in culture alters growth rate, matrix protease secretion and in vitro invasion. Clin. Exp. Metastasis 19: 727-733.

Hadley, C.W. and E.C. Miller, S.J. Schwartz, and S.K. Clinton. 2002. Tomatoes, lycopene, and prostate cancer: progress and promise. Exp. Biol. Med. (Maywood) 227: 869-880.

Heber, D. and Q.Y. Lu. 2002. Overview of mechanisms of action of lycopene. Exp. Biol. Med. (Maywood) 227: 920-923.

Hoosein, N.M. and D.D. Boyd, W.J. Hollas, A. Mazar, J. Henkin, and L.W.K. Chung. 1991. Involvement of urokinase and its receptor in the invasiveness of human prostatic carcinoma cell lines. Cancer Comm. 3: 255-264.

Jian, L. and C-J. Du, A.H. Lee, and C.W. Binns. 2005. Do dietary lycopene and other carotenoids protect against prostate cancer? Int. J. Cancer 113: 1010-1014.

Kucuk, O. and F.H. Sarkar, Z. Djuric, W. Sakr, M.N. Pollak, F. Khachik, M. Banerjee, J.S. Bertram, and D.P. Wood Jr. 2002. Effects of lycopene supplementation in patients with localized prostate cancer. Exp. Biol. Med. (Maywood) 227: 881-885.

Minorsky, P.V. 2002. Lycopene and the prevention of prostate cancer: the love apple lives up to its name. Plant. Physiol 130: 1077-1078.

Mohanty, N.K. and S. Saxena, U.P. Singh, N.K. Goyal, and R.P. Arora. 2005. Lycopene as a chemopreventative agent in the treatment of high-grade prostate intraepithelial neoplasia. Urol Oncol. 23: 383-385.

Pastori, M. and H. Pfander, D. Boscoboinik, and A. Azzi. 1998. Lycopene in association with α-tocopherol inhibits at physiological concentrations proliferation of prostate carcinoma cells. Biochem. Biophys. Res. Comm. 250: 582-585.

Pettaway, C.A. and S. Pathak, G. Greene, E. Ramirez, M.R. Wilson, J.J. Killion, and I.J. Fidler. 1996. Selection of highly metastatic variants of different human prostatic carcinomas using orthotopic implantation in nude mice. Clin. Cancer Res. 2: 1627-1636.

Rabbani, S.A. and J. Gladu. 2002. Urokinase receptor antibody can reduce tumor volume and detect the presence of occult tumor metastases in vivo. Cancer Res. 62: 2390-2397.

Rabbani, S.A. and R.H. Xing. 1998. Role of urokinase (uPA) and its receptor (uPAR) in invasion and metastasis of hormone-dependent malignancies. Int. J. Oncol. 12: 911-920.

Schuurman, A.G. and R.A. Goldbohm, H.A.M. Brants, and P.A. van den Brant. 2002. A prospective cohort study on intake of retinol, vitamins C and E, and carotenoids and prostate cancer risk (Netherlands). Cancer Causes Control 13: 573-582.

Sehgal, I. and K. Forbes, and M.A. Webb. 2003. Reduced expression of MMPs, plasminogen activators and TIMPS in prostate cancer cells during repeated metastatic selection. Anticancer Res. 23: 39-42.

Sidenius, N. and F. Blasi. 2003. The urokinase plasminogen activator system in cancer: recent advances and implications for prognosis and therapy. Cancer Meta. Rev. 22: 205-222.

Lycopene and Lung Cancer

Fuzhi Lian and Xiang-Dong Wang

Nutrition and Cancer Biology Laboratory, Jean Mayer United States Department of Agriculture Human Nutrition Research Center on Aging at Tufts University 711 Washington Street, Boston, MA 02111, USA

ABSTRACT

Although epidemiological studies have shown dietary intake of lycopene is associated with decreased risk of lung cancer, the effect of lycopene on lung carcinogenesis has not been well studied. A better understanding of lycopene metabolism and the mechanistic basis of lycopene chemoprevention must be elucidated under well-controlled experimental conditions. Particularly, knowledge of the lycopene dose administered *in vivo*, the accumulation of lycopene in a specific organ, the interaction of lycopene with tobacco exposure and alcohol drinking, and the effect of lycopene metabolites on cell signaling pathways and molecular targets must be addressed. Comprehensive research on the effects of lycopene metabolites is crucial in directing further research into identifying the active form(s) of lycopene and the mechanisms by which certain cancers are prevented by lycopene. This information is critically needed for future human studies involving lycopene for the prevention of cancer in the lung and at other tissue sites.

INTRODUCTION

Lung cancer is one of the most common cancers in the world and the leading cause of cancer death in the United States because of its poor prognosis. The total number of deaths attributed to lung and bronchial cancer has been estimated to exceed the combined total for breast, prostate, and colon cancer in the United States (American Cancer Society

---

A list of abbreviations is given before the references.

2007). Cigarette smoking is the main risk factor for the development of lung cancer and accounts for approximately 85% of lung cancer incidence in both women and men (American Cancer Society 2007). Although avoidance of tobacco products is the best way to prevent tobacco-related cancers, the addictive power of nicotine is strong, and exposure to environmental tobacco smoke persists. In addition, the 5 yr survival rate for lung cancer patients is only 13 to 16% (American Cancer Society 2007). Therefore, chemoprevention against smoke-induced earlier lung lesions by consumption of a healthy diet may be an effective way to reduce lung cancer risk.

Beneficial effects of carotenoid-rich fruits and vegetables on lung cancer risk have been found in many epidemiological studies (Ziegler et al. 1996). The failure of the β-carotene intervention trials to show a benefit against lung carcinogenesis in smokers indicates that carotenoids or phytonutrients other than β-carotene may account for the protective effects of fruits and vegetables on lung cancer risk shown in observational studies. Recent epidemiological studies provide supportive evidence that lycopene may have a chemopreventive effect against a broad range of epithelial cancers (Clinton et al. 1996, Giovannucci 1999a, b, 2002, Arab et al. 2001). This chapter reviews recent evidence regarding the protective effect of lycopene against lung carcinogenesis and possible mechanisms involved, including evidence for several potential chemopreventive activities of lycopene metabolites. In addition, the conversion of lycopene into lycopene metabolites and the interaction between lycopene metabolism and cigarette smoke exposure in lung tissue are discussed.

LYCOPENE AND LUNG CANCER RISK

Epidemiological Studies

As summarized in an earlier review by Giovannucci, a statistically significant or suggestive inverse association between tomato or lycopene intake and lung cancer risk was found in 10 out of 14 studies (Giovannucci 1999b). Since then, many investigations with different study designs have been published (Table 1) (Garcia-Closas et al. 1998, Knekt et al. 1999, Speizer et al. 1999, Stefani et al. 1999, Brennan et al. 2000, Michaud et al. 2000, Voorrips et al. 2000, Darby et al. 2001, Yuan et al. 2001, 2003, Holick et al. 2002, Rohan et al. 2002, Ito et al. 2003, 2005a, b, Smith-Warner et al. 2003, Wright et al. 2003, Mannisto et al. 2004). Among them, all five case-control studies suggest an inverse association between tomato or lycopene intake and lung cancer risk (Garcia-Closas et al. 1998, Stefani et al. 1999, Brennan et al. 2000, Darby et al. 2001, Wright et al. 2003). Of 11 analyses of

Table 1 Epidemiological studies of tomato and lycopene and risk for lung cancer, 1999-2007.

Author, year	Place of study	Type of study	No. of case subjects	Exposure	Relative risk (95% confidence)	Adjust factors
Garcia-Closas et al., 1998	Spain	Case-control	103, women	lycopene intake, tertile 3 vs. 1	0.56 (0.26-1.24), P for trend = 0.15	smoking habit and vitamin E, vitamin C, and total flavonoid intake
Stefani et al., 1999	Uruguay	Case-control	541	lycopene intake, quartile 4 vs. 1	no significant associate	
Knekt et al., 1999	Finland	Cohort study	138	lycopene intake, tertile 3 vs. 1	RR 1.00 (0.67-1.50), P for trend = 0.77	
Speizer et al., 1999	USA	Cohort study	593	lycopene and tomato intake, quintile 5 vs. 1	lycopene RR 0.8 (0.6-1.1), P for trend = 0.76; tomato RR 0.8, P for trend = 0.46	
Brennan et al., 2000	Europe	Case-control	506, non-smokers	tomato intake, upper third to bottom	0.5 (0.4-0.6), P = 0.02	
Michaud et al., 2000	USA	Cohort study	519 and 275	lycopene intake, quintile 5 vs. 1	pooled 0.8 (0.64-0.99), P for trend = 0.1; for current smoker, 0.63 (0.45-0.88)	
Voorrips et al., 2000	Netherlands	Cohort Study	939, men	lycopene intake, quintile 5 vs. 1	1.05 (0.75-1.46), P for trend = 0.14	age, smoking, educational level, and family history of lung cancer
Darby et al., 2001	UK	Case-control	982	tomato sauce intake or tomato sauce quartile	tomato 0.74 (0.57-0.96), P = 0.01; tomato sauce 0.69 (0.55-0.87), P = 0.001	adjusted for age, sex, and smoking
Yuan et al., 2001	China	Nested case-control	209	serum lycopene, quartile 4 vs 1	RR 0.46, (0.27-0.79), P for trend = 0.003; adjusted RR = 0.59 (0.31-1.14), P for trend 0.15	adjusted for age at starting to smoke, average no. of cigarettes smoked per day, and smoking status at the time of blood draw

Table contd.

Table contd.

Study	Country	Design	N	Exposure	Results	Adjustments
Holick et al., 2002	Finland	Cohort study	1644	lycopene intake, quintile 5 vs. 1	age-adjusted 0.63 (0.54-0.75), multivariate 0.72 (0.61-0.84)	
Rohan et al., 2002	Canada	Nested case-control	155	lycopene intake, quartile 4 vs. 1	1.04 (0.61-1.76)	
Yuan et al., 2003	Singapore	Cohort study	482	serum lycopene, quintile 5 vs 1	0.89 (0.65-1.21)	
Ito et al., 2003	Japan	Nested case-control	147, lung cancer death	serum lycopene, quartile 4 vs 1	RR for lung cancer death 0.46 (0.21-1.04)	
Wright et al., 2003	USA	Case-control	587	dietary lycopene intake 4 vs. 1	0.73 (0.48-1.1), $P = 0.11$	
Smith-Warner et al., 2003	Europe and North America	Meta-analysis	3206	tomato intake, quartile 4 vs. 1	RR 0.83 (0.74-1.02), P for trend = 0.10	
Mannisto et al., 2004	North America and Europe	Meta-analysis	3155	lycopene intake, quintile 5 vs. 1	age-adjusted RR 0.77 (0.68-0.88), $P = 0.03$; multivariate adjusted RR 0.91 (0.78-1.07), $P = 0.42$	adjusted for education, body mass index, alcohol consumption, energy intake, and smoking status
Ito et al., 2005a	Japan	Nested case-control	211, lung cancer death	serum lycopene, quartile 4 vs 1	male: 0.44 (0.19-1.05) $P = 0.03$; female: 0.63 (0.12-3.25), $P = 0.5$	
Ito et al., 2005b	Japan		31, lung cancer death	serum lycopene	no significant associate	age, smoke, alcohol, body mass index, serum cholesterol

cohort studies, three show that high dietary or serum lycopene is significantly associated with a 20-50% reduction in lung cancer risk (Michaud et al. 2000, Yuan et al. 2001, Holick et al. 2002), and two report that serum lycopene level is associated with decreased lung cancer mortality (Ito et al. 2003, 2005b). However, six of these studies find no association between dietary or serum lycopene level and lung cancer risk (Knekt et al. 1999, Speizer et al. 1999, Voorrips et al. 2000, Rohan et al. 2002, Yuan et al. 2003) or lung cancer mortality (Ito et al. 2005a). In a pooled analysis of eight cohort studies from Europe and North America, high intake of tomatoes was marginally significantly associated with an approximate 20% reduction of lung cancer risk (RR = 0.83; 95% CI: 0.74-1.02; P for trend = 0.1) (Smith-Warner et al. 2003), and dietary lycopene intake was inversely associated with lung cancer risk (age-adjusted RR = 0.77; 95% CI: 0.68-0.88; P for trend = 0.03); however, when the results are further adjusted for education, body mass index, alcohol consumption, energy intake, and smoking status, this association becomes insignificant (Mannisto et al. 2004). It seems that the epidemiological evidence regarding the protective effect of lycopene-enriched tomato and tomato products against lung cancer is inconclusive. In addition, whether lycopene alone provides a beneficial effect against lung cancer risk is still unclear.

Animal Studies

Only a few studies have been conducted to examine the chemopreventive activity of lycopene against lung cancer using animal models (Table 2) (Kim et al. 1997, Nishino 1997, Hecht et al. 1999, Guttenplan et al. 2001), and results have been inconsistent. Kim et al. (1997) showed that giving 50 ppm of lycopene in drinking water for 21 wk after carcinogenic initiation decreased the incidence and multiplicity of lung tumors induced by the combination of diethylnitrosamine, N-methyl-N-nitrosourea and 1,2-dimethylhydrazine in male B6C3F1 mice, but not in female mice. Similarly, Nishino (1997) reported that feeding with 0.2 mg lycopene mixed with olive oil three times a week for 25 wk in the promotion stage reduced both incidence (approximate 20% reduction) and multiplicity (approximate 55% reduction) of lung tumors induced by 4-nitro-quinoline-1-oxide plus glycerol, but this effect did not reach statistical significance. In contrast, Hecht et al. (1999) treated A/J mice with 10-550 ppm of lycopene in the form of lycopene-enriched tomato oleoresin (LTO) for 27 wk and showed no protective effect on benzo[a]pyrene (BaP) plus 4-(methylnitrosamino)-1-(3-pyridyl)-1-butanone (NNK)-induced lung tumors. Furthermore, Guttenplan et al. (2001) showed that treatment of

Table 2 Animal studies of lycopene supplementation and lung cancer, 1997–2006.

Author, year	Animal	Exposure	Lycopene dosage	Length of lycopene supplementation	Endpoint	Results
Nishino et al., 1997	Male ddY mice	two-stage carcinogenesis, 4NQO plus glycerol	0.2 mg lycopene/mouse in 0.2 ml olive oil and Tween 80, oral intubation three times a week	25 wk in promotion stage	lung tumor	20% decrease in incidence, not significant
Kim et al., 1997	B6C3F1 mice	combination of DEN, MNU and DMH	25, 50 ppm in drinking water	21 wk after carcinogen treatment	lung tumor	for male, control vs. 50 ppm lycopene, incidence: 75.0 vs. 18.8%, P < 0.02; multiplicity: 0.94 ± 0.17 vs. 0.25 ± 0.14, P < 0.001.
Hecht et al., 1999	A/J mice	BaP and NNK	10, 100, 550 ppm lycopene in the form of LTO	27 wk before and after carcinogen treatment	lung tumor	no effect of lung tumor multiplicity
Guttenplan et al., 2001	LacZ mice	short-term BaP-induced and long-term spontaneous mutagenesis	0.5 and 1.0 mmol/kg lycopene in the form of LTO	9 mon for long-term spontaneous mutagenesis 10 wk before and after BaP treatment for short-term experiment	in vivo mutagenesis	high lycopene increase BaP-induced mutagenesis, no effect on spontaneous mutagenesis
Liu et al., 2003	Ferrets	cigarette smoke exposure	1.1 and 4.3 mg lycopene/kg/d	9 wk	squamous metaplasia	lycopene supplementation inhibits smoke-induced lung squamous metaplasia

lycopene (536 ppm in diet in the form of LTO) actually enhanced BaP-induced mutagenesis in the lung and colon of LacZ mice. The inconsistency of these results may be attributed to the difference in routes of administration, mouse strains, forms of lycopene, and chemical carcinogen used to initiate carcinogenesis. In addition, the absorption of intact carotenoids by rodents is relatively low (Lee et al. 1999, Ferreira et al. 2000), which could also explain the outcome of these studies. Unfortunately, none of these studies recorded the levels of lycopene in blood and tissues. We have observed that lycopene supplementation inhibited lung squamous metaplasia induced by cigarette smoke exposure in ferret model, and the accumulation of lycopene in both ferret plasma and lung tissue to levels comparable to those in humans (Liu et al. 2003, 2006), suggesting a protective effect of lycopene against cigarette smoke–related lung carcinogenesis. While the inverse association between tomato or lycopene intake and lung cancer risk has been suggested in many population studies, a direct causal relationship in animal study has not yet been established.

Mechanisms of the Anticarcinogenic Activities of Lycopene

Although not fully understood, it has been believed that lycopene may function as a natural antioxidant (Di Mascio et al. 1989), enhance cellular gap junction communication (Zhang et al. 1991), induce phase II enzymes involved in activation of the antioxidant response element transcription system (Velmurugan et al. 2002), suppress insulin-like growth factor-1 (IGF-1)-stimulated cell proliferation by inducing IGF binding protein (IGFBP) (Karas et al. 2000, Liu et al. 2003), and inhibit neoplastic transformation of normal cells (Som et al. 1984). In this part, we discuss only evidence that directly links with lung carcinogenesis.

Effect of Lycopene on Lung Cell Growth and Apoptosis

The growth inhibitory activity of lycopene was first demonstrated by Levy et al. (1995), who showed that lycopene is a stronger cell growth inhibitor than β-carotene. The inhibition of cell proliferation was further observed in cell lines derived from breast cancer (Levy et al. 1995, Nahum et al. 2001), prostate cancer (Levy et al. 1995), lung cancer (Levy et al. 1995), and oral cavity cancer (Livny et al. 2002), as well as normal prostate epithelial cells (Obermuller-Jevic et al. 2003). The inhibition of MCF-7 breast cancer cell proliferation by lycopene was associated with inhibition of cell cycle progression from G1 to S phase, decreased cyclin D1 expression and the retention of p27 in cyclin E-CDK complex (Nahum et al. 2001, 2006). In addition, the growth inhibitory effect of lycopene may be attributed to the induction of apoptosis. Hwang et al. showed that 1 µM of water-soluble lycopene inhibited the growth of LNCaP prostate cancer cells, while lycopene at 5 µM blocked cells in G2/M phase and induced apoptosis, suggesting high concentrations of lycopene can induce DNA damage (Hwang and Bowen 2004). Palozza et al. reported that lycopene (0.5-2 µM) inhibited the growth of cigarette smoke condensate–exposed immortalized RAT-1 fibroblast cells by arresting cell cycle progression and inducing apoptosis (Palozza et al. 2005). Another study showed that lycopene at physiological concentrations (0.3-3 µM) did not affect the proliferation of LNCaP cells but rather affected mitochondrial function and induced apoptosis (Hantz et al. 2005).

Ferrets (*Mustela putorious furo*) offer an excellent model for mimicking the conditions of carotenoid intervention studies in humans because ferrets and humans are similar in terms of lycopene absorption, tissue distribution and concentrations, and metabolism (Wang 2005). We conducted a study to evaluate the effects of lycopene supplementation at both a low dose and a high dose on blood and lung tissue lycopene levels in ferrets with or without 9 wk of cigarette smoke exposure (Liu et al. 2003). Ferrets in the low-dose lycopene group were given lycopene at 1.1 mg/kg body weight per day, which is equivalent to an intake of 15 mg/d in humans. This dose is slightly higher than the mean intake of lycopene (9.4 ± 0.3 mg/d) in US men and women (Food and Nutrition Board 2002). Ferrets in the high-dose lycopene group were given lycopene at 4.3 mg/kg body weight per day, which is equivalent to 60 mg/d in humans and is achievable in a diet enriched with tomato products or supplements. The results show that lycopene supplementation at both a low dose and a high dose for 9 wk substantially increased the concentrations of lycopene in both plasma and lung tissue. The concentration of plasma lycopene (range 226-373 nmol/L) in ferrets after lycopene supplementation was similar to the lycopene concentration (range 290-350 nmol/L) reported in humans (Lu et al. 2001, Vogt et al. 2002). Furthermore, the lycopene concentrations in the lungs of ferrets that were given a low dose of lycopene (equivalent to 15 mg/d in humans) reached 342 nmol/kg, which is within the range of lung lycopene concentrations in normal humans (100 to 500 nmol/kg) (Schmitz et al. 1991). Lycopene concentrations in ferrets supplemented with a high dose of lycopene increased 3.4-fold in lung tissue and 1.6-fold in plasma compared with ferrets supplemented with a low dose of lycopene. We observed that squamous metaplastic lesions, which precede the appearance of lung cancer, were observed in the lung tissue of all six ferrets exposed to smoke alone but only in two of the six ferrets exposed to smoke and given a low dose of lycopene. No squamous metaplasia was observed in the lung tissue of ferrets in the control group, the low-dose lycopene group, the high-dose lycopene group, and the high dose plus smoke exposure group after 9 wk of intervention. Proliferating cellular nuclear antigen (PCNA) expression was increased 3-fold in the smoke-exposed group but did not differ between control ferrets and ferrets given either a low dose or a high dose of lycopene and exposed to smoke. Further, smoke exposure in ferrets for a 9 wk period significantly decreased apoptosis as indicated by a 74% reduction of cleaved caspase 3, compared with controls. However, there were no differences in cleaved caspase 3 levels in ferrets supplemented with either a low dose or a high dose of lycopene with or without smoke exposure as compared with control, suggesting that lycopene supplementation restored normal apoptosis reduced by smoke exposure. Our study indicates that lycopene supplementation for 9 wk at either a low dose or a high dose prevents smoke-induced squamous metaplasia, cell hyperproliferation, and the reduction of apoptosis in the lungs of smoke-exposed ferrets.

Effect of Lycopene on IGF Signaling

It has been suggested that the IGF signaling system may play a critical role in the biological action of lycopene (Levy et al. 1995, Karas et al. 2000). IGFs (including IGF-1 and 2) are the major mediators of the effects of the growth hormone, performing a fundamental role in the regulation of cellular proliferation, differentiation and apoptosis. By binding to membrane IGF-1 receptor, IGFs activate intracellular phosphatidylinositol 3'-kinase (PI3K)/Akt/protein kinase B and Ras/Raf/MAPK pathways, which regulate various biological processes such as cell cycle progression, survival, and transformation (Jones and Clemmons 1995). In circulation, IGFs are sequestered by a family of binding proteins (IGFBP1-IGFBP6), which regulate the availability of IGFs to bind to the IGF receptors (Jones and Clemmons 1995). Lycopene treatment was shown to inhibit IGF-1-stimulated insulin receptor substrate 1 phosphorylation and cyclin D1 expression, block IGF-1-stimulated cell cycle progression (Karas et al. 2000, Nahum et al. 2006), and increase membrane-associated IGFBPs in MCF-7 breast cancer cells (Karas et al. 2000), suggesting lycopene may inhibit cell proliferation by acting on IGF-1 signaling pathway. Consistent with *in vitro* studies that demonstrated lycopene as a modulator of IGF signaling (Levy et al. 1995, Karas et al. 2000), higher dietary intake of lycopene has been associated with lower circulating levels of IGF-1 (Mucci et al. 2001) and higher levels of IGFBP-3 (Holmes et al. 2002) in epidemiological studies. In addition, the modulation of IGF-1 signaling is thought to play an important role in lung carcinogenesis (Yu et al. 1999, London et al. 2002, Wakai et al. 2002).

We examined whether lycopene prevents cigarette smoke–related lung carcinogenesis through the modulation of IGF signaling (Fig. 1) in ferret model (Liu et al. 2003). We found that plasma concentrations of IGF-1 were not affected by either lycopene supplementation or cigarette smoke exposure. Interestingly, cigarette smoke exposure decreased ferret plasma

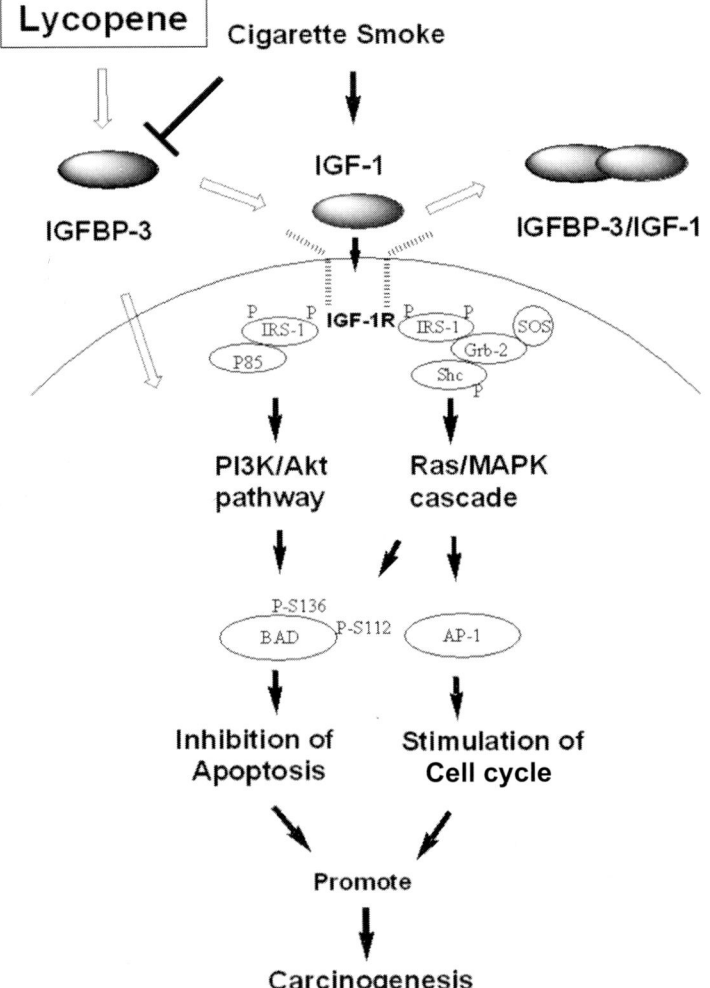

Fig. 1 Simplified illustration of IGF-1 signaling pathway and the effect of lycopene on IGF signaling. IGF binding protein-3 (IGFBP-3) regulates the bioactivity of IGF-1 by sequestering IGF-1 away from its receptor in the extracellular milieu, thereby inhibiting the mitogenic and anti-apoptotic action of IGF-1 and reducing cancer risk. Apart from modulating IGF-1 action, IGFBP-3 may exert intrinsic bioactivity via its interaction with other signaling pathways. The up-regulation of IGFBP-3 by lycopene may interrupt the signal transduction pathway of IGF-1, down-regulate phosphorylation of BAD, promote apoptosis and inhibit cell proliferation, thereby preventing smoke-induced lung carcinogenesis.

concentrations of IGFBP-3, while ferrets given both doses of lycopene had higher levels of IGFBP-3, regardless of cigarette smoke exposure. Furthermore, the ratio of IGF-1/IGFBP-3 was significantly decreased in ferrets given lycopene and exposed to smoke, compared with those exposed to smoke alone (Table 3) (Liu et al. 2003). The increased plasma IGFBP-3 by lycopene supplementation was associated with inhibition of cigarette smoke-induced lung squamous metaplasia, decreased PCNA and phosphorylated BAD protein levels, and restoration of cleaved caspase-3 levels in lung tissue (Liu et al. 2003), suggesting the inhibition of proliferation and induction of apoptosis. These results support our hypothesis that lycopene may affect IGF signaling, inhibit cell proliferation, and interfere with the development of cigarette smoke-related lung lesions. Further studies are needed to investigate whether the modulation of IGFBP-3 expression by lycopene plays a role in the prevention of lung tumor development.

Table 3 Plasma concentrations of IGF-1 and IGFBP-3 in six groups of ferrets after 9 wk lycopene supplementation.

Treatment	IGF-1 (ng/mL)	IGFBP-3 (ng/mL)	IGF-1/IGFBP-3 ratio
Control	754.07 ± 208.30[a]	2.33 ± 0.83[a]	360.01 ± 134.57[a]
Low-dose lycopene	1022.88 ± 121.83[a]	3.68 ± 0.63[b]	282.08 ± 41.98[a,b]
High-dose lycopene	928.36 ± 333.93[a]	3.85 ± 1.31[b]	238.27 ± 46.55[b]
Smoke	745.02 ± 141.36[a]	1.55 ± 1.15[c]	706.98 ± 40.65[c]
Smoke plus low-dose lycopene	929.04 ± 127.71[a]	3.34 ± 0.69[b]	281.89 ± 30.20[a,b]
Smoke plus high-dose lycopene	993.30 ± 80.25[a]	3.79 ± 0.86[b]	275.93 ± 76.58[a,b]

Ferrets were given lycopene at 1.1 mg/kg body weight per day (low dose) and 4.3 mg/kg body weight per day (high dose) and exposed or not exposed to cigarette smoke for 9 wk, and the plasma IGF-1 and IGFBP-3 levels were measured by ELISA. [a,b,c] For a given column, data not sharing a common superscript letter are statistically significantly different at $P < 0.05$. Values are expressed as mean ± SD. Adapted from Liu et al. (2003).

Lycopene Metabolism and the Biological Function of Lycopene Metabolites against Lung Carcinogenesis

Carotenoids are highly lipophilic compounds with polyisoprenoid structures, typically containing a series of conjugated double bonds in the central chain of the molecule, which make them susceptible to isomerization from *trans* to *cis* forms, oxidative cleavage, and the formation of potentially bioactive metabolites. The impact of oxidation on the biological effects of β-carotene and the potential for beneficial effects of small quantities or harmful effects of large quantities of β-carotene metabolic products in lung carcinogenesis has been reviewed (Wang 2004). Since lycopene metabolism involves both enzymatic and autoxidative cleavage, we focused our research efforts on the effects of smoke exposure on lycopene metabolism and the potential protective effect of lycopene metabolites against lung carcinogenesis.

Oxidative Cleavage Products

Lycopene is highly susceptible to degradation by chemical and physical factors, such as exposure to light, oxygen, temperature, and pH, because of the high reactivity of conjugated double bonds in the carbon chain of the molecule. A number of studies have identified the cleavage products of lycopene produced by *in vitro* oxidation (Kim et al. 2001, Caris-Veyrat et al. 2003). Kim et al. (2001) showed that after autoxidation of lycopene solubilized in toluene, aqueous Tween 40, or liposomal suspension at 37°C for 72 hr, a number of oxidative products were detected, which were identified as 3,7,11-trimethyl-2,4,6,10-dodecatetraen-1-al, 6,10,14-trimethyl-3,5,7,9,13-pentadecapentaen-2-one, acycloretinal, apo-14'-lycopenal, apo-12'-lycopenal, apo-10'-lycopenal, apo-8'-lycopenal, and apo-6'-lycopenal, as well as acycloretinoic acid. In addition to apo-lycopenals that contain one aldehyde end group, Caris-Veyrat et al. (2003) further identified a group of apo-carotendials that contain two aldehyde end groups. Using similar approaches, researchers have identified other polar metabolites of lycopene, such as 2,7,11-trimethyl-tetradecahexaene-1,14-dial (Aust et al. 2003) and (E,E,E)-4-methyl-8-oxo-2,4,6-nonatrienal (Zhang et al. 2003). Other lycopene metabolites have also been detected *in vivo* (Khachik et al. 2002, dos Anjos Ferreira et al. 2004, Gajic et al. 2006). Khachik et al. identified a group of lycopene oxidative products, 2,6-cyclolycopene-1,5-diol A and B, in human serum and many other tissues, including prostate, lung, and colon. Recently, Gajic and colleagues (2006) reported the detection of apo-8' and apo-12'-lycopenal as well as other unidentified polar metabolites of lycopene in the liver of rats given lycopene-enriched food. However, it is unknown whether these metabolites are the products of enzymatic cleavage or chemical oxidative cleavage.

Enzymatic Metabolism

For provitamin A carotenoids, such as β-carotene, α-carotene, and β-cryptoxanthin, central cleavage is a major pathway leading to vitamin A formation (Goodman and Huang 1965, Olson and Hayaishi 1965). This pathway is mediated by carotene 15,15'-monooxygenase 1 (CMO1), which cleaves carotenoids at their 15,15'-double bond (von Lintig and Vogt 2000, Wyss et al. 2000). CMO1 genes have recently been identified from different species, and their recombined enzymes have been biochemically characterized in different laboratories (von Lintig and Vogt 2000, Wyss et al. 2000, Leuenberger et al. 2001, Paik et al. 2001, Redmond et al. 2001, Lindqvist and Andersson 2002, Kloer et al. 2005, Poliakov et al. 2005). It was shown that mammalian CMO1 is highly expressed in liver, intestine, kidney, and testis (Redmond et al. 2001, Lindqvist and Andersson 2002). It specifically catalyzes the symmetric cleavage of b-ionone ring-containing carotenoids, such as β-carotene and β-cryptoxanthin, while lycopene is a good substrate for CMO1 (Redmond et al. 2001, Lindqvist and Andersson 2002).

In addition to central cleavage, an alternative pathway for carotenoid metabolism in mammals, the excentric cleavage pathway, was a controversial issue for decades. We previously demonstrated the random cleavage of the β-carotene molecule and identified a series of homologous carbonyl cleavage products, including β-apo-14'-, 12'-, 10'-, and 8'-carotenals, β-apo-13-carotenone and retinoic acid in tissue homogenates of humans, ferrets and rats (Tang et al. 1991, Wang et al. 1991, 1996). Kiefer et al. (2001) cloned and identified carotene-9',10'-monoxygenase (CMO2) gene in humans and mice, confirming the existence of this asymmetric cleavage pathway of carotenoids. They showed that CMO2 is highly expressed in intestine, liver, kidney, and testis of mice (Kiefer et al. 2001). When expressed in β-carotene-synthesizing and -accumulating *E. coli* strains, CMO2 cleaves β-carotene at its 9',10'-double bond to form β-apo-10'-carotenal, β-apo-10'-carotenol and its ester (Kiefer et al. 2001). Interestingly, it was also suggested that CMO2 cleaves lycopene in an *E. coli* strain able to synthesize and accumulate lycopene (Kiefer et al. 2001).

We cloned the CMO2 gene from ferret and characterized the CMO2 enzymatic properties using both β-carotene and lycopene as substrates (Hu et al. 2006). Consistent with a previous report from mouse (Kiefer et al. 2001), ferret CMO2 gene expression is detected in liver, testis, heart, spleen, lung, intestine, colon, stomach, kidney, bladder, and prostate tissues (Hu et al. 2006). Ferret CMO2 shares 87% sequence identity with human CMO2 and 81% with mouse CMO2. Kinetic studies show that recombinant ferret CMO2 cleaves *all-trans* β-carotene at 9',10'-double bond, generating β-apo-10'-lycopenal and β-ionone, with a K_m of 3.5 ± 1.1 mM and the V_{max} of 32.2 ± 2.9 pmol of β-apo-10'-carotenal/mg/hr, under the optimal pH of 8.0 to 8.5. This cleavage reaction is iron dependent (Hu et al. 2006). When using lycopene as substrate, we found that recombined ferret CMO2 does not significantly cleave *all-trans* lycopene (containing 97% as *all-trans*-lycopene); however, when incubated with the mixture of lycopene isomers (mainly 5-*cis*, and 13-*cis* lycopene), CMO2 was shown to cleave lycopene specifically at 9'10'-double bond, generating a cleavage product identified as apo-10'-lycopenal. Kinetic studies using the lycopene mixture containing 15-20% *cis*-isomers showed that the optimal pH for the reaction ranges between 8.0 and 8.5. Although we cannot calculate the exact values of K_m and V_{max} due to lycopene isomerization, our data suggested that CMO2 is an enzyme that preferentially cleaves *cis*-lycopene at 9',10'-double bond (Fig. 2).

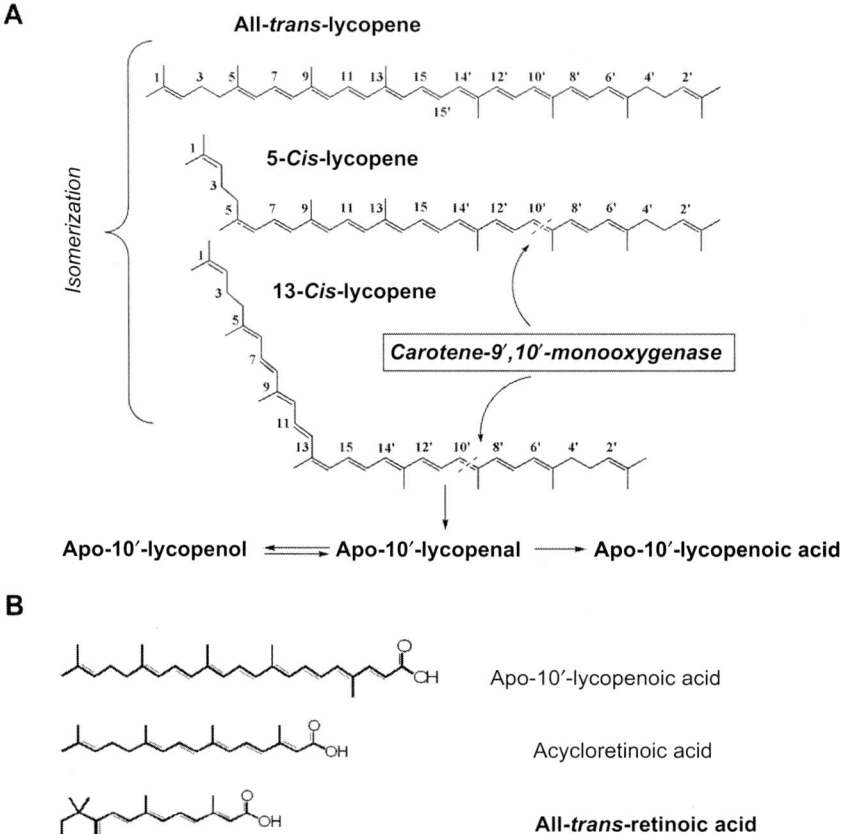

Fig. 2 Schematic illustration of lycopene metabolic pathway by CMO2. (A) 5-*cis* and 13-*cis* lycopene are preferentially cleaved by CMO2 at 9'10'-double bond. The cleavage product, apo-10'-lycopenal, can be further oxidized to apo-10'-lycopenol or reduced to apo-10'-lycopenoic acid, dependent on the presence of NADH. (B) Chemical structures of apo-10'-lycopenoic acid, acycloretinoic acid, and *all-trans* retinoic acid. Adapted from Hu et al. (2006).

The cleavage of lycopene was further confirmed in ferret lung tissue. After supplementation with all-*trans*-lycopene for 9 wk, ferrets preferentially accumulated more than 50% of *cis*-lycopene in the lung, mainly as 5-*cis* isomer (Liu et al. 2003). Among several metabolites detected in lung tissue of these lycopene-supplemented ferrets, one was identified as apo-10'-lycopenol, with the concentration of 8 ± 3 pmol/g wet weight lung tissue (Hu et al. 2006). However, no apo-10'-lycopenal was detected, suggesting that apo-10'-lycopenal, the primary cleavage product, might be an intermediate compound and could be either reduced to apo-10'-lycopenol or oxidized to apo-10'-lycopenoic acid. To confirm this hypothesis, we incubated apo-10'-lycopenal with the mixtures containing post-nuclear fraction of ferret liver homogenates and observed the conversion of apo-10'-lycopenal into apo-10'-lycopenoic acid in the presence of NAD^+, and to both apo-10'-lycopenoic acid and apo-10'-lycopenol in the presence of NADH (Hu et al. 2006) (Fig. 2). In addition, the expression of CMO2 mRNA in ferret lung was up-regulated 4-fold by lycopene supplementation for 9 wk, compared to animals not receiving lycopene supplementation (Hu et al. 2006). These results demonstrate that lycopene can be converted to apo-10'-lycopenoids in mammalian tissues both *in vitro* and *in vivo*, raising the question of whether apo-10'-lycopenoids have biological activity against lung carcinogenesis, or other important biological functions related to human health.

Lycopene Metabolism in Smoke-exposed Environment

Cigarette smoke accounts for approximately 85% of lung cancer incidence. It has consistently been associated with decreased concentration of lycopene as well as other carotenoids in plasma (Handelman et al. 1996, Alberg 2002, Dietrich et al. 2003). Meanwhile, clinical trials showed that supplementation with tomato products decreased levels of lymphocyte DNA damage (Briviba et al. 2004), and reduced plasma lipid peroxidation and susceptibility of low density lipoproteins to oxidation (Steinberg and Chait 1998) in smokers. However, the interaction between lycopene metabolism and cigarette smoke-generated free radicals in lung tissue has not been fully investigated.

We previously provided evidence that lycopene can protect against smoke-related lesions in the lungs of ferrets (Liu et al. 2003). In that study, ferrets with or without cigarette smoke exposure received low or high dose of lycopene. After 9 wk of supplementation, lycopene was detected in both plasma and lung tissue of ferrets; however, cigarette smoke exposure was associated with decreased lycopene levels, particularly in lung tissue (Table 4). The concentrations of plasma lycopene ranged from 226 to 373 nmol/L in the ferrets without cigarette smoke exposure, and the average levels of lycopene in the lungs were 342 nmol/kg and 1159 nmol/kg in ferrets supplemented with low and high dose of lycopene, respectively. Smoke exposure decreased lycopene concentrations by 40% and 90% in

Table 4 Plasma and lung concentrations of lycopene isomers in six groups of ferrets after 9 wk treatment.

Treatment	All-trans-lycopene and 5-cis-lycopene*	13-cis-lycopene	9-cis-lycopene	Total lycopene
Plasma (nmol/L)				
Control	ND	ND	ND	ND
Low-dose lycopene	128 ± 28[a]	48 ± 5[a,d]	52 ± 8[a]	226 ± 35[a]
High-dose lycopene	208 ± 38[b]	83 ± 14[b]	83 ± 11[b]	373 ± 60[b]
Smoke	ND	ND	ND	ND
Smoke plus low-dose lycopene	77 ± 17[c]	31 ± 15[a,c]	35 ± 9[c]	142 ± 36[c]
Smoke plus high-dose lycopene	110 ± 21[a]	53 ± 7[d]	65 ± 14[a,b]	228 ± 33[a]
Lung (nmol/kg)				
Control	ND	ND	ND	ND
Low-dose lycopene	240.2 ± 35.2[a]	61.8 ± 10.4[a]	40.2 ± 9.7[a]	342.2 ± 42.3[a]
High-dose lycopene	735.7 ± 123.7[b]	236.2 ± 49.4[b]	187.2 ± 36.2[b]	1159.2 ± 145[b]
Smoke	ND	ND	ND	ND
Smoke plus low-dose lycopene	15.6 ± 3.8[c]	6.8 ± 1.9[c]	6.6 ± 1.7[c]	29.2 ± 6.8[c]
Smoke plus high-dose lycopene	51.7 ± 8.7[d]	27.2 ± 4.8[d]	21.4 ± 4.2[d]	100.3 ± 16.5[d]

Ferrets were given lycopene at 1.1 mg/kg body weight per day (low dose) and 4.3 mg/kg body weight per day (high dose) and exposed or not exposed to cigarette smoke for 9 wk. The levels of lycopene in lung and plasma were measured by HPLC. [a,b,c,d] For the same tissue and a given column, data not sharing a common superscript letter are statistically significantly different at $P < 0.05$. Values are expressed as mean ± SD in nmol/kg wet weight. $n = 6$ ferrets in each group. ND, not detected. *At the time of HPLC analysis, we could not separate all-trans and 5-cis isomers of lycopene; however, later analysis showed that more than 50% of lycopene in this peak was 5-cis isomer. Adapted from Liu et al. (2003); details can be found in Hu et al. (2006).

plasma and in lung tissue, respectively, in groups receiving both the low dose and high dose of lycopene, as compared with non-supplemented groups. The ferret lycopene levels decreased by cigarette smoke exposure are consistent with findings from population studies showing smokers have lower serum levels of lycopene than non-smokers (Handelman et al. 1996, Wei et al. 2001, Alberg 2002, Dietrich et al. 2003). The decrease in lycopene levels after cigarette smoke exposure may be due to the enhanced lycopene degradation by either chemical or enzymatic oxidations. We have found that the expression of CMO2 in the lungs of ferrets was slightly induced by smoke exposure, and the production of apo-10'-lycopenol and other unidentified metabolites in the lungs of lycopene-supplemented ferrets was enhanced by smoke exposure, compared to ferrets receiving lycopene alone (Liu and Wang, unpublished data), indicating that lycopene degradation could be due to the autoxidation induced by smoke exposure-generated free radicals.

Previous studies have demonstrated that large amounts of oxidative metabolites of β-carotene account for the deleterious effect of high-dose β-carotene supplementation (Wang et al. 1999, Wang 2004). When the ferrets received β-carotene at a dose of 30 mg/d, the concentration of β-carotene in the lungs was 26,000 nmol/kg lung tissue, which was associated with an enhanced development of lung squamous metaplasia induced by cigarette smoke exposure (Wang et al. 1999). In contrast, when ferrets were given lycopene supplementation at a dose of 60 mg/d, the concentration of lycopene in the lungs was only 1,200 nmol/kg lung tissue, more than 10 times less than β-carotene accumulation (Liu et al. 2003). This amount of lycopene did not cause any harmful effects, but rather prevented the development of lung squamous metaplasia and cell proliferation induced by smoke exposure. The different outcomes between the lycopene and β-carotene studies in ferrets may be due to the differences in the levels of carotenoids that accumulated in lung tissue, and the generation of different levels of lycopene metabolites. It will be interesting to examine the biological activity of lycopene metabolites and the effect of lycopene metabolites on lung carcinogenesis.

Potential Chemopreventive Effect of Lycopene Metabolites against Lung Cancer

The identification of carotenoid metabolites *in vitro* and *in vivo* leads to the hypothesis that metabolites of carotenoids, which can possess either more or less activity than their parent compounds, or have entirely different functions, may contribute, at least in part, to the biological function of carotenoids (Wang 2004). Studies have shown that the mixture of lycopene oxidative products was able to inhibit the growth of HL-60 human promyelocytic leukemia cells (Nara et al. 2001), and induce antioxidant response element (ARE)-dependent transcription of phase II xenobiotic-metabolizing enzymes (Ben-Dor et al. 2005), suggesting lycopene metabolites may be biologically active components against initiation, promotion, or progression stages of carcinogenesis.

Among identified metabolites, acycloretinoic acid (Fig. 2B), the central cleavage product of lycopene at 15, 15'-double bond and an analog of retinoic acid, is the best-studied one. It has been shown to inhibit cell proliferation (Ben-Dor et al. 2001, Kotake-Nara et al. 2002, Nara et al. 2001), induce apoptosis (Kotake-Nara et al. 2002), and enhance gap junction communication (Stahl et al. 2000). As an analog of retinoic acid, the ability of acycloretinoic acid to activate retinoic acid receptor (RAR) has been examined (Stahl et al. 2000, Ben-Dor et al. 2001). Although acycloretinoic acid is able to activate retinoic acid response element (RARE)-driven luciferase gene transcription, the required concentration is much higher than that of *all-trans* retinoic acid, suggesting that acycloretinoic acid is a weak activator of RARs (Stahl et al. 2000, Ben-Dor et al. 2001).

The biological activities of other lycopene metabolites have also been investigated. For example, Aust et al. (2003) showed that 2,7,11-trimethyl-tetradecahexaene-1,14-dial, a metabolite of lycopene formed by a fragmentation at the 5,6 and 12',11' -positions, is able to enhance gap junction communication, while Zhang et al. (2003) demonstrated that (E,E,E)-4-methyl-8-oxo-2,4,6-nonatrienal, the product of oxidative cleavages at the 5, 6- and 13, 14-double bonds of lycopene, induced apoptosis, down-regulated Bcl-2 and Bcl-XL, and activated caspase cascades in HL-60 cells. However, the physiological roles of these lycopene products remain unknown, since none of these metabolites has been detected in biological systems.

We recently investigated the biological activity of apo-10'-lycopenoic acid, a relatively stable derivative of CMO2 cleavage of lycopene, as well as its potential chemopreventive effect against lung carcinogenesis (Lian et al. 2007). We first examined the effect of apo-10'-lycopenoic acid on cell growth. Three cell lines, which represent different stages of lung carcinogenesis, were used for the study: NHBE, a normal human bronchial epithelial cell line, BEAS-2B, an immortalized human bronchial epithelial cell line, and A549, a non-small cell lung cancer cell line. We showed that the growth of all three cells was inhibited by apo-10'-lycopenoic acid treatment, as measured by counting cell number (Fig. 3A) and 3-(4,5-dimethylthiazol-2-yl)-2,5-diphenyltetrazolium bromide (MTT) assay, albeit with different sensitivity. The growth inhibitory activity of

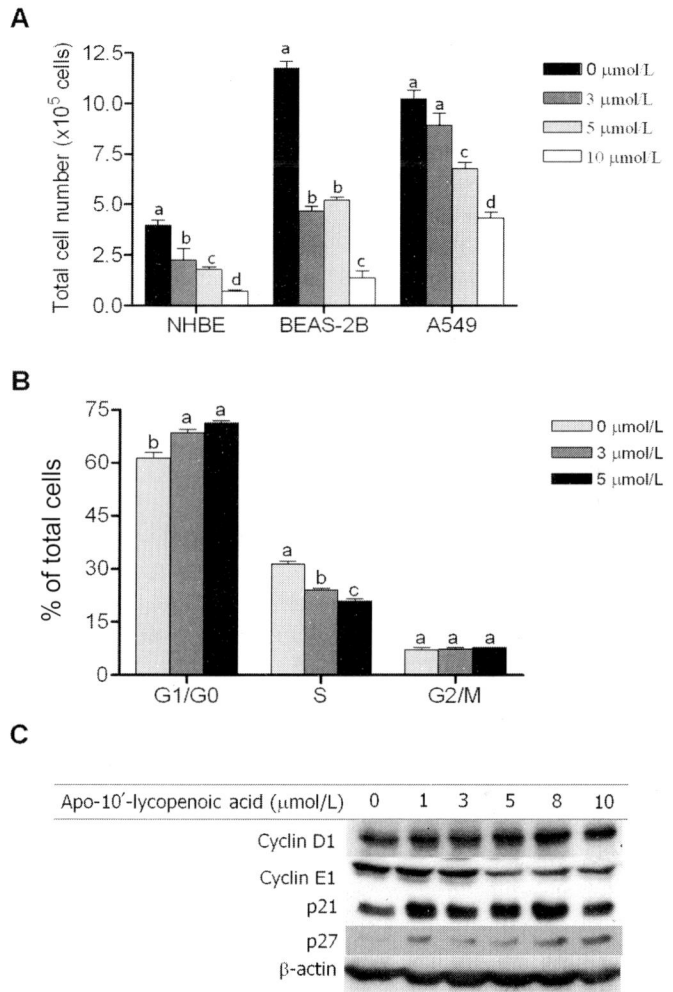

Fig. 3 Effect of apo-10′-lycopenoic acid on lung cell growth, cell cycle distribution, and the expression of cell cycle regulators. (A) Effect of apo-10′-lycopenoic acid on lung cell growth. 5×10^4 NHBE, BEAS-2B and A549 cells were treated with the indicated concentration of apo-10′-lycopenoic acid for 4 d. Cells were collected and cell numbers were counted using a hematometer. (B) Effect of apo-10′-lycopenoic acid on cell cycle distribution. 5×10^5 A549 cells were treated with indicated concentrations of apo-10′-lycopenoic acid for 48 hr. Then, cells were harvested, fixed, and analyzed for DNA content. Cell cycle distribution was analyzed using Modfit software. (A and B) Values are means ± SEM of three replicate assays. Means that do not share a letter differ, $P < 0.05$. (C) Effect of apo-10′-lycopenoic acid on protein levels of cell cycle regulators. 5×10^4 A549 cells were treated with the indicated concentration of apo-10′-lycopenoic acid for 4 d. Total cellular protein was separated on 12% SDS-polyacrylamide gels and blotted with anti-cyclin D1, cyclin E, p21, p27, and β-actin antibodies. The figure shows a representative Western blot photograph of three experiments. Adapted from Lian et al. (2007).

apo-10′-lycopenoic acid was largely due to the inhibition of cell proliferation, as we did not observe any induction of apoptosis. In addition, the lower sensitivity of A549 cells to apo-10′-lycopenoic acid treatment, as compared with NHBE and BEAS-2B cells, suggests that apo-10′-lycopenoic acid may be a better agent for cancer prevention than cancer therapy (Lian et al. 2007).

Cell proliferation is controlled by a series of cell cycle regulators, including cyclins, cyclin-dependent kinases (CDKs), and CDK inhibitors, which regulate cell cycle progress. Specifically, cyclin E regulates the transition of the cell from G1 phase to S phase; and p21 and p27 CDK inhibitors bind and inhibit the activity of cyclin E/CDK2 complex, blocking cell cycle progression in G1 phase. We showed that treatment with 3 mmol/L and 5 mmol/L of apo-10′-lycopenoic acid significantly decreased A549 cells in S phase and increased cells in G1 phase (Fig. 3B) and was associated with decreased cyclin E and increased p21 and p27 expression (Fig. 3C). These results further confirmed that apo-10′-lycopenoic acid is an inhibitor of cell proliferation (Lian et al. 2007).

Because of the similarity in chemical structures among apo-10′-lycopenoic acid, acycloretinoic acid, and *all-trans* retinoic acid (Fig. 2B), we questioned whether apo-10′-lycopenoic acid is an activator of RARs. We showed that treatment with 3 to 5 mmol/L of apo-10′-lycopenoic acid significantly increased the mRNA level of retinoic acid receptor β (RARβ), which is a transcriptional target of RARs (de The et al. 1990), in NHBE, BEAS-2B, and A549 cells (Lian et al. 2007). Similarly, IGFBP-3, another RAR target gene, was also induced by apo-10′-lycopenoic acid treatment in these cells (Lian and Wang, unpublished data). We then constructed a reporter vector containing the RARβ promoter fragment in the promoter region of luciferase gene. We showed that apo-10′-lycopenoic acid treatment increased the luciferase activity of HeLa cells transfected with this reporter vector. When the RARE in RARβ promoter was mutated, the ability of apo-10′-lycopenoic acid to transactivate RARβ promoter was abolished (Fig. 4). These results suggest that apo-10′-lycopenoic acid is a ligand for RARs, and that activation of RARs may account for the growth inhibitory effect of apo-10′-lycopenoic acid (Lian et al. 2007).

We further performed an *in vivo* study to determine whether apo-10′-lycopenoic acid could inhibit lung tumor development in the A/J mouse model of lung cancer. A/J mice were preloaded with control diet or diet containing 10, 40, and 120 mg/kg diet of apo-10′-lycopenoic acid for 2 wk before lung tumors were induced by injection of NNK. After the mice were on experimental diets for 14 wk, we found that the tumor number but not tumor incidence was significantly decreased by apo-10′-lycopenoic acid supplementation (Fig. 5). Supplementation with apo-10′-lycopenoic acid

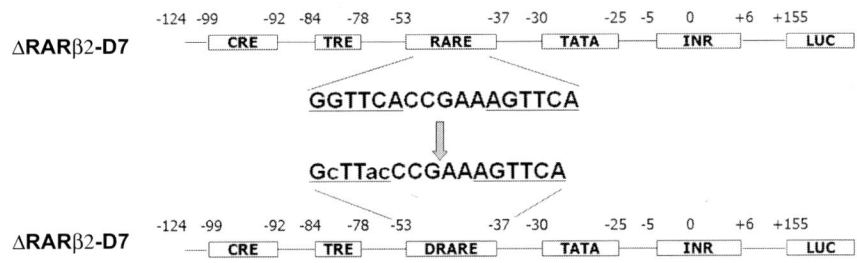

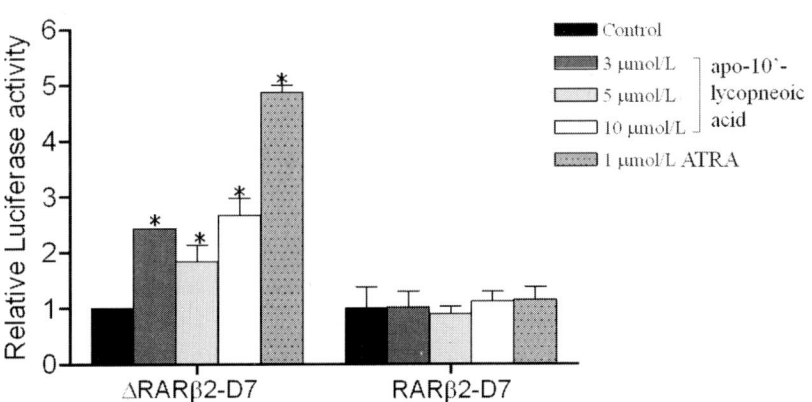

Fig. 4 The involvement of RARE on apo-10′-lycopenoic acid-transactivated RARβ expression. Upper panel shows a diagram of the construction of the RARβ reporter vector with wild-type or mutated RARE. Lower panel, HeLa cells transfected with the RARβ reporter vector and an internal control vector were treated with 5 μmol/L of apo-10′-lycopenoic acid or 1 μmol/L of *all-trans* retinoic acid for 24 hr. Luciferase activities were measured by Dual-luciferase reporter system. Values are means ± SEM of three replicate assays. *Statistically significantly different, as compared with control in the same group, $P < 0.05$. Adapted from Lian et al. (2007).

decreased tumor number in a dose-dependent manner from an average 16 tumors per mouse in the NNK injection alone group, to an average 10, 7, and 5 tumors per mouse in groups injected with NNK and given apo-10′-lycopenoic acid at doses of 10, 40, and 120 mg/kg, respectively, which represent 32.7%, 53.6%, and 65.4% declines in tumor number, as compared to the non-supplemented group ($P < 0.001$) (Fig. 5) (Lian et al. 2007).

In summary, both our *in vivo* and *in vitro* studies demonstrated that apo-10′-lycopenoic acid, an enzymatic metabolite of lycopene, is a potential chemopreventive agent against lung cancer. Furthermore, these results suggest that lycopene metabolites may, at least in part, mediate the chemopreventive effect of lycopene against lung cancer.

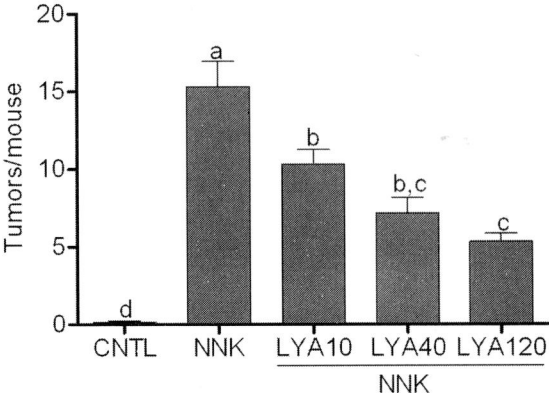

Fig. 5 Effect of apo-10′-lycopenoic acid supplementation on NNK-induced lung tumor development in A/J mice. A/J mice were fed control diet or diet supplemented with apo-10′-lycopenoic acid for 16 wk. Lung tumors were induced by injection of NNK at the third week of supplementation. Tumor nodules on the surface of mouse lung tissues were counted and recorded as tumor multiplicity (tumors per mouse). CNTL, non-supplementation plus sham injection; *NNK*, non-supplementation plus NNK injection; LYA10, LYA40, and LYA120, 10, 40, and 120 mg/kg diet of apo-10′-lycopenoic acid supplementation plus NNK injection. Values are presented as means ± SEM, n = 12-14. Groups that do not share a letter are significantly different, *P* < 0.05. Adapted from Lian et al. (2007).

ACKNOWLEDGMENTS

The authors thank Dr. Heather Mernitz for help in the preparation of the manuscript. This material is based on work supported by the NIH Grant R01CA104932 and US Department of Agriculture, under agreement No. 58-1950-7-707. Any opinions, findings, conclusion, or recommendations expressed in this publication are those of the authors and do not necessarily reflect the views of the US Department of Agriculture.

ABBREVIATIONS

BaP: benzo[a]pyrene; NNK: 4-(methylnitrosamino)-1-(3-pyridyl)-1-butanone; IGF-1: insulin-like growth factor-1; PCNA: proliferating cellular nuclear antigen; IGFBP: IGF-binding protein; CMO1: carotene 15,15′-monooxygenase; CMO2: carotene-9′,10′-monoxygenase; MTT: 3-(4,5-dimethylthiazol-2-yl)-2,5-diphenyltetrazolium bromide; CDK: cyclin-dependent kinase; RAR: retinoic acid receptor

References

Alberg, A. 2002. The influence of cigarette smoking on circulating concentrations of antioxidant micronutrients. Toxicology 180: 121-137.

American Cancer Society. 2007. Cancer Facts & Figures 2007. American Cancer Society, Atlanta.

Arab, L. and S. Steck-Scott, and P. Bowen. 2001. Participation of lycopene and beta-carotene in carcinogenesis: defenders, aggressors, or passive bystanders? Epidemiol. Rev. 23: 211-230.

Aust, O. and N. Ale-Agha, L. Zhang, H. Wollersen, H. Sies, and W. Stahl. 2003. Lycopene oxidation product enhances gap junctional communication. Food Chem. Toxicol. 41: 1399-1407.

Ben-Dor, A. and A. Nahum, M. Danilenko, Y. Giat, W. Stahl, H.D. Martin, T. Emmerich, N. Noy, J. Levy, and Y. Sharoni. 2001. Effects of acyclo-retinoic acid and lycopene on activation of the retinoic acid receptor and proliferation of mammary cancer cells. Arch. Biochem. Biophys. 391: 295-302.

Ben-Dor, A. and M. Steiner, L. Gheber, M. Danilenko, N. Dubi, K. Linnewiel, A. Zick, Y. Sharoni, and J. Levy. 2005. Carotenoids activate the antioxidant response element transcription system. Mol. Cancer Ther. 4: 177-186.

Brennan, P. and J. Butler, A. Agudo, S. Benhamou, S. Darby, C. Fortes, K.H. Jockel, M. Kreuzer, F. Nyberg, H. Pohlabeln, R. Saracci, H.R. Wichman, and P. Boffetta. 2000. Joint effect of diet and environmental tobacco smoke on risk of lung cancer among nonsmokers. J. Natl. Cancer Inst. 92: 426-427.

Briviba, K. and S.E. Kulling, J. Moseneder, B. Watzl, G. Rechkemmer, and A. Bub. 2004. Effects of supplementing a low-carotenoid diet with a tomato extract for 2 weeks on endogenous levels of DNA single strand breaks and immune functions in healthy nonsmokers and smokers. Carcinogenesis 25: 2373-2378.

Caris-Veyrat, C. and A. Schmid, M. Carail, and V. Bohm. 2003. Cleavage products of lycopene produced by in vitro oxidations: characterization and mechanisms of formation. J. Agric. Food Chem. 51: 7318-7325.

Clinton, S.K. and C. Emenhiser, S.J. Schwartz, D.G. Bostwick, A.W. Williams, B.J. Moore, and J.W. Erdman Jr. 1996. cis-trans lycopene isomers, carotenoids, and retinol in the human prostate. Cancer Epidemiol. Biomarkers Prev. 5: 823-833.

Darby, S. and E. Whitley, R. Doll, T. Key, and P. Silcocks. 2001. Diet, smoking and lung cancer: a case-control study of 1000 cases and 1500 controls in South-West England. Br. J. Cancer 84: 728-735.

de The, H. and M.M. Vivanco-Ruiz, P. Tiollais, H. Stunnenberg, and A. Dejean. 1990. Identification of a retinoic acid responsive element in the retinoic acid receptor beta gene. Nature 343: 177-180.

Di Mascio, P. and S. Kaiser, and H. Sies. 1989. Lycopene as the most efficient biological carotenoid singlet oxygen quencher. Arch. Biochem. Biophys. 274: 532-538.

Dietrich, M. and G. Block, E.P. Norkus, M. Hudes, M.G. Traber, C.E. Cross, and L. Packer. 2003. Smoking and exposure to environmental tobacco smoke decrease some plasma antioxidants and increase gamma-tocopherol in vivo after adjustment for dietary antioxidant intakes. Amer. J. Clin. Nutr. 77: 160-166.

dos Anjos Ferreira, A.L. and K.J. Yeum, R.M. Russell, N.I. Krinsky, and G. Tang. 2004. Enzymatic and oxidative metabolites of lycopene. J. Nutr. Biochem. 15: 493-502.

Ferreira, A.L. and K.J. Yeum, C. Liu, D. Smith, N.I. Krinsky, X.D. Wang, and R.M. Russell. 2000. Tissue distribution of lycopene in ferrets and rats after lycopene supplementation. J. Nutr. 130: 1256-1260.

Food and Nutrition Board, I.O.M. 2002. Dietary reference intakes for vitamin A, vitamin K, arsenic, boron, chromium, copper, iodine, iron, manganese, molybdenum, nickel, silicon, vanadium, and zinc. National Academy Press, Washington, DC.

Gajic, M. and S. Zaripheh, F. Sun, and J.W. Erdman Jr. 2006. Apo-8'-lycopenal and apo-12'-lycopenal are metabolic products of lycopene in rat liver. J. Nutr. 136: 1552-1557.

Garcia-Closas, R. and A. Agudo, C.A. Gonzalez, and E. Riboli. 1998. Intake of specific carotenoids and flavonoids and the risk of lung cancer in women in Barcelona, Spain. Nutr. Cancer 32: 154-158.

Giovannucci, E. 1999a. Nutritional factors in human cancers. Adv. Exp. Med. Biol. 472: 29-42.

Giovannucci, E. 1999b. Tomatoes, tomato-based products, lycopene, and cancer: review of the epidemiologic literature. J. Natl. Cancer Inst. 91: 317-331.

Giovannucci, E. 2002. A review of epidemiologic studies of tomatoes, lycopene, and prostate cancer. Exp. Biol. Med. (Maywood) 227: 852-859.

Goodman, D.S. and H.S. Huang. 1965. Biosynthesis of vitamin a with rat intestinal enzymes. Science 149: 879-880.

Guttenplan, J.B. and M. Chen, W. Kosinska, S. Thompson, Z. Zhao, and L.A. Cohen. 2001. Effects of a lycopene-rich diet on spontaneous and benzo[a]pyrene-induced mutagenesis in prostate, colon and lungs of the lacZ mouse. Cancer Lett. 164: 1-6.

Handelman, G.J. and L. Packer, and C.E. Cross. 1996. Destruction of tocopherols, carotenoids, and retinol in human plasma by cigarette smoke. Amer. J. Clin. Nutr. 63: 559-565.

Hantz, H.L. and L.F. Young, and K.R. Martin. 2005. Physiologically Attainable Concentrations of Lycopene Induce Mitochondrial Apoptosis in LNCaP Human Prostate Cancer Cells. Exp. Biol. Med. (Maywood) 230: 171-179.

Hecht, S.S. and P.M. Kenney, M. Wang, N. Trushin, S. Agarwal, A.V. Rao, and P. Upadhyaya. 1999. Evaluation of butylated hydroxyanisole, myo-inositol, curcumin, esculetin, resveratrol and lycopene as inhibitors of benzo[a]pyrene plus 4-(methylnitrosamino)-1-(3-pyridyl)-1-butanone-induced lung tumorigenesis in A/J mice. Cancer Lett. 137: 123-130.

Holick, C.N. and D.S. Michaud, R. Stolzenberg-Solomon, S.T. Mayne, P. Pietinen, P.R. Taylor, J. Virtamo, and D. Albanes. 2002. Dietary carotenoids, serum beta-carotene, and retinol and risk of lung cancer in the alpha-tocopherol, beta-carotene cohort study. Amer. J. Epidemiol. 156: 536-547.

Holmes, M.D. and M.N. Pollak, W.C. Willett, and S.E. Hankinson. 2002. Dietary correlates of plasma insulin-like growth factor I and insulin-like growth factor binding protein 3 concentrations. Cancer Epidemiol. Biomarkers Prev. 11: 852-861.

Hu, K.Q. and C. Liu, H. Ernst, N.I. Krinsky, R.M. Russell, and X.D. Wang. 2006. The biochemical characterization of ferret carotene-9',10'-monooxygenase catalyzing cleavage of carotenoids in vitro and in vivo. J. Biol. Chem. 281: 19327-19338.

Hwang, E.S. and P.E. Bowen. 2004. Cell cycle arrest and induction of apoptosis by lycopene in LNCaP human prostate cancer cells. J. Med. Food 7: 284-289.

Ito, Y. and M. Kurata, R. Hioki, K. Suzuki, J. Ochiai, and K. Aoki. 2005a. Cancer mortality and serum levels of carotenoids, retinol, and tocopherol: a population-based follow-up study of inhabitants of a rural area of Japan. Asian Pac. J. Cancer Prev. 6: 10-15.

Ito, Y. and K. Wakai, K. Suzuki, K. Ozasa, Y. Watanabe, N. Seki, M. Ando, Y. Nishino, T. Kondo, Y. Ohno, and A. Tamakoshi. 2005b. Lung cancer mortality and serum levels of carotenoids, retinol, tocopherols, and folic acid in men and women: a case-control study nested in the JACC Study. J. Epidemiol. 15 (Suppl 2): S140-149.

Ito, Y. and K. Wakai, K. Suzuki, A. Tamakoshi, N. Seki, M. Ando, Y. Nishino, T. Kondo, Y. Watanabe, K. Ozasa, and Y. Ohno. 2003. Serum carotenoids and mortality from lung cancer: a case-control study nested in the Japan Collaborative Cohort (JACC) study. Cancer Sci. 94: 57-63.

Jones, J.I. and D.R. Clemmons. 1995. Insulin-like growth factors and their binding proteins: biological actions. Endocr. Rev. 16: 3-34.

Karas, M. and H. Amir, D. Fishman, M. Danilenko, S. Segal, A. Nahum, A. Koifmann, Y. Giat, J. Levy, and Y. Sharoni. 2000. Lycopene interferes with cell cycle progression and insulin-like growth factor I signaling in mammary cancer cells. Nutr. Cancer 36: 101-111.

Khachik, F. and L. Carvalho, P.S. Bernstein, G.J. Muir, D.Y. Zhao, and N.B. Katz. 2002. Chemistry, distribution, and metabolism of tomato carotenoids and their impact on human health. Exp. Biol. Med. (Maywood) 227: 845-851.

Kiefer, C. and S. Hessel, J.M. Lampert, K. Vogt, M.O. Lederer, D.E. Breithaupt, and J. von Lintig. 2001. Identification and characterization of a mammalian enzyme catalyzing the asymmetric oxidative cleavage of provitamin A. J. Biol. Chem. 276: 14110-14116.

Kim, D.J. and N. Takasuka, J.M. Kim, K. Sekine, T. Ota, M. Asamoto, M. Murakoshi, H. Nishino, Z. Nir, and H. Tsuda. 1997. Chemoprevention by lycopene of mouse lung neoplasia after combined initiation treatment with DEN, MNU and DMH. Cancer Lett. 120: 15-22.

Kim, S.J. and E. Nara, H. Kobayashi, J. Terao, and A. Nagao. 2001. Formation of cleavage products by autoxidation of lycopene. Lipids 36: 191-199.

Kloer, D.P. and S. Ruch, S. Al-Babili, P. Beyer, and G.E. Schulz. 2005. The structure of a retinal-forming carotenoid oxygenase. Science 308: 267-269.

Knekt, P. and R. Jarvinen, L. Teppo, A. Aromaa, and R. Seppanen. 1999. Role of various carotenoids in lung cancer prevention. J. Nat. Cancer Inst. 91: 182-184.

Kotake-Nara, E. and S.J. Kim, M. Kobori, K. Miyashita, and A. Nagao. 2002. Acyclo-retinoic acid induces apoptosis in human prostate cancer cells. Anticancer Res. 22: 689-695.

Kotake-Nara, E. and M. Kushiro, H. Zhang, T. Sugawara, K. Miyashita, and A. Nagao. 2001. Carotenoids affect proliferation of human prostate cancer cells. J. Nutr. 131: 3303-3306.

Lee, C.M. and A.C. Boileau, T.W. Boileau, A.W. Williams, K.S. Swanson, K.A. Heintz, and J.W. Erdman Jr. 1999. Review of animal models in carotenoid research. J. Nutr. 129: 2271-2277.

Leuenberger, M.G. and C. Engeloch-Jarret, and W.D. Woggon. 2001. The reaction mechanism of the enzyme-catalyzed central cleavage of beta-carotene to retinal. Angew Chem. Int. Ed. Engl. 40: 2613-2617.

Levy, J. and E. Bosin, B. Feldman, Y. Giat, A. Miinster, M. Danilenko, and Y. Sharoni. 1995. Lycopene is a more potent inhibitor of human cancer cell proliferation than either alpha-carotene or beta-carotene. Nutr. Cancer 24: 257-266.

Lian, F. and D.E. Smith, H. Ernst, R.M. Russell, and X.D. Wang. 2007. Apo-10'-lycopenoic acid inhibits lung cancer growth in vitro, and suppresses lung tumorigenesis in the A/J mouse model in vivo. Carcinogenesis 28: 1567-1574.

Lindqvist, A. and S. Andersson. 2002. Biochemical properties of purified recombinant human beta-carotene 15,15'-monooxygenase. J. Biol. Chem. 277: 23942-23948.

Liu, C. and F. Lian, D.E. Smith, R.M. Russell, and X.D. Wang. 2003. Lycopene supplementation inhibits lung squamous metaplasia and induces apoptosis via up-regulating insulin-like growth factor-binding protein 3 in cigarette smoke-exposed ferrets. Cancer Res. 63: 3138-3144.

Liu, C. and R.M. Russell, and X.D. Wang. 2006. Lycopene supplementation prevents smoke-induced changes in p53, p53 phosphorylation, cell proliferation, and apoptosis in the gastric mucosa of ferrets. J. Nutr. 136: 106-111.

Livny, O. and I. Kaplan, R. Reifen, S. Polak-Charcon, Z. Madar, and B. Schwartz. 2002. Lycopene inhibits proliferation and enhances gap-junction communication of KB-1 human oral tumor cells. J. Nutr. 132: 3754-3759.

London, S.J. and J.M. Yuan, G.S. Travlos, Y.T. Gao, R.E. Wilson, R.K. Ross, and M.C. Yu. 2002. Insulin-like growth factor I, IGF-binding protein 3, and lung cancer risk in a prospective study of men in China. J. Nat. Cancer Inst. 94: 749-754.

Lu, Q.Y. and J.C. Hung, D. Heber, V.L. Go, V.E. Reuter, C. Cordon-Cardo, H.I. Scher, J.R. Marshall, and Z.F. Zhang. 2001. Inverse associations between plasma lycopene and other carotenoids and prostate cancer. Cancer Epidemiol. Biomarkers Prev. 10: 749-756.

Mannisto, S. and S.A. Smith-Warner, D. Spiegelman, D. Albanes, K. Anderson, P.A. van den Brandt, J.R. Cerhan, G. Colditz, D. Feskanich, J.L. Freudenheim, E. Giovannucci, R.A. Goldbohm, S. Graham, A.B. Miller, T.E. Rohan, J. virtamo, W.C. Willett, and D.J. Hunter. 2004. Dietary carotenoids and risk of lung cancer in a pooled analysis of seven cohort studies. Cancer Epidemiol. Biomarkers Prev. 13: 40-48.

Michaud, D.S. and D. Feskanich, E.B. Rimm, G.A. Colditz, F.E. Speizer, W.C. Willett, and E. Giovannucci. 2000. Intake of specific carotenoids and risk of lung cancer in 2 prospective US cohorts. Amer. J. Clin. Nutr. 72: 990-997.

Mucci, L.A. and R. Tamimi, P. Lagiou, A. Trichopoulou, V. Benetou, E. Spanos, and D. Trichopoulos. 2001. Are dietary influences on the risk of prostate cancer mediated through the insulin-like growth factor system? BJU Int. 87: 814-820.

Nahum, A. and K. Hirsch, M. Danilenko, C.K. Watts, O.W. Prall, J. Levy, and Y. Sharoni. 2001. Lycopene inhibition of cell cycle progression in breast and endometrial cancer cells is associated with reduction in cyclin D levels and retention of p27(Kip1) in the cyclin E-cdk2 complexes. Oncogene 20: 3428-3436.

Nahum, A. and L. Zeller, M. Danilenko, O.W. Prall, C.K. Watts, R.L. Sutherland, J. Levy, and Y. Sharoni. 2006. Lycopene inhibition of IGF-induced cancer cell growth depends on the level of cyclin D1. Eur J. Nutr. 45: 275-282.

Nara, E. and H. Hayashi, M. Kotake, K. Miyashita, and A. Nagao. 2001. Acyclic carotenoids and their oxidation mixtures inhibit the growth of HL-60 human promyelocytic leukemia cells. Nutr. Cancer 39: 273-283.

Nishino, H. 1997. Cancer prevention by natural carotenoids. J. Cell Biochem. 27 (Suppl): 86-91.

Obermuller-Jevic, U.C. and E. Olano-Martin, A.M. Corbacho, J.P. Eiserich, A. van der Vliet, G. Valacchi, C.E. Cross, and L. Packer. 2003. Lycopene inhibits the growth of normal human prostate epithelial cells in vitro. J. Nutr. 133: 3356-3360.

Olson, J.A. and O. Hayaishi. 1965. The enzymatic cleavage of beta-carotene into vitamin A by soluble enzymes of rat liver and intestine. Proc. Natl. Acad. Sci. USA 54: 1364-1370.

Paik, J. and A. During, E.H. Harrison, C.L. Mendelsohn, K. Lai, and W.S. Blaner. 2001. Expression and characterization of a murine enzyme able to cleave beta-carotene. The formation of retinoids. J. Biol. Chem. 276: 32160-32168.

Palozza, P. and A. Sheriff, S. Serini, A. Boninsegna, N. Maggiano, F.O. Ranelletti, G. Calviello, and A. Cittadini. 2005. Lycopene induces apoptosis in immortalized fibroblasts exposed to tobacco smoke condensate through arresting cell cycle and down-regulating cyclin D1, pAKT and pBad. Apoptosis 10: 1445-1456.

Poliakov, E. and S. Gentleman, F.X. Cunningham Jr., N.J. Miller-Ihli, and T.M. Redmond. 2005. Key role of conserved histidines in recombinant mouse beta-carotene 15,15'-monooxygenase-1 activity. J. Biol. Chem. 280: 29217-29223.

Redmond, T.M. and S. Gentleman, T. Duncan, S. Yu, B. Wiggert, E. Gantt, and F.X. Cunningham Jr. 2001. Identification, expression, and substrate specificity of a mammalian beta-carotene 15,15'-dioxygenase. J. Biol. Chem. 276: 6560-6565.

Rohan, T.E. and M. Jain, G.R. Howe, and A.B. Miller. 2002. A cohort study of dietary carotenoids and lung cancer risk in women (Canada). Cancer Causes Control 13: 231-237.

Schmitz, H.H. and C.L. Poor, R.B. Wellman, and J.W. Erdman Jr. 1991. Concentrations of selected carotenoids and vitamin A in human liver, kidney and lung tissue. J. Nutr. 121: 1613-1621.

Smith-Warner, S.A. and D. Spiegelman, S.S. Yaun, D. Albanes, W.L. Beeson, P.A. van den Brandt, D. Feskanich, A.R. Folsom, G.E. Fraser, J.L. Freudenheim, E. Giovannucci, R.A. Goldbohm, S. Graham, L.H. Kushi, A.B. Miller,

P. Pietinen, T.E. Rohan, F.E. Speizer, W.C. Willett, and D.J. Hunter. 2003. Fruits, vegetables and lung cancer: a pooled analysis of cohort studies. Int. J. Cancer 107: 1001-1011.

Som, S. and M. Chatterjee, and M.R. Banerjee. 1984. Beta-carotene inhibition of 7,12-dimethylbenz[a]anthracene-induced transformation of murine mammary cells in vitro. Carcinogenesis 5: 937-940.

Speizer, F.E. and G.A. Colditz, D.J. Hunter, B. Rosner, and C. Hennekens. 1999. Prospective study of smoking, antioxidant intake, and lung cancer in middle-aged women (USA). Cancer Causes Control 10: 475-482.

Stahl, W. and J. von Laar, H.D. Martin, T. Emmerich, and H. Sies. 2000. Stimulation of gap junctional communication: comparison of acyclo-retinoic acid and lycopene. Arch. Biochem. Biophys. 373: 271-274.

Stefani, E.D. and P. Boffetta, H. Deneo-Pellegrini, M. Mendilaharsu, J.C. Carzoglio, A. Ronco, and L. Olivera. 1999. Dietary antioxidants and lung cancer risk: a case-control study in Uruguay. Nutr. Cancer 34: 100-110.

Steinberg, F.M. and A. Chait. 1998. Antioxidant vitamin supplementation and lipid peroxidation in smokers. Amer. J. Clin. Nutr. 68: 319-327.

Tang, G.W. and X.D. Wang, R.M. Russell, and N.I. Krinsky. 1991. Characterization of beta-apo-13-carotenone and beta-apo-14'-carotenal as enzymatic products of the excentric cleavage of beta-carotene. Biochemistry 30: 9829-9834.

Velmurugan, B. and V. Bhuvaneswari, U.K. Burra, and S. Nagini. 2002. Prevention of N-methyl-N'-nitro-N-nitrosoguanidine and saturated sodium chloride-induced gastric carcinogenesis in Wistar rats by lycopene. Eur. J. Cancer Prev. 11: 19-26.

Vogt, T.M. and S.T. Mayne, B.I. Graubard, C.A. Swanson, A.L. Sowell, J.B. Schoenberg, G.M. Swanson, R.S. Greenberg, R.N. Hoover, R.B. Hayes, and R.G. Ziegler. 2002. Serum lycopene, other serum carotenoids, and risk of prostate cancer in US Blacks and Whites. Amer. J. Epidemiol. 155: 1023-1032.

von Lintig, J., and K. Vogt. 2000. Filling the gap in vitamin A research. Molecular identification of an enzyme cleaving beta-carotene to retinal. J. Biol. Chem. 275: 11915-11920.

Voorrips, L.E. and R.A. Goldbohm, H.A. Brants, G.A. van Poppel, F. Sturmans, R.J. Hermus, and P.A. van den Brandt. 2000. A prospective cohort study on antioxidant and folate intake and male lung cancer risk. Cancer Epidemiol. Biomarkers Prev. 9: 357-365.

Wakai, K. and Y. Ito, K. Suzuki, A. Tamakoshi, N. Seki, M. Ando, K. Ozasa, Y. Watanabe, T. Kondo, Y. Nishino, and Y. Ohno. 2002. Serum insulin-like growth factors, insulin-like growth factor-binding protein-3, and risk of lung cancer death: a case-control study nested in the Japan Collaborative Cohort (JACC) Study. Jpn J. Cancer Res. 93: 1279-1286.

Wang, X.D. 2004. Carotenoid oxidative/degradative products and their biological activities. pp. 313-335. *In:* N.I. Krinsky, S.T. Mayne and H. Sies [eds.]. Carotenoids in Health and Disease, Marcel Dekker, Inc., New York.

Wang, X.D. 2005. Can smoke-exposed ferrets be utilized to unravel the mechanisms of action of lycopene? J. Nutr. 135: 2053S-2056S.

Wang, X.D. and C. Liu, R.T. Bronson, D.E. Smith, N.I. Krinsky, and R.M. Russell. 1999. Retinoid signaling and activator protein-1 expression in ferrets given beta-carotene supplements and exposed to tobacco smoke. J. Nat. Cancer Inst. 91: 60-66.

Wang, X.D. and R.M. Russell, C. Liu, F. Stickel, D.E. Smith, and N.I. Krinsky. 1996. Beta-oxidation in rabbit liver in vitro and in the perfused ferret liver contributes to retinoic acid biosynthesis from beta-apocarotenoic acids. J. Biol. Chem. 271: 26490-26498.

Wang, X.D. and G.W. Tang, J.G. Fox, N.I. Krinsky, and R.M. Russell. 1991. Enzymatic conversion of beta-carotene into beta-apo-carotenals and retinoids by human, monkey, ferret, and rat tissues. Arch. Biochem. Biophys. 285: 8-16.

Wei, W. and Y. Kim, and N. Boudreau. 2001. Association of smoking with serum and dietary levels of antioxidants in adults: NHANES III, 1988-1994. Amer. J. Public Health 91: 258-264.

Wright, M.E. and S.T. Mayne, C.A. Swanson, R. Sinha, and M.C. Alavanja. 2003. Dietary carotenoids, vegetables, and lung cancer risk in women: the Missouri women's health study (United States). Cancer Causes Control 14: 85-96.

Wyss, A. and G. Wirtz, W. Woggon, R. Brugger, M. Wyss, A. Friedlein, H. Bachmann, and W. Hunziker. 2000. Cloning and expression of beta,beta-carotene 15,15'-dioxygenase. Biochem. Biophys. Res. Commun. 271: 334-336.

Yu, H. and M.R. Spitz, J. Mistry, J. Gu, W.K. Hong, and X. Wu. 1999. Plasma levels of insulin-like growth factor-I and lung cancer risk: a case-control analysis. J. Nat. Cancer Inst. 91: 151-156.

Yuan, J.M. and R.K. Ross, X.D. Chu, Y.T. Gao, and M.C. Yu. 2001. Prediagnostic levels of serum beta-cryptoxanthin and retinol predict smoking-related lung cancer risk in Shanghai, China. Cancer Epidemiol. Biomarkers Prev. 10: 767-773.

Yuan, J.M. and D.O. Stram, K. Arakawa, H.P. Lee, and M.C. Yu. 2003. Dietary cryptoxanthin and reduced risk of lung cancer: the Singapore Chinese Health Study. Cancer Epidemiol. Biomarkers Prev. 12: 890-898.

Zhang, H. and E. Kotake-Nara, H. Ono, and A. Nagao. 2003. A novel cleavage product formed by autoxidation of lycopene induces apoptosis in HL-60 cells. Free Radic. Biol. Med. 35: 1653-1663.

Zhang, L.X. and R.V. Cooney, and J.S. Bertram. 1991. Carotenoids enhance gap junctional communication and inhibit lipid peroxidation in C3H/10T1/2 cells: relationship to their cancer chemopreventive action. Carcinogenesis 12: 2109-2114.

Ziegler, R.G. and S.T. Mayne, and C.A. Swanson. 1996. Nutrition and lung cancer. Cancer Causes Control 7: 157-177.

3.4

Breast Cancer and Lycopene

[1,2,3]Nasséra Chalabi, [1,2,3,4,*]Yves-Jean Bignon and [1,2,3]Dominique J. Bernard-Gallon

[1]Département d'Oncogénétique, Centre Jean Perrin, 58 Rue Montalembert, BP 392 63011 Clermont-Ferrand Cedex 01, France
[2]INSERM UMR 484, rue Montalembert, BP 184, 63005 Clermont-Ferrand Cedex France
[3]Centre de Recherche en Nutrition Humaine (CRNH), 58 Rue Montalembert BP 321, 63009 Clermont-Ferrand Cedex 01, France
[4]Université d'Auvergne, 28 place Henri Dunant, BP 38, 63001 Clermont-Ferrand 1 France

ABSTRACT

Many studies have pointed out the benefits of carotenoids on breast cancer prevention. In this review, we provide a non-exhaustive list of *in vitro*, animal and human studies performed in order to improve an inverse correlation between lycopene and breast cancer incidence. Breast cancer is a major cause of death of women and the etiology is multifactorial. Among identified risk factors, dietary habits could play an important role, more precisely consumption of fruits and vegetables. Some evidence of this dietary preventive effect has come from the rising rate of breast cancer reported among immigrant Asian women. Fruits and vegetables are a source of a number of micronutrients such as vitamins, phyto-estrogens and carotenoids. During the past few years, tomatoes and tomato-derived products, which contain lycopene, appeared to be largely involved in breast cancer prevention. Numerous epidemiological studies have demonstrated a correlation between high serum level of lycopene and breast cancer prevention. This effect has been attributed in part to inhibition of breast cancer cell growth and antioxidant properties. In this review,

A list of abbreviations is given before the references.
Corresponding author

we focus on lycopene and its effects specifically on breast cancer prevention through *in vivo* and *in vitro* studies.

INTRODUCTION

In France, breast cancer is the leading cause of death among women. A majority of these cancers are hormone dependent. Moreover, *BRCA1* and *BRCA2* genes are the two major genes responsible for 5 to 10% of hereditary breast cancers. In sporadic breast cancer, their level of expression is modified. Nevertheless, this pathology could be influenced by environmental factors such as way of life and habits of food hygiene. Several epidemiological studies showed an opposite correlation between the consumption of lycopene, major carotenoid of tomatoes, and the incidence of breast cancer. Lycopene was discovered in 1873 by a researcher named Harsten. It is a carotenoid found in many fruits and vegetables such as grapefruit, watermelon, papaya, pink guava and rosehip. But lycopene is mainly a red fat-soluble pigment found essentially in tomatoes (*Lycopersicon esculentum*) and related products. Cooked tomatoes are richest in lycopene. Indeed, cooking and fatty acid addition increase lycopene bioavailability.

Lycopene is also known as the most potent antioxidant of carotenoid family. This property makes lycopene effective in preventing cancer and cardiac disorders. Among cancers sensitive to lycopene effects, numerous epidemiological studies have been performed on the most common genital cancers: prostate and breast (Bosetti et al. 2004, Bratt 2004, Etminan et al. 2004, Gomez-Aracena et al. 2003, Sato et al. 2002, Sesso et al. 2005, Tamimi et al. 2005, Wu et al. 2004). Moreover, it has been demonstrated that lycopene is the predominant carotenoid in human plasma and more precisely in low density and very low density lipoprotein fractions of the serum because of its lipophilic nature (Clinton 1998). In humans, lycopene is also found to concentrate in fat tissues, liver, testes, adrenal glands and prostate (Stahl et al. 1992). In breast tissue, lycopene level has been evaluated at 0.78 nmol/g wet weight (Agarwal and Rao 2000).

In this chapter, we focus on the role and effects of lycopene in breast cancer prevention according to *in vivo* and *in vitro* studies.

IN VIVO BREAST CANCER STUDIES

There are numerous epidemiological studies demonstrating that women with diet rich in fruits and vegetables have a decreased risk of breast cancer. But all these studies are not consistent with a preventive role of lycopene in breast cancer (Table 1).

Table 1 Summary of epidemiological studies in relation with lycopene and breast cancer.

Epidemiological studies	Countries	Breast cancer and lycopene relation	References
Case-control (304)	Australia	yes	Ching et al. 2002
Case-control (105)	United States	yes	Dorgan et al. 1998
Case-control (608)	United States	no	Freudenheim et al. 1996
Case-control (2,999)	United States	yes	Gaudet et al. 2004
Cross-sectional (51)	Spain	yes	Gomez-Aracena et al. 2003
Case-control (491)	Sweden	yes	Hulten et al. 2001
Case-control (417)	India	yes	Ito et al. 1999
Cohort (4,697)	Finland	no	Jarvinen et al. 1997
Case-control (11,571)	Italy	no	La Vecchia et al. 2002
Case-control (731)	Switzerland	yes	Levi et al. 2001
Case-control (550)	United States	no	London et al. 1992
Cohort (56)	United States	yes	Mc Eligot et al. 1999
Cohort (2,970)	United States	yes	Pierce et al. 2004
Case-control (196)	United States	no	Potischman et al. 1990
Cohort (79)	United States	yes	Rock et al. 1997
Case-control (805)	Uruguay	yes	Ronco et al. 1999
Case-control (590)	United States	yes	Sato et al. 2002
Cohort (39,876)	United States	no	Sesso et al. 2005
Case-control (56)	United States	yes	Simon et al. 2000
Case-control (1,947)	United States	yes	Tamimi et al. 2005
Cohort (56,837)	Canada	no	Terry et al. 2002
Case-control (540)	United States	yes	Toniolo et al. 2001
Cohort (302)	United States	no	Van Kappel et al. 2001
Case-control (90)	Korea	yes	Yeum et al. 1998
Case-control (109)	United States	no	Zhang et al. 1997

Lycopene Bioavailability and Breast Cancer

In 1996, Freudenheim et al. (1996) observed through a pre-menopausal breast cancer case-control study that there was a strong protective effect associated with habitual intake of vegetables; intake of tomatoes was correlated with a decreased risk of breast cancer, while no correlation was found with lycopene serum level. Van Kappel et al. (2001) observed the same result in a prospective study. One explanation of this phenomenon was that lycopene derived from tomatoes was more concentrated in tomato-rich products such as pizza, ketchup, and tomato paste, which were not correlated with high fruit and vegetable consumption. This result was also observed in a case-control study in Uruguay, where significant reduction risk of breast cancer associated with lycopene intake was due to a high consumption in this population of pasta and polenta eaten with abundant tomato sauce (Ronco et al. 1999). The reason appeared to be linked to bioavailability of *cis-* or *trans-*isomers of lycopene: *cis-*isomers most present in processed tomato products are more available because they are more soluble in biliary salts and more incorporated in chylomicrons (Giovannucci et al. 1995). This observation hypothesized that lycopene serum level should be associated with lipid metabolism. Indeed, a strong correlation of carotenoid concentrations has been reported between serum and breast adipose tissue, with the result that adipose tissue is a dynamic reservoir of carotenoids that reflects the circulating carotenoid concentrations (Gomez-Aracena et al. 2003, Kardinaal et al. 1995, Yeum et al. 1998). This relationship between high fruit and vegetable consumption and a decrease in breast cancer risk was corroborated by micronutrient serum level studies (Dorgan et al. 1998, Rock et al. 1997, Yeum et al. 1998).

Chen and Djuric (2002) showed that it would be more interesting to examine oxidation products of lycopene than lycopene alone. For that purpose, they investigated an HPLC study in order to check relationships between plasma and breast nipple aspirate fluid levels of 2,6-cyclolycopene-1,5-diol, which is the major oxidation product of lycopene. Moreover, this product has been identified in diet, breast milk and prostate tissue (Khachik et al. 1997, Kucuk et al. 2001). Results demonstrated an average ratio of the oxidation product to lycopene in breast nipple aspirate fluids 3-fold that in plasma, suggesting that 2,6-cyclolycopene-1,5-diol could be an interesting marker of oxidative stress in breast.

Lycopene and Hormonal Breast Cancer Status

In a North Sweden nested case-referent study, a significant association was found between lycopene and reduced breast cancer risk among post-menopausal women (Hulten et al. 2001). An inverse relationship was also observed between body mass index and plasma concentration of lycopene (Ascherio et al. 1992, Rock et al. 1997). These results were supported by other studies (Dorgan et al. 1998, Zhang et al. 1997) suggesting that age could be a non-negligible factor in lycopene bioavailability. Indeed, Rock et al. (1997) reported that factors such as age, baseline serum concentration, change in serum cholesterol concentration and change in alcohol intake were predictive of the degree of change in serum carotenoid concentrations in response to a high vegetable diet intervention. Another case-control study conducted by Ito et al. (1999) demonstrated that serum levels of carotenoids including lycopene were significantly lower among breast cancer cases and hospital controls (patients who had cancer at any site other than breast and gynecological organs) compared to healthy controls, especially in post-menopausal women. In addition, it was demonstrated that post-menopausal women consuming high level of lycopene had 34% lower incidence of breast cancer (Gaudet et al. 2004). In that study, the authors also found a stronger effect for post-menopausal women with ER+ (Estrogen Receptor Positive) tumors possibly due to a correlation between endogenous estrogen levels and oxidative stress. This hypothesis was in accordance with previous reports that DNA damage and oxidative stress enhanced the tendency of cells to metastasize (Malins et al. 1996), while carotenoids including lycopene may be able to reduce breast cancer cell proliferation (Ben-Dor et al. 2001, Chalabi et al. 2004, Nahum et al. 2001, Prakash et al. 2001). Nevertheless, Sesso et al. (2005) performed a large-scale prospective study of middle-aged and older women and observed that increasing the consumption of either dietary lycopene or tomato-based food products did not exhibit a dose-response relation with the risk of breast cancer, including ER+ and PR+ breast cancers.

In this context, an interesting study was conducted in 1992 to better understand problems of interpreting circulating biochemical indicators of nutritional status in case-control studies. The levels of selected nutrients including lycopene were examined before and after the diagnosis and treatment of breast cancer (Potischman et al. 1992). Results demonstrated that adjuvant therapy with tamoxifen was associated with decreased levels of β-carotene and increased levels of triglycerides and α-tocopherol, whereas chemotherapy was associated with elevation in levels of cholesterol, γ-tocopherol and retinol. The authors concluded that interpretations of case-control studies from nutrient values in blood samples may consider the different conditions of the blood collection (before or after surgery or treatment).

Lycopene, Environmental Factors and Breast Cancer

In 2000, Simon et al. investigated a pilot case-control study to evaluate the relationship between race and plasma lycopene levels and a significant interaction was observed with a higher breast cancer risk seen among African American cases in the lowest quintiles compared to Caucasian cases (Simon et al. 2000). This observation was also reported in a Korean case-control study where no differences in serum concentrations of carotenoids were found between benign breast tumor patients and breast cancer patients (Yeum et al. 1998). This was due to a bias in population recruitment revealed by the Korea Institute of Food Hygiene, where an extremely low intake of tomatoes and their products was reported in this population.

Lycopene and Breast Cancer in Animal Model Studies

All these observations suggested that there were many factors that could modify lycopene bioavailability and its effects on breast cancer prevention. However, even animal studies showed controversial results. In a study conducted on N-methylnitrosourea (NMU)-induced rat mammary tumor model, the authors compared the effects of pure lycopene with lycopene-rich tomato carotenoid oleoresin containing 66% lycopene, 22% β-carotene, 6% phytofluene, 5% phytoene and 0.007% ζ-carotene (Cohen et al. 1999). No effect of pure crystalline lycopene or lycopene-rich tomato carotenoid oleoresin (TCO) was observed on growth and development of NMU-induced mammary tumors. These results were not consistent with those found by Nagasawa et al. (1995) and Sharoni et al. (1997), where an inhibition of mammary tumors was noted. The first study was performed on spontaneous mammary tumor and the second on 7,12-dimethylbenz[a]anthracene (DMBA)-induced rat mammary tumors, which could in part explain these results contrary to those of Cohen et al. (1999). The authors pointed out that in all these studies there were many differences in lycopene preparation (pure crystalline lycopene or TCO), doses and administration method (e.g., intra-rectal, instillation gavage, intra-peritoneal injection) (Cohen 2002). Moreover, bioavailabilty of lycopene from a crystalline form or TCO could largely interfere in animal studies because lycopene, in the presence of other carotenoids, may be absorbed more efficiently than in the absence of other carotenoids (van den Berg 1999).

In conclusion, from these *in vivo* studies, it is difficult to say whether lycopene does or does not have beneficial effects on breast cancer development because of the numerous factors involved in human or animal studies. Nevertheless, these observations provided some food recommendations, as high intake of fruits and vegetables and low fat intake may help protect against breast cancer. In the case of lycopene, *in vitro* studies could provide more valid conclusions especially through mechanistic investigations.

IN VITRO BREAST CANCER STUDIES

Unlike *in vivo* studies, cell cultures could provide more reproducibility between different studies. Most disadvantages arise from lycopene concentration used, which is often supraphysiological (Table 2). Nevertheless, some studies were contradictory and one possible explanation came from the solubilization procedures used, which may induce carotenoid precipitation. A study conducted in 1992 determined relative solubility and stability of carotenoids in 18 organic solvents and results demonstrated that THF (Tetra Hydro Furane) was several hundred times as effective as ethanol and DMSO (Craft and Soares, 1992). The use of THF as a solubilizer has since been adopted in several studies (Bertram 1993, Chalabi et al. 2004, Levy et al. 1995).

To investigate the lycopene responsiveness of genes predisposing a person to breast cancer and to better understand the molecular mechanisms underlying the effects of lycopene, our group used a gene expression approach around *BRCA1* and *BRCA2*, oncosuppressor genes largely implicated in breast cancer using human breast cancer cell lines MCF-7 (*ERα+/ERβ+*) and MDA-MB-231 (*ERα–/ERβ+*) and a fibrocystic breast cell line MCF-10a (*ERα–/ERβ–*) either exposed or not to 10 µM lycopene for 48 hr. In Figs. 1 and 2, we illustrate our studies and results.

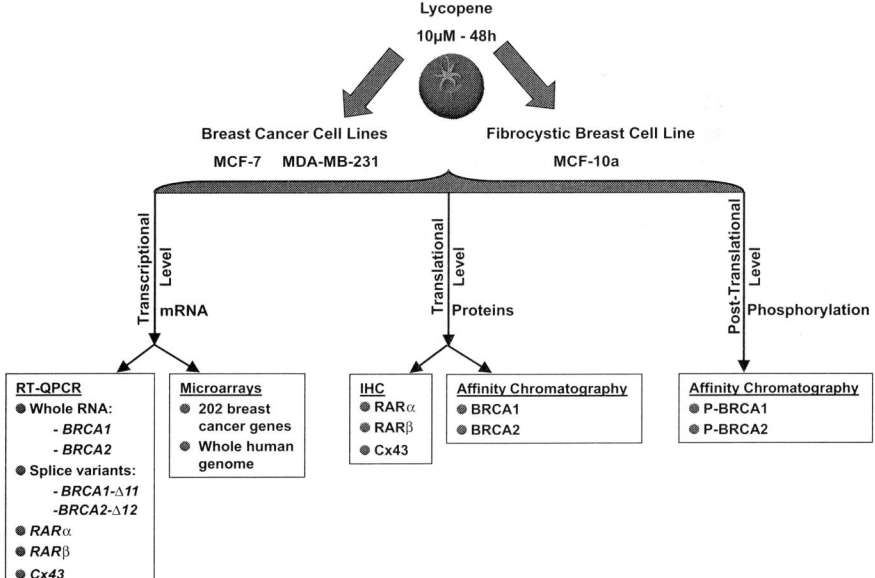

Fig. 1 Studies performed by Chalabi et al. (2004, 2005, 2006, 2007a, b).

Table 2 Summary of mechanistic studies in relation with lycopene and breast cancer.

Studied functions	Technologies used	Breast cancer and lycopene relation	References
Activation of RAR and proliferation	Transfection, flow cytometry, western blotting	yes	Ben-Dor et al. 2001
ARE expression	Transfection, real-time PCR, western blotting	yes	Ben-Dor et al. 2005
BRCA1 & *BRCA2* gene expression and cell proliferation	Flow cytometry, RT-QPCR	yes	Chalabi et al. 2004
Phosphorylated BRCA1 & BRCA2 protein expression	Affinity perfusion chromatography	yes	Chalabi et al. 2005
202 gene expression	Breast cancer dedicated-microarray	yes	Chalabi et al. 2006
Whole-genome gene expression	Pangenomic microarray	yes	Chalabi et al. 2007b
RAR and Cx43 protein expression	Immunohistochemistry	yes	Chalabi et al. 2007a
Mitogenic activity of IGF-1	Cell cycle analysis, western blotting	yes	Karas et al. 2000
Cancer cell growth	Cell proliferation assay and cell counting	yes	Levy et al. 1995
Apoptosis	Transfection	yes	Molnar et al. 2004
Cell cycle progression	Flow cytometry and western blotting	yes	Nahum et al. 2001
Proliferation	Flow cytometry, northern blotting	yes	Nahum et al. 2006
IGF-induced cancer cell growth	Cell cycle analysis, western blotting	yes	Prakash et al. 2001

RT-QPCR: Reverse Transcription-Quantitative Polymerase Chain Reaction.

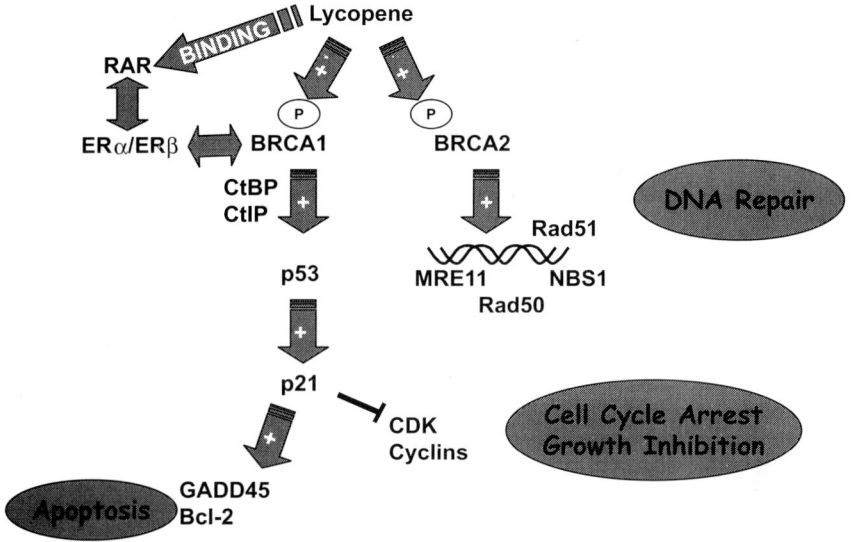

Fig. 2 Lycopene breast cancer molecular mechanism according to results of our group (Chalabi et al. 2006, 2007b). MRE11, meiotic recombination 11.

Among *in vitro* studies, it has been clearly demonstrated that lycopene effect on culture cells was in part due to its antioxidant ability. Several studies have concluded that lycopene was the most efficient singlet oxygen quencher among all carotenoids (Di Mascio et al. 1990, 1989, Sies et al. 1992). This was essentially due to its capacity to reduce oxidative damage of DNA, proteins and lipids. But other properties have also been pointed out, such as cell growth inhibition or gap junction communication induction.

Lycopene and Breast Cancer Cell Growth

One of the most widely studied effects of lycopene is its growth inhibitory ability on breast cancer cells. Prakash et al. (2001) observed this effect on MCF-7 and MDA-MB-231 cell lines suggesting that estrogen receptor status was an important but not essential factor for the responsiveness of breast cancer cells to lycopene treatment. Levy et al. (1995) demonstrated that lycopene was not only able to inhibit mammary cell growth but also was much more potent than α- and β-carotene. The authors concluded that this greater potency of lycopene may result from its larger number of conjugated double bonds and the absence of β-ionone rings.

Another explanation of this inhibitory effect of lycopene could be the reduction of IGF-1 level. Indeed, high blood levels of IGF-1 can predict increased risk for breast cancer (Hankinson et al. 1998). This result was consistent with those observed by Karas et al. (2000), where lycopene treatment in MCF-7 cells markedly reduced the IGF-1 stimulation of tyrosine phosphorylation of insulin receptor substrate 1 and binding capacity of the AP-1 transcription complex. Additionally, the authors observed an increase in membrane-associated IGF-binding proteins that negatively regulate IGF-1 receptor activation. This effect on growth cells was accompanied by a cell-cycle arrest through G_1 to S phase (Ben-Dor et al. 2001, Chalabi et al. 2004, Karas et al. 2000) mediated by the reduction of cyclin D1, an oncogene overexpressed in many breast cancer cell lines and tumors (Buckley et al. 1993) that led to a decrease of phosphorylated retinoblastoma phosphorylation (Nahum et al. 2001). This reduction of cyclin D1 has been explained by retention of p27 in the cyclin E/cdk2 complex resulting in inhibition of cdk 2 kinase activity. Moreover, it has been observed that IGF-1-mediated induction of p21 was also reduced by lycopene treatment because p21 and cyclin D1 inductions were tightly linked (Nahum et al. 2006).

Lycopene and Apoptosis in Breast Cancer

In breast cancer, apoptosis is one of the deregulated features that enhance anarchical cell proliferation leading to tumor development. Among the multiple properties of lycopene, its ability to induce apoptosis in breast cancer cell lines has been explored. Molnar et al. (2004) demonstrated that lycopene induced apoptosis of MDA-MB-231 breast cancer cell lines. We deserved similar results using microarray technology on MCF-7 and MDA-MB-231 breast cancer cell lines and we found up-regulation of different genes such as GADD45α, Bcl-2 and various caspases (Chalabi et al. 2006, 2007b). However, we also observed a specific cell line response to lycopene treatment that could be justified by the different estrogen receptor (ER) and retinoic acid receptor (RAR) cell status. Indeed, MCF-7 and MDA-MB-231 cells expressed basal levels of RARα and γ (Prakash et al. 2001) but only MCF-7 cell line expressed both ERα and β, while MDA-MB-231 cells are only ERβ+. Moreover, it has been demonstrated that RARα and γ could mediate programmed cell death in breast cell lines (Toma et al. 1998a, b) and some lycopene derivatives may interact with the retinoid receptors (Ben-Dor et al. 2001). Nevertheless, further investigations must be performed in this field.

Lycopene and Gap Junction Communication in Breast Cancer

Gap junction communication is an important factor in cancer prevention (Yamasaki 1990). Gap junctions allowed low molecular weight molecule exchange between cells. In the mammary gland, 15 connexins have been identified (Fishman et al. 1991, Kumar and Gilula 1996), including Cx43, Cx26 and Cx32. Cx43 is the predominant form in breast epithelium (Monaghan and Moss 1996). Nevertheless, loss of gap junction communication would be important for malignant transformation and its restoration may reverse the malignancy process (Hotz-Wagenblatt and Shalloway 1993, King et al. 2000, Loewenstein 1981). It has been demonstrated that lycopene stimulated the functionality of gap junction communication in a dose-dependent manner in MCF-7 breast cancer cells and could induce Cx43 mRNA and protein expression (Fornelli et al. 2007, Zhang et al. 1991, 1992). In a previous work, we also demonstrated an increase of Cx43 nuclear expression after lycopene treatment of MCF-7, MDA-MB-231 and MCF-10a breast cell lines (Chalabi et al. 2007a).

Lycopene and Xenobiotic Metabolism in Breast Cancer

Cytochrome P450 metabolizes both endogenous substrates (steroids, fatty acids, eicosanoids, retinoids) and exogenous substrates (xenobiotics) and is associated not only with tumor development and progression but also with efficacy of cancer treatment (Lewis 2004). In a previous work (Chalabi et al. 2007b), we demonstrated that lycopene up-regulated MGST1, a member of the glutathione S-transferase (GST) family of genes critical for certain life processes, as well as for detoxification via conjugation of reduced glutathione with numerous substrates, including pharmaceuticals and environmental pollutants (Fig. 3). The GST genes are up-regulated in response to oxidative stress and are overexpressed in many tumors, leading to problems during cancer chemotherapy (Nebert and Vasiliou 2004). Ben-Dor et al. (2005) found that lycopene increased glutathione levels in cancer cells including breast cancer MCF-7 cell line and induced phase II enzymes through activation of the antioxidant response element (ARE) transcription system and carotenoid-induced nuclear translocation of nuclear factor E_2-related factor 2, the major ARE-activating transcription factor implicated in induction of antioxidant and detoxifying genes.

In conclusion, lycopene seems to have a potential beneficial effect on breast cancer prevention. Nevertheless, further investigations must be performed in order specifically to determine molecular mechanisms

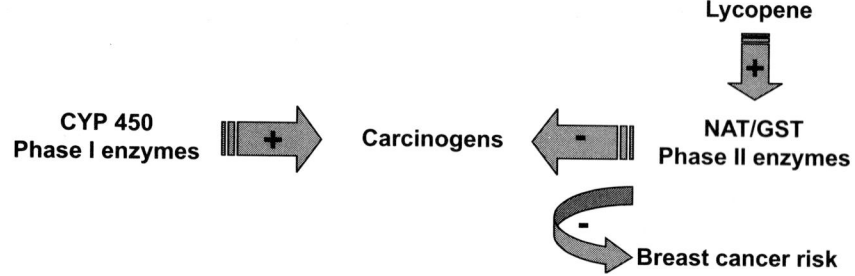

Fig. 3 Lycopene acts on xenobiotic metabolism in breast cancer. CYP 450, cytochrome P450; NAT, N-acetyl transferase.

involved in breast cancer regulation. For that, it would be most interesting for future investigations to focus on *in vitro* studies or animal control studies in which the various parameters such as bioavailability, source of lycopene or solubilization conditions could be controlled.

ACKNOWLEDGEMENTS

This work was supported by grants from Les Comités du Cantal et du Puy-de-Dôme de La Ligue Nationale Française de Lutte Contre le Cancer.

ABBREVIATIONS

ARE: Antioxidant response element; Bcl-2: B-cell leukemia/lymphoma-2; BRCA: Breast cancer gene; CDK: Cyclin-dependent kinase; CtBP: C-terminal binding protein; CtIP: CtBP-interacting protein; Cx43: Connexin 43; DMBA: 7,12-Dimethylbenz[*a*]anthracene; ER: Estrogen receptor; GADD45α: Growth arrest and DNA damage-inducible α; GST: Glutathione S-transferase; HPLC: High performance liquid chromatography; IGF-1: Insulin growth factor-1; NMU: N-Methylnitrosourea; P-BRCA1: Phosphorylated-breast cancer gene 1; PR: Progesterone receptor; RAR: Retinoic acid receptor; TCO: Tomato carotenoid oleoresin

References

Agarwal, S. and A.V. Rao. 2000. Tomato lycopene and its role in human health and chronic diseases. CMAJ 163: 739-744.

Ascherio, A. and M.J. Stampfer, G.A. Colditz, E.B. Rimm, L. Litin, and W.C. Willett. 1992. Correlations of vitamins A and E intakes with the plasma concentrations of carotenoids and tocopherols among American men and women. J. Nutr. 122: 1792-1801.

Ben-Dor, A. and A. Nahum, M. Danilenko, Y. Giat, W. Stahl, H.D. Martin, T. Emmerich, N. Noy, J. Levy, and Y. Sharoni. 2001. Effects of acyclo-retinoic acid and lycopene on activation of the retinoic acid receptor and proliferation of mammary cancer cells. Arch. Biochem. Biophys. 391: 295-302.

Ben-Dor, A. and M. Steiner, L. Gheber, M. Danilenko, N. Dubi, K. Linnewiel, A. Zick, Y. Sharoni, and J. Levy. 2005. Carotenoids activate the antioxidant response element transcription system. Mol. Cancer Ther. 4: 177-186.

Bertram, J.S. 1993. Cancer prevention by carotenoids. Mechanistic studies in cultured cells. Ann. NY Acad. Sci. 691: 177-191.

Bosetti, C. and R. Talamini, M. Montella, E. Negri, E. Conti, S. Franceschi, and C. La Vecchia. 2004. Retinol, carotenoids and the risk of prostate cancer: a case-control study from Italy. Int. J. Cancer 112: 689-692.

Bratt, O. 2004. A comparison of lycopene and orchidectomy vs orchidectomy alone in the management of advanced prostate cancer. BJU Int. 93: 637-638.

Buckley, M.F. and K.J. Sweeney, J.A. Hamilton, R.L. Sini, D.L. Manning, R.I. Nicholson, A. deFazio, C.K. Watts, E.A. Musgrove, and R.L. Sutherland. 1993. Expression and amplification of cyclin genes in human breast cancer. Oncogene 8: 2127-2133.

Chalabi, N. and L. Le Corre, J.C. Maurizis, Y.J. Bignon, and D.J. Bernard-Gallon. 2004. The effects of lycopene on the proliferation of human breast cells and BRCA1 and BRCA2 gene expression. Eur. J. Cancer 40: 1768-1775.

Chalabi, N. and J.C. Maurizis, L. Le Corre, L. Delort, Y.J. Bignon, and D.J. Bernard-Gallon. 2005. Quantification by affinity perfusion chromatography of phosphorylated BRCAl and BRCA2 proteins from tumor cells after lycopene treatment. J. Chromatog. B Analyt. Technol. Biomed. Life Sci. 821: 188-193.

Chalabi, N. and L. Delort, L. Le Corre, S. Satih, Y.J. Bignon, and D. Bernard-Gallon. 2006. Gene signature of breast cancer cell lines treated with lycopene. Pharmacogenomics 7: 663-672.

Chalabi, N. and L. Delort, S. Satih, P. Dechelotte, Y.J. Bignon, and D.J. Bernard-Gallon. 2007. immunohistochemical expression of RAR{alpha}, RAR{beta}, and Cx43 in breast tumor cell lines after treatment with lycopene and correlation with RT-QPCR. J. Histochem. Cytochem. 55: 877-883.

Chalabi, N. and S. Satih, L. Delort, Y.J. Bignon, and D.J. Bernard-Gallon. 2007b. Expression profiling by whole-genome microarray hybridization reveals differential gene expression in breast cancer cell lines after lycopene exposure. Biochim. Biophys. Acta 1769: 124-130.

Chen, G. and Z. Djuric. 2002. Detection of 2,6-cyclolycopene-1,5-diol in breast nipple aspirate fluids and plasma: a potential marker of oxidative stress. Cancer Epidemiol. Biomarkers Prev. 11: 1592-1596.

Ching, S. and D. Ingram, R. Hahnel, J. Beilby, and E. Rossi. 2002. Serum levels of micronutrients, antioxidants and total antioxidant status predict risk of breast cancer in a case control study. J. Nutr. 132: 303-306.

Clinton, S.K. 1998. Lycopene: chemistry, biology, and implications for human health and disease. Nutr. Rev. 56: 35-51.

Cohen, L.A. 2002. A review of animal model studies of tomato carotenoids, lycopene, and cancer chemoprevention. Exp. Biol. Med. (Maywood) 227: 864-868.

Cohen, L.A. and Z. Zhao, B. Pittman, and F. Khachik. 1999. Effect of dietary lycopene on N-methylnitrosourea-induced mammary tumorigenesis. Nutr. Cancer 34: 153-159.

Craft, N.E. and J.H. Soares. 1992. Relative solubility, stability, and absorptivity of lutein and beta-carotene in organic solvents. J. Agric. Food Chem. 40: 431-434.

Di Mascio, P. and T.P. Devasagayam, S. Kaiser, and H. Sies. 1990. Carotenoids, tocopherols and thiols as biological singlet molecular oxygen quenchers. Biochem. Soc. Trans. 18: 1054-1056.

Di Mascio, P. and S. Kaiser, and H. Sies. 1989. Lycopene as the most efficient biological carotenoid singlet oxygen quencher. Arch. Biochem. Biophys. 274: 532-538.

Dorgan, J.F. and A. Sowell, C.A. Swanson, N. Potischman, R. Miller, N. Schussler, and H.E. Stephenson Jr. 1998. Relationships of serum carotenoids, retinol, alpha-tocopherol, and selenium with breast cancer risk: results from a prospective study in Columbia, Missouri (United States). Cancer Causes Control 9: 89-97.

Etminan, M. and B. Takkouche, and F. Caamano-Isorna. 2004. The role of tomato products and lycopene in the prevention of prostate cancer: a meta-analysis of observational studies. Cancer Epidemiol. Biomarkers Prev. 13: 340-345.

Fishman, G.I. and R.L. Eddy, T.B. Shows, L. Rosenthal, and L.A. Leinwand. 1991. The human connexin gene family of gap junction proteins: distinct chromosomal locations but similar structures. Genomics 10: 250-256.

Fornelli, F. and A. Leone, I. Verdesca, F. Minervini, and G. Zacheo. 2007. The influence of lycopene on the proliferation of human breast cell line (MCF-7). Toxicol. In Vitro 21: 217-223.

Freudenheim, J.L. and J.R. Marshall, J.E. Vena, R. Laughlin, J.R. Brasure, M.K. Swanson, T. Nemoto, and S. Graham. 1996. Premenopausal breast cancer risk and intake of vegetables, fruits, and related nutrients. J. Nat. Cancer Inst. 88: 340-348.

Gaudet, M.M. and J.A. Britton, G.C. Kabat, S. Steck-Scott, S.M. Eng, S.L. Teitelbaum, M.B. Terry, A.I. Neugut, and M.D. Gammon. 2004. Fruits, vegetables, and micronutrients in relation to breast cancer modified by menopause and hormone receptor status. Cancer Epidemiol. Biomarkers Prev. 13: 1485-1494.

Giovannucci, E. and A. Ascherio, E.B. Rimm, M.J. Stampfer, G.A. Colditz, and W.C. Willett. 1995. Intake of carotenoids and retinol in relation to risk of prostate cancer. J. Natl. Cancer Inst. 87: 1767-1776.

Gomez-Aracena, J. and R. Bogers, P. Van't Veer, E. Gomez-Gracia, A. Garcia-Rodriguez, H. Wedel, and J. Fernandez-Crehuet Navajas. 2003. Vegetable consumption and carotenoids in plasma and adipose tissue in Malaga, Spain. Int. J. Vitam. Nutr. Res. 73: 24-31.

Hankinson, S.E. and W.C. Willett, G.A. Colditz, D.J. Hunter, D.S. Michaud, B. Deroo, B. Rosner, F.E. Speizer, and M. Pollak. 1998. Circulating concentrations of insulin-like growth factor-I and risk of breast cancer. Lancet 351: 1393-1396.

Hotz-Wagenblatt, A. and D. Shalloway. 1993. Gap junctional communication and neoplastic transformation. Crit. Rev. Oncog. 4: 541-558.

Hulten, K. and A.L. Van Kappel, A. Winkvist, R. Kaaks, G. Hallmans, P. Lenner, and E. Riboli. 2001. Carotenoids, alpha-tocopherols, and retinol in plasma and breast cancer risk in northern Sweden. Cancer Causes Control 12: 529-537.

Ito, Y. and K.C. Gajalakshmi, R. Sasaki, K. Suzuki, and V. Shanta. 1999. A study on serum carotenoid levels in breast cancer patients of Indian women in Chennai (Madras), India. J. Epidemiol. 9: 306-314.

Jarvinen, R. and P. Knekt, R. Seppanen, and L. Teppo. 1997. Diet and breast cancer risk in a cohort of Finnish women. Cancer Lett. 114: 251-253.

Karas, M. and H. Amir, D. Fishman, M. Danilenko, S. Segal, A. Nahum, A. Koifmann, Y. Giat, J. Levy, and Y. Sharoni. 2000. Lycopene interferes with cell cycle progression and insulin-like growth factor I signaling in mammary cancer cells. Nutr. Cancer 36: 101-111.

Kardinaal, A.F. and P. van't Veer, H.A. Brants, H. van den Berg, J. van Schoonhoven, and R.J. Hermus. 1995. Relations between antioxidant vitamins in adipose tissue, plasma, and diet. Amer. J. Epidemiol. 141: 440-450.

Khachik, F. and C.J. Spangler, J.C. Smith Jr., L.M. Canfield, A. Steck, and H. Pfander. 1997. Identification, quantification, and relative concentrations of carotenoids and their metabolites in human milk and serum. Anal. Chem. 69: 1873-1881.

King, T.J. and L.H. Fukushima, T.A. Donlon, A.D. Hieber, K.A. Shimabukuro, and J.S. Bertram. 2000. Correlation between growth control, neoplastic potential and endogenous connexin43 expression in HeLa cell lines: implications for tumor progression. Carcinogenesis 21: 311-315.

Kucuk, O. and F.H. Sarkar, W. Sakr, Z. Djuric, M.N. Pollak, F. Khachik, Y.W. Li, M. Banerjee, D. Grignon, J.S. Bertram, J.D. Crissman, E.J. Pontes, and D.P. Wood Jr. 2001. Phase II randomized clinical trial of lycopene supplementation before radical prostatectomy. Cancer Epidemiol. Biomarkers Prev. 10: 861-868.

Kumar, N.M. and N.B. Gilula. 1996. The gap junction communication channel. Cell 84: 381-388.

La Vecchia, C. 2002. Tomatoes, lycopene intake, and digestive tract and female hormone-related neoplasms. Exp. Biol. Med. (Maywood) 227: 860-863.

Levi, F. and C. Pasche, F. Lucchini, and C. La Vecchia. 2001. Dietary intake of selected micronutrients and breast-cancer risk. Int. J. Cancer 91: 260-263.

Levy, J. and E. Bosin, B. Feldman, Y. Giat, A. Miinster, M. Danilenko, and Y. Sharoni. 1995. Lycopene is a more potent inhibitor of human cancer cell proliferation than either alpha-carotene or beta-carotene. Nutr. Cancer 24: 257-266.

Lewis, D.F. 2004. 57 varieties: the human cytochromes P450. Pharmacogenomics 5: 305-318.

Loewenstein, W.R. 1981. Junctional intercellular communication: the cell-to-cell membrane channel. Physiol. Rev. 61: 829-913.

London, S.J. and E.A. Stein, I.C. Henderson, M.J. Stampfer, W.C. Wood, S. Remine, J.R. Dmochowski, N.J. Robert, and W.C. Willett. 1992. Carotenoids, retinol, and vitamin E and risk of proliferative benign breast disease and breast cancer. Cancer Causes Control 3: 503-512.

Malins, D.C. and N.L. Polissar, and S.J. Gunselman. 1996. Progression of human breast cancers to the metastatic state is linked to hydroxyl radical-induced DNA damage. Proc. Natl. Acad. Sci. USA 93: 2557-2563.

McEligot, A.J. and C.L. Rock, S.W. Flatt, V. Newman, S. Faerber, and J.P. Pierce. 1999. Plasma carotenoids are biomarkers of long-term high vegetable intake in women with breast cancer. J. Nutr. 129: 2258-2263.

Molnar, J. and N. Gyemant, I. Mucsi, A. Molnar, M. Szabo, T. Kortvelyesi, A. Varga, P. Molnar, and G. Toth. 2004. Modulation of multidrug resistance and apoptosis of cancer cells by selected carotenoids. In Vivo 18: 237-244.

Monaghan, P. and D. Moss. 1996. Connexin expression and gap junctions in the mammary gland. Cell Biol. Int. 20: 121-125.

Nagasawa, H. and T. Mitamura, S. Sakamoto, and K. Yamamoto. 1995. Effects of lycopene on spontaneous mammary tumour development in SHN virgin mice. Anticancer Res. 15: 1173-1178.

Nahum, A. and K. Hirsch, M. Danilenko, C.K. Watts, O.W. Prall, J. Levy, and Y. Sharoni. 2001. Lycopene inhibition of cell cycle progression in breast and endometrial cancer cells is associated with reduction in cyclin D levels and retention of p27(Kip1) in the cyclin E-cdk2 complexes. Oncogene 20: 3428-3436.

Nahum, A. and L. Zeller, M. Danilenko, O.W. Prall, C.K. Watts, R.L. Sutherland, J. Levy, and Y. Sharoni. 2006. Lycopene inhibition of IGF-induced cancer cell growth depends on the level of cyclin D1. Eur. J. Nutr. 45: 275-282.

Nebert, D.W. and V. Vasiliou. 2004. Analysis of the glutathione S-transferase (GST) gene family. Hum. Genomics 1: 460-464.

Pierce, J.P. and V.A. Newman, S.W. Flatt, S. Faerber, C.L. Rock, L. Natarajan, B.J. Caan, E.B. Gold, K.A. Hollenbach, L. Wasserman, L. Jones, C. Ritenbaugh, M.L. Stefanick, C.A. Thomson, and S. Kealey. 2004. Telephone counseling intervention increases intakes of micronutrient- and phytochemical-rich vegetables, fruit and fiber in breast cancer survivors. J. Nutr. 134: 452-458.

Potischman, N. and C.E. McCulloch, T. Byers, T. Nemoto, N. Stubbe, R. Milch, R. Parker, K.M. Rasmussen, M. Root, S. Graham, and T. Colin Campbell. 1990. Breast cancer and dietary and plasma concentrations of carotenoids and vitamin A. Amer. J. Clin. Nutr. 52: 909-915.

Potischman, N. and T. Byers, L. Houghton, M. Root, T. Nemoto, and T.C. Campbell. 1992. Effects of breast cancer treatments on plasma nutrient levels: implications for epidemiological studies. Cancer Epidemiol. Biomarkers Prev. 1: 555-559.

Prakash, P. and R.M. Russell, and N.I. Krinsky. 2001. In vitro inhibition of proliferation of estrogen-dependent and estrogen-independent human breast cancer cells treated with carotenoids or retinoids. J. Nutr. 131: 1574-1580.

Rock, C.L. and S.W. Flatt, F.A. Wright, S. Faerber, V. Newman, S. Kealey, and J.P. Pierce. 1997. Responsiveness of carotenoids to a high vegetable diet intervention designed to prevent breast cancer recurrence. Cancer Epidemiol. Biomarkers Prev. 6: 617-623.

Ronco, A. and E. De Stefani, P. Boffetta, H. Deneo-Pellegrini, M. Mendilaharsu, and F. Leborgne. 1999. Vegetables, fruits, and related nutrients and risk of breast cancer: a case-control study in Uruguay. Nutr. Cancer. 35: 111-119.

Sato, R. and K.J. Helzlsouer, A.J. Alberg, S.C. Hoffman, E.P. Norkus, and G.W. Comstock. 2002. Prospective study of carotenoids, tocopherols, and retinoid concentrations and the risk of breast cancer. Cancer Epidemiol. Biomarkers Prev. 11: 451-457.

Sesso, H.D. and J.E. Buring, S.M. Zhang, E.P. Norkus, and J.M. Gaziano. 2005. Dietary and plasma lycopene and the risk of breast cancer. Cancer Epidemiol. Biomarkers Prev. 14: 1074-1081.

Sharoni, Y. and E. Giron, M. Rise, and J. Levy. 1997. Effects of lycopene-enriched tomato oleoresin on 7,12-dimethyl-benz[a]anthracene-induced rat mammary tumors. Cancer Detect. Prev. 21: 118-123.

Sies, H. and W. Stahl, and A.R. Sundquist. 1992. Antioxidant functions of vitamins. Vitamins E and C, beta-carotene, and other carotenoids. Ann. NY Acad. Sci. 669: 7-20.

Simon, M.S. and Z. Djuric, B. Dunn, D. Stephens, S. Lababidi, and L.K. Heilbrun. 2000. An Evaluation of Plasma Antioxidant Levels and the Risk of Breast Cancer: A Pilot Case Control Study. Breast J. 6: 388-395.

Stahl, W. and W. Schwarz, A.R. Sundquist, and H. Sies. 1992. cis-trans isomers of lycopene and beta-carotene in human serum and tissues. Arch. Biochem. Biophys. 294: 173-177.

Tamimi, R.M. and S.E. Hankinson, H. Campos, D. Spiegelman, S. Zhang, G.A. Colditz, W.C. Willett, and D.J. Hunter. 2005. Plasma carotenoids, retinol, and tocopherols and risk of breast cancer. Amer. J. Epidemiol. 161: 153-160.

Terry, P. and M. Jain, A.B. Miller, G.R. Howe, and T.E. Rohan. 2002. Dietary carotenoids and risk of breast cancer. Amer. J. Clin. Nutr. 76: 883-888.

Toma, S. and L. Isnardi, P. Raffo, L. Riccardi, G. Dastoli, C. Apfel, P. LeMotte, and W. Bollag. 1998a. RARalpha antagonist Ro 41-5253 inhibits proliferation and induces apoptosis in breast-cancer cell lines. Int J. Cancer. 78: 86-94.

Toma, S. and L. Isnardi, L. Riccardi, and W. Bollag. 1998b. Induction of apoptosis in MCF-7 breast carcinoma cell line by RAR and RXR selective retinoids. Anticancer Res. 18: 935-942.

Toniolo, P. and A.L. Van Kappel, A. Akhmedkhanov, P. Ferrari, I. Kato, R.E. Shore, and E. Riboli. 2001. Serum carotenoids and breast cancer. Amer. J. Epidemiol. 153: 1142-1147.

van den Berg, H. 1999. Carotenoid interactions. Nutr. Rev. 57: 1-10.

van Kappel, A.L. and J.P. Steghens, A. Zeleniuch-Jacquotte, V. Chajes, P. Toniolo, and E. Riboli. 2001. Serum carotenoids as biomarkers of fruit and vegetable consumption in the New York Women's Health Study. Public Health Nutr. 4: 829-835.

Wu, K. and J.W. Erdman Jr. and S.J. Schwartz, E.A. Platz, M. Leitzmann, S.K. Clinton, V. DeGroff, W.C. Willett, and E. Giovannucci. 2004. Plasma and dietary carotenoids, and the risk of prostate cancer: a nested case-control study. Cancer Epidemiol. Biomarkers Prev. 13: 260-269.

Yamasaki, H. 1990. Gap junctional intercellular communication and carcinogenesis. Carcinogenesis 11: 1051-1058.

Yeum, K.J. and S.H. Ahn, S.A. Rupp de Paiva, Y.C. Lee-Kim, N.I. Krinsky, and R.M. Russell. 1998. Correlation between carotenoid concentrations in serum and normal breast adipose tissue of women with benign breast tumor or breast cancer. J. Nutr. 128: 1920-1926.

Zhang, L.X. and R.V. Cooney, and J.S. Bertram. 1991. Carotenoids enhance gap junctional communication and inhibit lipid peroxidation in C3H/10T1/2 cells: relationship to their cancer chemopreventive action. Carcinogenesis 12: 2109-2114.

Zhang, L.X. and R.V. Cooney, and J.S. Bertram. 1992. Carotenoids up-regulate connexin43 gene expression independent of their provitamin A or antioxidant properties. Cancer Res. 52: 5707-5712.

Zhang, S. and G. Tang, R.M. Russell, K.A. Mayzel, M.J. Stampfer, W.C. Willett, and D.J. Hunter. 1997. Measurement of retinoids and carotenoids in breast adipose tissue and a comparison of concentrations in breast cancer cases and control subjects. Amer. J. Clin. Nutr. 66: 626-632.

3.5

Lycopene and Colon Cancer

Martha Verghese, Judith Boateng, Louis Shackelford and Lloyd T. Walker

Nutritional Biochemistry, P.O. Box 1628, Alabama A&M University
Department of Food and Animal Sciences, Normal, AL 35762, USA

ABSTRACT

Considerable evidence from several epidemiological studies suggests that lycopene is a powerful antioxidant and may be inversely related to chronic diseases. These observations are based on findings that higher intakes of tomato and tomato products are associated with a reduced risk of several chronic diseases.

Colon cancer is one of the most commonly diagnosed cancers and perhaps one of the most preventable. Dietary habits are important modulatory factors in maintaining proper health. The hypothesis that there exists a relationship between westernization and colon cancer appears to strengthen a cascade of findings suggesting an increased incidence of colon cancer in westernized or developed countries. This chapter focuses on the role of lycopene in colon cancer risk reduction and/or prevention and the mechanisms involved in the proposed inhibitory effects, including its antioxidative property, gap junction communication, induction of detoxification enzymes such as glutathione-S-transferase and inhibition of insulin-like growth factor 1. A study was conducted to determine the effects of lycopene (tomato oleoresin containing 6% lycopene, Lyco-Mato, at 0, 200, and 400 ppm) and its interaction with fat levels (7 and 14%) on colonic aberrant crypt foci, which are preneoplastic lesions, and colon tumors in Fisher 344 male rats. The number of colonic aberrant crypt foci and tumors was larger in the control groups than in the groups fed dietary lycopene at 400 ppm with 7 and 14% fat. The control group (7% fat level) had an average

---

A list of abbreviations is given before the references.

tumor size of 1.02 cm, whereas the rats fed lycopene with 14% fat at I+P stage had an average tumor length of 0.23 cm. Since lycopene is a fat-soluble compound, the presence of high levels of dietary fat may increase its absorption and bioavailability. Incorporation of lycopene in the diet may have significant implications in reducing the risk of colon cancer.

INTRODUCTION

Lycopene is perhaps one of the most powerful antioxidants among dietary carotenoids. It has become evident from a preponderance of studies that lycopene is inversely related to chronic diseases such as cancer, cardiovascular disease, skin and eye health. These observations are based on epidemiological findings that higher intakes of tomato and tomato products are associated with a reduced risk of several chronic diseases (Giovannucci et al. 2002, Etminan et al. 2004).

Experimental observations on the characteristic mechanisms of lycopene have shown its biological and physicochemical properties as a natural antioxidant and its ability to act as a scavenger of free radicals (Kohlmeier et al. 1997, Giovannucci, 1999, Shi and Le Maguer, 2000, Böhm et al. 2003).

The best evidence relating to the increased scientific support for the role of lycopene as an important phytonutrient with important health benefits appears to be its protection against a broad range of epithelial cancers. Hence, the consensus based on decades of studies relays the significance of lycopene as it relates to the improvement of health.

COLON CANCER

Cancer is a worldwide problem that affects nearly 6.6 million lives every year. In the United States it is the second most common cause of death next to coronary heart disease, with an estimated 560,000 deaths annually (American Cancer Society (ACS) 2006) and almost 1.4 million new cases diagnosed each year.

Colon cancer is one of the most commonly diagnosed cancers and perhaps one of the most preventable. Current estimates dictate that more than 106,000 new cases (in both sexes) will be diagnosed and close to 55,000 deaths are expected in the United States. Worldwide, the disease affects more than a million people and claims more than 500,000 lives. This makes colon cancer the third most common cause of cancer deaths in the United States (ACS 2006) and the third worldwide (Boyle and Langman 2000). Since 5% of persons (1 in 20 persons) will likely develop colorectal cancer, this disease is an important public health issue (Calvert and Frucht 2002).

COLON CANCER DEVELOPMENT

The genesis of colon cancer is a multi-step process that develops as a result of the pathologic transformation of normal colonic epithelium. These changes involve a sequence of well-characterized pathological modifications ranging from discrete microscopic mucosal lesions such as aberrant crypt foci (ACF) to an adenomatous polyp and ultimately an invasive cancer (Gryfe et al. 1997, McLellan et al. 1991, Bird 1995). Colon cancer develops as a result of the pathologic transformation of normal colonic epithelium to an adenomatous polyp and ultimately to invasive cancer (Fig. 1). The adenoma-carcinoma sequence is defined by the accrual of multiple mutations in tumor suppressor genes and oncogenes that affect the balance between cell proliferation and apoptosis (Cruz-Bustillo 2004). The progression into malignant neoplasia (Fig. 2) takes years and possibly decades (Gryfe et al. 1997, Kinzler and Vogelstein 1996).

Colon cancer is categorized as hereditary or "familial" and non-hereditary or "sporadic" types. The main fact is that colon cancer whether familial or sporadic is caused by defects in genetic components that may be inherited or acquired (Bresalier and Kim 1985). The familial type of colon cancer accounts for 10-30% of colon cancer cases, while the sporadic form accounts for the remaining 70-90% (Cruz-Bustillo 2004). Mutations caused by sporadic cancers are a result of environmental factors including

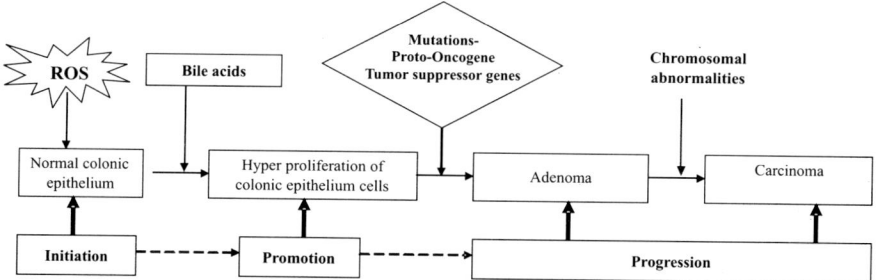

Fig. 1 Colon cancer development. Colon cancer is a multi-step process that develops as a result of the pathologic transformation of normal colonic epithelium by reactive oxygen species (ROS) resulting in initiation. Further changes result in hyperproliferation of colonic epithelial cells (promotion) resulting in discrete microscopic mucosal lesions developing from aberrant crypt foci (ACF) to an adenomatous polyp and ultimately an invasive cancer (progression). Adenoma-carcinoma sequence results by accumulation of multiple mutations in tumor suppressor genes and oncogenes.

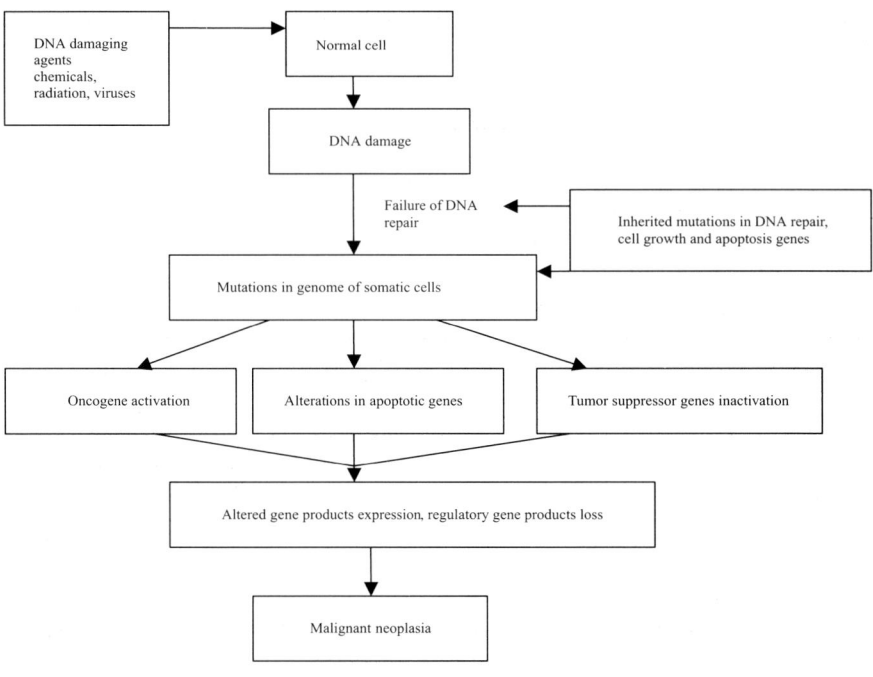

Fig. 2 Etiology of neoplasia. The etiology of neoplasia or cancer is a multi-step process that develops from the pathologic transformation of normal cells by DNA damage and failure of DNA repair due to inherited mutations in DNA repair, cell growth and apoptosis genes resulting in mutations in genome of somatic cells. The carcinoma sequence is defined by the accrual of multiple mutations in tumor suppressor genes, apoptotic genes and oncogenes that affect the balance between cell proliferation and apoptosis resulting in altered gene products expression and loss of regulatory gene products. The progression into malignant neoplasia may take years.

advanced age, high intake of red meat, a high fat diet, smoking, alcohol consumption, and obesity (Bresalier and Kim 1985).

DIET AND COLON CANCER

Dietary habits are important modulatory factors in maintaining proper health. The hypothesis that there exists a relationship between westernization and colon cancer (Giovannucci 1999) appears to strengthen a cascade of findings suggesting an increased incidence of colon cancer in westernized or developed countries (Burkitt 1971). Colorectal cancer is the third most common cancer in the world (IARC 2005). The highest incidence is seen in North America, Europe, and Australia and the lowest rates in Africa and Asia with a 20-fold variation in incidence rates between countries, mainly attributed to diet-related factors that could contribute to up to 80% of the differences between countries (Cummings and Bingham 1998). The major established diet-related risk factor is obesity; others include inadequate physical activity and some aspects of a Western diet, such as a higher intake of meat or fat, and lower intake of fiber, fruit, and vegetables (IARC 2002).

Meat

Research has shown that, overall, risk of colon cancer increases with increasing intake of red meat and processed meat but is not associated with total meat intake. Results from international studies (Armstrong and Doll 1975) have shown a strong association between consumption of meat per capita and deaths from colorectal cancer. The proposed mechanisms to explain how meat consumption might increase risk of colorectal cancer include mutagenic heterocyclic amines and polycyclic aromatic hydrocarbons formed when meat is cooked at high temperatures (Gooderham et al. 1997, Kazerouni et al. 2001) and nitrites and related compounds in smoked, salted, and processed meats that are converted to carcinogenic N-nitroso compounds in the colon (Bingham et al. 1996), and high concentrations of iron in the colon that could increase formation of free radicals that can be mutagenic. A high fat intake can increase sterols in the colonic lumen and colonic bacteria can convert these sterols into secondary bile acids and other toxic metabolic compounds that can damage the colonic mucosa, increase proliferation in the epithelium, activate ras potooncogene and also alter prostaglandin metabolism (Lund et al. 1999).

Fiber, Fruit and Vegetables

Results from case-control studies (Potter and Steinmetz 1996, Jacobs et al. 1998) have reported a lower risk of colon cancer with a high consumption of dietary fiber, fruits, and vegetables, compared to a low consumption; however, the findings of some prospective studies (Fuchs et al. 1999, Michels et al. 2000, Terry et al. 2001) have been inconsistent.

Burkitt (1971) suggested that the low rates of colorectal cancer in Africa were due to high consumption of dietary fiber. Fiber reduces the risk of colon cancer by increasing the stool bulk and speeding the transit of food through the colon, thereby diluting the gut contents and reducing exposure time and, therefore, absorption of carcinogens by the colonic mucosa. It also binds with potential bile carcinogens such as bile acids and decreases their concentrations. Fermentation of fiber (which escapes digestion by enzymes) in the large intestine produces short-chain fatty acids such as butyrate, which might protect against colorectal cancer through their ability to inhibit cell growth and proliferation and promote differentiation, induce apoptosis, modulate gene expression and inhibit production of secondary bile acids by inhibiting 7α-dehydroxylase by reduction of pH in the lumen (Hague 1995, Nagengast 1995). Fermentation of fiber also results in alterations in colonic microflora, thereby inhibiting microbial enzymes involved in carcinogen activation and changing bacterial species.

Folate

Several prospective studies (Chen et al. 1999, Giovannucci et al. 1995, Glynn et al. 1996) have reported the potential mechanisms of folate deficiency in increasing the risk of colorectal cancer. A low intake of folate, methionine, and a high intake of alcohol can impair the functions of folate. Folate mediates transfer of one carbon moieties and consequently plays an important part in DNA synthesis, methylation and metabolism. A deficiency of folate could increase the risk of cancer through mechanisms including reduced methylation of DNA and uracil incorporation instead of thymine into DNA, impaired DNA repair, increased mutagenesis, abnormal apoptosis, hyperproliferation and polymorphisms of genes involved in folate metabolic pathway and related gene-nutrient interaction (Chen et al. 1999, Giovannucci et al. 1995, Glynn et al. 1996).

Calcium and Vitamin D

High intakes of calcium, vitamin D, or both might also reduce risk of colorectal cancer (COMA 1998, Wu 2002, WCRF 1997). Calcium can bind free bile acids and fatty acids, forming insoluble calcium soaps, thereby reducing their carcinogenic effects on the colonic epithelium. It can also suppress proliferation and promote apoptosis of the colonic epithelium. Other mechanisms that may be affected by dietary calcium include reduction or suppression of molecular alterations implicated in colorectal cancer such as k-ras mutations, c-myc protooncogene expression, β-catenin transcriptional activation; suppressing the activation of secondary transduction signals (protein kinase C) and supplemental calcium may offer some protection against recurrence of colorectal adenomas (Bonithon-Kopp 2000, Baron 1999).

A Western diet (high consumption of saturated fat, red meat, refined sugars and carbohydrates) may be a risk factor for colon cancer, while a diet high in fruits and vegetables may reduce the risk of the disease. Plant-based foods are important sources of micronutrients and non-nutritive bioactive components. This fact has focused attention on the possible role of these phytonutrients in colon cancer prevention. Fruits, vegetables, legumes and cereals are by far the main sources of antioxidant nutrients such as ascorbic acid and vitamin E, carotenoids, including β-carotene and lycopene, flavonoids and polyphenolic compounds. The consumption of these antioxidants could play a vital role in protecting the body against reactive oxygen species (ROS) (Niki 1991, Rimm et al. 1993, Willet 1994, Negi et al. 2003), which are responsible for deleterious effects on critical macromolecules leading to the endogenous initiation of several chronic diseases such as cancer. Whether generated endogenously or from the diet, ROS can lead to the permanent modification of genetic material, thus resulting in the genesis of several chronic diseases such as colon cancer (Valko et al. 2006, Rao 2004).

While it has been argued by several investigators that suboptimal intake of micronutrients and phytonutrients is widespread in the West (Ames 1999, Ames and Wakimoto 2002), it is clear that this factor leads directly to DNA damage. Dietary intake of natural antioxidants could be an important factor in the body's defense against ROS (Negi et al. 2003). Moreover, there is strong epidemiological and experimental evidence that suggests that antioxidants that occur naturally in plant-based foods may play an important role as protectants against carcinogenesis (Manorama et al. 1993). Therefore, the consumption of dietary antioxidants such as lycopene could play a vital role in protecting the body against ROS (Niki 1991, Rimm et al. 1993, Willet 1994, Negi et al. 2003).

LYCOPENE AND COLON CANCER

Epidemiological and several case-control studies have suggested that dietary intake of tomato and tomato-based products that are rich sources of lycopene resulting in higher plasma levels of lycopene is protective against a variety of cancers (Giovannucci 1999). When it comes to lycopene's relation to cancer, the focus has been on prostate cancer; however, several studies have suggested that tomato juice rich in lycopene may have a protective effect against colon cancer (Narisawa et al. 1998). The majority of studies on the relationship between dietary lycopene intake and colon cancer risk have been investigated using data from case-control studies. Based on results obtained from these studies a number of *in vivo* and *in vitro* studies have been designed to test the efficacy and potential benefits of lycopene (Cohen 2002, Omoni and Aluko 2005).

Chemopreventive studies using animal models provide an excellent system to investigate *in vivo* biochemical functions of lycopene, although very few chemopreventive studies have been conducted using lycopene in animal models. Nevertheless, the results from these isolated studies indicate a relationship between lycopene and reduced colon cancer incidence.

The proposed mechanisms by which lycopene acts include its antioxidant properties, gap junction communication, induction of glutathione S-transferase, and anti-proliferation and apoptotic activities of insulin-like growth factor.

Antioxidant Properties of Lycopene in Cancer Prevention

Oxidation and the generation of free radicals are an essential part of basic human metabolism. In normal aerobic metabolism the body maintains an equilibrium between pro- and antioxidant systems by inducing enzymes such as superoxide dismutase (SOD) and various glutathione peroxidases (GPx), and endogenous or dietary antioxidants that scavenge or detoxify oxidants such as ROS that are constantly generated (Kanazawa et al. 2006).

Oxidative stress is linked to many diseases and may be the single most important causative factor in the etiology of human cancer (Hwang and Bowen 2007).

Lycopene, unlike α and β-carotenes, lacks provitamin A activity because of the absence of a β-ionone ring structure (Omoni and Aluko 2005, Lindshield et al. 2007). Despite this, it has been shown to be a powerful antioxidant and is 2- fold to 10-fold more effective at quenching singlet oxygen *in vitro* compared to β-carotene and α-tocopherol (Omoni and Aluko 2005, Lindshield et al. 2007).

The role of lycopene in colon cancer prevention and the mechanisms involved in their proposed inhibitory effects may include their antioxidative property. Modifications to DNA as a result of oxidative stress lead to adduct formation of base residues, mainly guanine residues (Kasai et al. 1984, 1991, Hamilton et al. 2001). The 8-oxo-deoxyguanosine is a product of the reaction between 2'-deoxyguanosine (dG) and ROS. Its formation leads to dG to T transversions and has been associated with diseases such as cancer (Kasai et al. 1984, 1991, Hamilton et al. 2001).

Lycopene has been reported to inhibit oxidative stress by potentially altering oxidation of protein thiols and 8-oxodeoxyguanosine contents of lymphocyte DNA (Rao and Agarwal, 1998). Rehman et al. (1999) indicated that consumption of a single serving of tomatoes by human volunteers was adequate to alter levels of oxidative DNA base damage.

The antioxidant defense includes enzymatic scavengers such as SOD, catalase and GPx (Ratnam et al. 2006). Lycopene has also been shown to increase antioxidant enzymes such as SOD, catalase and GPx (Pan et al. 2003). Previous work by Bhuvaneswari et al. (2001) indicated that lycopene significantly increased vitamins C and E and reduced glutathione (GSH) and GSH-dependent enzymes such as GPx, glutathione S-transferase (GST) and glutathione reductase. Decreased levels of these antioxidant enzymes are a causal factor for oxidative stress.

Reactive oxygen species have been reported to modify membrane-bound protein kinases, growth factors and their receptors and induce the expression of anti-apoptotic and proliferative transcription factors such as NF-κB and AP1 (Sahu 1990, Huang et al. 2007). They can activate oncogenes, such as c-fos and c-jun, and inactivate tumor suppressor genes/pathways (Hwang and Bowen 2007). The modifications of these genes indicate that ROS may impact cell cycle mechanisms and may eventually lead to the development of cancer.

Lycopene and Gap Junction Communication

First described by Loewenstein and Kanno (1966), gap junction communications (GJC) are channels connecting two neighboring cells. They enable exchange of molecules with weight of < 1-2 kD, such as nutrients or intracellular signaling molecules. Generally, GJC are essential for the regulation of growth control, differentiation and apoptosis of normal progenitor cells (Trosko and Ruch, 2002). A single gap junction channel is composed of water-filled pores created by the docking of two hemi-channels called connexons (King and Bertram 2005). Each connexon is composed of six subunits of proteins known as connexins (Saez et al. 2003). Connexins are a family of proteins comprising 21 members that are expressed in almost all tissues (Sohl and Willecke 2004). Of the proteins, the 43 kD protein connexin 43 is the most widely expressed (Kucuk et al. 2001). The decreased expression of the protein connexin 43 in precancerous tumors is an indication of neoplastic progression (Mesnil et al. 2005). Several reports have indicated that a disruption or loss of function of GJC often results in cancers and that its restoration or up-regulation is associated with decreased proliferation (Trosko and Ruch, 2002, Krinsky and Johnson, 2005, Valko et al. 2006).

The lack of function of GJC in malignant cells may be due to either down-regulated expression of connexins, altered trafficking to the cell membrane or the failure to form functional junctions (King and Bertram 2005). One of the chemopreventive actions of carotenoids regardless of their antioxidant activity is the ability to enhance GJC by up-regulating

C43 expression. According to Stahl and Sies (1998), the up-regulation of connexin mRNA might be a result of direct or indirect influences of carotenoids on the gene expression of connexins.

Lycopene has been shown to enhance gap junctional intercellular communication through stabilization and up-regulation of connexin 43 mRNA (Bertram and Bortkiewicz 1995, Krinsky and Johnson 2005, Valko et al. 2006). Previous studies by Zhang and colleagues (1991, 1995) demonstrated the ability of lycopene to inhibit chemically induced neoplastic transformation by their ability to enhance junctional communication in 10T1/2 cells via increased levels of connexin 43. Lycopene compared to α-carotene was effective in inducing junctional communication. Livny et al. (2002) showed that lycopene in a dose-dependent manner inhibited proliferation of KB-1 human oral tumor cells by up-regulation of both the transcription and the expression of connexin 43. This mechanism by lycopene may lend support to the proposition that lycopene may be an effective anticarcinogenic agent in colon carcinogenesis. Since the disruption of connexin 43 has been linked to proliferation of cancer cells, it has been suggested that connexins and GJC serve a tumor suppressor role (Omori and Yamasaki 1998). If this is the case, then this action by lycopene may have mechanistic significance by enabling the transfer of growth-regulatory signals between normal growth-inhibited cells and preneoplastic cells (Kucuk et al. 2001).

Lycopene and Induction of Glutathione-S-Transferase

In carcinogenesis, one of the key mechanisms of protection against carcinogens may be the induction of enzymes involved in their metabolism, specifically the phase II enzymes such as GSTs, UDP-glucuronosyl transferases, and quinone reductases (Kwak et al. 2001). These enzymes are responsible for the detoxification of endogenous toxins and xenobiotics (Table 2) by converting them to less toxic water-soluble metabolites that can be excreted readily from the body (Liska 1998).

Up-regulation of phase II detoxification might play a crucial role in cancer prevention, and several bioactive compounds including carotenoids have been reported to be potent inducers of these enzymes. Previous studies have shown that the expression of phase II enzymes is governed by the antioxidant response element (ARE). Transcription factor Nrf2 (nuclear factor E2-related factor 2) binds to the ARE found in these genes (Jaiswal 2000, Nguyen et al. 2000, Ben-Dor et al. 2005). The binding of Nrf2 to ARE sequence leads to the induction of Phase II enzymes in response to a variety of stimuli including antioxidants, xenobiotics and UV

Table 1 Detoxification enzymes.

Phase I enzymes

Cytochrome P450 (CYP 450)

- Cyp1A2
- Cyp2E1
- Cyp2A6
- Cyp2C9
- Cyp3A4
- Cyp2D6

Phase II enzymes

GST: Glutathione S-transferases

UDP-GT: Urinidine-3′5′-diphosphoglucuronic acid-glucuronosyl transferases

QR: Quinone Reductases

MnSOD: Manganese superoxide dismutase

ST: Sulfotransferase

EH: Epoxide hydrolase

λ GCS: λ Glutamylcysteine synthatase

One of the key mechanisms of protection against carcinogens may be the induction of enzymes involved in their metabolism, specifically the phase II enzymes such as glutathione S-transferases, UDP-glucuronosyl transferases, and quinone reductases, which are responsible for the detoxification of endogenous toxins and xenobiotics by converting them to less toxic water-soluble metabolites that can be excreted readily from the body. Up-regulation of phase II detoxification might play a crucial role in cancer prevention, and several bioactive compounds including carotenoids have been reported to be potent inducers of these enzymes.

irradiation (Jaiswal 2000, Nguyen et al. 2000, Ben-Dor et al. 2005, Wang and Jaiswal, 2006).

There are indications that lycopene enhances the expression of phase II enzymes through the transcription factor Nrf2 and ARE (Velmurugan et al. 2001, Astorg et al. 1994). Ben-Dor and colleagues (2005) observed that lycopene compared to Astaxanthin, phytoene and β-carotene was more potent in activating ARE, which led to the increased expression of phase II enzymes. The authors suggested that the ARE activation was not related directly to the antioxidant function of the carotenoids. Lycopene-induced elevation and stimulation of phase II detoxification enzymes such as GPx, GST and glutathione reductase may be one of the mechanisms by which it offers protection against colon cancer.

Lycopene and Insulin-like Growth Factor 1: Antiproliferation and Apoptotic Activities

Insulin-like growth factor 1 (IGF-1) has been reported as an important risk factor for colon cancer (Giovannucci 1999, Khandwala et al. 2000). Insulin-like growth factors are mainly synthesized by the liver, although several other organs have been reported to produce this protein (Jones and Clemmons, 1995, Khandwala et al. 2000). The effects of circulating IGFs are mediated through its interaction with its receptor proteins, IGF-binding proteins (IGFBPs). There are six classes of IGFBP (IGFBP 1-6) but the most prevalent is IGFBP-3, since it is present at the highest levels in the blood (Jones and Clemmons, 1995, Khandwala et al. 2000). The binding of IGF-1 to IGFBP-3 leads to a decrease in the bioavailability of IGF-1, and this has been reported to increase proapoptotic activity (Giovannucci 2001, Chan et al. 2002). IGFBP-3 can induce apoptosis independent of IGF-1 by inhibiting PI3K/Akt/PKB and MAPK signaling pathways (Lee et al. 2002, Liu et al. 2003). There is also evidence linking IGFBP-3 to increased P-53-dependent apoptosis (Williams et al. 2000). Thus, circulating levels of IGFBP-3 and IGF-1 may be a determining factor in the risk of some cancers such as colon cancer.

In vitro experiments have indicated that increased levels of IGFs may stimulate the development of cancer by regulating cell proliferation, replication, inhibiting apoptosis and stimulating DNA synthesis by causing cells to navigate through the successive phases of the cell cycle (Dunn et al. 1997, Jones and Clemmons 1995, Khandwala et al. 2000). Liu et al (2003) explained that IGF-1 can reduce apoptosis and increase cell survival through phosphorylation of BAD, a member of BH3, the sole subfamily of Bcl-2. Phosphorylated BAD is found in the cytosol bound to 14-3-3 proteins rather than Bcl-xL and hence does not induce apoptosis (Liu et al. 2003).

Lycopene has been reported to suppress IGF-1-stimulated growth in cell culture and animal models (Siler et al. 2005). Growth suppression was related to a delayed G_1-S cell cycle progression, thus inhibiting IGF signaling (Siler et al. 2005, Karas et al. 2000). Lycopene was found to potentially up-regulate IGFBP-3 and down-regulate the phosphorylation of BAD, which promotes apoptosis and inhibits cell proliferation (Liu et al. 2003). It was shown that lycopene at both low and high dosage overturned the reduction in plasma IGFBP-3 and apoptosis as well as reduced hyperproliferation (Liu et al. 2003). A review by Giovannucci (2001) suggested that hyperinsulinemia might be associated with colon cancer risk. Hyperinsulinemia has been implicated in increased IGF-1 levels. There is evidence suggesting that IGF-1 may increase epithelial energy, leading to proliferation of colon epithelium tissues, which correlates with colon cancer (Bruce et al. 2000, Giovannucci 2001). Hence, a decrease in IGF-1 might be one of the mechanisms by which lycopene may reduce or inhibit colon cancer (Fig. 3).

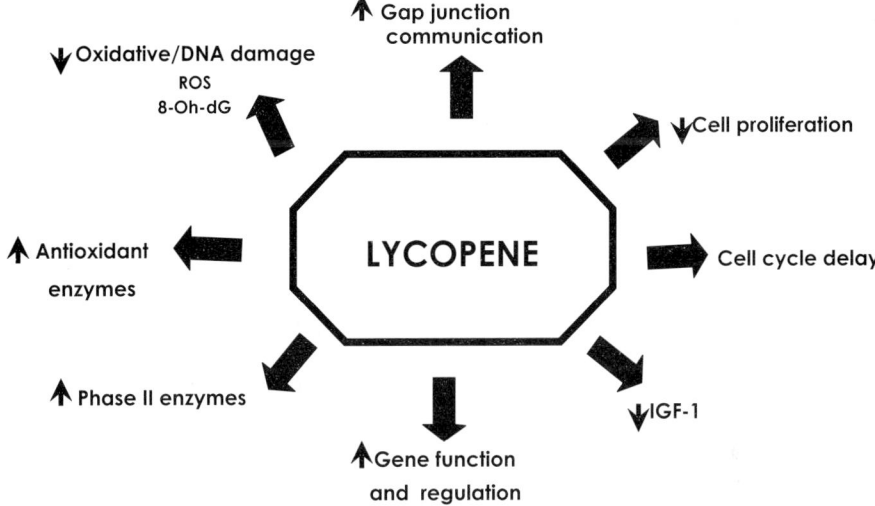

Fig. 3 Some proposed mechanisms of the chemopreventive properties of lycopene. IGF: insulin-like growth factor.

EXPERIMENTAL STUDIES

Lycopene Reduced Incidence of Azoxymethane-induced Colon Tumors in Fisher 344 Rats

Studies have shown that about 50% of all cancers are due to poor diet (Agarwal and Rao 2000a, b). We conducted a study to determine the effects of lycopene (tomato oleoresin containing 6% lycopene — Lyco-Mato, Beersheeba, Israel — at levels of 0, 200, and 400 ppm) and its interaction with fat levels (7 and 14%) (Table 2) on azoxymethane (AOM)-induced colon carcinogenesis using a Fisher 344 rat model.

Azoxymethane, a metabolite of 1,2-dimethylhydrazine (DMH) (Fig. 4), an organotropic colon carcinogen, has been extensively used to induce colon carcinogenesis in susceptible laboratory animals (Ma et al. 1996, Papanikolaou et al. 1998, Dommels et al. 2003). It is generally preferred to DMH because it is a more potent carcinogen based on molarity and it has enhanced chemical stability in solutions (Papanikolaou et al. 1998). Most studies have reported that two successive injections of AOM, 1 wk apart, are adequate to induce colon cancer in rats. This results in a much shorter initiation phase than in a DMH model, thus making AOM models more suitable for studying events taking place during either the initiation or promotion phases of carcinogenesis (Wijnands et al. 1999). Induction by

Table 2 Composition of the diet AIN 93G/M.[1]

Ingredient (g/kg diet)	Control		Lycopene			
			200 ppm		400 ppm	
	7% fat	14% fat	7% fat	14% fat	7% fat	14% fat
Cornstarch	397.50	327.50	397.47	327.47	397.54	327.54
Lycopene	0.00	0.00	3.33	3.33	6.66	6.66
Soybean oil	70.00	140.00	66.70	136.70	63.30	133.30
Common ingredients[2/3]	532.50	532.50	532.50	532.50	532.50	532.50

[1]Formulation of diets based on American Institute of Nutrition 93 Growth (AIN 93G) (Reeves et al. 1993a, b).
[2]Common ingredients for AIN 93G: casein, 200; dextrose, 132; sucrose, 100; fiber, 50; AIN 93G-MX Mineral Mix, 35; AIN 93-VX VITAMIN MIX, 10; L-cystine, 3; choline bitartate, 2.5.
[3]Common ingredients for AIN 93M: casein, 140; dextrose, 155; sucrose, 100; fiber, 50; AIN 93M-MX Mineral Mix, 35; AIN 93-VX VITAMIN MIX, 10; L-cystine, 1.8; choline bitartate, 2.5.

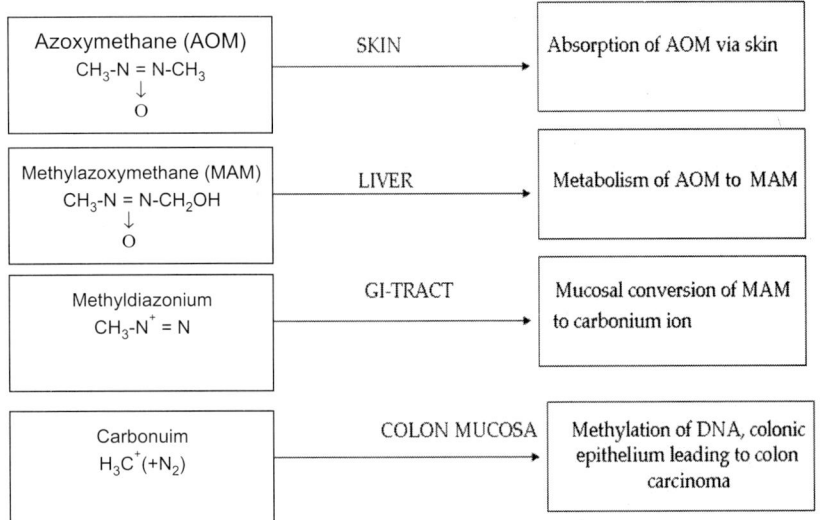

Fig. 4 Azoxymethane (AOM) metabolizes to carbonium ion, the ultimate carcinogen that causes DNA methylation (Ma et al. 1996, Papanikolaou et al. 1998, Dommels et al. 2003). Azoxymethane, a metabolite of 1,2-dimethylhydrazine (DMH), an organotropic colon carcinogen, is used to induce colon carcinogenesis in susceptible laboratory animals. Colon cancer chemically induced in rodents by a colon-specific carcinogen such as AOM is often used as an intermediate endpoint to evaluate the natural history of colon carcinogenesis in rats using several nutritional factors. Treatments with AOM result in histopathological characteristics similar to human tumors.

AOM is the most popular experimental model used to identify dietary modulations of colon cancer (Dommels et al. 2003). Chemically induced colon cancer in rodents by a colon-specific carcinogen such as AOM is often used as an intermediate endpoint to evaluate the natural history of colon carcinogenesis in rats using several nutritional factors (Zalatnai et al. 2001, Vanamala et al. 2006, Xiao et al. 2005).

Treatments with AOM share many histopathological characteristics with human tumors (Reddy 2004, Corpet and Pierre 2003). In laboratory models (Fisher 344 male rats), AOM induces colon tumors predominantly in the distal colon, which is similar to the regional distribution of tumors in the human colon (Reddy 2004). Reports by Singh et al. (1994, 1997) have indicated that AOM treatment also induced several genetic mutations in critical genes such as oncogenes and tumor suppressor genes that have been causally associated with colon tumor development in addition to enhanced expression of cyclooxygenase-2, which are also present in human colon tumors (Kawamori et al. 1999, Corpet and Pierre 2003).

Aberrant Crypt Foci (ACF)

ACF are the earliest recognizable putative preneoplastic precursors of adenoma/carcinoma that appear on the surface of rodents after treatment with chemically induced colon carcinogens such as AOM (Bird 1995, Suh et al. 2007) and are quantified as number per animal or per colon. First described by Bird (1987, 1995), ACF are defined as having altered luminal openings, exhibiting thickened epithelia, and being larger than adjacent normal crypts (Fig. 5).

According to Pretlow and Pretlow (2005), lesions that are identified microscopically in the intact mucosa and meet the above criteria are, by definition, ACF. Aberrant crypt foci appear a few weeks after treatment with a carcinogen and become larger with time, with more distinct nuclear atypia or dysplasia (Takayama et al. 2005). They are specific to colon carcinogenesis and the number and growth features of ACF numerically increase with time as the disease develops (Bird and Good 2000). Additionally, since cancer is a multi-step process whereby only a small number of lesions at each stage evolve to the next, the number of crypts per precursor lesion should increase with time (McKellan et al. 1991).

Aberrant crypt foci consisting of ≥ 4 crypts have been reported as putative premalignant lesions and as an intermediate biomarker for colon cancer development (Seraj et al. 1997).

Aberrant crypt foci appearing in the colon of rodents chemically induced with carcinogens are enzyme-altered and have also been reported in colonic mucosa in humans. Aberrant crypt foci as well as colon

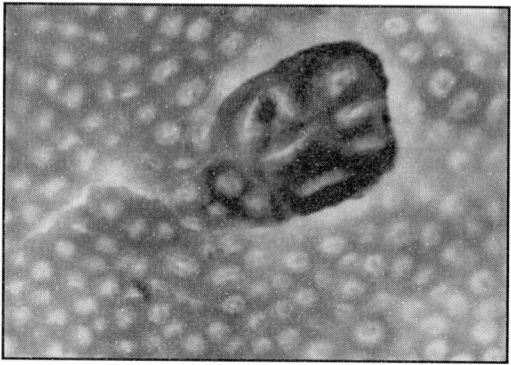

Fig. 5 Aberrant crypt foci (ACF): foci with multiple crypts. Aberrant crypt foci have thick epithelium and stain darker using methylene blue compared to the normal colon epithelium. They are the earliest recognizable putative preneoplastic precursors of adenoma/carcinoma that appear on the surface of rodents after subsequent treatments with AOM. They appear a few weeks after treatment with a carcinogen and become larger with time with more distinct nuclear atypia or dysplasia (Bird 1985).

adenomas in humans contain mutations of *Kras* and *Apc*, two genes that are important in colon tumorigenesis (Takayama et al. 2005, Suh et al. 2007). This supports several observations that indicate that ACF may thus provide an ideal model system in which to study the role of oncogene expression in the early stages of human colon cancer (Stopera et al. 1992).

Aberrant crypt foci are perhaps the only system that provides a quantitative assessment of the stepwise development of the disease (Bird and Good 2000). Using ACF as a model for screening assay for colon tumorigenesis in laboratory rodents has so far proven to be a reliable biomarker (Bird 1987, Pretlow et al. 1991, Bird and Good 2000).

Glutathione S-transferase activity in the liver of rats was analyzed in order to study the potential of lycopene in increasing levels of detoxifying enzymes. Phase II enzymes, including GST, are involved in the detoxification of carcinogen metabolites such as polycyclic aromatic hydrocarbons and ROS (González et al. 2001). Glutathione S-transferase also modulates the induction of other enzymes and proteins important for cellular functions, such as DNA repair (Hayes and Pulford 1995). These classes of enzymes are important for maintaining cellular genomic integrity and possibly play an important role in cancer susceptibility.

Fisher 344 male rats (Harlan IN) approximately 4 wk old were randomly assigned into groups (Fig. 7) and fed diets containing AIN 93G/M (Control)-7 and 14% fat and AIN 93G/M with 200, 400 ppm (7 and 14% fat) (Table 2). The animals remained on their respective diets for 13 wk in the ACF study and 42 wk in the endpoint tumor study.

In the ACF study we observed that feeding lycopene at 200 and 400 ppm with fat (7% level) significantly ($P \leq 0.05$) reduced the incidence of ACF by 12 and 33.4% compared to the control (7% fat). At 14% fat levels, ACF was reduced by 30 and 48% after feeding with 200 and 400 ppm (14%) (Table 3) compared to the control (14% fat). With increasing levels of lycopene, a reduction in ACF was seen, but the reduction was even more pronounced in the rats fed higher fat (14%) compared to the 7% fat level.

For the endpoint tumor model study, following an acclimatization period of 1 wk, six groups of 15 Fisher 344 weanling male rats (4 wk old) were fed AIN 93-G diets (Reeves et al. 1993a, b) containing 0, 200 and 400 ppm lycopene/kg diet (each diet at 7% fat and 14% fat) for 46 wk (Fig. 6). In experiment 2, beginning at age 4 wk, eight groups of 15 rats each

Table 3. Effect of dietary lycopene on ACF in colon of Fisher 344 rats.

Groups	Proximal colon*	Distal colon*	Total*
Control (7% fat)	39.9 ± 0.94[b]	118.2 ± 1.7[b]	154.4 ± 1.8[b]
Control (14% fat)	53.75 ± 2.8[a]	125.62 ± 3.2[a]	178.25 ± 2.1[a]
Lycopene 200 ppm (7% fat)	20.15 ± 0.89[c]	110.2 ± 1.11[b]	135.9 ± 2.9[c]
Lycopene 200 ppm (14% fat)	18.22 ± 1.3[c]	107.9 ± 2.0[b]	125.0 ± 1.2[d]
Lycopene 400 ppm (7% fat)	18.0 ± 0.79[c]	85.1 ± 1.5[c]	103.1 ± 1.4[e]
Lycopene 400 ppm (14% fat)	16.5 ± 1.2[c]	68.6 ± 1.4[d]	85.18 ± 2.0[f]
Lycopene 800 ppm (7% fat)	18.9 ± 1.2[c]	57.0 ± 1.0[e]	76.0 ± 1.2[f]

*Values are means ± SEM, n = 10 for all groups.

Means within the columns with different letters are significantly different ($P < 0.05$) by Tukey's studentized range test.

Aberrant crypt foci were counted in the distal and proximal colon after rats were fed control and lycopene diets for 13 wk (until 17 wk of age). Feeding with lycopene at 200 and 400 ppm with fat (7% level) significantly ($P \leq 0.05$) reduced the incidence of ACF by 12 and 33.4% compared to the control (7% fat); at 14% fat levels, ACF was reduced by 30 and 48% after feeding 200 and 400 ppm compared to the control. With increasing levels of lycopene a reduction in ACF was seen, but the reduction was even more pronounced in the rats fed higher fat (14%) compared to the 7% fat level.

were assigned to eight dietary treatments: (1) control diet (AIN 93G, 0% lycopene, 7% fat), (2) control diet (AIN 93G, 0% lycopene, 14% fat), (3) lycopene diet (400 ppm, 7% fat, I), (4) lycopene diet (400 ppm, 7% fat, P), (5) lycopene diet (400 ppm, 7% fat, I+P), (6) lycopene diet (400 ppm, 14% fat, I), (7) lycopene diet (400 ppm, 14% fat, P), and (8) lycopene diet (400 ppm, 14% fat, I+P) (Fig. 7). Each lycopene diet was fed during initiation (I), promotion (P) and initiation plus promotion (I+P) stages of carcinogenesis (Fig. 8).

In the endpoint tumor study, number of tumors in the rats fed lycopene (200 and 400 ppm) was significantly ($P \leq .05$) lower and tumor size was significantly ($P \leq .05$) smaller compared to the control (Table 4). We

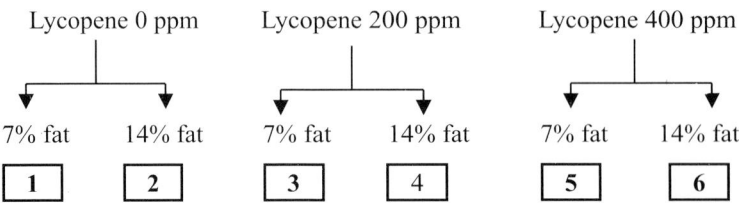

Fig. 6 Dietary treatments fed Fisher 344 male rats (Experiment 1). Following an acclimatization period of 1 wk, six groups of 15 Fisher 344 weanling male rats (4 wk old) were fed AIN 93-G diets containing 0, 200 and 400 ppm lycopene/kg diet (each diet at 7 and 14% fat) for 41 wk.

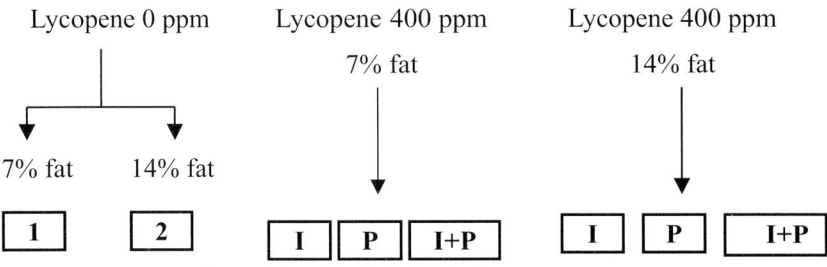

Fig. 7 Diets consumed by Fisher 344 male rats (I = Initiation, P = Promotion) (Experiment 2). Eight groups of 15 rats each were assigned to eight dietary treatments: (1) control diet (AIN 93G, 0% lycopene, 7% fat), (2) control diet (AIN 93G, 0% lycopene, 14% fat), (3) lycopene diet (400 ppm, 7% fat, I), (4) lycopene diet (400 ppm, 7% fat, P), (5) lycopene diet (400 ppm, 7% fat, I+P), (6) lycopene diet (400 ppm, 14% fat, I), (7) lycopene diet (400 ppm, 14% fat, P), and (8) lycopene diet (400 ppm, 14% fat, I+P).

observed a negative correlation between the dose of lycopene in the diet and the number of colon tumors in the 7% and 14% fat groups; the lowest incidence of tumors was seen in rats fed a diet with lycopene 400 ppm (7% fat) (Fig. 9). However, in the short-term study (ACF study), total number of ACF was decreased with increased fat level. Studies have shown that the presence of dietary fat at the initial stage of ingestion can help to ensure optimum absorption of lycopene in the body. This could be due to the presence of an increased number of carbon-carbon double bonds in lycopene's structure that makes it more soluble in fats and lipids and more readily absorbable (Nguyen and Schwartz 1999).

Martinez-Ferrer et al. (2006) reported significant ($P < 0.05$) increases in GST activity in rats fed different doses of lycopene (200-400 ppm) compared to a control group (Table 5). The increase in Phase II enzyme activity may demonstrate one of the proposed mechanisms by which lycopene prevents tumorigenesis. A possible mechanism for the inhibition

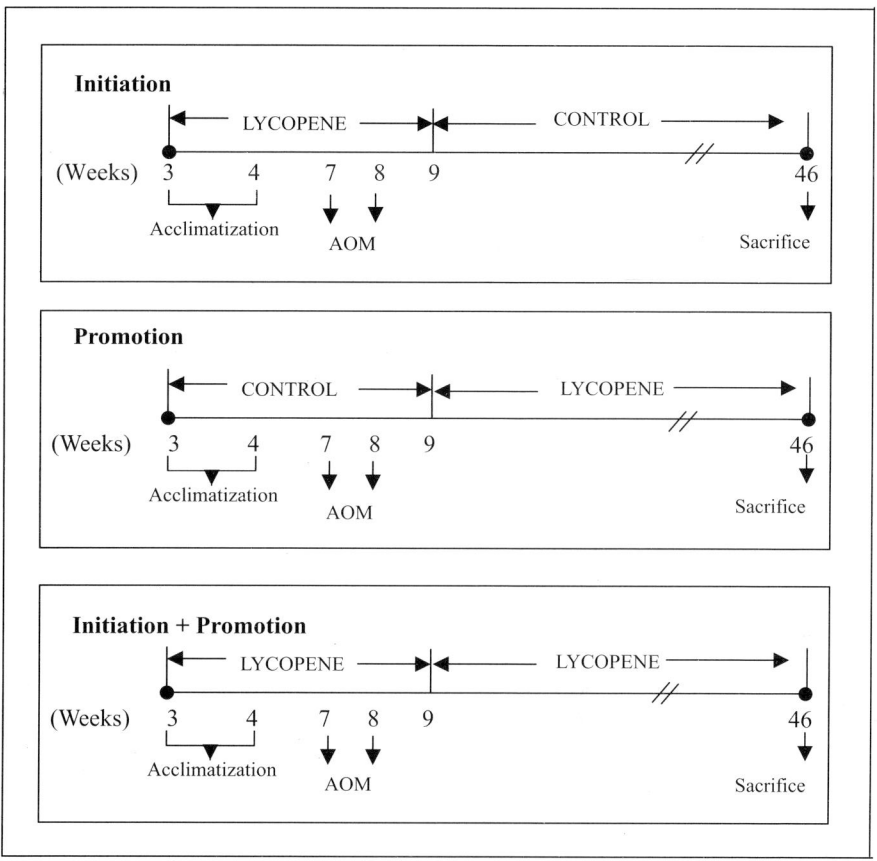

Fig. 8. Potential of lycopene in inhibiting colon tumors at the I and P stages of colon carcinogenesis. In experiment 2, beginning at 4 wk of age, eight groups of 15 rats each were assigned to eight dietary treatments: (1) control diet (AIN 93G, 0% lycopene, 7% fat), (2) control diet (AIN 93G, 0% lycopene, 14% fat), (3) lycopene diet (400 ppm, 7% fat, I), (4) lycopene diet (400 ppm, 7% fat, P), (5) lycopene diet (400 ppm, 7% fat, I+P), (6) lycopene diet (400 ppm, 14% fat, I), (7) lycopene diet (400 ppm, 14% fat, P), and (8) lycopene diet (400 ppm, 14% fat, I+P) (Figs. 6 and 7). Each lycopene diet was fed during I, P and I+P stages of carcinogenesis (Fig. 8). In the I group, rats were fed lycopene in the diet for 5 wk (3 wk prior to first AOM injection and 1 wk after second AOM injection); rats were then switched to the control diet. In the P group, the rats received control diet up to 10 wk of age (2 wk after the second AOM injection) followed by lycopene diets for the rest of the experiment. In the I+P group, rats received lycopene diet throughout the experiment.

of colon tumors by these agents is through an increase in the levels of detoxifying enzymes in the liver. Our present study demonstrated that dietary lycopene increased liver GST activity. The increase in enzyme activity was greater in rats fed diets with higher fat (14%) content. Since

Table 4 Incidence of small intestinal tumors, colon tumors and size of tumors in Fisher 344 male weanling rats fed dietary lycopene[1]

Treatment	n_1/n_0 [2]	Incidence (% of animals with tumors)				
		Small intestinal tumors (%)	Colon tumors (%)	Proximal (%)	Distal (%)	TBR[3]
Control, 7% fat	13/15	31	87	8	92	1.9[b]
Control, 14% fat	12/12	29	100	8	92	4.4[a]
Lycopene 200 ppm, 7% fat	5/14	0	36	0	100	1.4 [d]
Lycopene 200 ppm, 14% fat	4/12	0	33	0	100	1.5[c]
Lycopene 400 ppm, 7% fat	1/13	0	8	0	100	1.0[e]
Lycopene 400 ppm, 14% fat	3/14	0	21	0	100	1.0[e]
Pool SD						1.18

[1]Values are means or %.
[2]n_1, number of rats with tumors; n_0, total number of rats at the end of the experiment.
[3]TBR: Tumor-bearing rat.
Means in the same column with the same letter are not significantly different by Tukey's Studentized Range Test ($P < 0.05$).
The incidence of tumors in the distal section of the colon was significantly higher than in the proximal sections. Lycopene groups showed 100% incidence in the distal section and control groups showed 92%. Previous studies showed that distal segments of the colon had significantly higher number of crypts than the proximal segments. Decrease in TBR was directly proportional to increasing dietary lycopene and was higher in groups fed diets containing 14% fat, as compared with 7% fat diets. Rats fed the control diet (14% fat) showed the highest (4.4) TBR value and rats fed lycopene at 400 ppm levels (7 and 14% fat) showed the lowest (1.0). TBR values are considered better indicators for studying the effect of phytochemicals on endpoint tumors because they give a more precise picture of tumor inhibition, specifically the number of tumors induced per TBR.

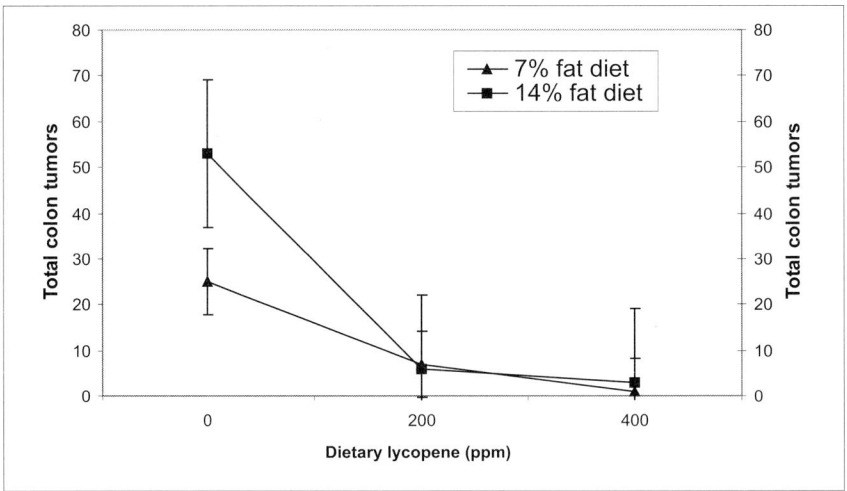

Fig. 9 Dietary lycopene and number of total colon tumors in rats. Six groups of 15 Fisher 344 weanling male rats (4 wk old) were fed AIN 93-G diets containing 0, 200 and 400 ppm lycopene/ kg diet (each diet at 7 and 14% fat) for 41 wk. There was a negative correlation between the dose of lycopene in the diet and the number of colon tumors in the 7 and 14% fat groups; the lowest incidence of tumors was seen in rats fed a diet with lycopene 400 ppm (7% fat).

Table 5 Specific activity of glutathione S-transferase enzyme in liver of male azoxymethane-induced Fisher 344 rats.

Treatment	Specific activity ($\mu moles\ min^{-1}\ mg^{-1}$)
Lycopene 400 ppm 14% fat	193.93 ± 19.86^a
Lycopene 400 ppm 7% fat	170.20 ± 29.25^b
Lycopene 200 ppm 14% fat	137.76 ± 11.85^c
Lycopene 200 ppm 7% fat	131.44 ± 3.59^c
Control 14% fat	75.94 ± 6.19^d
Control 7% fat	52.22 ± 15.60^e

Means with the same letter are not significantly different by Tukey's Studentized Range (HSD) Test ($P < 0.05$). Values are means $\pm$ SD.

Feeding different doses of lycopene (200-400 ppm) resulted in significant ($P < 0.05$) increases in GST activity over the level in control rats. The increase in Phase II enzyme activity may demonstrate one of the proposed mechanisms by which lycopene prevents tumorigenesis.. The increase in enzyme activity was greater in rats fed diets with higher fat (14%) content. Since lycopene is a fat-soluble compound, presence of high levels of dietary fat increases its absorption and makes it more bioavailable.

lycopene is a fat-soluble compound, presence of high levels of dietary fat increases its absorption and makes it more bioavailable.

We hypothesized that the induction of GST activity may have led to the detoxification of AOM, thus leading to the reduced incidence of tumors observed in the lycopene-treated groups.

At the I, P and I+P stages of carcinogenesis, colon tumors were significantly ($P \leq .05$) lower in the rats fed lycopene (400 ppm with 7 and 14% fat) than in the rats fed control diet (AIN-93 G/M) with 7 and 14% fat. A significant ($P \leq .05$) reduction in colon tumors was observed in the rats fed lycopene 400 ppm (7% fat) at P stage (Table 6). The incidence of tumors in the distal segment of the colon in the control (7 and 14% fat) was 92%. There were no tumors in the proximal colon in the rats fed lycopene (except those fed lycopene at 400 ppm at 14% fat level). The rats fed dietary lycopene (14% fat) at the P stage showed 96% tumor induction in the distal segment, whereas the other dietary lycopene groups had 100% tumor induction in the distal section of the colon. An important attribute of colorectal cancer is its anatomical site of origin (González et al. 2001). Results from this experiment are consistent with data available in humans, where colon cancer is more prevalent in the distal colon.

Tumors were larger in the control groups than in the groups fed dietary lycopene at 400 ppm with 7 and 14% fat. The control group fed 7% fat level had an average tumor size of 1.02 cm, whereas the rats fed lycopene with 14% fat at I+P stage had an average tumor length of 0.23 cm. Because lycopene is a fat-soluble compound, the presence of high levels of dietary fat may increase its absorption and make it more bioavailable.

Table 6 Effect of dietary lycopene on small intestine tumors, colon tumors and size of tumors in Fisher 344 male weanling rats during initiation and post-initiation stages (Experiment 2).[1]

Treatment	n_1/n_0[2]	Incidence (% of animals with tumors)				Tumor size		
		Colon tumors (%)	Proximal (%)	Distal (%)	TBR[3]	Diameter (cm)	Length (cm)	Volume (cm²)
Control 7%	13/15	87	8	92	1.9^c	0.41^a	1.02^a	6.703
Control 14%	12/12	100	8	92	4.4^a	0.35^c	0.70^b	0.642
Lycopene 400 ppm 7% (I)	10/13	76	0	100	1.4^f	0.30^d	0.60^c	1.051
Lycopene 400 ppm 7% (P)	9/15	60	0	100	1.7^e	0.35^c	0.44^e	1.869
Lycopene 400 ppm 7% (I+P)	8/9	88	0	100	1.0^h	0.34^c	0.39^f	0.723
Lycopene 400 ppm 14% (I)	12/14	86	0	100	1.8^d	0.34^c	0.58^d	3.216
Lycopene 400 ppm 14% (P)	11/15	73	4	96	2.1^b	0.38^b	0.43^e	5.648
Lycopene 400 ppm 14% (I+P)	7/10	70	0	100	1.1^g	0.21^e	0.23^g	0.468
Pool SD					1.01	0.17	0.28	1.2

[1]Values are means or %.

[2]n_1, rats with tumors; n_0, total number of rats at the end of the experiment.

[3]TBR: Tumor-bearing rat.

Means with the same letter are not significantly different by Tukey's Studentized Range (HSD) Test ($P < 0.05$).

The total number of colon tumors induced was significantly lower in the lycopene 400 ppm, 7% fat P group (60%). Among the lycopene-fed groups, lycopene 400 ppm, 7% fat I+P showed the highest (88%) incidence of colon tumors. Incidence of colon tumors was the highest in the control 14% fat group (100%) and was the lowest in the lycopene 400 ppm 7% fat P group (60%). Feeding lycopene (at both fat levels) at the P stage resulted in substantial reduction in colon tumorigenesis induced by AOM. The incidence of colon tumors was lower at P as compared to the I stages. In the I and P groups, rats consuming diets containing 14% fat showed a higher colon tumor incidence than their respective 7% fat groups. Dietary fat seemed to have an effect on tumor promotion in those groups. Rats that were fed diets with 14% fat I+P showed lower (70%) tumor incidence than the respective 7% fat group (88%). Since lycopene is a fat-soluble compound, dietary fat increases its absorption and may make it more bioavailable, with the reduction being more pronounced with feeding lycopene at I+P stages as compared to I or P alone.

The findings of this study suggest that dietary lycopene suppressed AOM-induced colon tumors in Fisher 344 rats. The results also showed a proportional relation with GST activity in the liver. Studying the relationship between dietary lycopene in AOM-induced colon carcinogenesis at the I, P and progression stages is an important finding in providing recommendations to incorporate dietary lycopene in the prevention or treatment of colon cancer.

Lycopene-rich Juices Offer Protection against Chemically Induced ACF in Fisher 344 Male Rats

Tomato and watermelon are rich sources of carotenoids, particularly lycopene, which is presently considered one of the most powerful carotenoids in protecting against free radical damage. The experiment was carried out to determine the inhibitory effects of watermelon juice (WMJ) and tomato juice (TJ) on the incidence of AOM-induced ACF in Fisher 344 male rats. Fisher 344 male rats were randomly divided into five groups (n = 6) and fed control diet (C) AIN-93G, 25% WMJ, 50% WMJ, 25% TJ and 50% TJ.

Formation of ACF (means ± SEM) in the proximal and distal colons of rats fed C, 25% WMJ, 50% WMJ, 25% TJ, and 50% TJ ranged from 23.33 ± 0.6 to 70.67 ± 51.2 and 8.67 ± 1.4 to 101.67 ± 14.2. Rats fed WMJ and TJ had significantly ($P < 0.05$) higher GST activity (mg/μmol) than rats fed C.

The results showed that WMJ and TJ when given at 25 and 50% levels significantly reduced the incidence of AOM-induced ACF in Fisher 344 rats; hence, the inclusion of WMJ and TJ in the diet may play a vital role in alleviating the occurrence of colon cancer.

ABBREVIATIONS

ACF: Aberrant crypt foci; AIN 93 G/M: American Institute of Nutrition 93 Growth/Maintenance; AOM: Azoxymethane; ARE: Antioxidant response element; DMH: 1,2-Dimethylhydrazine; GST: Glutathione-S-transferase; GSHPx/GPx: Glutathione Peroxidase; IGF-1: Insulin-like growth factor 1; IGF-BP: IGF-binding protein; ROS: Reactive oxygen species; SOD: Superoxide dismutase

References

Agarwal, S. and A.V. Rao. 2000a. Carotenoids and chronic diseases. Drug Metabol. Drug Interact. 17: 189-210.

Agarwal, S. and A.V. Rao. 2000b. Tomato lycopene and its role in human health and chronic diseases. CMAJ 163: 739-744.

American Cancer Society. 2006. Colorectal Cancer Facts and Figures—Special Edition. American Cancer Society. Available at www.acs.gov.

Ames, B.N. 1999. Micronutrient deficiencies. A major cause of DNA damage. Ann. NY Acad. Sci. 889: 87-106.

Ames, B.N. and P. Wakimoto. 2002. Are vitamin and mineral deficiencies a major cancer risk? Natl. Rev. Cancer 2: 694-704.

Armstrong, B. and R. Doll. 1975. Environmental factors and cancer incidence and mortality in different countries, with special reference to dietary practices. Int. J. Cancer 15: 617-631.

Astorg, P. and S. Gradelet, J. Leclerc, M.C. Canivenc, and M.H. Siess. 1994. Effects of beta-carotene and canthaxanthin on liver xenobiotic-metabolizing enzymes in the rat. Food Chem. Toxicol. 32: 735-742.

Baron, J.A. and M. Beach, and J.S. Mandel. 1999. Calcium supplements and colorectal adenomas. Polyp Prevention Study Group, Ann. NY Acad. Sci. 889: 138-145.

Ben-Dor, A. and M. Steiner, L. Gheber, M. Danilenko, N. Dubi, K. Linnewiel, A. Zick, Y. Sharoni, and J. Levy. 2005. Carotenoids activate the antioxidant response element transcription system. Mol. Cancer Ther. 4: 177-186.

Bertram, J.S. and H. Bortkiewicz. 1995. Dietary carotenoids inhibit neoplastic transformation and modulate gene expression in mouse and human cells. Amer. J. Clin. Nutr. 62 (Suppl 6): 1327S-1336S.

Besalier, R.S. and Y.S. Kim. 1985. Diet and colonic cancer: putting the puzzle together. N. Engl. J. Med. 1313: 1412-1414.

Bhuvaneswari, V. and B. Velmurugan, S. Balasenthil, C.R. Ramachandran, and S. Nagini. 2001. Chemopreventive efficacy of lycopene on 7,12-dimethylbenz[a]anthracene-induced hamster buccal pouch carcinogenesis. Fitoterapia 72: 865-874.

Bingham, S.A. and B. Pignatelli, and J.R. Pollock. 1996. Does increased endogenous formation of N-nitroso compounds in the human colon explain the association between red meat and colon cancer? Carcinogenesis 17: 515-523.

Bird, R.P. 1987. Observation and quantification of aberrant crypt foci in murine colon treated with a colon carcinogen: preliminary findings. Cancer Lett. 37: 147-151.

Bird, R.P. 1995. Role of aberrant crypt foci in understanding the pathogenesis of colon cancer. Cancer Lett. 93: 55-71.

Bird, R.P. and C.K. Good. 2000. The significance of aberrant crypt foci in understanding the pathogenesis of colon cancer. Toxicol. Lett. 112-113: 395-402.

Bohm, V. and K. Frohlich, and R. Bitsch. 2003. Rosehip—a "new" source of lycopene? Mol. Aspects Med. 24: 385-389.

Bonithon-Kopp, C. and O. Kronborg, A. Giacosa, U. Rath, and J. Faivre. 2000. European Cancer Prevention Organisation Study Group, Calcium and fibre supplementation in prevention of colorectal adenoma recurrence: a randomised intervention trial. Lancet 356: 1300-1306.

Boyle, P. and J.S. Langman. 2000. ABC of colorectal cancer: Epidemiology. BMJ 321(7264): 805-808.

Burkitt, D.P. 1971. Epidemiology of cancer of the colon and rectum. Cancer 28: 3-13.

Calvert, P.M. and H. Frucht. 2002. The genetics of colorectal cancer. Ann. Intern. Med. 137: 603-612.

Chan, J.M. and M.J. Stampfer, J. Ma, P. Gann, J.M. Gaziano, M. Pollak, and E. Giovannucci. 2002. Insulin-like growth factor-I (IGF-I) and IGF binding protein-3 as predictors of advanced-stage prostate cancer. J. Natl. Cancer Inst. 94: 1099-1106.

Chen, J. and E.L. Giovannucci, and D.J. Hunter. 1999. MTHFR polymorphism, methyl-replete diets and the risk of colorectal carcinoma and adenoma among US men and women: an example of gene-environment interactions in colorectal tumorigenesis. J. Nutr. 129(suppl): 560S-564S.

Cohen, L.A. 2002. A review of animal model studies of tomato carotenoids, lycopene, and cancer chemoprevention. Exp. Biol. Med. (Maywood) 227: 864-868.

COMA. 1998. Nutritional aspects of the development of cancer (Report of the Working Group on Diet and Cancer of the Committee on Medical Aspects of Food and Nutrition Policy). Stationery Office, London.

Corpet, D.E. and F. Pierre. 2003. Point: From animal models to prevention of colon cancer. Systematic review of chemoprevention in min mice and choice of the model system. Cancer Epidemiol. Biomarkers Prev. 12: 391-400.

Cruz-Bustillo Clarens, D. 2004. Molecular genetics of colorectal cancer. Rev. Esp. Enferm. Dig. 96: 48-59.

Cummings, J.H. and S.A. Bingham. 1998. Diet and the prevention of cancer. BMJ 317: 1636-1640.

Dommels, Y.E.M. and S. Heemskerk, H. van den Berg, G.A. Alink, P.J. van Bladeren, and B. van Ommen. 2003. Effects of high fat fish oil and high fat corn oil diets on initiation of AOM-induced colonic aberrant crypt foci in male F344 rats. Food Chem. Toxicol. 41: 1739-1747.

Dunn, S.E. and F.W. Kari, J. French, J.R. Leininger, G. Travlos, R. Wilson, and J.C. Barrett. 1997. Dietary restriction reduces insulin-like growth factor I levels, which modulates apoptosis, cell proliferation, and tumor progression in p53-deficient mice. Cancer Res. 57: 4667-4672.

Etminan, M. and B. Takkouche, and F. Caamano-Isorna. 2004. The role of tomato products and lycopene in the prevention of prostate cancer: a meta-analysis of observational studies. Cancer Epidemiol. Biomarkers Prev. 13: 340-345.

Fuchs, C.S. and E.L. Giovannucci, and G.A. Colditz. 1999. Dietary fiber and the risk of colorectal cancer and adenoma in women. N. Engl. J. Med. 340: 169-176.

Giovannucci, E. 1999a. Insulin-like growth factor-I and binding protein-3 and risk of cancer. Hormone Res. 51 (Suppl 3): 34-41.

Giovannucci, E. 1999b. Tomatoes, tomato-based products, lycopene, and cancer: review of the epidemiologic literature. J. Natl. Cancer Inst. 91: 317-331.

Giovannucci, E. 2001. Insulin, insulin-like growth factors and colon cancer: a review of the evidence. J. Nutr. 131 (Suppl 11): 3109S-3120S.

Giovannucci, E. 2002. A review of epidemiologic studies of tomatoes, lycopene, and prostate cancer. Exp. Biol. Med. (Maywood) 227: 852-859.

Giovannucci, E. and E.B. Rimm, A. Ascherio, M.J. Stampfer, G.A. Colditz and W.C. Willett. 1995. Alcohol, low-methionine, low-folate diets, and risk of colon cancer in men. J. Natl. Cancer Inst. 87: 265-273.

Glynn, S.A. and D. Albanes, and P. Pietinen. 1996. Alcohol consumption and risk of colorectal cancer in a cohort of Finnish men. Cancer Causes and Control 7: 214-223.

González, E.C. and R.G. Roetzheim, J.M. Ferrante, and R. Campbell. 2001. Predictors of proximal vs. distal colorectal cancers. Dis. Colon Rectum 44: 251-258.

Gooderham, N.J. and S. Murray, and A.M. Lynch. 1997. Assessing human risk to heterocyclic amines. Mutat. Res. 376: 53-60.

Gryfe, R. and C. Swallow, B. Bapat, M. Redston, S. Gallinger, and J. Couture. 1997. Molecular biology of colorectal cancer. Curr. Prob. Cancer 21(5): 233-300.

Hague, A. and D.J. Elder, D.J. Hicks, and C. Paraskeva. 1995. Apoptosis in colorectal tumour cells: induction by the short chain fatty acids butyrate, propionate and acetate and by the bile salt deoxycholate. Int. J. Cancer 60: 400-406.

Hamilton, M.L. and Z. Guo, C.D. Fuller, H. Van Remmen, W.F. Ward, S.N. Austad, D.A. Troyer, I. Thompson, and A. Richardson. 2001. A reliable assessment of 8-oxo-2-deoxyguanosine levels in nuclear and mitochondrial DNA using the sodium iodide method to isolate DNA. Nucleic Acids Res. 29: 2117-2126.

Hayes, J. and D. Pulford. 1995. The glutathione S-transferase supergene family: regulation of GST and the contribution of the isoenzymes to cancer chemoprevention and drug resistance. Crit. Rev. Biochem. Mol. Biol. 30: 445-600.

Huang, C.S., Y.E. Fan, C.Y. Lin and M.L. Hu. 2007. Lycopenen inhibits matrix metalloproteinase-9 expression and down-regulates the binding activity of nuclear factor-kappa B and stimulatory protein-1. J. Nutr. Biochem. 18(7): 449-456.

Hwang, E.S. and P.E. Bowen. 2007. DNA damage, a biomarker of carcinogenesis: its measurement and modulation by diet and environment. Crit. Rev. Food Sci. Nutr. 47: 27-50.

International Agency for Research on Cancer, Globocan (http: // www.dep.iarc.fr/globocan/globocan.html).

International Agency for Research on Cancer. 2002. Overweight and lack of exercise linked to increased cancer risk. IARC Handbooks of Cancer Prevention, vol. 6., IARC Press, Lyon.

Jacobs, D.R.J. and L. Marquart, J. Slavin and L.H. Kushi. 1998. Whole-grain intake and cancer: an expanded review and meta-analysis. Nutr. Cancer 30: 85-96.

Jaiswal, A.K. 2000. Regulation of genes encoding NAD(P)H: quinone oxidoreductases. Free Radical Biol. Med. 29: 254-262.

Jones, J.I. and D.R. Clemmons. 1995. Insulin-like growth factors and their binding proteins: biological actions. Endocr. Rev. 16: 3-34.

Kanazawa, K. and M. Uehara, H. Yanagitani, and T. Hashimoto. 2006. Bioavailable flavonoids to suppress the formation of 8-OHdG in HepG2 cells. Arch. Biochem. Biophys. 455: 197-203.

Karas, M. and H. Amir, D. Fishman, M. Danilenko, S. Segal, A. Nahum, A. Koifmann, Y. Giat, J. Levy, and Y. Sharoni. 2000. Lycopene interferes with cell cycle progression and insulin-like growth factor I signaling in mammary cancer cells. Nutr. Cancer 36: 101-111.

Kasai, H. and H. Hayami, Z. Yamaizumi, H. Saito, and S. Nishimura. 1984. Detection and identification of mutagens and carcinogens as their adducts with guanosine derivatives. Nucleic Acids Res. 12: 2127-2136.

Kasai, H. and M.H. Chung, D.S. Jones, H. Inoue, H. Ishikawa, H. Kamiya, E. Ohtsuka, and S. Nishimura. 1991. 8-Hydroxyguanine, a DNA adduct formed by oxygen radicals: its implication on oxygen radical-involved mutagenesis/carcinogenesis. J. Toxicol. Sci. 16 (Suppl 1): 95-105.

Kawamori, T. and R. Lubet, V.E. Steele, G.J. Kelloff, R.B. Kaskey, C.V. Rao, and B.S. Reddy. 1999. Chemopreventive effect of curcumin, a naturally occurring anti-inflammatory agent, during the promotion/progression stages of colon cancer. Cancer Res. 59: 597-601.

Kazerouni, N. and R. Sinha, C.H. Hsu, A. Greenberg, and N. Rothman. 2001. Analysis of 200 food items for benzo[a]pyrene and estimation of its intake in an epidemiologic study. Food Chem. Toxicol. 39: 423-436.

Khandwala, H.M. and I.E. McCutcheon, A. Flyvbjerg, and K.E. Friend. 2000. The effects of insulin-like growth factors on tumorigenesis and neoplastic growth. Endocr. Rev. 21: 215-244.

King, T.J. and J.S. Bertram. 2005. Connexins as targets for cancer chemoprevention and chemotherapy. Biochim. Biophys. Acta 1719: 146-160.

Kinzler, K.W. and B. Vogelstein. 1996. Lessons from hereditary colorectal cancer. Cell 87: 159-170.

Kohlmeier, L. and J.D. Kark, E. Gomez-Gracia, B.C. Martin, S.E. Steck, A.F. Kardinaal, J. Ringstad, M. Thamm, V. Masaev, R. Riemersma, J.M. Martin-Moreno, J.K. Huttunen, and F.J. Kok. 1997. Lycopene and myocardial infarction risk in the EURAMIC Study. Amer. J. Epidemiol. 146: 618-626.

Krinsky, N.I. and E.J. Johnson. 2005. Carotenoid actions and their relation to health and disease. Mol. Aspects Med. 26: 459-516.

Kucuk, O. and F.H. Sarkar, W. Sakr, Z. Djuric, M.N. Pollak, F. Khachik, Y.W. Li, M. Banerjee, D. Grignon, J.S. Bertram, J.D. Crissman, E.J. Pontes, and D.P. Wood Jr. 2001. Phase II randomized clinical trial of lycopene supplementation before radical prostatectomy. Cancer Epidemiol. Biomarkers Prev. 10: 861-868.

Kwak, M.K. Itoh, M. Yamamato, T.R. Sutter, and T.W. Kensler. 2001. Role of transcription factor Nrf2 in the induction of hepatic phase 2 and antioxidative enzymes in vivo by the cancer chemoreventive agent, 3H-1,2-dimethiole-3-thione. Mol. Med. 7: 135-145.

Lee, H.Y. and K.H. Chun, B. Liu, S.A. Wiehle, R.J. Cristiano, W.K. Hong, P. Cohen, and J.M. Kurie. 2002. Insulin-like growth factor binding protein-3 inhibits the growth of non-small cell lung cancer. Cancer Res. 62: 3530-3537.

Lindshield, B.L. and K. Canene-Adams, and J.W. Erdman Jr. 2007. Lycopenoids: are lycopene metabolites bioactive? Arch. Biochem. Biophys. 458: 136-140.

Liska, D.J. 1998. The detoxification enzyme systems. Altern. Med. Rev. 3: 187-198. Review.

Liu, E. and H.K. Law, and Y.L. Lau. 2003. Insulin-like growth factor I promotes maturation and inhibits apoptosis of immature cord blood monocyte-derived dendritic cells through MEK and PI 3-kinase pathways. Pediatr. Res. 54: 919-925.

Livny, O. and I. Kaplan, R. Reifen, S. Polak-Charcon, Z. Madar, and B. Schwartz. 2002. Lycopene inhibits proliferation and enhances gap-junction communication of KB-1 human oral tumor cells. J. Nutr. 132: 3754-3759.

Loewenstein, W.R. and Y. Kanno. 1966. Intercellular communication and the control of tissue growth: lack of communication between cancer cells. Nature 209: 1248-1249.

Lund, E.K. and S.G. Wharf, S.J. Fairweather-Tait, and I.T. Johnson. 1999. Oral ferrous sulfate supplements increase the free radical-generating capacity of feces from healthy volunteers. Amer. J. Clin. Nutr. 69: 250-255.

Ma, Q. and H. Hoper, I. Halliday, and B.J. Rowlands. 1996. Diet and experimental colorectal cancer. Nutr. Res. 16: 413-426.

Manorama, R. and N. Chinnasamy, and C. Rukmini. 1993. Effect of red palm oil on some hepatic drug-metabolizing enzymes in rats. Food Chem. Toxicol. 31: 583-588.

Martínez-Ferrer, M. and M. Verghese, L.T. Walker, L. Shackelford, C.B. Chawan, and N. Jhala. 2006. Lycopene reduces azoxymethane-induced colon tumors in Fisher 344 rats. Nutr. Res. 26: 84-91.

McLellan, E.A. and A. Medline, and R.P. Bird. 1991. Dose response and proliferative characteristics of aberrant crypt foci: putative preneoplastic lesions in rat colon. Carcinogenesis 12: 2093-2098.

McLellan, E.A. and A. Medline, and R.P. Bird. 1991. Sequential analyses of the growth and morphological characteristics of aberrant crypt foci: putative preneoplastic lesions. Cancer Res. 51: 5270-5274.

Mesnil, M. and S. Crespin, J.L. Avanzo, and M.L. Zaidan-Dagli. 2005. Defective gap junctional intercellular communication in the carcinogenic process. Biochim. Biophys. Acta 1719: 125-145.

Michels, K.B. and G. Edward, and K.J. Joshipura. 2000. Prospective study of fruit and vegetable consumption and incidence of colon and rectal cancers. J. Natl. Cancer Inst. 92: 1740-1752.

Nagengast, F.M. and M.J. Grubben, and I.P. van Munster. 1995. Role of bile acids in colorectal carcinogenesis. Eur. J. Cancer 31: 1067-1070.

Narisawa, T. and Y. Fukaura, M. Hasebe, S. Nomura, S. Oshima, H. Sakamoto, T. Inakuma, Y. Ishiguro, J. Takayasu, and H. Nishino. 1998. Prevention of

N-methylnitrosourea-induced colon carcinogenesis in F344 rats by lycopene and tomato juice rich in lycopene. Jpn. J. Cancer Res. 89: 1003-1008.

Negi, P.S and G.K. Jayaprakasha, and B.S Jena. 2003. Antioxidant and antimutagenic activities of pomegranate peel extracts. Food Chem. 80: 393-397.

Nguyen, M.L. and S.J. Schwartz. 1999. Lycopene: chemical and biological properties. Food Technol. 53: 38-48.

Nguyen, T. and H.C. Huang, and C.B. Pickett. 2000. Transcriptional regulation of the antioxidant response element. Activation by Nrf2 and repression by MafK. J. Biol Chem. 275: 15466-15473.

Niki, E. 1991. Vitamin C as an antioxidant. World Rev. Nutr. Diet. 64: 1-30.

Omoni, A.O. and R.E. Aluko. 2005. The anticarcinogenic and anti-atherogenic effects of lycopene: a review. Trends Food Sci. Technol. 16: 344-350.

Omori, Y. and H. Yamasaki. 1998. Mutated connexin43 proteins inhibit rat glioma cell growth suppression mediated by wild-type connexin43 in a dominant-negative manner. Int. J. Cancer 78: 446-453.

Pan, H. and G. Shi, W. Chen, and D. Wang. 2003. Effect of lycopene on the function of antioxidative enzyme system in rats. Wei Sheng Yan Jiu 32: 441-442.

Papanikolaou, A. and O. Wang, D.A. Delker, and D.W. Rosenberg. 1998. Azoxymethane-induced colon tumors and aberrant crypt foci in mice of different genetic susceptibility. Cancer Lett. 130: 29-34.

Potter, J.D. and K. Steinmetz. 1996.Vegetables, fruit and phytoestrogens as preventive agents. pp. 61-90. *In:* B.W. Stewart and D. McGregor [eds.]. Principles of Chemoprevention. IARC Scientific Publication No. 139.

Pretlow, T.P. and T.G. Pretlow. 2005. Mutant KRAS in aberrant crypt foci (ACF): initiation of colorectal cancer? Biochim. Biophys. Acta. 1756: 83-96.

Pretlow, T.P. and B.J. Barrow, W.S. Ashton, M.A. O'riordan, T.G. Pretlow, J.A. Incisek, and T.A. Stellato. 1991. Aberrant crypts: putative preneoplastic foci in human colonic mucosa. Cancer Res. 51: 1564-1567.

Rao, A.V. and S. Agarwal. 1998. Bioavailability and in vivo antioxidant properties of lycopene from tomato products and their possible role in the prevention of cancer. Nutr. Cancer 31: 199-203.

Rao, C.V. 2004. Nitric oxide signaling in colon cancer chemoprevention. Mutat. Res. 555: 107-119.

Ratnam, D.V. and D.D. Ankola, V. Bhardwaj, D.K. Sahana, and M.N. Kumar. 2006. Role of antioxidants in prophylaxis and therapy: A pharmaceutical perspective. J. Control Release 113: 189-207.

Reddy, B.S. 2004. Studies with the azoxymethane-rat preclinical model for assessing colon tumor development and chemoprevention. Environ. Mol. Mutagen. 44: 26-35. Review.

Reeves, P.G. and F.H. Nielsen, and G.C. Fahey Jr. 1993a. AIN-93 purified diets for laboratory rodents: final report of the American Institute of Nutrition ad hoc writing committee on the reformulation of the AIN-76A rodent diet. J. Nutr. 123: 1939-1951.

Reeves, P.G. and K.L. Rossow, and J. Lindlauf. 1993b. Development and testing of the AIN-93 purified diets for rodents: results on growth, kidney calcification and bone mineralization in rats and mice. J. Nutr. 123: 1923-1931.

Rehman, A. and L.C. Bourne, B. Halliwell, and C.A. Rice-Evans. 1999. Tomato consumption modulates oxidative DNA damage in humans. Biochem. Biophys. Res. Commun. 262: 828-831.

Rimm, E.B. and M.J. Stampfer, A. Ascherio, E. Giovannucci, G.A. Colditz, and W.C. Willett. 1993. Vitamin E consumption and the risk of coronary heart disease in men. N. Engl. J. Med. 328: 1450-1456.

Saez, J.C. and V.M. Berthoud, M.C. Branes, A.D. Martinez, and E.C. Beyer. 2003. Plasma membrane channels formed by connexins: their regulation and functions. Physiol. Rev. 83: 1359-1400.

Sahu, S.C. 1990. Oncogenes, oncogenesis, and oxygen radicals. Biomed. Environ. Sci. 3: 183-201.

Seraj, M.J. and A. Umemoto, A. Kajikawa, S. Mimura, T. Kinouchi, Y. Ohnishi, and Y. Monden. 1997. Effects of dietary bile acids on formation of azoxymethane-induced aberrant crypt foci in F344 rats. Cancer Lett. 115: 97-103.

Shi, J. and M. Le Maguer. 2000. Lycopene in tomatoes: chemical and physical properties affected by food processing. Crit. Rev. Biotechnol. 20: 293-334.

Siler, U. and A. Herzog, V. Spitzer, N. Seifert, A. Denelavas, P.B. Hunziker, L. Barella, W. Hunziker, M. Lein, R. Goralczyk, and K. Wertz. 2005. Lycopene effects on rat normal prostate and prostate tumor tissue. J. Nutr. 135: 2050S-2052S.

Singh, J. and N. Kulkarni, G. Kelloff, and B.S. Reddy. 1994. Modulation of azoxymethane-induced mutational activation of ras protooncogenes by chemopreventive agents in colon carcinogenesis. Carcinogenesis 15: 1317-1323.

Singh, J. and R. Hamid, and B.S. Reddy. 1997. Dietary fat and colon cancer: modulating effect of types and amount of dietary fat on ras-p21 function during promotion and progression stages of colon cancer. Cancer Res. 57: 253-258.

Sohl, G. and K. Willecke. 2004. Gap junctions and the connexin protein family. Cardiovasc. Res. 62: 228-232.

Stahl, W. and H. Sies. 1998. The role of carotenoids and retinoids in gap junctional communication. Int. J. Vitam. Nutr. Res. 68: 354-359.

Stopera, S.A. and J.R. Davie, and R.P. Bird. 1992. Colonic aberrant crypt foci are associated with increased expression of c-fos: the possible role of modified c-fos expression in preneoplastic lesions in colon cancer. Carcinogenesis 13: 573-578.

Suh, N. and S. Paul, X. Hao, B. Simi, H. Xiao, A.M. Rimando, and B.S. Reddy. 2007. Pterostilbene, an active constituent of blueberries, suppresses aberrant crypt foci formation in the azoxymethane-induced colon carcinogenesis model in rats. Clin. Cancer Res. 13: 350-355.

Takayama, T. and K. Miyanishi, T. Hayashi, T. Kukitsu, K. Takanashi, H. Ishiwatari, T. Kogawa, T. Abe, and Y. Niitsu. 2005. Aberrant crypt foci: detection, gene abnormalities, and clinical usefulness. Clin. Gastroenterol. Hepatol. 3 (Suppl 7): S42-5.

Terry, P. and E. Giovannucci, and K.B. Michels. 2001. Fruit, vegetables, dietary fiber, and risk of colorectal cancer. J. Natl. Cancer Inst. 93: 525-533.

Trosko, J.E. and R.J. Ruch. 2002. Gap junctions as targets for cancer chemoprevention and chemotherapy. Curr. Drug Targets 3: 465-482.

Valko, M. and C.J. Rhodes, J. Moncol, M. Izakovic, and M. Mazur. 2006. Free radicals, metals and antioxidants in oxidative stress-induced cancer. Chem. Biol. Interact. 160: 1-40.

Vanamala, J. and T. Leonardi, B.S. Patil, S.S. Taddeo, M.E. Murphy, L.M. Pike, R.S. Chapkin, J.R. Lupton, and N.D. Turner. 2006. Suppression of colon carcinogenesis by bioactive compounds in grapefruit. Carcinogenesis 27: 1257-1265.

Velmurugan, B. and V. Bhuvaneswari, S. Balasenthil, and S. Nagini. 2001. Lycopene, an antioxidant carotenoid modulates glutathione-dependent hepatic biotransformation enzymes during experimental gastric carcinogenesis. Nutr. Res. 21: 1117-1124.

Wang, W. and A.K. Jaiswal. 2006. Nuclear factor Nrf2 and antioxidant response element regulate NRH:quinone oxidoreductase 2 (NQO2) gene expression and antioxidant induction. Free Radical Biol. Med. 40: 1119-1130.

Wijnands, M.V. and M.J. Appel, V.M. Hollanders, and R.A. Woutersen. 1999. A comparison of the effects of dietary cellulose and fermentable galacto-oligosaccharide, in a rat model of colorectal carcinogenesis: fermentable fibre confers greater protection than non-fermentable fibre in both high and low fat backgrounds. Carcinogenesis 20: 651-656.

Willett, W.C. 1994. Micronutrients and cancer risk. Amer. J. Clin. Nutr. 59(Suppl 5): 1162S-1165S.

Williams, A.C. and T.P. Collard, C.M. Perks, P. Newcomb, M. Moorghen, J.M. Holly, and C. Paraskeva. 2000. Increased p53-dependent apoptosis by the insulin-like growth factor binding protein IGFBP-3 in human colonic adenoma-derived cells. Cancer Res. 60: 22-27.

World Cancer Research Fund. 1997. Food, nutrition, and the prevention of cancer: a global perspective. American Institute for Cancer Research, Washington, DC.

Wu, K. and W.C. Willett, C.S. Fuchs, G.A. Colditz, and E.L. Giovannucci. 2002. Calcium intake and risk of colon cancer in women and men. J. Natl. Cancer Inst. 94: 437-446.

Xiao, R. and T.M. Badger, and F.A. Simmen. 2005. Dietary exposure to soy or whey proteins alters colonic global gene expression profiles during rat colon tumorigenesis. Mol. Cancer 4: 1.

Zalatnai, A. and K. Lapis, B. Szende, E. Raso, A. Telekes, A. Resetar, and M. Hidvegi. 2001. Wheat germ extract inhibits experimental colon carcinogenesis in F-344 rats. Carcinogenesis 22: 1649-1652.

Zhang, L.X. and R.V. Cooney, and J.S. Bertram. 1991. Carotenoids enhance gap junctional communication and inhibit lipid peroxidation in C3H/10T1/2 cells: relationship to their cancer chemopreventive action. Carcinogenesis 12: 2109-2114.

Zhang, L.X. and P. Acevedo, H. Guo, and J.S. Bertram. 1995. Upregulation of gap junctional communication and connexin43 gene expression by carotenoids in human dermal fibroblasts but not in human keratinocytes. Mol. Carcinogenesis 12: 50-58.

Index

α-carotene 39, 85, 153, 202, 247, 283, 284, 376, 422

α-Crystallin 294

α-tocopherol 19, 47, 117, 225, 228, 231, 233, 238, 239, 255, 284, 341, 359, 399, 420

β-carotene 19, 39, 42, 44-48, 50, 53, 55, 56, 57, 66, 68-71, 73, 85, 93, 106, 119, 120, 122, 123, 134, 135, 147, 166, 168-170, 176, 184, 197, 202, 225, 238, 247, 252, 257-259, 261, 262, 313, 316, 366, 371, 375-377, 381, 399, 400, 403, 419, 420, 423

1,2-dimethylhydrazine (DMH) 425, 426

7α-dehydroxylase 418

8-oxo-deoxyguanosine 420

A

A/J mouse model of lung cancer 384

Aberrant crypt foci (ACF) 415, 427, 428

Acetoacetyl-CoA 254

Acetyl-CoA 254

Adenoma/carcinoma 427, 428

Adenomatous polyp 415

Age 169, 246, 318, 369, 431

Aggregation 72

Aging 111, 117, 118, 122, 169, 187, 211, 212, 216, 261, 294

All-*trans* lycopene 5, 32

Antiatherogenic activity 10

Anticarcinogenic potential of tomatoes 13

Antioxidant response elements (ARE) 371, 382, 405, 406, 422, 435

Antioxidant(s) 3, 4, 6, 9, 10, 25, 47, 65, 67, 76, 106, 117, 123, 135, 145, 159, 160, 163, 169-173, 175, 183-186, 193, 194, 196-198, 201, 203, 207, 208, 210, 211, 213-218, 225, 227-232, 234, 238, 243, 244, 247, 248, 251-253, 255, 258-262, 273, 277, 280, 287, 288, 293, 294, 296, 297, 309-313, 316, 317, 320, 333, 334, 338, 339, 341, 344, 345, 353, 355, 395, 396, 403, 405, 413, 414, 419-423

–activity 10, 225, 227, 238, 260, 277, 280, 299, 421

–capacity 225, 229

Apo-lycopenals 37, 39-44, 46, 49-51, 54-57, 59, 71

Apo-10'-lycopenoic acid 72, 379

Apoptosis 49, 54, 55, 57, 133, 135-137, 141-145, 201, 203, 334, 337, 342, 372, 374, 404, 415, 416, 418, 421, 424

Atherosclerosis 10, 18, 134, 135, 146, 159, 160, 162, 170, 171, 173, 244, 246, 247, 253, 255, 257-261, 275, 297, 305

Atherosclerosis 10, 18, 134, 135, 146, 159, 160, 162, 170, 173, 244, 246, 247, 253, 255, 257-261, 275, 297, 305

Azoxymethane (AOM) 425, 426